Economic
Nematology

Economic Nematology

Edited by
JOHN M. WEBSTER
Pestology Centre,
Department of Biological Sciences,
Simon Fraser University,
Burnaby, Vancouver, B.C., Canada

1972

Academic Press   ·   London   ·   New York

ACADEMIC PRESS INC. (LONDON) LTD.
24/28 Oval Road,
London NW1

United States Edition published by
ACADEMIC PRESS INC.
111 Fifth Avenue
New York, New York 10003

Library of Congress Catalog Card Number: 77–1894–80
ISBN: 0–12–741050–3

PRINTED IN GREAT BRITAIN BY
T. &. A. CONSTABLE LTD., EDINBURGH

Contributors

C. D. BLAKE
Riverina College of Advanced Education, P.O. Box 588, Wagga Wagga, N.S.W. 2650, Australia

E. COHN
The Volcani Institute of Agricultural Research, Bet Dagan, Israel

K. B. ERIKSSON
Department of Plant Pathology and Entomology, Agricultural College of Sweden, Uppsala 7, Sweden

N. G. M. HAGUE
Department of Zoology, University of Reading, Reading, England

J. J. HESLING
Glasshouse Crops Research Institute, Rustington, Littlehampton, Sussex, England

M. ICHINOHE
National Institute of Agricultural Sciences, Nishigahara, Kita-Ku, Tokyo, Japan

H. J. JENSEN
Department of Botany and Plant Pathology, Oregon State University, Corvallis, Oregon, U.S.A.

M. H. KHAN
Department of Economics and Commerce, Simon Fraser University, Burnaby, Vancouver, B.C., Canada

J. KORT
Plantenziektenkundige Dienst, Geertjesweg 15, Wageningen, The Netherlands

L. G. E. LORDELLO
Department of Zoology, Escola Superior de Agricultura "Luiz de Queiroz", University of São Paulo, Piracicaba, Brazil

F. D. MCELROY
Research Station, Canada Department of Agriculture 6660 N.W. Marine Drive, Vancouver, B.C., Canada

D. L. MILNE
Citrus and Sub-Tropical Fruit Research Institute, Nelspruit, E. Transvaal, South Africa

M. OOSTENBRINK
Laboratorium voor Nematologie, Landbouwhogeschool, Wageningen, The Netherlands

S. K. PRASAD
Department of Nematology, Indian Agricultural Research Institute, New Delhi, India

J. L. RUEHLE
Forestry Sciences Laboratory, U.S.D.A, Carlton Street, Athens, Georgia, U.S.A.

J. N. SASSER
Department of Plant Pathology, North Carolina State University, Raleigh, North Carolina, U.S.A.

P. SIVAPALAN
Tea Research Institute of Ceylon, Talawakele, Ceylon

W. STEUDEL
Biologische Bundesanstalt, Institut für Hackfruchkrankheiten und Nematodenforschung, Münster, Germany

J. M. WEBSTER
Pestology Centre, Department of Biological Sciences, Simon Fraser University, Burnaby, Vancouver, B.C., Canada

B. WEISCHER
Biologische Bundesanstalt, Institut für Hackfruchkrankheiten und Nematodenforschung, Münster, Germany

R. J. WILLIS
Ministry of Agriculture, Nematology Laboratory, Felden House, Newtownabbey, Northern Ireland

R. D. WINSLOW
Ministry of Agriculture, Nematology Laboratory, Felden House, Newtownabbey, Northern Ireland

Preface

The aim of this book is to satisfy the needs of agriculturalists, nematologists, researchers and students for a reference text to the important nematode pests of the world's major crops. It was believed there was an urgent need to produce such a textbook to complement the taxonomic and physiological nematology texts. Information on the cultural practices used in the production of each of the crops was essential to show how these practices could influence the presence and development of the specific nematode pests of the crops. One of the major concerns of the book was to emphasize not only the nematode pests of crops and how they are controlled, but also to give some estimate of the economic loss currently caused by nematode pests.

This breadth of coverage meant that no one person could write authoritatively on such diverse crops and so twenty-two experts have contributed the various chapters of this book. Although the coverage is wide it is inevitable that many crops are included only briefly or not at all. In choosing crops for inclusion, cognizance was taken of the volume and value of the world production of the crop as well as any special growing conditions. It is nevertheless regrettable that separate chapters could not be devoted to such crops as pineapple, maize, grapes or hops, but space was at a premium.

One of the possible hazards of several authors writing on a common theme is the chance of considerable overlap of chapter content. This has been kept to a minimum, and as each chapter probably will be read as a self-contained unit such overlap as occurs should afford completion and emphasis of the topic rather than cause confusion.

Except for the first, the penultimate and the final chapters, which themselves serve to draw together and place new perspective on the content of the seventeen "crop chapters", the general format for each chapter follows the same pattern: namely, introduction to the crop and its production, nematode pests, cultural and environmental influences on the nematode pests, control and, finally, economics of the nematode pest problems. It is this last section in each chapter which really is the *raison d'être* for the book. Hence, the reason, also, for the first chapter on the economics of crop diseases. I venture to suggest that the completed text indicates that we are still unable, for many crops, to make dependable predictions on crop losses. Therefore, as well as serving as a useful reference work on nematode pests this book should also encourage researchers to consider in economic terms some of the data that current investigations provide.

I wish to thank Mrs Janie Collins, Mrs Angela Hamlett, Mrs Carole Thompson, Mrs Sandi Lauer, Miss Karen Ballinger and Miss Shirley Vander Molen for typing and proofreading; Mrs Catherine Fockler for assistance with the subject index; the Simon Fraser University Audio-Visual Centre for redrawing many of the figures; the other twenty-one contributors whose enthusiasm and advice has helped to make this book possible. During its preparation I held a research grant from the National Research Council of Canada.

January, 1972   JOHN M. WEBSTER

Contents

A*

1

Economic Aspects of Crop Losses and Disease Control

Mahmood Hasan Khan

Department of Economics and Commerce
Simon Fraser University
Burnaby, Vancouver, B.C., Canada

I. Introduction

Plant diseases and the consequent crop losses are as much a subject of concern to the economist as they are to the plant pathologist, the nematologist, the entomologist, the agronomist and the rest. The economic consequences of crop losses due to diseases are many and varied.* First, there is the loss of revenue or the increased cost of production to the farm producer because of the reduction in crop yield or the loss in quality of the marketable produce. In addition, there is the factor of uncertainty which the farmer must face. Second, there is the loss to the consumer, household and industry, either through the price increase or through the deterioration in quality of the consumable product or a combination of these two. Third, there is the loss to the society, of which the producer and consumer are a part, in the form of wasted

---

* For this study, crop diseases include those which are caused by the attacks of insect pests, fungi, viruses, nematodes and weeds.

1

resources which have direct and implicit costs. Finally, there is the loss sustained by the world community, especially in those parts of the world that are struggling against food and raw material shortages, population growth and slow development. These losses are not confined to food crops. They affect also the cash crops which form a major source of revenue to the farmers and to the industry.

There is yet another dimension to this problem: the economic aspects of disease control. If controls are technically feasible and effective, they normally involve additional costs which have to be compared with the revenue saved or the benefits derived. It may be that the costs of control exceed their benefits. If that is so, the economic problem is clear and a decision must be made as to whether to control the pest or to shift the given crop resources to some other use. In other words, resource reallocation may be considered.

Current literature on crop diseases and their control, while it is voluminous and impressive in the field of biology, is scarce and incomplete in that of economic analysis, as can be seen from Cramer (1967), Anon. (1967), and a recent study published by Anon. (1971). There seem to be many reasons for this and they need no restatement here. In this study, those deficiencies of analysis which have economic foundation will be examined. This examination will take the following form:

1. An attempt will be made to discuss available data on the extent and type of crop losses, in quantity and value, caused by diseases.

2. The economic principle will be defined and its relevance to the subject of crop diseases and their control analysed.

3. The economic aspects of crop losses due to diseases will be examined.

4. The economic aspects of disease control will be analysed.

5. A research proposal on the methods of economic evaluation of crop diseases and their control will be presented.

II. The Extent and Type of Crop Losses due to Diseases

It is well recognized that there is a paucity of information about the extent and type of crop losses due to various diseases. Even in those few countries where statistics are collected and published, they are often incomplete and inaccurate. There are several reasons for this. First, in the majority of cases, there are conceptual problems arising from the definition of losses. Second, there are the problems of estimation of losses. Third, in many parts of the world there is no system of record collection, in which case sound interpretation of the available data is almost impossible.

Notwithstanding the above reservations, one recent study of the extent and type of crop losses due to plant diseases is quite comprehensive (Cramer, 1967). With some changes in the arrangement of data, the annual world crop losses in quantity and value are reproduced in Tables I and II.

TABLE I. Annual world crop losses (in million tons; 1 ton = 1·02 metric tons

			Crop losses due to			
Commodity	Actual production	Potential production	Insect pests	Diseases	Weeds	Total
Cereals	961·1	1467·5	203·7	135·3	167·4	506·4
Potatoes	270·8	400·0	23·8	88·9	16·5	129·2
Sugar-beet and Sugar-cane	694·6	1330·4	228·4	232·3	175·1	635·8
Vegetables	201·7	279·9	23·4	31·3	23·7	78·2
Fruits	141·7	197·2	11·3	32·6	11·6	55·5
Stimulants	10·2	16·5	1·9	2·6	1·8	6·3
Oils	94·7	137·0	14·5	13·5	14·3	42·3
Fibres	16·0	23·2	3·0	2·6	1·6	7·2
Natural Rubber	2·3	3·0	0·1	0·5	0·1	0·7
All Crops	2393·1	3854·7	510·1	539·6	412·1	1461·6

After Cramer (1967).

TABLE II. Annual world crop losses (in 1000 million U.S. dollars)

			Crop losses due to			
Commodity	Actual value	Potential value	Insect pests	Diseases	Weeds	Total
Cereals	63·9	98·0	14·4	8·7	11·0	34·1
Potatoes	10·6	15·6	1·0	3·4	0·6	5·0
Sugar-beet and Sugar-cane	7·6	13·9	2·3	2·3	1·7	6·3
Vegetables	16·7	23·1	2·0	2·3	2·0	6·3
Fruits	14·3	20·1	1·2	3·3	1·2	5·7
Stimulants	7·2	11·4	1·3	1·7	1·2	4·2
Oils	10·6	15·7	1·8	1·6	1·7	5·1
Fibres and Rubber	8·6	12·7	1·8	1·5	0·8	4·1
All Crops	139·7	210·5	25·8	24·8	20·2	70·8

After Cramer (1967).

As Cramer (1967) admits in his study, the estimates given in Tables I and II are no more than the dimension of magnitudes they represent. Yet these approximations emphasize the enormity of the problem in that about 33% of the potential agricultural value is lost annually. These are, however, not the only losses which result from crop diseases. Virtually no data are available on the value of losses to the consumer and to the society. Assuming that such losses occur, the total loss due to crop diseases must exceed the figures given in the studies by Cramer (1967), Anon. (1965) and Anon. (1971).*

III. The Economic Principle

Economic science deals with a particular relationship between ends and means. Ends or objectives may deal with physical production, consumption or profits. Means or resources are concerned with physical resources, funds or organizations which can be used in achieving the objectives. However, this relationship between ends and means is not *per se* an economic problem. It is an economic problem only if there are *many* ends that need satisfying, and that the means to achieve these ends are *limited*. Given this condition, the central problem in economics is the problem of choice between alternatives. To resolve the problem of choosing between alternatives, economics deals with the maximizing and minimizing conditions. The maximizing condition deals with the maximization of ends, like physical output, consumer satisfaction and resource allocation. The minimizing condition involves the minimization of means, like the use of land, labour, capital and organization. Economics, therefore, deals with either the maximization of ends with the given means or the minimization of the use of means for the given ends.

Since at the heart of economics lies the problem of choice, it seems appropriate to demonstrate this problem of choice. Suppose that a farmer has one unit of resources which can be used in the production of crop A or B. His goal is profit maximization. Specifying the production conditions of crops A and B, that one unit of resources will yield 20 bushels of crop A or 50 bushels of crop B, should he grow A or B? His choice principle is given by the crop price ratios. If the price of crop A is $2.00 per bushel and that of crop B is $1.00 per bushel, the choice ratio of A to B is 2 : 1. Given this, the farmer should obviously select B and not A. But if this ratio changes in favour of A, because either the price of A goes up and the price of B remains unchanged or

* This study by a committee chaired by Dr J. Feldmesser provides extensive data, perhaps for the first time, on crop losses due to nematodes in agriculture in the U.S.A.

the price of A remains unchanged and the price of B goes down, he must select A rather than B.

Of course, in practice the problem of choice is not as simple as stated above, but this example does illustrate the choice principle clearly. The choice indicator may be different, depending on the objective(s) selected, but the principle remains intact. The economist has come to rely more and more on mathematics as a tool of developing these choice indicators.

Having so defined the economic principle, its relevance to the problems of crop protection becomes at once obvious. This principle is equally applicable to the farmer, the consumer and the society. To the farmer, its relevance lies in the fact that, at any given moment, he has limited resources which he can use in different farm enterprises. Crop losses due to diseases, whether they affect the crop yield or the price because of the deterioration in the quality of the product or the residual effect on land, give rise to either losses in revenue or increases in the cost of production. To the consumer, the adverse effects of crop losses result in price increases which reduce his real income or through the loss in quality of the product they affect his total satisfaction. To the society, these losses result in the non-optimum use of national resources. It is obvious that in all these cases the maximizing and minimizing conditions are being violated. Using the economic principle, the economist can demonstrate more accurately the consequences of crop losses. It seems quite appropriate to pause and look at some of these consequences through the economist's eye.

A. The Consequences to the Farmer

Crop losses due to plant diseases appear in different forms on the farm. The more common of these forms would be the loss in yield of the product and the deterioration in the quality of the product. Both of these are likely to affect the price and total revenue—total revenue being the product of the price per unit multiplied by the quantity sold. The yield and quality losses also increase the cost of production per unit of resources. There may be yet another type of loss that the farmer faces if the disease is of a persistent character, in the sense that it leaves residual effects in soil or seed, which will affect farm planning for several seasons or years. This could force the farmer to reallocate his resources. In addition, if the farmer undertakes disease control, his cost of production might increase to the extent that, on per unit basis, the additional cost is greater than the additional revenue. In almost every case, controls add to the current or variable costs (like labour) of production. If these additional costs, while kept to a minimum, cannot save much additional value or return, these controls cannot be regarded as profitable. In fact, the farmer may find it more profitable to invest his resources in another crop or farm enterprise. The discussion on the economic consequences of controls will be postponed until a later section.

Returning to the price and revenue effects, assume that there are no effective controls available and that the yield and quality of the crop are adversely affected in one season only. The initial effect is that there is a decrease in marketable supply. This reduction in supply, if it occurs only on one farm and if the affected producer is not an influential supplier (i.e. he is only one among many), is most likely to reduce the total revenue of this producer because he cannot determine the price of the product by himself. But if the disease affects the majority of the suppliers in a market, the economic scene changes. As for the price and revenue effects on the farmers, much will depend on the nature of demand for the product so affected. This can be demonstrated by a simple example.

Assume that there are only two crops, wheat and apples. Also assume that the supply conditions for these two are similar. On the demand side, assume that it is dissimilar for these crops. The conditions of demand may differ for several reasons. First, wheat and apples occupy different positions in the consumer's budget and preference scale. Wheat is a staple crop, while apples are not. Second, apples are likely to have more close substitutes than wheat has. Following these assumptions of supply and demand, it can be shown that the decrease in the supply of the two crops caused by disease will have different results.

In Figs 1 and 2 the familiar demand and supply conditions for wheat and apples are depicted. The demand line, DD, shows the combinations of various quantities that the consumers would be willing to buy and the price levels at which these quantities will be bought. The supply line, SS, indicates the combination of various quantities that the producers or suppliers would be willing to sell and the price levels at which these quantities will be sold. The demand lines or curves for wheat and apples have different slopes, which reflects the fact that the two products occupy different places in the consumer's basket. The demand line for wheat has a steeper slope. In the economist's jargon, wheat has a relatively "less elastic" demand and apples have a relatively "more elastic" demand.

The concept of "elasticity" has a specific meaning in economics. The ratio of the percentage change in quantity to the percentage change in price, when the price change is small, gives the elasticity of demand for a particular product. If, in response to a percentage change in the price of a product, the percentage change in the quantity demand is greater, the demand is said to be relatively more elastic. Conversely,

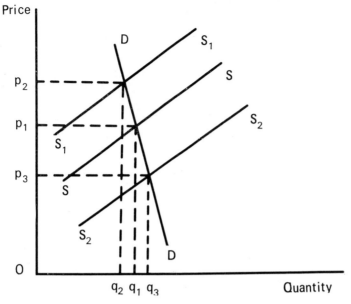

FIG. 1. Wheat: price and quantity effects.

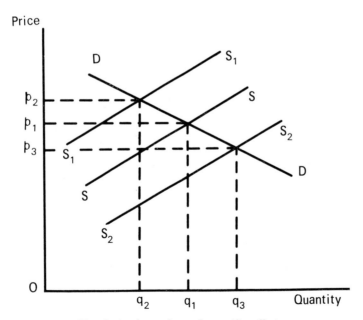

FIG. 2. Apples: price and quantity effects.

if the percentage change in the quantity demanded in response to a percentage change in the price is smaller, the demand is said to be relatively less elastic. In the example used here, the demand for wheat is relatively less elastic and for apples it is relatively more elastic. Buyers of wheat are, in other words, less responsive in terms of the change in quantity than to a change in the price. In the example of apples, buyers are more responsive.

Returning to Figs. 1 and 2, say the original positions of supply and demand are shown by SS and DD lines. The prices and quantities which are agreeable to the buyers and sellers are at the point of intersection of supply and demand curves. These are labelled as p_1 and q_1 for wheat and apples. Assume now that owing to diseases on both wheat and apples, the supply at each given price is reduced: less of these two products is offered for the same price. This means that the supply curves shift upward to the left, from SS to S_1S_1. Also assume that the demand conditions remain unchanged. The result is obvious: that the prices of wheat and apples will increase and the quantities bought and sold will decrease, p_1 to p_2 and q_1 to q_2. The changes in the price and quantity of wheat are likely to increase the total revenue and decrease the costs. However, for apples, these changes will reduce the total revenue and perhaps will increase the costs. While the price and quantity changes are similar in both instances, in one instance the farmers stand to gain and in the other they may lose.

In reality it is not as simple as that. The reduction in supply may also mean that farm resources have not been used optimally or at the lowest cost per unit of output. First, in instances where additional revenue has accrued through price increases that may not equal the cost that the waste in resources imply. Second, the wasted resources could have been used optimally had they been put to some other crops. Finally, to a single farmer, who may be the only one affected adversely by crop losses due to diseases, the total loss may even be greater.

Modifying the foregoing example, assume now that the farmers have been growing wheat and apples under disease conditions but they now find controls which save the yield and the quality of products to be marketed. What will this increased supply do to the farmers? Going back to Figs 1 and 2, the immediate effect is that the supply curves shift outward to the right from SS to S_2S_2: more is supplied at each given price. The price received will decrease, from p_1 to p_3, but the quantity sold will increase, from q_1 to q_3. In the case of wheat, the decrease in price is likely to result in a loss of total revenue to the farmers, while for apples, this may increase the total revenue. So it is not always true that disease control will increase the farmers' total revenue. Further, the costs of control may be excessive, despite the fact that the farmers save their revenue.

B. The Consequences to the Consumer

There are two classes of consumers of agricultural products: (a) the consumers who use these products for final consumption, and (b) the consumers who use them for producing other goods. In other words, agricultural products can be used for final and intermediate consumption. The effects on these two classes of consumers of crop losses due to diseases tend to differ in their consequences.

To the final consumer, the price increase of the diseased crop may mean that, with the given total budget, he is left with less income for other needs. So if he decides to consume a constant amount of this crop, the real income of the consumer is reduced. Alternatively, the price increase may result in a reduction of the amount of the crop he consumes because he has a fixed budget for this item. In addition, if the quality of the crop is also affected adversely, the consumer may find his total satisfaction reduced. In some instances, the price effect on the consumer may force him to shift to inferior substitutes, which will also reduce his total satisfaction.

For intermediate consumers, who use the raw material for further processing or manufacturing of other goods, the price increase will increase the cost of production and may also affect the quality of the final product. However, this consumer as a producer is more likely to pass the increased cost on to the final consumer in terms of increased prices of final goods.

C. The Consequences to the Society

Since the producers and consumers are part of the general community, obviously the losses incurred by these groups are eventually the losses to the society. Losses to the society may appear in other forms such as the payment of relief and subsidies to the farmers and in some instances to the consumers. The national resources so diverted could have been utilized for some other purposes. This often leads to the non-optimum use of a society's economic resources. In countries which are economically weak and agriculturally backward, these losses could be critical. Crop losses due to diseases in these countries may defeat the efforts to increase production and efficiency.

IV. The Evaluation of Losses

In the previous section, the relevance of the economic principle to crop losses was demonstrated at length. Now the stage is set to analyse some economic aspects of evaluating the losses to the farmer, the consumer and the society.

A. Evaluating the Economic Loss to the Farmer

As was mentioned earlier, crop diseases lead to at least three types of losses, namely, (*a*) the reduction in the yield of crop to be marketed, (*b*) the deterioration in the quality of the crop which affects its market value, and (*c*) the residual effect which the disease may leave in the soil or seed, thus necessitating either replacement or change in the cropping pattern. In current literature, there is at least one sound method of evaluation which includes all these effects (Grainger, 1967). It deserves further consideration and analysis. The formulation of this method is:

$$\text{percentage of disease} \times k = \text{percentage of loss}$$

where the percentage of disease can be estimated by its three components, namely, (1) severity: percentage of disease on a single plant, (2) incidence: percentage of diseased plants in a crop, and (3) prevalence: percentage of diseased crops in a region. It is suggested that the value of the factor "k" is made up of three parts, namely, (1) k_y: the effect on yield, (2) k_m: the effect on market value, and (3) k_r: the replacement factor. By multiplying k_y, k_m, and k_r, the value of the constant "k" can be derived.

The procedure adopted by Grainger improves the estimation of crop losses on the farm. It also makes the identification of the sources and forms of loss possible. However, it is not the complete answer. Even if it is granted that the extent of loss by this method can be determined accurately, it is only partial in its coverage. It does not tell the farmer the extent of loss in terms of revenue and cost. To put it differently: the farmer, or the farming community in general, is not necessarily interested in how much of the product is lost or saved, but rather how much is the extra revenue lost and the extra cost incurred. The farm entrepreneur is a profit maximizer. It is the economic test that he uses to evaluate his losses. The losses expressed through the values of "k" do not mean much to the farmer, except that they may provide the raw data which has to be translated into economic terms. They have to be made relevant to the fulfilment of the economic objectives.

The calculation that the farmer understands is the value of the lost revenue or the increased cost per unit of resources used. For instance, if a disease or a combination of diseases affects only the yield of the crop, it may have two simultaneous economic consequences. First, the loss of yield results in a reduced supply to the market, and with given prices, a reduced total revenue. Even in those instances where the reduced supply increases the price, it may still reduce the total revenue. Second, the reduced yield is reflected in the increased cost per unit of the crop. This aspect, important though it is, is not always so obvious. For example, the farmer decides to grow a crop for which he requires a certain quantity of resources. The resource requirement will be determined partly by the technology available; the resource mix will be determined partly by the price of these resources. The cost of these resources can be determined by the pricing system and the quantities required to produce one unit of the crop. Now if as a result of a particular disease the yield of the crop is reduced in a given season, it means that the farmer gets less output with the given cost per unit: the cost per unit of this crop has increased. In addition, these costs have yet another dimension. The resources which are wasted through disease could have been used for an alternative purpose. This means that some income is foregone by not using the resources differently. If this is also included in the increased costs, the loss picture becomes even more serious.

Thus, in general, the economic aspect of crop losses to the farmer does not get a fair and complete treatment by the methods used by biologists. It is in the expression of additional revenue and additional cost that the adequate treatment lies.

B. Evaluating the Economic Loss to the Consumer

In the preceding section the economic consequences of crop losses to the consumers were indicated. Consumers are affected in at least two ways, namely: (a) by price increases which either reduce their real incomes or increase the cost of production, and (b) by the deterioration in the quality of some products, they are not able to maximize their satisfaction. In the literature one finds almost no evaluation of these losses. The methods of assessment of these losses are a part of demand analysis or market research. In fact, in economics there are standard criteria which are used for such problems. Undoubtedly the quantification of losses accruing to the consumers poses tricky questions. It is quite difficult to isolate a particular cause behind price increases, especially when there may be several factors influencing the price of a product. Also there is the question of quantifying the qualitative effect on consumer satisfaction. In the final section of this study an attempt will be made to resolve these difficulties.

C. Evaluating the Economic Loss to the Community

As was stated earlier, crop losses due to diseases may reduce the income of the society and also misdirect the utilization of its resources.

In the literature on crop losses, this is another unexplored territory. In economic science, there are standard methods, although not without limitations, which could be used in the estimation of the extent and nature of such losses. If to the losses accruing to the producers and consumers are added the direct costs to the society, such as subsidies, relief payments, research expenditures, the total estimates become more accurate. In addition to these direct costs, there are the implicit or indirect costs, which can be derived from the alternative methods of allocation of national resources.

V. Economic Aspects of Disease Control

The basic objective of farm planning is the efficient and profitable use of the productive resources. Disease control methods offer to the farm planner, the consumer and the society the opportunity to avoid losses due to plant diseases either before the first signs of disease appear or after it spreads. From an economic point of view, control methods can be classified into two groups. First, there are methods which may prevent the occurrence of some diseases. Second, there are other methods which are only effective once the disease agent is actually causing damage.

In both of these control methods, there are different degrees of effectiveness. The simple test of effectiveness is not *how many* disease agents have been eliminated or prevented from multiplying but *how few* have been left to exist and multiply. It is the *best degree* of control and not *some degree* of control which it is hoped to achieve. Take an example. Say there are two chemicals, X and Y, for controlling a particular disease. Assume that the performance of X is 100% and that of Y is 60%. The test of their performance would be the number of disease agents which survive to multiply. Also assume that the cost of application of X and Y is the same, say 30 units. The performance cost of X is $\frac{100}{100} \times 30 = 30$ and that of Y is $\frac{100}{60} \times 30 = 50$. From this calculation, it is quite obvious that Y is not only less efficient but also has a higher performance cost. However, the more appropriate test of the value of controls lies in what the economist calls the cost–benefit analysis.

The choice criterion in cost–benefit analysis is the cost–benefit ratio. This ratio could be interpreted in two ways. First, the ratio of cost to benefit is an average concept: per unit cost of control must be less than the per unit benefit of that control. Second, this ratio is a marginal concept: one additional unit of cost should not exceed the one additional unit of benefit derived from the control method used.

Costs and benefits of controls depend on many factors, like the nature of the crop, the control method, the concept of costs and benefits used, and the time period for which the costs and benefits are computed. From the conceptual point of view, the inclusion or exclusion of certain costs and benefits play an important role. For instance, in computing the costs and benefits involved in a certain control method both the *direct* and *indirect* costs and benefits should be included. However, in practice it is not quite so easy to compute the indirect costs and benefits.

The economics of disease control does not end here. There are controls which are either low in efficiency or economically unprofitable. In both of these instances, if a control method adds more to the per unit cost than to the per unit benefit, it should not be applied. Possibly the farmer would benefit most by replacing one farm enterprise by another. In farm planning, geared to the maximization of objectives and the minimization of resource use, the alternative should be considered on the basis of the criterion of choice just stated. However, the replacement problem is not always an easy one. There may be several problems involved in growing a different crop. On the supply side, the soil and climatic conditions and capital resources could be the major constraints. On the demand side, the market conditions may be such that either the value of the new crop is low or the elasticity of demand is unfavourable.

VI. An Economic Approach to the Evaluation of Crop Losses and Disease Control

The preceding sections have indicated the importance of the economic aspects of crop losses and disease control. The stage is now set to propose an economic framework of inquiry. This framework will highlight the areas of interaction and cooperation between the biologist and the economist. The advantages inherent in this approach will become obvious as the analysis proceeds.

The biological and economic aspects of diseases and their control brought together should not only improve the existing methods of evaluation of losses but by so doing should also benefit the parties for whom the research work is done. From the discussion contained in the foregoing sections, it is clear that the economic approach should be an integral part of all scientific research in this area. With the use of the economic principles and tools of analysis, the problems can be defined more precisely and their solutions made more realistic.

From the viewpoint of good management, the economist can help devise some definite procedures for analysing the micro (farm level) and macro (economy level) problems. On the farm level, it seems that an approach such as the following or a variant of it is worth considering.

1. Identify the disease agent and determine the degree of infestation on the crop.

2. Determine the forms that the losses take on the crop and quantify them where possible.

3. Translate these losses into economic terms, such as the loss in revenue or the increase in cost or both.

4. Determine the available means of disease control and the degree of their effectiveness, considering individual and collective methods.

5. Determine all costs, fixed and variable, including new investment and annual fixed cost for additional resources, under the alternative control methods which are technically feasible and efficient.

6. Determine the indirect costs of control, including major changes in crop rotations, management practices, etc.

7. Determine the benefits by evaluating the net effects on profits from the affected crops and the total farm under alternative methods of control.

8. Compare costs with benefits and derive ratios to determine the profitability or otherwise of the control methods to be considered.

9. Compare the above with the alternative, i.e. eliminating the crop altogether and using the released resources for some other purpose.

The nine-step procedure outlined above has several advantages. Above all, it integrates the interests of the biologist and the economist. Given the vast body of literature on the economics of production and farm management together with the studies by biologists, a formal design of research can be evolved such that it is suited to the achievement of the objectives. It must, however, be recognized that such a design cannot be too general in its form. Therefore, the presentation of such a general research design based on the procedure outlined above lies beyond the scope of this study. In what follows, each of the nine points will be elaborated.

1. The first step is basically the field and laboratory work of the biologist. In this area there is very little room, if any, for suggestions by the economist. One approach which this author finds particularly appealing is the one suggested by Grainger (1967).

2. The second step also belongs to the biologist. However, here the economist may be of some help in identifying the type and extent of the loss in yield and quality. The procedure suggested by Grainger (1967) seems to have great merit.

3. It is in the third stage that the economist is needed. The losses already identified should be transformed into cost and revenue terms.

The economic principle is as clear as the methods of estimation are definite. The cost effect can be determined by the figures of the value of the product reduced by all costs. The revenue lost, if any, can be computed by price changes and quantity changes.

4. The fourth step is primarily a technical matter, which should be left to the biologist to determine. Again the method suggested by Grainger (1967) would seem to be quite satisfactory.

5. In the fifth stage, the economist should calculate all costs, original and additional, of control methods. These should include direct and indirect costs, which could be explicit or implicit.

6. The point is taken care of by the foregoing step.

7. Here farm profits or benefits from controls will have to be determined. The direct and indirect benefits will have to be computed and their effect on total farm profits analysed.

8. After determining all the quantifiable costs and benefits, cost–benefit ratios can be established for each control and crop and compared with each other. These ratios have to be related to the value of the crop to the farm and also to its alternatives.

9. Finally, if the cost–benefit ratios are not favourable, a comparison may have to be made between the situation with no control and the alternative use of farm resources. In some instances, it may be found that the costs and revenues will be favourably affected by shifting the resources to some other use. For this the total farm picture will have to be analysed.

After this brief discussion of the suggested approach to evaluate the economic aspects of crop losses and disease control, there still remains the question of incorporating the economic consequences to the consumers and the society. While it is true that without the inclusion of these effects of crop losses and control the study remains incomplete, it is this area of analysis which will confront the economist with several tricky problems. First, in these areas the economic tools are conceptually quite unclear. Second, because of the nature of indirect and often intangible effects which seem to dominate, quantitative analysis will at best be inadequate. Finally, the researchers may find that, because of these problems and the lack of sufficient funds, a comprehensive study is beyond them. However, it would still be rewarding to initiate studies in this direction for the simple reason that almost nothing is known about these economic consequences which may be quite significant.

VII. Conclusion

In conclusion it is hoped that this brief study, despite its many limitations, has succeeded in bringing into sharper focus the significance of economic aspects of plant diseases and their control. After all, the primary purpose of research in the area of crop losses due to diseases and their control is to improve the present state of the art of production and in so doing to maximize economic welfare. The quantity of food and cash crops that can be saved by this type of research may in the end be equal to, if not greater than, the increase in production which will otherwise have to come about through the additional use of land, labour, fertilizer, water and machines. That which is so lost to the society is no doubt of great value, especially in those parts of the world where the struggle for development has just begun.

References

Anon. (1965). "Losses in Agriculture." United States Department of Agriculture, Washington, D.C.

Anon. (1967). *FAO Symposium on Crop Losses*, Food and Agriculture Organization, Rome.

Anon. (1971). Estimated crop losses due to plant-parasitic nematodes in the United States. Special publication No. 1. 7 pp. Suppl. *J. Nematol.*

Cramer, H. H. (1967). "Plant Protection and World Crop Production." Bayer Pflanzenschutz, Leverkusen.

Grainger, J. (1967). Economic aspects of crop losses caused by diseases. In *FAO Symposium on Crop Losses*, Rome, 55–98.

2

Nematode Diseases of Potatoes

R. D. Winslow and R. J. Willis

Ministry of Agriculture, Nematology Laboratory
Felden House, Newtownabbey, Northern Ireland

I. The Crop and its Nematode Problems

A. Crop importance

The world potato crop occupies some 20 million hectares (50 million acres) annually, producing some 300 million metric tons, about 35% of it in the U.S.S.R. and some 50% elsewhere in Europe (Table I). Producer prices vary greatly with country, season and potato usage (human or animal food or industrial use), but may be as high as £20 sterling ($50 U.S.) or more per ton in western Europe and North America.

Conservatively estimated at £4 ($10) per ton, gross returns to the world's producers may total some £1,200,000,000 ($3,000,000,000) annually.

It is difficult to forecast changes in annual crop area, but it will probably decrease in the developed countries and increase or remain static in the developing countries. Any overall decrease in area may be compensated for, at least partly, by increased yields per unit area, resulting from the use of improved varieties. Thus potatoes are likely to remain an important crop, particularly in Europe and the U.S.S.R.

TABLE I. Annual world potato production (approximate)

Region	Crop area (thousand ha)	Production (million metric tons)	Yield (metric tons/ha)
U.S.S.R.	8000	100	12
Eastern Europe*	5000	80	16
Western Europe	3000	60	20
North America	700	16	23
South and Central America	1000	8	8
Asia†	1000	13	13
Africa	300	2	7
Australasia	50	1	20
World‡	19,050	280	15

* Excluding U.S.S.R.
† Excluding U.S.S.R., China, Indonesia.
‡ Excluding China, Indonesia, for which countries recent data are lacking.

B. The Nematodes

The potato cyst nematode, *Heterodera rostochiensis*, is by far the most important nematode pathogen of potatoes. Other major pathogens include species of *Ditylenchus*, *Meloidogyne*, *Pratylenchus* and *Trichodorus*, and minor pathogens include species of *Longidorus*, *Hexatylus* and *Neotylenchus*.

II. Potato Cyst Nematode, *Heterodera rostochiensis*

A. History and Distribution

The potato cyst nematode (potato root eelworm, golden nematode) undoubtedly evolved, along with its chief food plant, the cultivated potato, in the Andes Mountain regions in South America. It was introduced to Europe, probably in the middle or latter half of the 19th century, on potatoes imported for breeding purposes and it has since spread throughout most of Europe and to Asia (Israel, India), North Africa, Canary Islands, Iceland, and North and Central America (Jones, 1970a).

Its arrival in North America had the semblance of a well-planned military invasion, with the establishment of three offshore bases on Long Island (detected 1941), Newfoundland (1962) and Vancouver Island (1965). The United States Department of Agriculture undertook a vigorous and costly campaign to prevent its establishment on the mainland but the recent discoveries of infestations in western New York State and in Delaware indicate that the campaign was not wholly successful.

B. Morphology and Biology

1. Life-cycle

The persistent or dormant stage found in the soil is the so-called cyst, the toughened integument of a dead female, filled with embryonated eggs. In the absence of a host crop such as potatoes, tomatoes or egg-plant, a small proportion of the juvenile worms or larvae hatch each year but die for want of adequate nutrition. When a host crop is grown most of the remaining juveniles are stimulated to hatch by a chemical factor, of unknown composition, emanating from the host roots. These juveniles invade the roots and mature into vermiform males or rotund females. The fertilized females, pearly-white and pear-shaped at this stage, increase in size to become spherical (Fig. 1), and, on dying, change colour, through golden yellow or cream (Guile, 1970), to the final brown cyst—a dead structure filled with embryonated eggs. The cysts are left in the soil when the crop is lifted, completing the life-cycle.

2. Host-parasite Relationships

After entering a potato rootlet the nematode larva usually settles to feed with its head end in the vascular tissue. Presumably in response to glandular secretions from the nematode, certain cells in the stele around the head of the nematode enlarge and coalesce to form "giant cells" or syncytia, of which an adequate supply is essential for normal female development. Sex is environmentally determined, the relatively sexless larvae developing to adult females or males, or perishing in the roots, depending on the production of giant cells in response to invasion.

3. Longevity and Dispersal

Viable embryonated eggs within cysts persist in the soil in the absence of hosts for up to thirty years. Eggs within cysts are also highly resistant

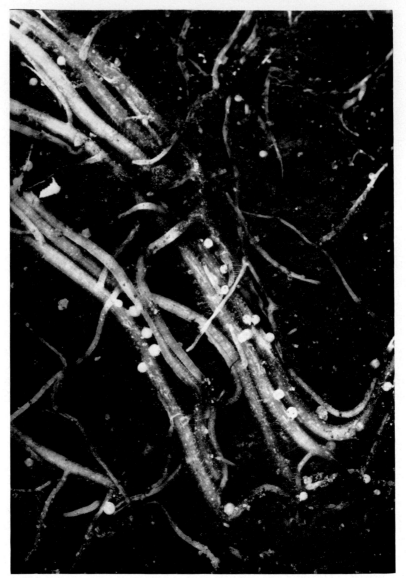

FIG. 1. *Heterodera rostochiensis* white females attached to potato roots. Magnified ×2 approximately.

to desiccation and are readily dispersed, through the movement of cyst-bearing soil on farm implements, flood water and, occasionally, as a result of wind-blow. Transportation over greater distances normally occurs through the movement of infested soil adhering to seed potato tubers or to the roots of other forms of plant propagating material.

C. Ecology

The environmental conditions associated with extensive commercial potato production provide optimum conditions for eelworm survival and multiplication. Free-draining, aerated sands, silts and peat soils with a moisture content of 50–75% water capacity are more suitable for potato production—and survival and movement of the free-living stages of the eelworm in the soil—than are heavy clay soils.

Growth of the potato crop will occur in spring at lower soil temperatures than that required to initiate eelworm activity. A minimum soil temperature of 10°C is necessary before eelworm activity occurs at Long Island, U.S.A. (Chitwood and Feldmesser, 1948), but a somewhat lower minimum is probably necessary in the cooler British soils. In some early potato growing areas in Britain this discrepancy between the minimum temperature requirements of the host and the eelworm can be exploited to minimize damage by the pest. At the other extreme, tuberization of the potato plant is inhibited at high soil temperatures and, in tropical areas such as Peru and India, potato production is, of necessity, confined to the cooler, high altitude areas. On tomato plants under glasshouse conditions the eelworm tolerates higher soil temperatures. Jones and Parrott (1969) concluded, from experiments conducted out of doors in England, that except under extreme temperature variations, an accumulation of approximately 1100 day-°C above a threshold of 4·4°C was required after planting potatoes for the successful completion of the life-cycle.

The eelworm apparently tolerates pH values tolerable to the potato crop and is unaffected by the nutrient status of the soil, though the latter, by affecting crop performance, can influence the eelworm multiplication rate.

D. Population Dynamics

Population dynamics, especially population increase, vary greatly with climate and season. Also, while the rate of decrease may be relatively density-independent, rate of increase is density-dependent, varying inversely with pre-crop infestation. The multiplication factor may range from perhaps 50 or more for very low pre-crop infestations to unity or less (nematode density maintained or decreased) at saturation levels.

Decrease in infestation, when the land is rested from potatoes, is some 30–33% per year for the British Isles (various authors), although it may be greater in the first year, and may vary with cultivations and the type of non-host crops grown. Groundkeepers may help to maintain low infestations but have little effect on high infestations (Ouden, 1967).

Assessment of infestation levels and changes is essential in studies on control. Therefore, in areas where this nematode is, or threatens to become, a pest, annual increase and decrease rates should be determined. They may well be different from those in neighbouring areas.

Population estimates are best conducted outside the growing season so that the cysts are relatively full of viable eggs. Some workers have classified cysts into empty (without live contents) and "viable" cysts or cysts with contents (embryonated eggs). Others have classified viable cysts into "full" cysts (with 50 or more eggs), and "half-full" (less than 50 eggs), two "half-full" cysts being regarded as equivalent to one "full" cyst. Estimates of viable or full cysts seem adequate for most purposes, but in precise investigations all the eggs in a sample of cysts are counted or estimated, the infestation level being expressed as eggs/g of soil.

E. Potato Resistance and Nematode Pathogenicity

1. S. andigena Resistance

Ellenby (1948, 1952) examined wild and cultivated species in the British Commonwealth Potato Collection and found good resistance in certain lines of two South American species, *Solanum vernei* and *S. andigena* (strictly *S. tuberosum* subsp. *andigena*). Later, resistance was discovered in several other species. Resistance of *S. andigena* was the first to be commercially exploited, since this species, being tetraploid like the commercial European *S. tuberosum*, is easily crossed with it. Also, the resistance behaves as a single-gene dominant character and is readily transferred to the hybrids. Thirdly, *S. andigena* produces edible tubers and thus is much closer, agronomically, to *S. tuberosum* than are other resistant species. Fourthly, *S. andigena* behaves as a trap-crop. Its roots produce a hatching factor and are readily invaded by the larvae, which, however, fail to mature and reproduce (Williams, 1956). This failure is because the root tissues are not induced to produce giant cells, but instead the cells near the nematode's head become necrotic and there is inadequate nutrition for the female to develop.

Dutch workers first realized the commercial possibilities of resistance from *S. andigena* (Toxopeus and Huijsman, 1953), but work in other countries soon followed. The fodder variety, Antinema, and the table variety Ulster Glade were among the first commercial resistant varieties to be introduced, and there are now more than a dozen registered potato varieties with this resistance. Such varieties are of limited use, however, because they may have some undesirable features such as susceptibility to slugs, tuber blight, and resistance-breaking strains of the nematode.

Strains capable of breaking *andigena*-resistance were first reported from Peru (Quevedo Diaz *et al.*, 1956) and Scotland (Dunnett, 1957). In preliminary surveys of England and Wales (Jones, 1958) more than 50% of populations tested were classified as "resistance-breaking" or aggressive. Fortunately, aggressive populations seem comparatively rare in the main seed-producing areas of the British Isles (Scotland, Ireland) and in Europe generally. In Holland they form 24% of the total (Kort, 1962), and are rare in Germany and Scandinavia. They seem rare in North America, also, although some exist in Newfoundland.

2. Strain Nomenclature and Other Sources of Resistance

At least two strains of the nematode were identified initially, using the *S. andigena* type of resistance: (i) the "common" or "non-aggressive" strain, A, which parasitized most commercial varieties but not potatoes with resistance from *andigena*, and (ii) the "aggressive" strain, B, which parasitized both above types of potato. The resistance of *S. vernei* was studied at several centres in Europe and North America. Other species with promising resistance were *S. kurtzianum, S. multidissectum, S. sanctae-rosae* and *S. famatinae*.

Using resistant lines of some of the above species as test plants, Kort (1962) recognized four strains in the Netherlands, defined as tabulated below:

Test plant	Nematode strain			
	A	B	C	D
Solanum tuberosum	+	+	+	+
Resistant *S. andigena*	−	+	+	+
Resistant *S. kurtzianum*	−	−	+	+
Resistant *S. vernei*	−	−	−	+

+ = plant susceptible, nematode strain reproduces freely.

− = plant resistant, nematode strain reproduces poorly or not at all.

In Britain a different classification evolved. Dunnett (1961) found that some lines of *S. multidissectum* which had no resistance to the common strain A were resistant to a population, from Duddingston in Scotland, which was aggressive to resistant *S. andigena*. This population became known as strain B in Britain. By crossing *S. multidissectum* with *S. andigena*, resistance to both strains could be combined in the same plant, which, however, might be susceptible to a third strain

B

widely distributed in northern and western parts of England and Wales. This third strain was named strain E by Dunnett, to avoid confusion with the Dutch strains C and D. *S. vernei* appears to have good general resistance to all three strains known to occur in the British Isles. Thus the British classification in its simplest form reads as follows:

Test plant	Nematode strain		
	A	B	E
Solanum tuberosum	+	+	+
S. andigena	−	+	+
S. multidissectum	+	−	+
S. mult. × *S. and.*	−	−	+
S. vernei	−	−	−

The situation is complicated, however, by the existence of additional resistance in some *S. multidissectum* lines which, when crossed with *S. andigena*, may produce progeny with some resistance to E.

3. Inheritance and Nature of Resistance

Resistance to a single strain (e.g. resistance of *S. andigena* to A; of *S. multidissectum* to B) is attributed to a single dominant gene, but minor genes may also be involved, which would explain the discrepancies sometimes found. Resistance to two strains is attributed to two independent, dominant genes; examples are resistance of *S. kurtzianum*, *S. famatinae*, and *S. andigena* × *S. multidissectum*. The general resistance exhibited by *S. vernei* is less well understood but may be polygenic.

As stated above, the nature of *S. andigena* resistance is such that the nematodes hatch and invade the roots normally, but the roots fail to form the giant cells essential for nematode reproduction. The resistance mechanism of other *Solanum* species has been less well studied, but they appear generally more hostile to the nematodes because not only do they provide inadequate nutrition, but they fail to stimulate normal hatching and root invasion (Williams, 1956). Hence they are likely to prove less efficient trap crops than *S. andigena*.

4. Strains or Species?

Although the potato cyst nematode has hitherto been regarded as a single species, *Heterodera rostochiensis*, certain morphological and physiological differences between strains have been detected. Guile (1970) has studied the transient colour changes undergone by the cystic stage of British strains of the nematode in changing from a living, white female to a dead, brown cyst. In strain A, the sequence is a short white phase, a short pale yellow phase, then the well-known golden-yellow phase lasting for several weeks, before turning brown. This strain is the "golden nematode" of Long Island.

In strain B there is a short white phase, a short cream phase, a long creamy-yellow phase, then brown. In strain E there is a prolonged white phase, a short creamy-yellow phase, a short tan phase, then brown. Thus the predominant colour of the developing cyst in strain A is golden-yellow, in strain B creamy-yellow, and in strain E white.

Some British advisory nematologists have used these colour differences to determine which strains are present on the roots of growing potato crops. Such a method, if successful, would save much time and effort, compared with the established method of growing a selection of resistant and susceptible potatoes in pots of soil from the infested field and counting the cysts formed on the root-balls, the whole test occupying several weeks.

The new method, however, has its limitations and drawbacks. Firstly, only fields currently growing potatoes can be assessed. Secondly, the roots of the crop must be inspected at the correct stage of nematode development, i.e. when most of the strain A females are in the transient golden stage. Thirdly, whilst a relatively pure strain A population (most cysts golden) may be distinguished readily from a non-A population (no golden cysts), populations of strain mixtures between these extremes are difficult to analyse. Fourthly, the method is not universally applicable, because in other countries populations with golden-yellow cysts may parasitize potatoes resistant to British strain A, thus behaving like British strains B or E (Trudgill *et al.*, 1970).

The distinctive golden-yellow phase is not the only feature distinguishing British strain A worms, which also differ distinctly in larval and male characters from B and E (Webley, 1970; Trudgill *et al.*, 1970). Moreover, strain A worms do not interbreed freely with B and E worms, which, however, interbreed freely with each other (Jones, 1968). Strain A worms also differ from B and E in their protein content, as determined by polyacrylamide gel electrophoresis (Jones, 1970b). These differences have led Jones and his colleagues at Rothamsted to conclude that two species are involved, the golden nematode (British strain A) representing *Heterodera rostochiensis sensu stricto*, and British strains B and E representing an undescribed species (Jones, 1970b).

F. Interactions

In pot tests conducted at Auchincruive in southwestern Scotland, Grainger and Clark (1963) found that potato cyst nematode and the black-scurf fungus *Rhizoctonia solani* together caused much greater reduction in yield than did the organisms separately. Indeed they concluded that at the dosage used (0·9 viable cyst/g soil) the nematode alone had no effect. Stelter and Meinl (1967) in Germany, however, found the nematode to be the more serious pathogen, the fungus alone causing no damage. From pot tests conducted at Edinburgh, Dunn and Hughes (1967) concluded that the two organisms together or separately had little effect on potato growth, there being no evidence of enhanced effect when they were used jointly. Dunn and Hughes used a nematode dosage of 20 larvae/g soil, which was probably much lower than that used by Grainger and Clark. This may help to explain the discrepancy in findings, but another reason may be the climatic differences of the two localities. The fungus thrives best in cool, damp conditions, more likely to apply at Auchincruive. The Auchincruive work was repeated in later years, confirming the early findings (Anon. 1966), therefore it seems that, given suitable conditions, the nematode and fungus interact and so enhance disease and yield depression in potatoes.

H. rostochiensis also interacts with *Verticillium dahliae*, enhancing the severity of the wilt caused by the fungus. The leaf symptoms appeared earlier and the tuber yields were depressed in the presence of the nematode (Jones, 1968).

G. Control

Five main control agents are used in practice, namely hygiene, crop rotation, disease escape, resistant varieties, and nematicidal treatments. (These, which are rarely used singly, may be enforced or regulated by legislation.)

1. Hygiene

The main objective of hygiene, frequently enforced by legislation, is to prevent or delay the introduction of the pest. Costly programmes of inspections and sanctions may be undertaken to minimize introductions of the pest with imported commodities.

Imported seed potatoes with adherent soil are the most likely means of introducing potato cyst nematode to potato fields in new countries, but the nematode is found frequently in soil on tubers, rhizomes, corms, bulbs and rootstocks of other plants. Washing, brushing or warm-water or other nematicidal treatment may greatly decrease the level of contamination, but may not eliminate it. Indeed, with modern international commerce, the avenues of entry for the pest are legion, and its entry is at best merely delayed.

2. Crop Rotation

The decline in soil infestation level in the absence of a host crop is regarded as density-independent. Thus, given similar environmental conditions each year, infestation level in terms of numbers of living eelworms would show a steady asymptotic fall. The type of non-host crop, season, cultivations, soil type, organic content and micro-organisms, may affect the rate of fall, but these effects are generally small and do not greatly alter the overall picture (Fig. 2). The annual decrease determined for the British Isles is some 30–33%, or a survival rate of 4% after 8–9 years, as shown in Table II.

In contrast, the increase in infestation level is density-dependent. For main-crop potatoes in Northern Ireland, multiplication factors of 13–25× have been recorded. Even greater multiplication rates may apply to infestations below normal detection levels. Hence an interval of at least 8 or 9 years between potato crops may be required to com-pletely prevent any possible build-up of infestation (Table II).

TABLE II. Progressive decrease in soil infestation level of potato cyst nematode in the British Isles, when host crops are not grown. (Assuming an asymptotic annual decrease of 30% or 33%.)

	% Infestation remaining								
Annual decrease = 30%	70	49	34	24	17	12	8	6	4
Annual decrease = 33%	67	45	30	20	14	9	6	4	3
Years since host crop	1	2	3	4	5	6	7	8	9

Early potato crops and also canning and seed crops that require a shorter growth period will restrict nematode increase, and shorter rota-tions will suffice. In warmer climates the annual decrease in infestation level may be greater and increase per potato crop less, again decreasing the interval necessary between potato crops.

3. Disease Escape

Complete disease escape by potatoes in infested soil does not occur, but partial disease escape by first early varieties has been reported from Long Island, Scotland and Belgium. These varieties start growth in spring at soil temperatures too low for much nematode activity, and may be harvested before many of the nematodes can reproduce. Thus these crops, because they escape serious damage and restrict multiplication of the pest, can be profitably grown frequently or continuously, on infested land. Strain A worms develop more quickly than do B or E worms, at least in Northern Ireland (unpublished), and would better survive such cropping. This may explain why B and E strains are rare in traditional first-early areas.

Crops of potatoes for canning, which also need a short growth period,

FIG. 2. *Heterodera rostochiensis*; reduced growth of potatoes (right foreground) planted in infested soil where potatoes have been grown too frequently—once every 3 years. In contrast good growth of potatoes (left and background) planted in infested soil where potatoes have been grown once every 6 years and infestation has not reached a damaging level.

similarly restrict nematode increase. In Northern Ireland two profitable canning crops were grown, one immediately after the other on the same infested (strain A) land in 1969. The overall increase in eelworm was 9-fold against 25-fold for a conventional ware crop. Disease escape also was found to result from very late planting (August, instead of June or July) in Lithuania.

4. Resistant Varieties

Existing resistant varieties are of limited use, being resistant only to strain A. Such varieties may safely be grown *ad lib* in soil infested with strain A only, where they will decrease the infestation level to undetectable levels. Many infestations, however, contain traces of other strains and if existing resistant varieties are grown repeatedly in such soils, these other strains multiply unchecked, until the soils are appreciably infested with a strain or strains to which no commercially acceptable resistant variety is yet available.

Clearly, then, existing resistant varieties must be used judiciously. Jones (1970a) studied the problem in England, where strains other than A are frequent; he suggested how resistant varieties might replace some normal potato crops in rotations, thus lengthening the interval between fully susceptible crops and preventing eelworm populations increasing to damaging levels.

In seed-producing and other ostensibly "clean" fields, alternating resistant and susceptible crops may be equally beneficial, provided that any eelworm which might arise or be introduced is of strain A. This would apply to growers, in areas where only strain A worms are known, who purchase their seed from similar areas. In such areas in Northern Ireland, where present legislation requires a 4-year interval between potato crops, the period between susceptible crops would then be at least 9 years. The intervening resistant crop would act as a trap crop, however, causing a decrease in infestation equivalent to 2 years without potatoes, thus making the interval equivalent to at least 10 years. This utopian state of affairs can only be realized, however, with a massive rise in demand for resistant varieties, which is unlikely in the foreseeable future.

Potato varieties with resistance to more than one strain of the pest are desirable and Dr J. M. Dunnett in Scotland is producing material with good, albeit incomplete, resistance, derived from *Solanum vernei*, to all populations of the pest against which it has been tested.

5. Nematicides

Generally, nematicidal treatment of main-crop potato land for control of potato cyst nematode has not been economic, for several reasons.

Effective nematicides are costly to acquire or apply. They are usually phytotoxic and so must be applied outside the growing season. At that time nematodes are most resistant, being dormant within cysts, and the soils may be cold and wet, decreasing the degree of soil permeation and the nematicidal activity of the chemical. Even under the most favourable conditions the kill may be inadequate; given a nematode increase rate of some 20-fold per potato crop, a 95% kill will merely remove the worms produced in one season, and kills of this order are rarely achieved.

DD (dichloropropane-dichloropropene) mixture was the most used of the earlier nematicides, probably because it was cheaper or more effective than its rivals. Its use was profitable in certain instances, for example in Ayrshire (Grainger, 1951) where reduced multiplication of the pest (occasioned by disease escape) and enhanced prices for Epicure first-early crops combined to make a moderate kill economically worthwhile.

It has been liberally used on Long Island, where soil and climatic conditions for its application are generally more favourable than those in northern Europe. The ultimate purpose was undoubtedly to protect the mainland from invasion, and immediate crop responses form only part of the benefits from the treatment. A parallel situation exists on Vancouver Island, where the closely similar Vidden-D is being used. Apart from these special instances, DD has been little used in potato land, although interest in it has been revived recently, especially in the Netherlands.

Possible nematicides of the future are Dazomet, Temik, and yellow oxide of mercury. Dazomet, which liberates methyl isothiocyanate, is more nematicidal than DD in peats and clay soils, and often gives large yield increases (Jones, 1970a). Temik (2-methyl-2-(methylthio) pro-prionaldehyde O-(methyl-carbamoyl) oxime) is promising but largely untried. Yellow oxide of mercury is said to give good, inexpensive control when used at 5·6 kg per hectare, thoroughly mixed with the topsoil. The fact, however, that none of the above nematicides found general acceptance even before the recent public outcry against pesticide pollution suggests that their use is still problematic.

Nevertheless the full range of the beneficial side effects of nematicides is not generally appreciated. Some of them kill wireworms (DD), aphids (Temik), weeds and fungi (Dazomet), and other plant-parasitic nematodes in addition to *H. rostochiensis* (Jones, 1970a). These effects may benefit not only the potato crop, but also other crops in the rotation.

6. Legislation

Legislative control varies greatly between countries, the stringency of the measures tending to vary inversely with the degree of infestation.

Thus the U.S.A., where infestation is very restricted, has very rigorous legislation, whereas in England, where the nematode is widespread, regulations are minimal. Four main areas of legislation may be recognized, namely, import, internal traffic, cropping frequency and export regulations.

7. Integrated Control

The combined use of two or more methods of control, which has been heralded as a new approach to potato nematode control, has in fact been practised for decades. Examples already mentioned include the combination of disease escape and DD treatment in Ayrshire, and hygiene, in the form of seed certification, combined with crop rotation in seed-growing areas.

New variations on the theme involve combinations of nematicide, rotation, and resistant variety (Jones, 1970a). Use of a variety resistant to 90% of the worms in an infestation, followed by a nematicide, killing 90% of the post-crop population, could be equivalent to a 99% kill. And a post-crop nematicidal treatment killing 75% or 90% of the worms could shorten a safe rotation by 4 or 6 years (Table II). Natural enemies of the pest, such as predatory fungi and tardigrades, may have a limited role in integrated control.

H. Economic Importance

The main factors which contribute to the status of the potato cyst nematode as probably the most important pest of the potato crop are the substantial reductions in crop yields which occur when susceptible potato varieties are grown in heavily infested land and the lack, at present, of an inexpensive nematicidal treatment capable of providing an adequate level of control of the pest under field conditions. Other contributory factors are the relative ease with which the nematode may be spread from area to area, its persistence in a viable state in soil for many years and its high reproductive capacity in the presence of a growing potato crop.

In the absence of generally acceptable chemical control measures, avoidance or limitation of crop losses resulting from nematode damage is dependent on the employment of various other control measures. These may include official inspection procedures designed to minimize the risk of introducing the nematode to agricultural land and the voluntary or compulsory adoption by growers of restrictions on the cropping or treatment of their land in order to avoid or delay the development of any incipient infestation to a level where crop damage may occur. The nature and extent of these preventive measures reflect the local importance attached to the nematode as an agricultural pest and this,

B*

in turn, is largely dependent on the prevalence of existing nematode infestations within the area.

1. Nematode Widespread

In areas where the nematode is already established and widely distributed as a result of intensive potato production the pest status of the nematode is largely determined by the importance of the potato crop in the local agricultural economy. In England and Wales, for example, the widespread occurrence of the nematode throughout the traditional potato-growing areas is generally regarded as being of little practical significance as a limiting factor in the production of the national potato crop. The acreage of ware potato crops required to be grown each year has steadily declined over the past 20 years (it is now less than 5% of the tillage area available), and, under these conditions, the national importance of the nematode as a pest problem has also declined. The National Agricultural Advisory Service of England and Wales (N.A.A.S.)* provides advice to farmers regarding the nematode density of land and indicates the level of crop loss likely to occur if infested land is cropped with a susceptible potato variety. This loss has been estimated, in fen peat soil in eastern England, to approximate to a reduction in yield of 2·5 metric tons/hectare for each 20 nematode eggs present/g of soil prior to potato cropping (Brown, 1961). Such advice enables each individual grower to assess whether or not a profitable potato crop can be produced on infested land, and growers of ware potato crops are not subjected to any State-imposed restrictions on the frequency with which infested land may be cropped with potatoes.

Experience has shown that the N.A.A.S. advisory scheme has resulted in a marked reduction in the occurrence of obvious and severe crop losses (Winfield, 1965), though insidious and widely distributed losses in potato crops due to nematode attack are still common. Recent estimates (Jones, 1970a) suggest that, in England and Wales, these may total some £5 million annually.

2. Nematode Present but not Widely Distributed

Typical examples of such areas are the seed-producing areas of Scotland, Ireland and the Netherlands. In these districts, the direct crop losses due to nematode attack are generally of negligible importance and the economic significance of the nematode is largely derived from the need to protect valuable seed-producing land from nematode invasion and establishment. The distribution of seed potatoes contaminated with cysts of the nematode is the principal means by which the nematode spreads nationally and internationally and it is now a

* Now the Agricultural Development and Advisory Service (A.D.A.S.).

normal requirement that seed crops must be certified by the country of origin as having been produced in land which has been shown to be free from detectable eelworm infestation. Stringent measures to prevent any spread of the nematode to valuable seed-producing areas are, therefore, essential. These measures are normally enforced by legislation and generally include a prohibition on the cropping of infested land with potatoes, a limitation on the frequency with which non-infested land may be cropped with potatoes, and various restrictions on the movement of infested soil, plant produce, equipment, etc. The adverse economic consequences of such restrictions on the farming community may be considerable. The market value of land found to be infested is substantially reduced since it ceases to be available for potato production, and the restrictions on the frequency of cropping any land with potatoes restricts the acreage grown and the choice of fields. Restrictions on the types of crops which may be grown in infested land may be such that, in practice, the infested land is normally sown to grass. The cost of operating the official inspection services (soil sampling and laboratory services) may be borne, in part, by the growers; such services may involve an expenditure of £5 or more per hectare.

3. Nematode with Foothold Only

Where the nematode has gained a foothold but is not yet established, rigorous measures may be taken to prevent or delay its establishment, as in the U.S.A. and Canada. The campaign by the U.S. Department of Agriculture, to confine the nematode to Long Island, must rank as one of the most vigorous and costly ever mounted against a crop pest. Infested land is taken out of production, with compensation to growers, and movement of infested or suspect soil, plants, produce, containers, machinery, etc., is severely restricted. Thousands of hectares are treated with DD mixture costing millions of dollars. Machinery, containers and plants for propagating are also treated with nematicides, and attempts have been made to disinfest seed potato tubers.

The nematode has now been detected parasitizing mainland crops, some 28 years after detection on Long Island. It is impossible to estimate the value of this respite to the mainland, but it may well outweigh the cost of the Long Island campaign.

Canada is now in much the same plight, with the nematode established offshore, in Newfoundland and in Vancouver Island. Because Newfoundland is more remote and has less traffic with the mainland than has Long Island, the threat is perhaps less. Vancouver Island, like Long Island, is close to and has much commuting with the mainland, but is more remote from the main potato growing areas. Therefore the Canada Department of Agriculture, in attempting to protect its potato-growing industry, would seem to be more favourably placed than was the U.S. Department of Agriculture in 1941. This is assuming, of course, that the nematode is not already established in undetected foci on the Canadian mainland.

4. Nematode Apparently Absent

Some countries, e.g. Australia, New Zealand, are remote from known infested areas and import relatively little produce likely to be contaminated; thus their chance of remaining uninfested is high. Other countries, like South Africa, import considerable quantities of seed potatoes annually from Europe, and run a much greater risk of becoming infested. The risk is reduced by extra precautions such as washing or brushing of seed tubers; nevertheless the pest probably has been or will be introduced, but it will not necessarily become an established pest. Although introduced to the coastal plain of Peru, it apparently failed to establish, the climate and/or soil type being somehow unsuitable (Simon F., 1955). Current costs to South Africa are mainly the extra cost of importing washed or brushed seed potatoes.

III. Genus *Ditylenchus*

A. History and Distribution

Two distinct types of damage to potatoes were reported from Europe in 1888, namely tuber rot associated with distorted top growth, and tuber rot without above-ground symptoms. Both types of damage were attributed to the stem nematode, *Ditylenchus dipsaci*, until Thorne (1945) described nematodes causing tuber rot in Idaho, U.S.A., as a new species, *D. destructor*, the potato rot or potato tuber nematode.

D. destructor soon became recognized as a troublesome pest of potatoes in North America, especially in the northwestern U.S.A. and Prince Edward Island, Canada. Its importance in Canada soon waned, owing to the removal of infested land from potato cultivation (Baker, 1952). More recently this species has become recognized as a major pest of potato crops in the western and southern borders of the U.S.S.R., from Estonia to Kazakh. Elsewhere it is causing concern only in a few localities in Europe.

D. dipsaci is not a serious pest of potatoes except perhaps in Germany and the Netherlands.

B. Morphology and Biology

1. Life-cycle

All stages of *Ditylenchus* may be found in the host tissues or in the surrounding soil, the worms moving into, out of and through the tissues at will. They feed mainly on stems and foliage (*D. dipsaci*) or on underground stem structures (*D. destructor*). Morphologically, the two species are very similar, but biologically they differ in some important ways. *D. dipsaci* apparently is an obligate parasite of living stem and leaf tissue of higher plants, incapable of existing on decayed plant tissue or associated fungi. *D. destructor*, on the other hand, thrives both on higher plants and on fungi. Moreover, *D. dipsaci* is a complex species comprising many "races" distinguished only by their host preference; whereas only limited race specialization occurs in *D. destructor*. Also, *D. dipsaci* is very tolerant of desiccation, persisting anabiotically in the dried state for several years; *D. destructor* is thought incapable of such anabiosis.

2. Host–parasite Relationships

(a) *Host Ranges*. The limited occurrence of serious attacks by *D. dipsaci* on potato is puzzling, as potatoes are reportedly susceptible to many races or isolates of this pest. Perhaps, despite these findings, potatoes are normally damaged only by specialized races of limited distribution and host range.

Damage by *D. destructor* has been much more frequently and widely reported. This species has a wide host range amongst crops and weeds (particularly those with swollen, or rhizomatous underground parts) and common fungi. Beet, mangold, carrot, dahlia, gladiolus, iris, tulip, clovers, vetches, grasses and several common weeds of arable land are hosts.

(b) *Interactions and Plant Symptoms*. *D. destructor* infestation starts as small glistening, whitish "feeding pockets" in the cortex, detected by removing the peel (Duggan and Moore, 1963). As the pockets enlarge they may coalesce, the affected tissue darkening gradually through greyish to dark brown or black in colour, as secondary organisms— fungi, bacteria, microbivorous nematodes, etc.—enter. *D. destructor* is seldom found in the dark tissue, being concentrated in the whitish "mealy" tissue at the advancing edge of the lesion (Fig. 3a). The tuber skin is not attacked and may remain intact but papery thin over the lesions, or may crack as stresses are set up in the tuber (Fig. 3b). Affected tissues are soft and mealy, unlike the hardness of tissue attacked by blight.

Unlike the above lesions which tend to be superficial, lesions caused by *D. dipsaci* may extend throughout the tuber, the affected tissues being spongy and yellowish to brown in colour. Also, *D. dipsaci* attacks the above-ground parts, causing typical stunting, thickening and distortion which may occur without tuber symptoms. *D. destructor* rarely occasions foliage symptoms, but confines its attacks to underground stem structures such as stolons, tubers and stem bases.

Both species are primary pathogens, invading tubers via eyes, lenti-cels, or stolons. The rot caused by *D. destructor* is accentuated, however, when fungi also are present, and damage by *D. dipsaci* may accelerate invasion by *Phoma solanicola* (Hijink, 1963).

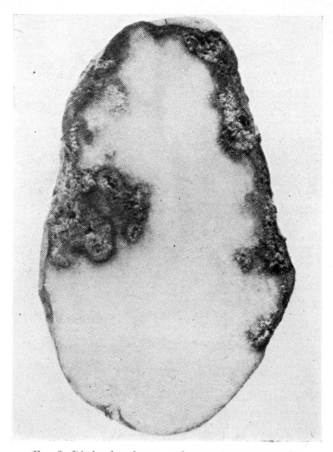

FIG. 3. *Ditylenchus destructor* damage to a potato tuber.

(a) Potato tuber cut to show internal lesions. (Crown Copyright, permission from H.M.S.O.)

C. Ecology

Ladigina (1956) found that *D. destructor* survived at −28°C and developed in 60–18 days at temperatures varying from 5° to 34°C; greatest infestation occurred at 15–20°C and 90–100% relative humid-ity, the worms not surviving relative humidities below 40%. Hence this pest is well able to parasitize growing and stored tubers under most climatic conditions. Its intolerance of drought rather than heat may be the reason why it is not a potato pest of any consequence in some warm climates—India, Africa, South America. Russian workers found that the nematode overwintered mainly in unharvested, infested tubers

FIG. 3. *Ditylenchus destructor* damage to a potato tuber.
(b) Surface view showing cracked skin. (Crown Copyright, permission from H.M.S.O.)

or in seed in stores. Less is known about other environmental factors but Safyanov (1966) found that ammonium fertilizers increased yields and decreased storage losses in Kazakh.

Incidence of *D. dipsaci* on potatoes in Germany seemed independent of temperature, rainfall and pH. Otherwise little is known of its ecology, except that most races overwinter better in clay soils than in the lighter soils in which potatoes are normally grown (Seinhorst, 1957).

D. Control

1 Hygiene

Unharvested, infested tubers should be removed and destroyed. Seed should come from uninfested crops, or should be hand-picked. The modern tendency to use pre-emergence and haulm-destroying herbicides should greatly decrease nematode propagation on weeds.

2. Seed Treatments

Although hot-water treatment has been successfully used to control *D. destructor* in iris bulbs, no equally effective treatment has been evolved for potato tubers. Formalin dips and low temperature storage have been tried in the U.S.S.R., but visual sorting of seed tubers before planting seems the main method practised (Ismailov, 1967).

3. Soil Nematicides

Ethylene dibromide has been used successfully against *D. destructor* in the U.S.A. when applied as two split applications, the soil being turned over at the second application, to bury loose topsoil and infested tubers (Dallimore, 1955). Elsewhere soil nematicides have been deemed uneconomic and ineffective against *D. destructor* (Ismailov, 1967), and generally unsuited to the heavier soil types preferred by *D. dipsaci* (Seinhorst, 1957).

4. Crop Rotation

Crop rotation is one of the chief means of controlling *D. destructor* in the U.S.S.R.; although Thorne (1961) thought it would be ineffective in the U.S.A., because of the wide host range of the pest.

5. Disease Escape

Russian workers found that propagating from early-harvested seed gave healthy stocks in one or two years, and advocated late planting and early harvesting, especially for seed crops. Presumably disease escape occurs, late plantings perhaps coinciding with low soil moisture, insufficient for maximum invasion. Likewise early harvesting may not allow sufficient time for maximum nematode multiplication. Disease escape may also be promoted by cool, dry storage, which limits multiplication within tubers and spread of infestation to other tubers.

6. Resistance

Deliberate attempts to breed potatoes resistant to *Ditylenchus* have not been made, and existing varieties tested have generally been found susceptible. Varieties vary, however, in degree of susceptibility to rotting caused by *D. destructor*, the Dutch variety Rode Star being found resistant by Seinhorst (1949), who also found resistance to *D. dipsaci* in wild potatoes and hybrids. A comprehensive search for resistance might well be rewarding.

7. Integrated Control

None of the above control measures, taken singly, is likely to be effective where *Ditylenchus* is prevalent in potato crops. Therefore combinations of them are recommended, such as: healthy seed, planted late, harvested early, stored as cool and dry as possible; proper rotation —potatoes not more frequently than once in 3 or 4 years, avoid other susceptible crops; field hygiene—removal of old infested tubers, weed control.

E. Economic Importance

1. D. dipsaci

Seinhorst (1957) regarded *D. dipsaci* as a serious pest of potatoes in Germany and the Netherlands, but relatively little has been published in other countries. It seems to be a potato pest of only local importance. Estimates of actual losses caused are lacking, although Kaai (1964) published a mathematical formula linking nematode population density and crop damage.

2. D. destructor

Generally, this is a much more important pest of potatoes. Judging from the literature it reaches its greatest importance in parts of the U.S.S.R., infesting 10–20% of potato crops in Armenia and causing great losses; and causing losses in western Ukraine estimated at some 162,500 metric tons annually. Elsewhere in Europe it is of only local importance, although it is said to be one of the nematode species spreading into new polders in the Netherlands. It is currently a local problem in Ireland.

Importance of *D. destructor* in North America seems to have declined, especially in Canada as noted above. In the U.S.A., successful soil fumigation with ethylene dibromide may have contributed to the decline.

IV. Other Nematodes

A. General Assessment

Species of *Meloidogyne*, *Pratylenchus* and *Trichodorus* are associated with crop losses in potatoes in many countries, and are considered as major pathogens in the sections below. *Longidorus maximus* caused depressed growth, associated with typical root galling, in patches in potato crops in Germany (Sprau, 1960). *Hexatylus vigissi* attacked leaves, stems and tubers, causing withering and reduced yield of a potato crop in the U.S.S.R. (Skarbilovich, 1959), *Neotylenchus abulbosus* caused rotting of stored tubers in Estonia (Krall, 1958), and *Tylenchorhynchus dubius* was associated with reduced potato growth in pot trials in England (Kyrou, 1969).

Species of *Tylenchorhynchus*, *Rotylenchus* and *Belonolaimus* are parasitic on potatoes, high numbers of them being sometimes associated with the crop, but their pathogenic status is not clear. Species of *Helicotylenchus*, *Hemicycliophora*, *Nacobbus*, *Paratylenchus*, *Rotylenchulus*, *Xiphinema* and other spear nematodes have been found associated with potato roots or tubers, but they are probably of no economic importance. Species of *Rhabditis*, *Diplogaster* and other non-spear (microbivorous) nematodes often occur as secondary organisms in rotting tissues, e.g. in tubers attacked by *Ditylenchus*. They are presumed to be harmless, but may help to spread bacterial and fungal pathogens.

B. Root-knot Nematodes, *Meloidogyne* spp.

1. Distribution

The dominant species on potato in Europe and North America is *M. hapla*, followed by *M. incognita* and *M. incognita acrita*. In Africa and Asia *M. javanica* and *M. incognita* are dominant, followed by *M. incognita acrita* and *M. hapla*, the last being the species found on potatoes in Hokkaido, Japan. Data for South America is sparse, the species most often cited being *M. incognita* and *M. javanica*. *M. arenaria* has occasionally been recorded from potatoes in most continents, although some African populations assigned to it are now regarded as representing a new species, *M. ethiopica* (Whitehead, 1968). *M. thamesi* is mentioned occasionally from potatoes in America and Africa, and *M. acronea* and *M. africana* have been recorded on potatoes in Africa.

2. Morphology and Biology

(a) *Life-cycle*. Species of *Meloidogyne* are sedentary root and tuber endoparasites, the life-cycle being similar to that of *H. rostochiensis*, with some important differences. The white, swollen females do not develop into persistent, brown cysts, but deposit all their eggs in an external egg-mass. Hatching occurs when the physical conditions are suitable, and is not dependent on a hatching factor from host roots. Males may be present but are not necessary for reproduction. Sex is environmentally determined as for *H. rostochiensis*.

(b) *Host Ranges*. Species of *Meloidogyne* usually have wide host ranges amongst crops and weeds in many botanical families.

(c) *Plant Symptoms and Interactions.* The diagnostic symptoms of a *Meloidogyne* attack on potatoes are the galls found on roots and tubers. Like *Heterodera*, the nematodes feed on giant cells formed in the stele in response to the nematodes' presence. The galling, mainly an enlargement of plant cortex, is probably not essential for nematode nutrition, but may protect the female worms from predators and unfavourable soil conditions. Galling on tubers may render them unsaleable.

Meloidogyne spp. are associated with the fungi *Rhizoctonia solani* and *Verticillium* sp. and the bacterium *Pseudomonas solanacearum* in disease complexes (Hoyman and Dingman, 1967; Hickey, 1969; Feldmesser and Goth, 1970).

3. Ecology

Meloidogyne damage to potatoes is usually associated with light soils or peats, partly because the nematodes may prefer such soils, but mainly because such soils are preferred for potato growing.

Soil moisture has little effect on the degree of tuber galling by *M. hapla* (Griffin and Jorgenson, 1969b). In wetter soils, however, tubers have swollen lenticels, possibly providing easy ingress for nematodes. Aggravated attack by *Meloidogyne* on potatoes under irrigation is usually ascribed to spread of the nematodes in the irrigation water.

Soil temperature requirements of these nematodes vary with species. Griffin and Jorgenson (1969a) found an optimum of 25°C for *M. hapla* in Utah, at which temperature a generation occupied 10 weeks. Other species have even higher temperature requirements, which obviously limits their effective geographical range. Hence they reach their greatest importance in tropical and warm temperate climates, and are relatively unimportant in cool temperate regions such as northern Europe.

Other soil factors have been little studied and are little understood. Cultivations and organic additives, which sometimes cause unexplained suppression of *Meloidogyne* populations, are discussed under control measures below.

C. Root lesion Nematodes, *Pratylenchus* spp.

1. Distribution

Recent changes in nomenclature (Seinhorst, 1968) cast doubt on the identity of many species in publications, but undoubtedly *P. penetrans* is the dominant species on potatoes in Europe and North America. Others recorded are *P. crenatus, P. minyus, P. thornei* and *P. scribneri* in Europe; *P. crenatus, P. brachyurus* and *P. scribneri* in N. America; *P. andinus* and *P. scribneri* in S. America; *P. brachyurus* and *P. scribneri* in Africa; and *P. vulnus* and *P. coffeae* in Japan.

2. Morphology and Biology

(a) *Life-cycle.* Species of *Pratylenchus* are migratory or wandering root endoparasites, the vermiform juveniles and adults moving freely into and out of roots. Males are common in some species, but not in others. Eggs are laid in soil or in roots, and hatch when the physical conditions are suitable.

(b) *Host Ranges.* The host ranges of species extensively studied are usually very wide indeed. Oostenbrink *et al.* (1957) found *P. penetrans* in the roots of 164 crop plants and weeds. Host efficiency, however, varies greatly and is discussed below under control by rotation.

(c) *Plant Symptoms and Interactions.* *Pratylenchus* spp. cause root necrosis, sometimes visible externally as darkened lesions. Some species, e.g. *P. brachyurus* and *P. scribneri*, also cause lesions—scabs, pustules or

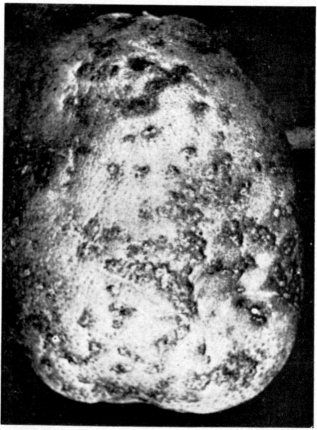

Fig. 4. The surface damage to a potato tuber caused by *Pratylenchus brachyurus*. Reproduced with permission from Koen and Hogewind (1967) and under copyright Authority 4500 of the Government Printer of the Republic of South Africa.

pimples—on tubers (Fig. 4), reducing their market value. Above-ground symptoms may appear as circular patches of stunted, chlorotic plants (Dickerson *et al.*, 1964).

P. penetrans and possibly other species are associated with *Rhizoctonia solani*, *Verticillium* and *Fusarium* in disease complexes of potato.

3. Ecology

Pratylenchus damage to potatoes is usually associated with light soils, partly because some of the species involved, e.g. *P. penetrans*, prefer sandy soils, and partly because such soils are preferred to heavier soils for potato culture.

Species of *Pratylenchus* depend on optimum moisture for migration from plant to plant, although Koen (1967) found *P. brachyurus* remarkably resistant to drought in South Africa.

Soil temperature requirements vary greatly with species. The optimum for *P. penetrans* in Wisconsin was 16–20°C (Dickerson *et al.*, 1964), and for invasion of potatoes by *P. coffeae* in Japan, 25–28°C (Yokoo and Kuroda, 1966). These differing soil temperature requirements obviously limit the geographical ranges of the various species. Hence *P. penetrans* is an important pest in regions of Europe but is unimportant or absent in warmer climates where species with higher temperature optima, such as *P. brachyurus* in Africa and *P. coffeae* in southern Japan, replace it.

D. Stubby-root Nematodes, *Trichodorus* spp.

1. Distribution

Species originally involved in transmitting tobacco rattle virus to potato were *T. pachydermus* and *T. primitivus* in Europe and *T. christiei* in North America. More recently, nine species in Europe (Hoof, 1968) and three in the U.S.A. have been proven vectors. Probably most members of the genus are potential vectors.

2. Morphology and Biology

(a) *Life-cycle.* Species of *Trichodorus* are migratory root ectoparasites, the vermiform juveniles and adults wandering freely from root to root. Males are known in some species but not in others. Eggs are laid in the soil.

(b) *Host Ranges.* Host ranges of a few species have been studied and they are generally wide, but very little is known about the host preferences of many other species.

(c) *Plant Symptoms and Interactions.* *Trichodorus* spp. cause stunting of roots ("stubby-root"), but their main significance as potato pests is that they transmit tobacco rattle virus, the cause of stem "mottle" and tuber "spraing" or corky ringspot in potato.

3. Ecology

Trichodorus spp. associated with rattle virus in potato crops usually occur in light, sandy soils, where they may be found at great depths. Other factors affecting distribution of the disease are that potato cultivars vary greatly in susceptibility, and that the right combination of nematode species and virus strain must be present (Hoof, 1968). Little is known of the effects of other ecological factors. The disease appears to be most prevalent in northwestern Europe, probably because this is where it has been most studied; relatively little is known of its distribution in many other potato-growing regions.

E. Control

1. Hygiene

General precautions, such as minimizing spread of infested soil and plant material and controlling weeds, restrict the spread of these nematodes. Tubers from *Meloidogyne* or *Pratylenchus* infested crops should not be used as seed, as even outwardly clean seed may be infested. Tobacco rattle virus is not readily spread by infested seed, as it is unlikely that the virus strain will suit the *Trichodorus* population encountered in the new habitat; each population transmits its own strain of the virus (Hoof, 1968). In several countries plant health regulations prohibit the use of seed infested by these nematodes.

2. Rotation

Cereals, grasses, asparagus and onion were not attacked by *M. hapla* in the U.S.A. (Faulkner and McElroy, 1964), and should be useful rotation crops against this species. Attack on some other crops (soybean, lucerne, *Phaseolus vulgaris*) varied greatly with variety. *Tagetes*, the roots of which contain nematicidal substances, and *Eragrostis curvula*, which greatly reduces *Meloidogyne* populations, are sometimes grown in rotations.

Oostenbrink *et al.* (1957) found *P. penetrans* in the roots of 164 crops and weeds tested, the density of infestation varying from 2·3 in *Tagetes patula* to over 10,000/g of root in *Vicia faba*. Since host efficiency is a product of root size and nematode density in the roots, crops like cereals, grasses and legumes with large root systems and moderate to high nematode densities are efficient hosts and unsuitable rotation crops. Beets and mangold, on the other hand, with smaller, lightly infested root systems are inefficient hosts and suitable rotation crops.

The host ranges of many *Trichodorus* vectors of rattle virus have not been studied but they may well be wide. A study of the host range of both *Trichodorus* and the virus may be rewarding, as use of crops immune to them combined with weed control may usefully decrease virus incidence. Asparagus decreases *Trichodorus* populations (Rohde and Jenkins, 1958).

3. Resistant Varieties

Resistance to *Meloidogyne* in potatoes has not been much exploited, despite the serious crop losses ascribed to these nematodes, notably in the U.S.A. and South Africa (see p. 46). The resistance of wild potatoes and hybrids to *Meloidogyne* is being investigated in India (Khanna, 1966) and Argentina (Brücher, 1967). Khanna investigated multiple resistance—to drought, heat, frost and *Meloidogyne*.

Useful resistance to *Pratylenchus* and *Trichodorus* is not known, but potato varieties differ in susceptibility to rattle virus.

4. Direct Control

(a) *Seed Treatments.* *Meloidogyne* in tubers was controlled by hot-water treatment at 46–47·5°C for 2 h, and by dry heat at 45°C for 48 h. Hot-water treatment at 50°C for 30–60 min. has been suggested for *Pratylenchus.*

(b) *Soil Treatments.* DD mixture, ethylene dibromide, dibromochloropropane (DBCP), and methyl isothiocyanate (MIC) compounds such as Vapam are the chemicals most used to control *Meloidogyne* and *Pratylenchus* in potato fields, although DBCP may be phytotoxic and DD occasionally causes taint. MIC compounds also control harmful fungi such as *Verticillium* and *Fusarium.* Use of nematicides to control *Pratylenchus* in potato fields in Europe is, however, uneconomic.

Heating soil by open or smouldering fires controlled *M. incognita* (Nirula *et al.*, 1968) but clearly this is only applicable on a small scale.

(c) *Treatment of Growing Crop.* Post-emergence spraying of potatoes, growing from infested seed, with dimethoate eliminated *M. hapla* under glass in U.S.A., but applying another systemic organophosphorus compound, Disyston, caused no reduction in *Meloidogyne* attack in Germany.

5. Cultural and Other Methods

Cultivations occasionally cause unexplained suppression of these nematodes, as do farmyard and green manures and crop residues. Cultivations may alter soil structure, temperature and moisture sufficiently to interfere with nematode activity and multiplication, particularly in hot dry climates. The addition of organic matter to soils may promote fungal and other enemies of nematodes, or release toxic compounds.

Such phenomena merit close investigation, especially in developing countries where nematicide use may be uneconomic.

A measure of disease escape from *Meloidogyne* may result from early spring planting; conversely, an unusually heavy infestation of *M. incognita acrita* in South Carolina in 1964 was attributed to late planting occasioned by inclement spring weather (Sitterly and Fassuliotis, 1965). Early lifting and cool storage conditions prevent deterioration of tubers infested with *P. brachyurus* (Koen and Hogewind, 1967).

F. Economic Importance

1. Meloidogyne

Serious crop losses attributed to root-knot have been reported from many countries. In South Africa, soil fumigation with ethylene dibromide before the potato crop in a four-year rotation controlled *M. javanica* and doubled the value of the crop, increasing the net return by £124/hectare (Koen, 1966). *M. hapla* is a limiting factor and a hazard to potato production in the western U.S.A., and unusually heavy attacks by *M. incognita acrita* in South Carolina in 1964 caused losses of up to 2500 dollars/hectare (Sitterly and Fassuliotis, 1965).

2. Pratylenchus

Oostenbrink (1961) found potato crop losses of one-third or more in the Netherlands correlated with density of *P. penetrans* in the roots. The same nematode caused patches of stunted, chlorotic growth and reduced yields in potato crops in Wisconsin (Dickerson *et al.*, 1964). Koen (1967) found *P. brachyurus* an economic pathogen of potatoes in South Africa, the visible lesions on tubers reducing their market value as table potatoes and making them worthless as seed.

3. Trichodorus

Although neither the nematodes nor tobacco rattle virus are considered seed-borne in the commercial sense, presence of affected tubers in a seed crop may lead to its rejection. Published data of such rejections are not available, but at least one Northern Irish seed crop failed certification, with considerable loss to the grower.

References

Anon. (1966). *Rep. W. Scotl. agric. Coll.* Year 1965–66, pp. 63–68.
Baker, A. D. (1952). *Can. Insect Pest Rev.* **30**, 118–120.
Brown, E. B. (1961). *Nature, Lond.* **191**, 937–938.

Brücher, H. (1967). *Am. Potato J.* **44**, 370–375.

Chitwood, B. G. and Feldmesser, J. (1948). *Proc. helminth. Soc. Wash.* **15**, 43–55.

Dallimore, C. E. (1955). *Pl. Dis. Reptr.* **39**, 511–515.

Dickerson, O. J., Darling, H. M. and Griffin, G. D. (1964). *Phytopathology* **54**, 317–322.

Duggan, J. J. and Moore, J. F. (1963). *Ir. J. agric. Res.* **2**, 75–86.

Dunn, E. and Hughes, W. A. (1967). *Eur. Potato J.* **10**, 327–328.

Dunnett, J. M. (1957). *Euphytica* **6**, 77–89.

Dunnett, J. M. (1961). *Rep. Scott. Pl. Breed. Stn* Year 1961, 39–46.

Ellenby, C. (1948). *Nature, Lond.* **162**, 704.

Ellenby, C. (1952). *Nature, Lond.* **170**, 1016.

Faulkner, L. R. and McElroy, F. D. (1964). *Pl. Dis. Reptr* **48**, 190–193.

Feldmesser, J. and Goth, R. W. (1970). *Phytopathology* **60**, 1014.

Grainger, J. (1951). *Res. Bull. W. Scotl. agric. Coll.* **10**.

Grainger, J. and Clark, M. R. M. (1963). *Eur. Potato J.* **6**, 131.

Griffin, G. D. and Jorgenson, E. C. (1969a). *Phytopathology* **59**, 11.

Griffin, G. D. and Jorgenson, E. C. (1969b). *Pl. Dis. Reptr* **53**, 259–261.

Guile, C. T. (1970). *Pl. Path.* **19**, 1–6.

Hickey, K. D. (1969). Fungicide and nematicide tests. Results of 1968. *Winchester, Va: America Phytopathological Society* Vol. 24.

Hijink, M. J. (1963). *Neth. J. Pl. Path.* **69**, 318–321.

Hoof, H. A. van (1968). *Nematologica* **14**, 20–24.

Hoyman, W. G. and Dingman, E. (1967). *Am. Potato J.* **44**, 165–173.

Ismailov, G. M. (1967). *In* "Nematode Diseases of Crops" (N. M. Sveshnikova, ed.), pp. 138–144. Izdatelstvo "Kolos", Moscow. [In Russian]

Jones, F. G. W. (1958). *Pl. Path.* **7**, 24–25.

Jones, F. G. W. (1968). *Rep. Rothamsted exp. Stn for 1967*, pp. 141–162.

Jones, F. G. W. (1970a). *Jl. R. Soc. Arts* **118**, 179–196.

Jones, F. G. W. (1970b). *Rep. Rothamsted exp. Stn. for 1969*, Part 1, pp. 176–204.

Jones, F. G. W. and Parrott, D. M. (1969). *Ann. appl. Biol.* **63**, 175–181.

Kaai, C. (1964). *Nematologica* **10**, 72.

Khanna, M. L. (1966). *Curr. Sci.* **35**, 494–496.

Koen, H. (1966). *Nematologica* **12**, 109–112.

Koen, H. (1967). *Nematologica* **13**, 118–124.

Koen, H. and Hogewind, W. L. (1967). *S. Afr. J. agric. Sci.* **10**, 543–550.

Kort, J. (1962). *Meded. LandbHogesch. OpzoekStns Gent* **27**, 754–759.

Krall, E. (1958). *Izv. Akad. Nauk est. SSR ser. biol.* **3**, 187–193.

Kyrou, N. C. (1969). *Eur. Potato J.* **12**, 215–218.

Ladigina, N. M. (1956). *Probleme Parazit.*, 2nd, 298–299.

Nirula, K. K., Kumar, R. and Bassi, K. K. (1968). *Am. Potato J.* **45**, 262–263.

Oostenbrink, M. (1961). *Neth. J. agric. Sci.* **9**, 188–209.

Oostenbrink, M., s'Jacob, J. J. and Kuiper, K. (1957). *Tijdschr. PlZiekt.* **63**, 345–360.

Ouden, H. den (1967). *Nematologica* **13**, 325–335.

Quevedo Diaz, A., Simon F., J. E. and Toxopeus, H. J. (1956). *Infme. mens. Estac. exp. agric. La Molina* **30**, 10–15.

Rohde, R. A. and Jenkins, W. R. (1958). *Phytopathology* **48**, 463.

Safyanov, S. (1966). *Zashch. Rast. Vredit. Bolez.* No. 3, pp. 53–54.

Seinhorst, J. W. (1949). *Landbouwk. Tijdschr., s-Grav.* **61**, 638–641.

Seinhorst, J. W. (1957). "Phytonematology in western Europe." *Technical Committee, Southern Regional Nematology Project (S–19)*, Auburn, Alabama.

Seinhorst, J. W. (1968). *Nematologica* **14**, 497–510.

Simon F., J. E. (1955). *Infme mens. Estac. exp. agric. La Molina* **29**, 8–14.

Sitterly, W. R. and Fassuliotis, G. (1965). *Pl Dis. Reptr* **49**, 723.

Skarbilovich, T. S. (1959). *Trudy vses. Inst. Gel'mint.* **6**, 395–400.

Sprau, F. (1960). *Nematologica Suppl.* **2**, 49–55.

Stelter, H. and Meinl, G. (1967). *Z. PflKrankh. PflPath. PflSchutz* **74**, 671–675.

Thorne, G. (1945). *Proc. helminth. Soc. Wash.* **12**, 27–34.

Thorne, G. (1961). "Principles of Nematology." McGraw-Hill, New York.

Toxopeus, H. J. and Huijsman, C. A. (1953). *Euphytica* **2**, 180–186.

Trudgill, D. L., Parrott, D. M. and Stone, A. R. (1970). *Nematologica* **16**, 410–416.

Webley, D. (1970). *Nematologica* **16**, 107–112.

Whitehead, A. G. (1968). *Trans. zool. Soc. Lond.* **31**, 263–401.

Williams, T. D. (1956). *Nematologica* **1**, 88–93.

Winfield, A. L. (1965). *N.A.A.S. q. Rev.* **67**, 110–117.

Yokoo, T. and Kuroda, Y. (1966). *Agric. Bull. Saga Univ.* No. 22, pp. 93–103.

3

Nematode Diseases of Sugar-beet

B. Weischer and W. Steudel

Biologische Bundesanstalt
Münster, Germany

I. Introduction

The sugar-beet, *Beta vulgaris saccharifera,* is a comparatively young cultivated plant, which was bred by selecting from white strains of fodder beet, originating from the Mediterranean wild beet, *Beta maritima.* After A. S. Markgraf demonstrated the similar identity of beet-sugar and cane-sugar in 1747 sugar production from beet was started by F. C. Achard. France and Germany were the first to develop this sugar industry and were later followed by other European countries. In spite of several backward steps, not the least being that caused by the sugar-beet cyst eelworm, the European sugar-beet industry developed rapidly. Most European countries, the Middle East, several Asian countries, some parts of Africa and several countries of North and South America currently produce sugar-beet providing that the climatic and agricultural conditions are suitable for the cultivation of this demanding crop. The beet growing area is still extending and certainly will continue to do so for some time. The world sugar production in 1970/71 amounted to roughly 73 million tons (Licht, 1971), and about 30·4 million tons (about 42%) of this is produced from sugar-beet. 81·2% of the beet-sugar comes from the U.S.S.R. and Europe. Western Europe alone produces 11·5 million tons and eastern Europe 13·3 million tons. The remainder is produced elsewhere, but not in Australia where no sugar-beet is grown. The U.S.A. produces 3·9% of the world's beet-sugar. Apart from sugar production, sugar-beet is very important as a forage, as not only the crown leaves but also beet pulp are used to feed cattle in many countries. Additionally, the molasses can be used as forage and for the production of alcohol and yeast.

II. Nematodes Attacking Sugar-beet

A. Sugar-beet Cyst Eelworm, *Heterodera schachtii*

H. schachtii is the most important nemic pest of sugar-beet production. Kühn (1881) demonstrated that this nematode was the cause of the so called "Rübenmüdigkeit" in the region of Madeburg, Germany. At present the sugar-beet cyst eelworm is known from almost every country that has a long tradition of sugar-beet production. The nematode is distributed throughout most of Europe including the U.S.S.R., parts of the Middle East, of North America and of Australia and probably will be discovered eventually in sugar-beet growing countries at present considered to be nematode-free.

H. schachtii shows pronounced sexual dimorphism, the adult females being sedentary, lemon-shaped bodies full of eggs and larvae, whereas the males are typically vermiform. The hatching of the second stage larvae is stimulated by root secretions of many plants, hosts as well as non-hosts, and by other soil chemical or physical factors (Shepherd, 1962). At a soil humidity of 10–20% and at a soil particle size of 150–250 μm the larvae are active at 8–10°C and reach maximum activity at 20–24°C (Wallace, 1958; Raski and Johnson, 1959). The life-cycle is completed within 57 days at 17·8°C and within 23 days at 29°C (Ladigina, 1960). The number of generations per year therefore, depends on the average soil temperature during the period of active host growth. In temperate zones like central Europe two complete generations per year are produced on sugar-beet. A third generation may start to develop but fails to complete the life-cycle because of lower temperatures in the autumn. In warmer regions like the Mediterranean area and the Middle East more generations can be expected and they already have been recorded from California (Raski and Johnson, 1959). The greatest multiplication of the nematode takes place on sugar-beet roots at 5–40 cm depth. Deeper in the soil the number of cysts decreases rapidly although single cysts with viable larvae can be found as deep as 1·5 m in porous soils. The vertical distribution of *H. schachtii* is influenced by the age of infestation. Thus in recently infested fields the cysts are more or less confined to the topsoil whereas in older infestations the cysts occur at much greater depth (Table I). These differences in vertical distribution have a remarkable influence on the effect of control measures as will be discussed later.

TABLE I. Number of *Heterodera schachtii* cysts
per 100 g of soil in young and old foci of natural
infestation (after Goffart, 1954)

Depth in cm	Number of viable cysts	
	Young focus	Older focus
0–10	228	31
10–20	31	41
20–30	11	22
30–40	2	27
40–50	4	35

The host range of *H. schachtii* includes mainly species of Chenopodiaceae and Cruciferae. The size of the host root system affects the nematode's ability to reproduce (Jones, 1965). Consequently, the nematode multiplication rate is higher the lower the initial population density, within certain limits (Table II). The damage caused by this nematode

TABLE II. Multiplication rate of *Heterodera schactii* on
sugar-beet (modified after Goffart, 1953)

Number of initial cysts per pot	Number of new cysts per pot	Multiplication factor
1	1641	1641
2	1206	605
3	1563	521
5	1324	265
10	1826	183
20	2050	102
50	2162	43

depends on the initial nematode population density, and on the general soil and climatic conditions which influence the growth of the host plant and the nematode survival. In areas of intensive sugar-beet production in central Europe sugar losses of 25% and more can be expected when nematodes are favoured by one-sided crop rotations (Goffart, 1952). In warmer climates the losses can be higher because the nematode damage is often increased by secondary pathogens. Recent experiments have shown that young beet seedlings are particularly sensitive to nematode attack (Steudel, 1970). Provided that beet seedlings develop a good root system prior to the main attack by the nematodes the losses are comparatively low even at high infestation rates, and in spite of the remarkable multiplication rate of the nematode. In areas with a temperate climate where beets are drilled at low temperatures the early sown beets are less damaged than beets sown in May because the larvae are able to invade the roots only at higher temperatures. The earlier germinated plants have already developed an efficient root system by the time of nematode attack and can tolerate and compensate for the attack. When beets are drilled at higher temperatures (subtropical regions or May drilling in a temperate climate) the young plants are attacked and heavily damaged whilst they are germinating (Raski and Johnson, 1959; Polychronopoulos and Lownsbery, 1968). Under such conditions the losses can amount to 50% and more. In temperate climates the greater part of these losses is due to a reduction in weight whereas in warmer climates the sugar content of the beets also can be greatly decreased.

Infection with *H. schachtii* often aggravates losses caused by other pests, e.g. *Cercospora beticola* (Veselý, 1970), *Rhizoctonia solani* (Polychronopoulos *et al.*, 1969), *Beta* virus 4 (Goffart, 1956) and others.

B. Stem Eelworm, *Ditylenchus dipsaci*

Although *Ditylenchus dipsaci* was one of the earliest recognized plant-parasitic nematodes its damage to sugar-beet was not noted till 1900 when it was found in England, Germany and the Netherlands. Recently, remarkable damage was reported from Belgium, C.S.S.R., Poland, U.S.A. and U.S.S.R. (Decker, 1969). Initially mainly fodder beets were attacked but now an increasing number of heavy losses on sugar-beets are recorded.

Soon after seed germination the nematodes invade the hypocotyl, commence feeding and multiply. The action of the nematode and its secretions change the plant tissue histologically and chemically, and heavily attacked plants soon show the symptoms of twisted stems and petioles and swollen leaves. The main damage occurs later in the season when secondary pathogens enter the damaged tissue and cause rotting that completely destroys the storage root (Fig. 1). Since neither the vascular system nor the adventitious roots are significantly affected, either by the nematodes or by the rot, the foliar system retains an almost healthy appearance. Frequently, therefore, the damage is detected only late in the season. Heavily infected plants can be killed, and attacked beets should be removed from the soil. During storage rotting continues and the sugar factories sometimes refuse to accept sugar-beet infected with *D. dipsaci*.

FIG. 1. Sugar-beet damaged by stem eelworm, *Ditylenchus dipsaci*, and subsequent rot

The invasion of the young beet plant by stem eelworms is favoured by cool and rainy weather during May whilst higher temperatures and dry conditions decrease the nematode's activity and lessen the chance of heavy losses. During summer practically no invasion of the plant occurs and so the spring weather determines the amount of damage to the crop under central European conditions.

D. dipsaci has many races that are differentiated from each other by their respective host ranges. The host ranges overlap in many instances and at least six different races have been found in sugar-beet. However, there is a high degree of variability within races, within populations of a single race and even within the offspring of a single female (Sturhan, 1971). Furthermore, new races can emerge when populations of different races meet and mate in a common host. These phenomena make it difficult to control *D. dipsaci* by means of crop rotation. The races attacking sugar-beet also can attack rye, oats, maize, onion, carrot, bean, cucumber, and sunflower crops and many weeds.

The yield losses in sugar-beet caused by *D. dipsaci* can amount to 50% and more (Graf *et al.*, 1963). Apart from the reduction in quantity, the quality of the crop also is affected. Thus, the sugar content is diminished in infected plants whereas the concentration of undesirable by-products like reducing sugars, noxious nitrogen and soluble ash is increased (Goffart and Heiling, 1959).

C. Root-knot Nematodes, *Meloidogyne* spp.

The general biology and pathogenicity of the root-knot nematodes on sugar-beet is similar to that described in other chapters of this book. Out of the about 30 valid species of *Meloidogyne* at least five are economically important in sugar-beet production namely: *M. arenaria*, *M. incognita*, *M. javanica*, *M. hapla* and *M. naasi*. All have virtually world-wide distribution and are known to infect a wide range of plants. Sugar-beet is only one of the many important crops that suffer heavy losses from these nematodes. *M. arenaria*, *M. incognita* and *M. javanica* are essentially hot-weather organisms and are most important where beets are grown in regions with long, hot summers and short, mild winters, e.g. the Mediterranean area, California, Chile, etc. *M. hapla* and *M. naasi* withstand frost and are able to develop at lower temperatures although the degree of gall formation they cause varies according to the temperature. In Germany, beets and other crops in infested fields remain practically free of galls by *M. hapla* if spring and early summer are cool and wet, whereas high spring temperatures may evoke severe damage at the same level of infestation. These root-knot nematodes thrive best in fairly loose, well aerated, moderately dry soils. Although nematode attack is favoured by sandy soils damage may occur in any type of soil in which sugar-beets are grown.

M. arenaria, *M. incognita* and *M. javanica* attack the smaller roots and the tap root as well as the storage root and they induce large galls. The galls frequently fuse together and result in irregularly shaped, thickened structures. The number of feeder roots is decreased. *M. hapla* and *M. naasi* attack mainly the smaller roots and induce the development of numerous small galls. Those of *M. hapla* are spherical and often bear many small rootlets giving an appearance similar to the root galls caused by *Nacobbus aberrans*. *M. naasi* causes elongate and sometimes spiral galls at the root tips.

Attack by root-knot nematodes generally inhibits the development of the tap root thereby making the plant particularly sensitive to shortage of water. In such conditions the damaged root system cannot supply the plant with the required amount of water and nutrients and the main root and shoot remains small. In addition, the sugar content may be affected. D'Herde (1965) reports from Belgium a reduction in yield of about 60% in fields heavily infested with *M. naasi* as compared to fields free of these nematodes. The sugar content of the storage root was 14·1% in uninfested and 12·2% in nematode-infected sugar-beet. It may be noted that according to Gooris (1968) beet suffers severely from an attack of *M. naasi* but also it decreases the larval population in the soil.

D. False Root-knot Nematode, *Nacobbus aberrans*

This nematode, with which Sher (1970) synonymized *N. batatiformis*, *N. batatiformis bolivianus* and *N. serendipiticus*, is native to the western parts of North and South America, where it parasitizes a number of wild and cultivated plants. It causes galls on the roots that are similar to those produced by the root-knot nematodes of the genus *Meloidogyne*. Previously, therefore, damage by *N. aberrans* (formerly as *N. batatiformis* and *N. serendipiticus*) has often been mistakenly attributed to *Meloidogyne*. The species also has been found on greenhouse tomatoes in England and in the Netherlands where it obviously was introduced. *N. aberrans* shows a marked sexual dimorphism. Larvae, males and young females are typical filiform nematodes whereas the adult females are transformed into swollen, more or less irregularly shaped bodies. The host–parasite relationship of this nematode on sugar-beet has recently been studied by Schuster *et al.* (1965).

In the root tissue the young larvae live intracellularly and eventually establish a feeding site. Extended necrosis usually occurs and as a consequence small rootlets often die. The maturing sedentary females cause hypertrophy and hyperplasia of root cells resulting in large galls. Within the galls next to the stele a spindle-like structure develops formed by thin-walled enlarged cells. The root-galls caused by *Nacobbus* bear numerous small rootlets. The galls induced by *Nacobbus* differ from those caused by *Meloidogyne* mainly by the presence of a "spindle", the lack of giant cells disorganizing the vascular tissue and the formation of starch in the gall cells. The number of generations varies with temperature and host. Prasad and Webster (1967) have shown that at 25°C the nematode completes several generations per year on a favourable host.

The nematode is confined mainly to warmer soils with a sandy, coarse texture. The geographical distribution is incompletely known

C

and is probably much wider than so far recorded since the nematode can be carried easily in seedling plants, nursery stock and soil. The wide host range of *Nacobbus* includes sugar-beet, tomato, carrot, many brassicas, lettuce, several cucurbits, egg-plant and pea.

E. Migratory Root Nematodes

Of the numerous species of endo- and ectoparasitic migratory root nematodes only few are of economic importance in sugar-beet production. These are *Trichodorus pachydermus*, *T. primitivus*, *T. viruliferus*, *T. similis*, *T. cylindricus*, *T. teres*, *T. anemones*, *Longidorus attenuatus* and *L. elongatus*. They are known to cause heavy damage particularly to sugar-beet seedlings and young plants in England (Whitehead, 1965; Whitehead and Hooper, 1970) and in the Netherlands (Kuiper and Loof, 1961). In England the disease caused by these migratory nematodes is called "Docking Disorder" of sugar-beet.

The above species of nematode feed ectoparasitically on the developing roots of sugar-beet and impair their function or kill them. *Trichodorus* spp. cause stubby-ended lateral roots which turn brown and often die. New roots are in turn attacked and damaged. The tap root is scurfy and may be killed by a heavy attack and other roots growing horizontally or diagonally replace its function, producing a fangy storage root. When attacked by *Longidorus* spp. the lateral roots appear shortened and sometimes with swollen and darkened tips.

In most instances of reported injury the different nematode species mentioned above occur in different combinations so that the root system shows a variable combination of symptoms. The root stunting and fanging caused by these nematodes has a certain similarity to the symptoms caused by waterlogging, compact soil layers, etc. Attacked plants are stunted and can show symptoms of nutrient deficiencies due to the deterioration of the feeder roots. Damage is greatest in light sandy soils. Since many of the nematode species involved are known to be vectors of either the tobacco rattle virus (some *Trichodorus* spp.) or certain ringspot viruses (*Longidorus attenuatus* and *L. elongatus*) it was thought that viruses could play a role in causing certain disease symptoms. However, the studies by Heathcote (1965) have shown that little if any of the stunting associated with Docking Disorder of sugar-beet can be attributed to virus infection. The retardation of top growth in attacked plants is most pronounced in spring, and heavy rainfall during May seems to increase the damage. In July the affected beet plants start to grow again and they may achieve an almost healthy foliage but the root yield will be much reduced. This relative recovery of the plants is most marked in dry summers probably due to the decreased mobility of the nematode in dry soils. The severity of the damage varies from year to year, according to the climate.

F. Other Nematodes

Apart from the nematode species mentioned above there are several others reported to attack sugar-beet (for references see Goodey *et al.*, 1965). They may occasionally or locally cause some damage but are of no general economic importance in sugar-beet production.

III. Control

A. Cultural Practices

Of the many possible cultural practices early sowing and a suitable crop rotation are the most efficient and economical way to avoid nematode damage. Early sowing and crop rotation can be used both as prevention and as cure. As already pointed out beets that are sown at temperatures at which nematodes are more or less inactive have a good chance of escaping severe damage. This applies in particular to *H. schachtii, D. dipsaci* and the root-knot nematodes.

Crop rotation is most effective against those nematode species that are relatively host specific. Alfalfa, cereals and potatoes are non-hosts of *H. schachtii* and are used successfully in crop rotation to reduce the nematode population. Könnecke (1967) recommends long-term crop rotations like: sugar-beet + stable manure – green oats + alfalfa as undercrop – alfalfa – potatoes + stable manure – winter wheat – winter rye – sugar-beet – etc. or sugar-beet + stable manure – pea – winter barley + alfalfa as undercrop – alfalfa – alfalfa – potato – rye – sugar-beet – etc. Experience shows that such a sequence of crops decreases even the severe infestations, but a few cysts always survive particularly in the older infestation sites.

The use of trap crops as suggested by earlier workers is risky because nematodes can multiply if the crop is not completely destroyed before they finish their life-cycle. Furthermore it is normally uneconomical since the grower has the expense of planting and destroying a crop that brings no revenue.

Crop rotation is not very effective in controlling *D. dipsaci* because of the wide host range and broad genetical variability of this species. In Switzerland attempts are being made to decrease the populations of *D. dipsaci* by weed-free rotations with as many non-hosts as possible but reliable results are not yet available (Meyer, 1971). Fodder beets may be sown in seedbeds free of stem eelworms and later transplanted into infested fields without being damaged by the nematodes. This procedure cannot be applied to sugar-beet because they develop a fangy root system when transplanted even though they will not be attacked by the nematode.

The control of the *Meloidogyne* spp. on sugar-beet by crop rotation is very difficult and is virtually impossible if mixed populations are present. About 150 hosts are recorded for *M. arenaria* and only some varieties of cotton, some species of *Nicotiana, Crotalaria,* and strawberry are known non-hosts. About 280 hosts are reported hosts for *M. incognita* and of the known non-hosts peanuts, *Panicum* and *Sorghum, Lycopersicum peruvianum,* a few *Nicotiana* spp. and strawberry could be used in a crop rotation with sugar-beet. *M. javanica* attacks about 400 plant species including almost all economically important crops. Only a few varieties of oats and barley, some species of *Crotalaria* and *Nicotiana, Gossypium barbadense* and *Capsicum annuum* are known nonhosts. At least 350 plant species are recorded hosts of *M. hapla,* but cereals and other gramineae, cotton, onion, *Asparagus* and a few cucurbits are non-hosts. Some 60 hosts are known for the recently described *M. naasi* and among the non-hosts are crop plants like oats, maize, sorghum, alfalfa, carrot, celery, tomato and strawberry.

Although *Nacobbus aberrans* has many host plants, crop rotation can be used successfully to decrease populations below the danger level. Crop plants like alfalfa, clover, wheat, oats, barley, maize, onion and potato are suitable non-hosts for rotation.

No general recommendations for suitable crop rotations against *Trichodorus* spp. and *Longidorus* spp. attacking sugar-beet can be given because natural populations usually contain more than one species, each with a different host range. Efficient rotation could be worked out on a local base according to the species present but the prospects are not very favourable because most of the species are polyphagous. Furthermore the host ranges are incompletely known.

It must be stressed that many common weeds can harbour one or more of these important nematode pests of sugar-beet. Therefore, efficient weed control is essential to make a crop rotation successful.

B. Resistant Varieties

Until now no sugar-beet varieties resistant to one or more of the important nematode species have been produced and attempts to introduce resistance from wild *Beta* species into the commercial varieties have failed. Some progress has been made in breeding varieties tolerant to *H. schachtii* but the results are not yet of practical value to the farmer.

C. Chemical Control

As the available nematicides are capable of efficiently decreasing nematode populations, the chemical control of noxious nematode species in sugar-beet production is not a technical problem but merely an economic one. Of course, a complete eradication of a field infestation cannot be achieved under normal conditions.

Heterodera schachtii can be controlled by the common soil fumigants DCP (1,3-dichloropropene-1), EDB (ethylene dibromide), chloropicrin and DBCP (1-2 dibromo-3 chloropropane) although it must be mentioned that sugar-beet shows some sensitivity to DBCP. Recently the non-phytotoxic carbamoyloximes gave satisfactory results.

Against the stem eelworm, *Ditylenchus dipsaci*, organic phosphorus compounds like fensulfothion and parathion are highly effective and are used widely in sugar-beet production in infested areas in Europe.

Nematicides were developed first as a means to control root-knot, and these nematodes still are used as the principal test-organism in screening for new nematicides. Hence, all recognized preparations available on the market give satisfactory control of *Meloidogyne* spp. and *Nacobbus aberrans*. Losses caused by migratory eelworms are frequently more easily avoided than those caused by the endoparasitic, cyst- and gall-forming species.

IV. Economics

Plant-parasitic nematodes influence the profitability of sugar-beet cultivation in various ways, e.g. reduction in quantity and quality of yield, need for additional fertilizer and water, application of nematicides, limitation in soil use by enforced, less profitable crop rotations and impediment of production and trade by phytosanitary regulations (Weischer, 1967). Sugar-beet cultivation is profitable to the grower but is particularly sensitive to any reduction in revenue. It is, therefore, important for the grower to know what damage he can expect in the presence of nematodes. Some data are available which predict yield reduction according to nematode population density, and for the economics of chemical control. However, the financial losses caused indirectly by nematodes in one way or the other cannot be estimated due to lack of reliable information.

Taylor (1967) compiled the results of 94 field experiments where sugar-beet had been treated with different nematicides and calculated an overall yield increase of 61·4% as compared with the untreated plots. Warren (1963) applied 250–400 kg DCP/ha and observed a yield increase of 200–300% whilst Zambelli and De Leonardis (1970) reported similar yield increases after application of DCP and carbamoyloxim.

Such results seem to justify quite high expenditure for control measures but it must be stressed that these increases cannot be expected in every instance. The increased yields are due to the elimination of not only the parasitic nematodes but also the other noxious organisms since most nematicides are merely biocides. Also soil fumigation usually increases the amount of plant-available nitrogen in the soil (Peachey and Chapman, 1966) and can decrease the readiness of beet plants to wilt, thereby enabling them to assimilate and grow under dry conditions in which untreated plants would suffer even in the absence of parasitic nematodes (Goffart and Heiling, 1958). The nematode damage as well as the effect of control varies with climate, soil type and predominant nematode species. All these observations emphasize that reliable data on losses can be obtained on only a local or regional basis. Unfortunately there is insufficient information on the harmful population densities of most of the nematodes attacking sugar-beet. However, some data are available for the most important species, *H. schachtii*. Schemes of damage prediction for the use of farmers and advisory services in Germany (Goffart, 1952) and in the Netherlands (Heijbroek, 1971) are available (Table III).

TABLE III. Evaluation of initial infestation by *Heterodera schachtii* and damage prediction (based on Goffart, 1952 and Heijbroek, 1971)

Number of eggs and larvae per 100 cm³ of soil	Degree of damage	Advice
< 150		no danger with normal crop rotation
150– 300	light	small chance of damage only, one year waiting recommended
300– 800		chance of damage, 2–3 years waiting recommended
800–1750	moderate	good chance of damage, 4–5 years waiting recommended
1750–2750	severe	big chance of damage, 5–6 years waiting recommended
> 2750		serious damage to be expected, 7–8 years waiting recommended

The relationships between the number of larvae and degree of damage to be expected are given in Table III and are valid for timely drilling and normal weather conditions. Beet drilled later in the season is much more sensitive to nematodes for the reasons mentioned earlier. Steudel (1970) studied the effect of *H. schachtii* on sugar yield in relation to initial population density and drilling time. The field experiments were done in the region of Cologne, Germany and clearly illustrated the beneficial effect of early drilling. In early drilled beets significant damage occurred at an initial population density of about 2000 viable larvae per 100 cm³ of soil whereas in late drilled beets even 250 larvae caused a noticeable decrease in sugar yield (Fig. 2). The weather during the period of germination plays an important role because it influences the speed of plant growth as well as the hatching and activity of the nematodes. The planting of non-hosts decreases the infestation and Hijner (1956) estimated an annual decline at the rate of 40% of the density of the previous year. Moriarty (1966) assumed an annual decline of 50% but finally concluded that there is no constant decline and that the decline in population is less in the first year than in the subsequent years.

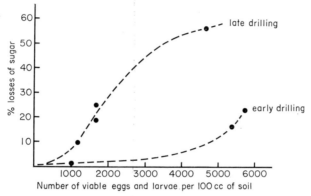

Fig. 2. The effect of sugar-beet cyst eelworm, *Heterodera schachtii*, on sugar yield in relation to initial infestation of eggs and larvae and to drilling time. Reproduced with permission from Steudel (1970).

The beneficial effect of crop rotation on beet yield has been demonstrated by Könnecke (1967). Under the conditions of the well known "Magdeburger Börde" where *H. schachtii* was observed for the first time, the average beet yield in a rotation with sugar-beet every third year was 14% less than in rotations with sugar-beet every eighth year. Crop rotation is not so effective and important in controlling other nematode species as it is for *H. schachtii* on sugar-beet. Furthermore, few reliable data about the economics of such procedures are available.

Efficiency and economics of chemical control of nematodes on sugar-beet depends greatly on local conditions. High infestation levels, presence of other noxious organisms which are also controlled, favourable side effects on plant growth and the chance for the farmer to avoid less profitable crop rotations lessen the high costs of nematicide application. Furthermore a less costly row treatment may be used instead of the broadcast application. Owing to the decreasing sensitivity of the seedlings with age the row treatment can protect the young plants sufficiently to avoid economic damage. Steudel and Thielemann (1968) obtained normal yield by row application of 5 kg active material per hectare of a carbamoyloxim preparation at infestation levels up to 5000 larvae of *H. schachtii* per 100 cm³ of soil. The average yield increase amounted to 13 tons/ha. The same amount of nematicide had little effect on yield when applied in June.

Row treatment with fensulfothion (3 kg active material per hectare) on the drilled seed was proved to be very effective against the stem eelworm on sugar-beet (Löcher, 1964). The nematicide prevents the invasion of the seedlings by nematodes and its systemic action protects the plants also from aphids and other noxious insects. In Switzerland fensulphothion gave insufficient control because of the higher rainfalls, and repeated application of parathion would be the best way to avoid damage. 3·5 tons of beet are considered to be the economic equivalent of the treatment and this yield increase can easily be obtained in infested areas when the weather favours nematode attack. Damage prediction on the basis of nematode numbers frequently fails in Switzerland owing to the influence of environmental conditions. Therefore, at all levels of infestation of the stem nematode in the soil either chemical treatment or a non-host crop is recommended (Meyer, 1971).

To control ectoparasitic nematodes attacking sugar-beet Whitehead *et al.* (1970) found row application of smaller amounts of DCP (< 135 litres/ha) than customarily used (> 370 litres/ha) for soil disinfection gave very good results. The nematode populations were greatly decreased and the increase in beet yield more than repaid the costs of treatment. Other nematicides showed a similar beneficial effect. This very efficient and economic control, however, can be achieved only in alkaline sandy soils which are well drained but not too dry. This limitation demonstrates once again the importance of local situations and the difficulty in formulating general rules about the economics of chemical nematode control.

One more aspect of chemical control should be mentioned although it may not be directly related to the economics of sugar-beet production. Nematicides contribute to the increasing pollution of the environment particularly when used on the broad scale necessary in sugar-beet fields, and this may limit their use. Hence, instances may arise where crop losses may have to be tolerated because nematicides would endanger the health of human beings.

Although none of the nematode species attacking sugar-beet has a phytosanitary importance comparable to the potato cyst eelworm, *Heterodera rostochiensis*, some of them are subject to regulations which

influence the economics. In England crop rotation is enforced by legislation (Anon., 1960 and 1962) on all land infested by *H. schachtii* and
in some areas which include infested and "clean" land. This "clean"
land is subject to a 3-course rotation of host plants and moderately
infested land is subject to a 4-course rotation. Heavily infested land
must be rested from host crops for at least 5 years. Soil tests are done
subsequently to ascertain whether or not the nematode population has
declined before sanction is given for a continued 4-course rotation.

In some countries *H. schachtii, Meloidogyne* spp. and *Nacobbus
aberrans* are subject to strict quarantine regulations which can influence
the economics of sugar-beet production by impeding transport and trade.

All the data and remarks on economics discussed above are based
on the traditional cultural practices as they are commonly used. Sugar-
beet production, however, is at present undergoing a radical change,
particularly in countries with an intensive agriculture. Labour-saving
practices like the use of monogerm seed instead of seed balls, drilling
to a stand instead of continuous-row drilling with subsequent singling,
and weed control by means of herbicides instead of mechanical weeding
are increasingly used. These changes will certainly influence the nematode populations, and the yield and damage of the crop. However, the
present knowledge gives no satisfactory indication as to whether or
not nematode populations will increase or decrease and if so, by how
much. Furthermore there is a tendency towards an increasing monotony
of crop rotation and to a concentration of sugar-beet production in the
vicinity of sugar factories, for economic reasons. This intensive cropping
with sugar-beet which is now occurring is thus a repeat of what induced
the first severe outbreaks of nematode pests of the crop. Present changes
occurring in sugar-beet production necessitate a more thorough examination of some aspects of the problem of nematode pests of this crop.
There is an urgent need for more research in spite of the fact that nematode diseases of sugar-beet have been studied for more than a hundred
years.

References

Anon. (1960). The Beet Eelworm Order, 1960. Statutory Instr., No. 2147,
    H.M.S.O., London.
Anon. (1962). The Beet Eelworm (Amendment) Order, 1962. Statutory Instr.,
    No. 2497, H.M.S.O., London.
Decker, H. (1969). "Phytonematologie." VEB Deutscher Landwirtschafts-
    verlag, Berlin.
D'Herde, J. (1965). *Meded. LandbHogesch. OpzoekStns Gent* **30**, 1429–1436.
Goffart, H. (1952). *Zucker* **5**, 315–317.
Goffart, H. (1953). *Z.ZuckInd.* **3**, 229–231.
    C*

Goffart, H. (1954). *Zucker* **7**, 130–137.

Goffart, H. (1956). *Meded. LandbHogesch. OpzoekStns Gent* **21**, 351–360.

Goffart, H. and Heiling, A. (1958). *Nematologica* **3**, 213–228.

Goffart, H. and Heiling, A. (1959). *Z.ZuckInd.* **9**, 349–351.

Goodey, J. B., Franklin, M. T. and Hooper, D. J. (1965). "T. Goodey's The nematode parasites of plants catalogued under their hosts." 3rd Edn Tech. Commun. No. 30, Commonw. Bur. Helminth. Farnham Royal, England.

Gooris, J. (1968). *Meded. Rijksfakulteit Landbouwwetensch. Gent* **33**, 85–100.

Graf, A. E., Joseph, E., Keller, E., Liechti, H. and Savary, A. (1963). *Mitt. Schweiz. Landw.* **11**, 134–141.

Heathcote, G. D. (1965). *Pl. Path.* **14**, 154–157.

Heijbroek, W. (1971). 34th I.I.R.B. Winter Congress, 1971, Sessions IV, Report No. 4.

Hijner, J. A. (1956). *Rapp. Inst. Int. Rech. Betteravières* XVIe Assembl. Bruxelles, 129–139.

Jones, F. G. W. (1965). Beet eelworm. *In* "Plant Nematology" (J. F. Southey, ed.), No. 7 (2nd Edn), 189–198. Tech. Bull. Minist. Agric., London.

Könnecke, G. (1967). "Fruchtfolgen." VEB Deutscher Landwirtschafts-verlag, Berlin.

Kühn, J. (1881). *Ber. physiol. Lab. Univ. Halle* **3**, 1–153.

Kuiper, K. and Loof, P. A. A. (1961). *Versl. Meded. Plziektenk. Dienst* No. 136 (Jaarboek 1961), 193–200.

Ladigina, N. M. (1960). *Tezisy dokladov, Samarkand,* 57-58.

Licht, F. O. (1971). *Zucker* **24**, 203.

Löcher, F. (1964). *Z. PflKrankh. PflPath. PflSchutz* **70**, 413–417.

Meyer, H. (1971). *Die Zuckerrübe* **20**, 17–19.

Moriarty, F. (1966). *Nematologica* **12**, 349–354.

Peachey, J. E. and Chapman, M. R. (1966). Chemical control of plant nematodes. Tech. Commun. No. 36, Commonw. Bur. Helminth. Farnham Royal, England.

Polychronopoulos, A. G. and Lownsbery, B. F. (1968). *Nematologica* **14**, 526–534.

Polychronopoulos, A. G., Houston, B. R. and Lownsbery, B. F. (1969). *Phytopathology* **59**, 482–485.

Prasad, S. K. and Webster, J. M. (1967). *Nematologica* **13**, 85–90.

Raski, D. J. and Johnson, R. T. (1959). *Nematologica* **4**, 136–141.

Schuster, M. L., Sandstedt, R. and Estes, L. W. (1965). *J. Am. Soc. Sug. Beet Technol.* **13**, 523–537.

Shepherd, A. M. (1962). The emergence of larvae from cysts in the genus *Heterodera*. Tech. Commun. No. 32, Commonw. Bur. Helminth. Farnham Royal, England.

Sher, S. A. (1970). *J. Nematol.* **2**, 228–235.

Steudel, W. (1970). *Proc. Third. Int. Symp. Sugar Beet Protection*, Novi Sad 1970, Savremena Poljoprivredna *XVII*, 13–18.

Steudel, W. and Thielemann, R. (1968). *Meded. Rijksfakulteit Landbouw-wetensch. Gent* **33**, 707–718.

Sturhan, D. (1971). Biological races. *In* "Plant Parasitic Nematodes" (B. M. Zuckerman, W. F. Mai and R. A. Rohde, eds), vol. 2, pp. 51–71. Academic Press, London, New York.

Taylor, A. L. (1967). Principles of measurement of crop losses: nematodes. *Proc. FAO Symp. Crop Losses*, Rome 1967, 225–232.

Veselý, D. (1970). *Zucker* **24**, 10–14.

Wallace, H. R. (1958). *Ann. appl. Biol.* **46**, 86–94.

Warren, L. E. (1963). *J. Am. Soc. Sug. Beet Technol.* **12**, 348–358.

Weischer, B. (1967). Types of losses caused by nematodes. *Proc. FAO Symp. Crop Losses*, Rome 1967, 181–187.

Whitehead, A. G. (1965). *Br. Sug. Beet Rev.* **34**, 77–78, 83–84, 92.

Whitehead, A. G. and Hooper, D. J. (1970). *Ann. appl. Biol.* **65**, 339–350.

Whitehead, A. G., Tite, D. J. and Fraser, J. E. (1970). *Ann. appl. Biol.* **65**, 361–375.

Zambelli, N. H. and De Leonardis, A. (1970). *Abstr. 10th Nematology Symp. Europ. Soc. Nem.* Pescara 1970, 185–187.

4

Nematode Diseases of Pasture Legumes and Turf Grasses

K. B. Eriksson*

Department of Plant Pathology and Entomology
Agricultural College of Sweden
Uppsala 7, Sweden

I. Introduction

Pasture and turf are indispensable to us all. Pasture is not only the cheapest source of nitrogen and palatable nutrients for livestock, but it is valuable in restoring and maintaining soil fertility and structure. One of the results of the greater leisure time now available is an increasing interest in turf management, especially the utility and aesthetic values of high quality turf.

Pests hamper the establishment of pasture, decrease forage and seed yields, and impair the quality of turf. Nematode diseases of pasture legumes, forage grasses and turf have been recognized as a major agricultural problem especially in temperate regions.

II. Crop Production

Mixtures of legumes and grasses are used for grazing, and for hay and forage production, while turf grasses are the chief ingredient of home

* The author gratefully acknowledges valuable advice from Dr S. Bingefors, private information from many other persons, and photographic help from Mrs Birgit Wallentinsson.

lawns, parks, and the various types of sports fields. Most of the important pasture legumes and turf grasses are native to Europe and temperate Asia, and have been introduced to other parts of the world. They have adapted to many climatic and soil conditions (Table I).

III. Nematode Diseases

A. Pasture Legumes

The most important nematode problems occur in temperate legumes, but a few tropical legumes will be considered also.

TABLE I. Characteristics of some important pasture legumes and turf grasses

Species	Area of cultivation; adaptation; utilization
Legumes	
Medicago sativa lucerne, alfalfa	World-wide, extensively under irrigation. Drought-resistant perennial that grows best in relatively dry climates.
Trifolium pratense red clover	Cool moist climates all over the world. Variable perennial; thrives in fertile soils.
T. repens white clover	Cool moist areas all over the world. Variable perennial, including the large-leaved Ladino type. A nuisance in sports turfs.
T. hybridum alsike clover	Temperate zones of Europe, Asia, America. Less pretentious than red clover. Perennial.
T. incarnatum crimson clover	Important winter-annual in cool humid climates with moderate winter temperatures.
T. resupinatum Persian clover	South Australia, southern U.S.A., India. Winter-annual that prefers deep heavy soils.
T. subterraneum subterranean clover	Areas with long warm winters and dry summers. Important in Australia. Winter-annual.
Melilotus alba, M. officinalis sweet clover	Widespread in temperate zones, important in the U.S.A. and Canada. Drought-tolerant. Used for hay, pasture and green manure.
Lotus corniculatus birdsfoot trefoil	Adapted to cool climates and less-fertile soils. Perennial, used for hay and pasture.
Lespedeza spp. *striata* common; *stipulacea* Korean; *cuneata* sericea	Warm-season pasture plants, widely grown on moist soils in the eastern half of the U.S.A. *L. striata*, with the variety Kobe, and *L. stipulacea* are annuals; *L. cuneata* is perennial.
Alysicarpus vaginalis alyce-clover; *Desmodium* spp. beggarweeds; *Phaseolus atropurpureus* Siratro; *Stylosanthes humilis* Townsville lucerne	Key legumes in the tropics and subtropics, however, with a short history as pasture plants (Hutton, 1969). Lucerne and white clover fill special ecological niches in subtropical areas.

TABLE I (*continued*)

Species	Area of cultivation; adaptation; utilization
Grasses	
Agrostis spp. bent grass; *alba* redtop; *canina* velvet bent; *palustris* (*stolonifera*) creeping bent; *tenuis* browntop, colonial bent grass	Cool humid regions. Cold-hardy, thrives on poor wet soils, *A. tenuis*, however, in dry sandy situations. Fine-leaved, variable perennials for lawns, bowling, putting, and golf greens. Spread by rhizomes or stolons.
Festuca spp. fescue; *ovina* sheep fescue; *rubra* red fescue; *rubra* var. *commutata* Chewing fescue; *arundinacea* (*elatior*) tall fescue; *pratensis* (*elatior*) meadow fescue	Cool-season, fine-textured and highly variable perennials, adapted to most soil types. Red and Chewing fescues major turf grass species, tall and meadow fescues valuable pasture and forage grasses used also as general-purpose turf grasses.
Poa pratensis smooth-stalked meadow grass, Kentucky bluegrass	Cool-season, rhizomatous perennial with worldwide distribution. Cold-hardy, withstands wear and tear. Valuable pasture species and a major turf grass.
P. trivialis rough-stalked meadow grass	Cool moist climates. Fine-textured perennial for second-quality lawns.
P. annua annual meadow grass	Cool-season annual. Ingredient in many swards, weed problem in fine turf.
Lolium multiflorum Italian ryegrass; *L. perenne* perennial ryegrass, English ryegrass	Mild, humid-temperate climates and moist fertile soils. Valuable forage grasses, often seeded in lawns. *L. perenne* suitable for sports grounds.
Bromus inermis smooth brome grass; *Dactylis glomerata* cocksfoot, orchard grass; *Phleum pratense* timothy	Temperate areas throughout the world. Cool-season, perennial hay and pasture species.
Cynodon dactylon Bermuda grass	Tropical and subtropical areas. Warm-season, perennial species, valuable for pasture and turf.
Eremochloa ophiuroides centipede grass	Warm-season, dwarf, stoloniferous plant. General-purpose turf and lawn grass.
Stenotaphrum secundatum St Augustine grass	Humid tropics and subtropics and rich moist soils. Valuable perennial for pastures and lawns.
Zoysia spp. zoysia grass	Warm-season grasses for turf of differing quality.

1. Lucerne, *Medicago sativa*

The most extensive nematode damage to lucerne is probably caused by the stem and bulb nematode, *Ditylenchus dipsaci*. Among the approximately twenty biological races within this cosmopolitan species

(Decker, 1969), there are three which have, respectively, lucerne, red clover or white clover as their main host plant. *D. dipsaci* is frequently reported as a serious pest in lucerne growing areas in the U.S.A., especially in heavier soils, in regions of high rainfall and in areas under irrigation. More or less severe damage to lucerne is reported from many other countries (Table II).

TABLE II. Nematode infestations of lucerne and their geographical locations

Nematode species	Geographical locations	References
Ditylenchus dipsaci stem and bulb nematode	Europe (Denmark, England, France, Hungary, Sweden, U.S.S.R., and others); U.S.A. (several of the eastern and western states); Canada; S. America (Argentina, Brazil, Chile); S. Africa; New Zealand; New South Wales	Brown (1957) Thorne (1961) Bingefors (1969) Decker (1969) Grandison (person. communic.)
Meloidogyne spp. (*arenaria, hapla, incognita, javanica*) root-knot nematodes	U.S.A. (southern and south-western states); S. America (Chile)	Reynolds (1955) Allison (1956) Guinez (1965)
M. hapla	U.S.A. (Oregon, California, Iowa, Maryland, N. Carolina)	Norton (1969)

D. dipsaci attacks the stem and foliage, causing susceptible plants to become stunted and distorted with swollen buds, shoots and stem bases (Fig. 1). If the inflorescences are infected nematodes may be dispersed in the seed (Brown, 1957). Stem nematodes may occasionally infect lucerne roots, causing internal cavities (Krusberg, 1961), or gall-like outgrowths which girdle the crown and sub-crown portions of tap roots (Hawn, 1969). A field infection appears as round or irregular areas of sparse growth. If severely and extensively attacked, the whole lucerne stand may have to be ploughed. However, the deep root system and drought-resistance of lucerne may facilitate recovery later in the season to give an acceptable yield.

Stem nematodes seem to play an important role in the development of bacterial wilt of lucerne, caused by *Corynebacterium insidiosum* (Hawn and Hanna, 1967); in a field test the nematode infection broke the bacterial wilt resistance of one variety, rendering it very susceptible. Hawn (personal communication, 1970) has found evidence that *D. dipsaci* may be involved also in the epidemiology of the crown bud rot of lucerne.

Root-knot on lucerne, caused by *Meloidogyne* spp., is prevalent in southwestern and southern U.S.A., especially in light, sandy soils where it can be economically important. Species frequently found associated with lucerne are *M. arenaria*, *M. incognita*, *M. javanica*, and also *M. hapla*, to which it is most susceptible (Allison, 1956). Reports are conflicting concerning its suitability as a host for *M. naasi* (Gooris, 1968; Radewald *et al.*, 1970).

Fig. 1. Disease symptoms caused by *Ditylenchus dipsaci* in pasture legumes; left, lucerne; centre two, red clover; right, white clover. Note stunted plants, shortened internodal spacing, swollen leafstalks, and bulb-like swelling of shoot base.

Root-knot nematode attacks on lucerne result in a shallow, profusely-branched root system with only a few long tap roots; a severe infestation may cause very heavy galling (Reynolds, 1955; Stanford *et al.*, 1958). The infections cause slow seedling growth and decreased hay yield. Attacked plants, however, may recover and give a satisfactory stand in following years (Stanford *et al.*, 1958).

Root-knot nematode and *Fusarium* wilt interact, and according to McGuire *et al.* (1958) the degree of wilting in lucerne is proportional to the severity of root-knot infection. In greenhouse tests *M. hapla* increased the incidence of bacterial wilt in both wilt-resistant and susceptible lucerne varieties (Norton, 1969; Hunt *et al.*, 1971).

Pratylenchus penetrans reproduces readily in lucerne and decreases yield (Chapman, 1959; Willis and Thompson, 1969). *Ditylenchus medicaginis*, described from lucerne in Poland, infects roots and aerial plant parts without causing specific symptoms (Decker, 1969). Reproduction of a number of ectoparasitic nematodes has been reported on lucerne, but they cause little or no damage (Coursen *et al.*, 1958; McGlohon *et al.*, 1961). *Belonolaimus gracilis* may cause severe injury on several lucerne varieties (Good and Thornton, 1956). Lucerne appears not to be a host to *Heterodera* spp. (Mankau and Linford, 1960).

Pasture legumes are hosts of several viruses some of which are nematode-transmitted. Lucerne is a host of pea early-browning virus, transmitted by *Trichodorus* spp. (Gibbs and Harrison, 1964), and nematodes, potential virus vectors, frequently have been found associated with lucerne. *Xiphinema americanum* is definitely associated with poor growth of lucerne in a few states of the U.S.A. (Ward, 1960; Norton, 1965). *Trichodorus christiei* caused significant reduction in lucerne shoot and root weights in greenhouse studies (McGlohon *et al.*, 1961), and *Longidorus maximus* produced typical symptoms of attacks on lucerne (Sturhan, 1963).

2. Red Clover, *Trifolium pratense*

Ditylenchus dipsaci was reported on red clover, under the name "Stock", in Germany during the early 19th century (Decker, 1969), and is now known to attack, or to be a potential threat to this crop almost wherever it is cultivated (Table III).

Stem nematode damage in the field is seen best in the early summer or in the autumn, as small circular, or larger irregular patches with dwarfed and mis-shaped plants. Diseased red clover plants exhibit shortened internodal spacing, and very characteristic, spongy-textured swollen shoots (Fig. 1), which are easily loosened from the root. Flower-buds, leaves and leaf-stalks are frequently thickened and twisted. If infected plants are harvested for seed there is a risk of dissemination of stem nematodes in the seed or plant debris.

Red clover varies in its degree of resistance to *D. dipsaci*. Both resistant and susceptible plants are invaded (Bingefors, 1957), but further nematode development depends on the plant response. The nematode causes cells to separate and enlarge, producing swelling of the tissue in a susceptible plant. Resistant plants are stunted and develop necrotic spots, but are not typically swollen.

Heterodera trifolii, clover cyst nematode, parasitizes the clovers and many other legumes, and is of major importance in the field in white clover. The nematode was first found parasitizing red clover roots in Germany in 1932 (Decker, 1969) and has since been recorded in red clover from various parts of the world (Table III). Reports are conflicting regarding the ability of *H. trifolii* to infect and produce cysts on clovers, and significant differences between nematode isolates in their cyst production, indicating different biological forms within the nematode, were reported by Singh and Norton (1970). Though pathogenicity to red clover has been proved (Mankau and Linford, 1960), great variability in susceptibility was observed among red clover accessions (Norton and Isely, 1967). According to Chapman (1964) *H. trifolii*, under greenhouse conditions attacks more rapidly and injures red clover more severely than white clover. Though the amount of damage to red clover under field conditions generally appears to be rather unimportant (D. C. Norton, personal communication, 1970), this crop was severely damaged in field microplots (McGlohon *et al.*, 1961), and Norton (1967) reported significant yield losses.

TABLE III. Nematode infestations of red and white clovers and their geographical locations

Nematode species	Geographical locations	References
Ditylenchus dipsaci stem and bulb nematode	Europe (England, Germany, the Netherlands, Scandinavia, Switzerland, U.S.S.R., and others); U.S.A. (western states and the Pacific northwest); New Zealand	Thorne (1961) Bingefors (1969) Decker (1969) Oostenbrink (person. communic.) Grandison (person. communic.)
Heterodera trifolii clover cyst nematode	Several European countries; U.S.A. (at least 25 states); Canada; New Zealand	Grandison (1963) Decker (1969) Stehman and Norton (1969)
Meloidogyne spp. (*arenaria, hapla, incognita, incognita acrita, javanica*) root-knot nematodes	Southeastern U.S.A.; Chile (red clover); New South Wales	Allison (1956) McGlohon and Baxter (1958) Colman (1964) Guinez (1965)
M. hapla	Europe (Denmark, Germany, the Netherlands)	Lindhardt (1963) Hijink and Kuiper (1964) Decker (1969)

Root-knot occurs on the clovers mainly in the warmer regions, where several *Meloidogyne* species affect the establishment and persistence of these crops (Table III). *M. hapla* on clovers is known from a few European countries, though not of economic importance. The species survives the winters of Denmark, where extensive galling on wild red clover has been observed (Lindhardt, 1963). Heavy infestations of *M. hapla* on individual clover plants in the field have been found occasionally in Estonia (E. Krall, personal communication, 1970). Martin (1961) tested a great number of red clover varieties from different countries and found all of them to be heavily infested by *M. javanica*. Red clover is more susceptible to *M. incognita* than is white clover (Bain, 1959), and appears to suffer more from root-knot injury than does lucerne (Allison, 1956).

Pratylenchus penetrans causes a significant reduction in red clover growth (Chapman, 1959) and foliage yield (Willis and Thompson, 1969). The plant favours reproduction of *Tylenchorhynchus martini* and *T. claytoni* (Chapman, 1959; Krusberg, 1959). Red clover is a host for *Ditylenchus destructor* (Thorne, 1961; Moore, 1971), and large numbers of these nematodes were associated with root necrosis of this plant as well as of white and alsike clovers (S. Andersen, personal communication, 1971).

Xiphinema americanum is common in red clover fields in some states in the U.S.A., and is frequently associated with stunting and patchiness of the crop (McGlohon *et al.*, 1961; Norton, 1965). *X. diversicaudatum*, vector of the arabis mosaic virus, reproduces on red clover (Thomas, 1970). Red clover is an "excellent" host for *Trichodorus christiei* (Rohde and Jenkins, 1957), and a "good" host of *Longidorus elongatus* which can transmit raspberry ringspot and tomato black ring viruses to red clover (Thomas, 1969). *L. elongatus* causes galling of red clover roots where it multiplies better than with white clover (Thomas, 1969). *L. maximus* is a significant pest in southern Germany, causing thickening and contortions of red clover root tips (Sturhan, 1963).

3. White Clover, *Trifolium repens*

Heterodera trifolii is widely distributed in lawns and in long-term pastures in Europe and in the U.S.A. (Table III), mainly associated with white clover. It appears to be a factor in white clover decline in north-western U.S.S.R. (Krall, 1965), and in the dying out of white clover in New Zealand pastures (Grandison, 1963). Reports of damage, however, are inconsistent (Dijkstra, 1971). While Ennik *et al.* (1965) considered *H. trifolii* to be an important factor in the disappearance of white clover in grasslands, Seinhorst and Sen (1966) found no obvious damage by the nematode in a number of pastures. It has been amply demonstrated that *H. trifolii* is pathogenic to white clover (Mankau and Linford, 1960; Chapman, 1964; Norton, 1967). Norton and Isely (1967) stated that white clover is generally more susceptible than red clover, and in their tests with different nematode isolates Singh and Norton (1970) found greater varietal susceptibility in red than in white clover. Secondary effects of nematode attack may be a decreased number of *Rhizobium* root nodules (Wardojo *et al.*, 1963).

The incidence of root-knot damage to white clover coincides with that of red clover. Several *Meloidogyne* spp. are associated with unthrifty growth of the clovers in the U.S.A. (Table III), and pot experiments confirmed field observations on root-knot associated with considerable injury to white clover in New South Wales (Colman, 1964). Hijink and Kuiper (1964) believe that *M. hapla* may be a factor in white clover decline in meadows of the Netherlands.

Stem nematode attacks on white clover were first described in England, but are known also from Denmark and were recently observed in New Zealand (Grandison, 1965; Bingefors, 1969). It is considered to be a distinct white clover race of *D. dipsaci* which causes most damage (Fig. 1), but both the red clover and lucerne stem nematode races reproduce slightly on white clover (Böning, 1964; Bingefors, 1967). *Pratylenchus penetrans* decreases white clover yields (Wardojo *et al.*, 1963; Willis and Thompson, 1969), and large numbers of *P. scribneri* were found in declining stands of white clover in Alabama (Minton, 1965), the yield being significantly reduced. McGlohon *et al.* (1961) report that several ectoparasitic nematodes reproduce readily on Ladino clover.

Xiphinema americanum was found in large numbers associated with white clover in New York state (Ward, 1960), and the plant favours reproduction of *X. diversicaudatum* (Thomas, 1970). The latter species, which is a vector of arabis mosaic virus, is widely distributed in southern Britain, mainly in heavy soils, and patchiness in grasslands containing much white clover was associated with a frequent occurrence of the nematode (Harrison and Winslow, 1961). The authors found that the white clover plants were often infected with arabis mosaic virus, and that patches of virus-infected plants coincided with patches of soil infected with *X. diversicaudatum*. *Longidorus elongatus* attacks white clover (Thomas, 1969) and *L. maximus* causes thickening and contortions of white clover root tips (Sturhan, 1963). Rohde and Jenkins (1957) listed the plant as an "excellent" host for *Trichodorus christiei*.

4. Miscellaneous Pasture Legumes

Lotus corniculatus is resistant to the lucerne and red clover races of *Ditylenchus dipsaci* (Bingefors, 1967), and is a non-host for *Heterodera trifolii* (Norton and Isely, 1967). One variety each of big trefoil and birdsfoot trefoil are very susceptible to *Meloidogyne arenaria*, *M. hapla*, *M. incognita*, *M. incognita acrita* and *M. javanica* (Allison, 1956). Martin (1961) observed that some varieties of *L. corniculatus* were only lightly to moderately infested by *M. javanica*, while *Pratylenchus penetrans* decreased foliage yield significantly (Willis and Thompson, 1969). Large numbers of *Xiphinema americanum* were associated with *L. corniculatus* (Ward, 1960), and Rohde and Jenkins (1957) found that the plant was an "excellent" host for *Trichodorus christiei*.

A number of other temperate legumes vary in their suitability as hosts for various plant nematodes (Table IV).

All three cultivated *Lespedeza* species are susceptible to root-knot nematodes, *L. stipulacea* apparently being more susceptible than Kobe lespedeza (*L. striata*) (Offutt and Riggs, 1970). Allison (1956) reported that five accessions of sericea lespedeza were moderately susceptible to four *Meloidogyne* spp. but moderately resistant to *M. hapla*. *H. glycines* occurs wherever Kobe lespedeza is grown in the U.S.A. and the crop was highly susceptible to nine isolates of this nematode from six states (Epps and Golden, 1967). *L. cuneata* and *L. stipulacea* were listed as highly favourable for reproduction of *Tylenchorhynchus claytoni* by Krusberg (1959). Populations of *Belonolaimus gracilis* can be maintained or carried over on Kobe lespedeza (Holdeman and Graham, 1953).

As might have been expected, *Meloidogyne* spp. are the most prevalent nematode problem of tropical pasture legumes, but Hutton (1969) believes that important pasture species are not badly affected. In an investigation with five *Meloidogyne* spp. Minton *et al.* (1967) found that three *Desmodium* species differed in their response to the nematodes, *D. intortum* having the highest level of resistance; varying degrees of root-knot resistance were observed in *Stylosanthes humilis*, *Indigofera hirsuta*, *Alysicarpus vaginalis* and Siratro. All four plants were susceptible to *M. arenaria*, and *A. vaginalis* was highly susceptible to four root-knot species, but had some resistance to *M. hapla*. Siratro seems to have a very high level of root-knot resistance (Hutton, 1969). Good *et al.* (1965) reported *D. tortuosum* as susceptible to *M. hapla*, but resistant to four other *Meloidogyne* spp. These authors found also that the plant greatly increased populations of *Belonolaimus longicaudatus*, and supported moderate populations of *X. americanum*, *T. christiei* and *Criconemoides ornata*. *B. gracilis* can also build up under *D. tortuosum* (Holdeman and Graham, 1953).

B. Turf Grasses

It is mainly during the past two decades that nematodes have been recognized as important pests of turf grasses. A number of nematodes, most of them ectoparasites, cause considerable damage to golf course greens, home lawns and turf nurseries (Good *et al.*, 1959; Rhoades, 1962). A. W. Johnson (personal communication, 1970), referring to the situation in the southeastern U.S.A., lists the following nematodes:

TABLE IV. Host status and reactions of some pasture legumes to attacks of plant nematodes

Crop species	Nematode species	Host status, plant reaction	References
Trifolium hybridum alsike clover	*Ditylenchus dipsaci*	slightly susceptible	Bingefors (1967)
	Heterodera trifolii	host – very poor host – non-host	Norton and Isely (1967)
	Paratylenchus projectus	non-host	Coursen et al. (1958)
	Trichodorus christiei	excellent host	Rohde and Jenkins (1957)
T. incarnatum crimson clover	*Ditylenchus dipsaci*	resistant to lucerne and red clover races	Bingefors (1967)
	Heterodera trifolii	non-host	Norton and Isely (1967)
	Meloidogyne incognita acrita	severely galled	McGlohon and Baxter (1958)
	Belonolaimus gracilis	significant increase of population	Good and Thornton (1956)
	Paratylenchus projectus	host	Coursen et al. (1958)
	Tylenchorhynchus claytoni	favourable for reproduction	Krusberg (1959)
	Trichodorus christiei	excellent host	Rohde and Jenkins (1957)
T. resupinatum Persian clover	*Ditylenchus dipsaci*	resistant to red clover race	Bingefors (1967)
	Heterodera trifolii	host	Norton and Isely (1967)
	Meloidogyne incognita acrita	severely galled	McGlohon and Baxter (1958)
T. subterraneum subterranean clover	*Meloidogyne incognita acrita*	severely galled	McGlohon and Baxter (1958)
	Meloidogyne hapla	susceptible to attacks	Colman (1964)
	Meloidogyne javanica	susceptible to attacks	Colman (1964)
Melilotus alba white sweet clover	*Ditylenchus dipsaci*	slightly susceptible to lucerne race, resistant to red clover race	Bingefors (1967)
	Heterodera trifolii	resistant or slightly susceptible	Gerdemann and Linford (1953) Norton and Isely (1967)
	Paratylenchus projectus	host	Coursen et al. (1958)
M. officinalis yellow sweet clover	*Ditylenchus dipsaci*	susceptible to lucerne race, resistant to red clover race	Bingefors (1967)
	Heterodera trifolii	resistant or slightly susceptible	Gerdemann and Linford (1953) Norton and Isely (1967)

Belonolaimus longicaudatus, Meloidogyne graminis, Hoplolaimus galeatus, Trichodorus christiei, Tylenchorhynchus martini, Pratylenchus zeae, Helicotylenchus spp., *Criconemoides ornata,* and *Xiphinema americanum.* Grasses are generally poor hosts to root-knot nematodes, with the exception of *M. naasi* and the pseudo root-knot nematode, *M. graminis* (syn. *Hypsoperine graminis,* but recently synonymized as *M. graminis* by Whitehead, 1968). Grass for seed production and forage grasses are attacked by nematodes, causing galls in the inflorescences (*Anguina* spp.), foliage (*A. graminis, Ditylenchus graminophilus*), and on the roots (*D. radicicola*).

General symptoms of nematode injury to turf grasses are chlorotic foliage and patches with declining growth, especially in hot weather, moisture stress and poor soils. Root symptoms may appear as sparseness of roots, discoloration, stubby roots, and galled or slightly swollen rootlets (Heald and Perry, 1969).

1. Bent grass, *Agrostis* spp.

One of the oldest records of plant nematode damage is that by Steinbuch in 1799 who described galls in the flowers of *A. capillaris* caused by a nematode now known as *Anguina agrostis* (Goodey, 1959). This nematode attacks several *Agrostis* spp., and is a serious pest of bent grass grown for seed (Apt *et al.,* 1960). The nematode is dependent on the developing flowers for its reproduction and suppressing the heading of bent grass for a year, thereby causing a break in the nematode's life-cycle, effectively controls the nematode (Apt *et al.,* 1960). Leaf galling on the various bent grasses is produced by *A. graminis* and *D. graminophilus* (Goodey, 1959). *D. radicicola* causes galls on the roots of grasses such as *A. stolonifera* and *A. tenuis* (Kuiper, 1953). *Meloidogyne* spp. are infrequently reported as parasites of bent grass. *M. naasi* is recorded from Europe and as an associate of turf grasses in a few states in the U.S.A. (Gooris, 1968; Radewald *et al.,* 1970). The nematode reproduces extremely well on *A. palustris* and *A. tenuis; A. canina* supports reproduction while the host status of *A. alba* appears uncertain (Gaskin, 1965; Gooris, 1968; Radewald *et al.,* 1970). *A. stolonifera* is a good host for *Longidorus macrosoma* but a poor host for *Xiphinema diversicaudatum;* both nematode species are important virus vectors (Fritzsche and Hofferek, 1969a, 1969b).

2. Fescue, *Festuca* spp.

Chewing and sheep fescues are hosts for *Anguina agrostis* (Goodey, 1959), and the galls caused on chewing fescue are apparently poisonous to cattle and sheep (Decker, 1969). *A. graminis* causes leaf galling on several *Festuca* spp. (Goodey, 1959). *F. pratensis* is a host of *D. radicicola* (Kuiper, 1953).

Many grass species are attacked by *Heterodera avenae* (*H. major*) (Bovien, 1953), but are less efficient hosts than the cereals, and several workers have reported that populations of this nematode fall rapidly under grasses, such as *F. pratensis* (Gair, 1968; Andersen and Andersen, 1970). R. D. Winslow (personal communication, 1970) made some interesting observations on "strains" and morphological "variants" in the *H. avenae* complex. Two "variants", which appear to be undescribed species, reproduce poorly on cereals and on some grasses, differ in their host range with regard to some other grasses and reproduce on three *Festuca* spp.

Several *Festuca* species and varieties are hosts for *Meloidogyne naasi* (Gooris, 1968; Radewald *et al.*, 1970). *F. pratensis* and *F. rubra* have been recorded as hosts for *Pratylenchus neglectus* (Wetzel, 1969), *F. rubra* is host for *Paratylenchus projectus* and *Trichodorus christiei* (Coursen *et al.*, 1958), and *Longidorus macrosoma* reproduces on *F. ovina* (Fritzsche and Hofferek, 1969b).

Numerous investigations concern nematodes associated with tall fescue. "Kentucky 31" tall fescue was highly resistant to five *Meloidogyne* spp. in the investigations of McGlohon *et al.* (1961). They also found that it produced lower top yield and fewer roots than control plants when infested with various ectoparasitic nematodes. Extremely high nematode populations were recovered from the soil with tall fescue. *Pratylenchus scribneri* reproduced well on *F. elatior* and in greenhouse experiments, infected plants yielded significantly less forage than control plants (Minton, 1965). Tall fescue was listed as a very suitable host for *T. christiei* by Rohde and Jenkins (1957), and has also been recorded as host for *Paratylenchus projectus* (Coursen *et al.*, 1958), and *Hypsoperine graminis* (Williams and Laughlin, 1968).

3. Meadow Grass, *Poa* spp.

A number of different nematodes have been recorded as associated with or pathogenic on *P. pratensis* (Table V). It is known as a host for *Ditylenchus radicicola* in several countries (Kuiper, 1953). Lewis and Webley (1966) considered this nematode capable of causing damage to pasture under certain climatic conditions, and according to R. D. Winslow (personal communication, 1970) the nematode has been detected in pasture, associated with shallow or poorly-anchored grass root systems at a few sites in N. Ireland. L. Bumbulucz (personal communication, 1970) observed heavy attacks of *D. radicicola* causing some damage in meadows locally in western Norway. Oostenbrink (1953) reported that the nematode occurred in 75% of old meadows in the Netherlands, though not causing obvious damage.

Spiral nematode genera cause significant injury to turf grasses (Heald and Perry, 1969). *Helicotylenchus* spp., especially *H. digonicus,* caused unthrifty growth and a disease called "summer dormancy" in Kentucky bluegrass turf in Wisconsin (Perry *et al.,* 1959). The pale or chlorotic plants with shallow, necrotic, brown root systems showed an appreciable loss of production.

TABLE V. Host status and reactions of *Poa pratensis* to attacks of plant nematodes

Nematode species	Host status, reaction	References
Anguina agrostis	host	Goodey (1959)
Ditylenchus radicicola	host	Kuiper (1953)
Heterodera avenae	light attack	Bovien (1953)
Meloidogyne spp. (*arenaria, incognita, incognita acrita, javanica*)	infection and repro- duction	McGlohon *et al.* (1961)
Meloidogyne graminis	host (var. "Merion")	Williams and Laughlin (1968)
Meloidogyne hapla	no reproduction	McGlohon *et al.* (1961)
Meloidogyne incognita	successful invasion, galling and repro- duction	Potter *et al.* (1969)
Meloidogyne naasi	host	Gooris (1968)
Paratylenchus projectus	host ("Common" and "Merion" varieties)	Coursen *et al.* (1958)
Trichodorus christiei	good host '('Merion")	Rohde and Jenkins (1957)
Longidorus macrosoma	host	Fritzsche and Hofferek (1969b)
Xiphinema diversi- caudatum	good host	Fritzsche and Hofferek (1969a)

P. annua and *P. trivialis* are hosts for *Anguina agrostis* (Goodey, 1959; Goffart, 1962), *M. naasi* (Gooris, 1968), and for *D. radicicola,* the latter plant being attacked in permanent pastures in England (Lewis and Webley, 1966). *P. trivialis* was attacked by *H. avenae* in Denmark (Bovien, 1953). *P. annua* appears to be a good host for *L. elongatus,* causing galls on the roots (Thomas, 1969), and Thomas (1970) found that populations of *X. diversicaudatum* increased under this grass species.

4. Bermuda Grass, *Cynodon dactylon*

The nematode pests of this important warm-season turf grass have attracted much attention. Root-knot has appeared sporadically as "trouble spots" in Bermuda grass pastures and turf, and Riggs *et al.* (1962) believe that the problem may become more important with the trend toward better lawns. These authors tested pasture and lawn types of Bermuda grass against five *Meloidogyne* spp. *M. incognita acrita* was more universally damaging than any other species, while *M. hapla* was the least damaging. McGlohon *et al.* (1961) stated that "Coastal" Bermuda grass, a pasture type, was highly resistant to five *Meloidogyne* spp., including *M. hapla*. Good *et al.* (1965) concluded that this variety may be used to control root-knot, but that it can increase populations of some destructive ectoparasitic nematodes including *Belonolaimus longicaudatus*, which is considered the most important pest of turf grasses in Florida (Heald and Perry, 1969). Brodie *et al.* (1969) found that "Coastal" Bermuda grass suppressed populations of *Pratylenchus zeae, T. christiei, Helicotylenchus dihystera* and *X. americanum*, and concluded that this, besides its root-knot resistance, increases the value of this grass as a nematode-reducing crop. *M. graminis* causes severe damage to some turf varieties of Bermuda grass in that it prevents normal root development and suppresses vegetative growth (Heald, 1969; Maur and Perry, 1969), and Williams and Laughlin (1968) listed several varieties of *C. dactylon*, and the species *C. transvaalensis*, as hosts for this nematode.

There are several ectoparasitic nematodes of great importance in Bermuda grass turf (Good *et al.*, 1959; Heald and Perry, 1969). Populations of *B. longicaudatus* and *Hoplolaimus coronatus* increase on this grass and reduce the yield (Di Edwardo, 1963). Johnson (1970) found that *B. longicaudatus, Criconemoides ornata* and *Tylenchorhynchus martini* reproduced readily on six varieties of Bermuda grass and that the plants showed less growth than control plants. Bermuda grass top and root weights were significantly decreased by *T. christiei* according to McGlohon *et al.* (1961).

5. St Augustine Grass, *Stenotaphrum secundatum*

Several plant parasitic nematodes have been reported as parasites of this grass (Good *et al.*, 1959). *Hypsoperine graminis*, originally described from St Augustine grass in Florida, caused considerable loss in average root fresh weight in pathogenicity tests (Maur and Perry, 1969). *Heterodera leuceilyma* parasitized exclusively St Augustine grass in Florida (Di Edwardo and Perry, 1964). Top and root growth were reduced and the roots became necrotic, distorted and with abnormal swellings, sometimes with proliferation. A pronounced interveinal chlorosis resembled the symptoms of iron deficiency.

Criconemoides lobatum reproduces readily on St Augustine grass (Johnson and Powell, 1968). *B. longicaudatus* and *T. christiei* are serious pathogens and economically important pests of St Augustine grass in Florida and Georgia (Rhoades, 1962; Heald and Perry, 1969).

6. Miscellaneous Grasses

A number of ectoparasitic nematodes were listed by Good *et al.* (1959) as detrimental parasites of the important warm-season turf grasses *Zoysia japonica* and *Eremochloa ophiuroides*. Parasitism of *Criconemoides lobatum* on centipede grass and "Emerald" zoysia caused sufficient damage to grass roots, mainly through the reduction of fibrous roots, to become a problem under certain field conditions (Johnson and Powell, 1968). The deleterious effect of *C. citri* on centipede grass growth was clearly demonstrated by nematicide treatment in greenhouse pot tests (Tarjan, 1964). "Meyer" zoysia is host of *Hypsoperine graminis* (Williams and Laughlin, 1968). *Digitaria decumbens*, pangola grass, a warm-season pasture and forage grass, is interesting because of its inhibitory effect on *Meloidogyne incognita acrita*, apparently by a root secretion that kills the nematode larvae (Ayala *et al.*, 1967).

The ryegrasses seem to be very suitable hosts for a number of plant-parasitic nematodes (Table VI), including several ectoparasites (McGlohon *et al.*, 1961). *Longidorus elongatus* increases substantially on *Lolium multiflorum* and *L. perenne*, causing distortion or galling of the subterminal portion of the roots, and was found to transmit the tomato black ring virus to both grass species (Taylor, 1967; Thomas, 1969). *L. perenne* is a very suitable host for *Longidorus macrosoma* and *X. diversicaudatum*, and a host also of the brome grass mosaic virus transmitted by both of these nematode species (Fritzsche and Hofferek, 1969a, 1969b; Fritzsche, 1970).

Pasture, forage and cereal grasses are parasitized by *Anguina* spp. and closely related genera (Goodey, 1959; Goffart, 1962) as well as by a number of other nematodes (Table VI). The seed-galling species *A. agrostis* attacks, amongst others, cocksfoot, timothy and *Trisetum flavescens*; Wagner (1969) observed noticeable infestations in *T. flavescens* seed crops in Germany and in seed imported from Czechoslovakia. Recently, Southey (1969) reported that an *Anguina* sp. produced galls on both leaf and flower tissues and in the seed of cocksfoot in pot tests. *Nothanguina cecidoplastes*, producing galls on the leaves, stem and flowers of its grass host (Thorne, 1961), has been recorded from the pasture grass *Andropogon pertusus* in India (Goodey, 1959). An interesting association of the leaf-galling nematode *Ditylenchus graminophilus* with the fungus *Dilophospora alopecuri*, mentioned by Goodey (1959), was recorded also by Goffart (1962) from *Calamagrostis* sp.

Compared with oats and barley the fodder grasses are generally poor or very poor hosts for *Heterodera avenae* (Bovien, 1953; Gair, 1968), and Andersen and Andersen (1970) reported a 58% decrease of *H. avenae* populations under several fodder grasses, no difference being found between species or varieties.

TABLE VI. Host status and reactions of some grasses to attacks of plant nematodes

Crop species	Nematode species	Host status, plant reaction	References
Lolium multiflorum Italian ryegrass	*Meloidogyne arenaria, incognita, incogn. acr., javanica*	infection and reproduction	McGlohon et al. (1961)
	Meloidogyne hapla	resistant to infection	McGlohon et al. (1961)
	Meloidogyne graminis	non-host, no reproduction	Williams and Laughlin (1968)
	Meloidogyne naasi	host	Gooris (1968)
	Pratylenchus neglectus	heavily attacked but not suffering	Wetzel (1969)
	Longidorus elongatus	one of the best hosts	Thomas (1969)
	Xiphinema diversicaudatum	population increase	Thomas (1970)
L. perenne perennial ryegrass	*Ditylenchus radicicola*	host	Kuiper (1953)
	Meloidogyne naasi	host	Gooris (1968)
	Pratylenchus neglectus	heavily attacked but not suffering	Wetzel (1969)
	Trichodorus christiei	host	Rohde and Jenkins (1957)
	Longidorus elongatus	rapid population increase	Taylor (1967)
	Longidorus macrosoma	very suitable host	Fritzsche and Hofferek (1969b)
	Xiphinema diversicaudatum	very suitable host	Fritzsche and Hofferek (1969a)
Bromus inermis smooth brome grass	*Meloidogyne incognita*	successful invasion, galling, reproduction	Potter et al. (1969)
	Pratylenchus neglectus	moderately attacked	Wetzel (1969)
	Paratylenchus projectus	host	Coursen et al. (1958)
	Trichodorus christiei	good host	Rohde and Jenkins (1957)

Host plant	Nematode	Reaction	Reference
Dactylis glomerata cocksfoot	*Ditylenchus radicicola*	galls observed	Lewis and Webley (1966)
	Meloidogyne graminis	host	Williams and Laughlin (1968)
	Meloidogyne naasi	host	Gooris (1968)
	Meloidogyne incognita	successful invasion, galling, reproduction	Potter *et al.* (1969)
	Pratylenchus neglectus	heavily attacked	Wetzel (1969)
	Paratylenchus projectus	host	Coursen *et al.* (1958)
	Tylenchorhynchus claytoni	favourable for reproduction	Krusberg (1959)
	Trichodorus christiei	poor host	Rohde and Jenkins (1957)
	Longidorus elongatus	host	Thomas (1969)
	Longidorus macrosoma	very suitable host	Fritzsche and Hofferek (1969b)
	Xiphinema diversicaudatum	very suitable host	Fritzsche and Hofferek (1969a)
Phleum pratense timothy	*Ditylenchus radicicola*	host	Kuiper (1953)
	Meloidogyne naasi	host	Gooris (1968)
	Meloidogyne incognita	successful invasion, galling, reproduction	Potter *et al.* (1969)
	Pratylenchus neglectus	heavily attacked	Wetzel (1969)
	Paratylenchus projectus	host	Coursen *et al.* (1958)
	Trichodorus christiei	good host	Rohde and Jenkins (1957)
	Longidorus macrosoma	host	Fritzsche and Hofferek (1969b)
	Xiphinema diversicaudatum	good host	Fritzsche and Hofferek (1969a)

A number of forage and cereal grasses are hosts for *L. macrosoma* and *X. diversicaudatum*, and are hosts also for the brome grass mosaic virus transmitted by these nematodes (Fritzsche and Hofferek, 1969a, 1969b; Fritzsche, 1970).

IV. Cultural and Environmental Influences

In pastures and turfs the soil is not cultivated and disturbed every year as with annual cultures. These crops tend to improve soil structure and fertility, and this, together with the perennial nature of the crops, probably favours the build-up of various plant nematode populations. To prevent an expected severe damage it is often recommended to maintain a high level of soil fertility for good plant growth. The nematodes apparently also benefit from this, and in established well-tended pastures and turfs the plants may tolerate very high numbers of nematodes. The disease cycle of root ectoparasitic nematodes in Kentucky bluegrass turf, and the effects of a cool moist environment on root growth and nematode build-up was discussed by Perry *et al.* (1959). Norton and Wilsie (1967) stressed the importance of maintaining a high level of soil fertility, where other means of nematode control are not feasible. This was illustrated by Riispere and Krall (1970), who showed that the tolerance of white clover to attacks by *Heterodera trifolii* increased considerably if the plants were supplied with optimum amounts of water and nutrients. Many reports of severe root-knot injury to lucerne in coarse-textured, light or sandy soils exemplify the influence of soil texture; stem nematodes, on the other hand, prefer heavier soils.

Several cases of seed dispersal emphasize the importance of careful cleaning of the seed harvest. Brown (1957) and Bingefors (1960) described evident long-range spread of stem nematode infestations in lucerne to different countries in the 1950s, by nematodes attached to the seed or with harvest debris. Thorne (1961) believes that the lucerne stem nematode originated from an infested locality in southeast Asia from where it has spread throughout the world with lucerne seed. Böning (1964) maintains that seed dispersal of nematodes in the red clover crop is of an accidental occurrence only, and not of any general importance. Irrigation water seemed to be the only logical source of dissemination of root-knot and lucerne stem nematode in infested fields of the Columbia Basin in the U.S.A. (Faulkner and Bolander, 1967). Stem nematode damage following the direction of slopes or the direction of ploughing and cutting in red clover and lucerne fields indicates spread by surface drainage water (Fig. 2).

Weed control is also an important cultural practice as several common weeds are hosts for many of the plant nematodes. *Poa annua*, generally considered a weed in good lawns and pastures, is a host for many nematodes, as accounted for under the *Poa* section. *Agropyron repens* is host for *Meloidogyne incognita* (Potter *et al.*, 1969), *Heterodera avenae* (Bovien, 1953), and *Ditylenchus radicicola* (Kuiper, 1953). It appears to be a poor host for *Longidorus elongatus* and *Xiphinema diversicaudatum* (Thomas, 1969, 1970).

FIG. 2. Spread of stem nematode (*Ditylenchus dipsaci*) infection with surface water in a sloping field. (By courtesy of S. Bingefors.)

Leguminous pastures supply the soil with nitrogen, with the aid of the *Rhizobium* bacteria living in nodules on the plant roots. Several workers have reported that nematodes interfere with bacterial nodulation (Taha and Raski, 1969). Wardojo *et al.* (1963) found that *Heterodera trifolii* infections caused a strong decline of *Rhizobium* nodules on white clover roots. However, Taha and Raski (1969) showed that the introduction of nematodes with the bacteria did not hinder nodule formation.

V. Control

Nematodes in pasture and turf can be controlled by crop rotation, the use of resistant varieties, or with chemicals. Seed crops cause special problems; they must be harvested and cleaned with great care to prevent the dispersal of adhering nematodes.

A. Crop Rotation

Nematode control by crop rotation is largely implicated with the cultural practices discussed above. For the control of stem nematodes in lucerne and red clover it is advisable not to grow these crops in infested soil for 8–10 years, but rather to grow sugar-beet, potatoes, cereals, and other non-hosts (Decker, 1969). Wagner (1967), however, reported stem nematode attacks in red clover where this crop had not been grown for 15 years, and recommends the cultivation of *Trifolium resupinatum*, which is resistant to this nematode, where red clover is damaged by stem nematodes. Forage grasses are not very efficient hosts for *Heterodera avenae*, and a grass break in the rotation can be of value in decreasing levels of this nematode (Gair, 1968; Andersen and Andersen, 1970). Hijink and Kuiper (1964) discussed the cultivation of cereals and grasses to prevent the build-up of *Meloidogyne hapla* populations.

Pastures and leys are often needed in crop rotations because of their value as soil improvers. Cotton is rotated with lucerne in the southwest U.S.A. (Reynolds, 1955), and the increased use of forage grasses and legumes in rotation with row-crops was mentioned by Brodie *et al.* (1969). An important aspect of this cropping system is that plant species and varieties are chosen so that they do not increase, but depress nematode populations. Brodie *et al.* (1969) presented evidence that nematodes are involved in crop rotation effects, and recommended "Coastal" Bermuda grass as a nematode-reducing crop in sod-based rotations. Ayala *et al.* (1967) found pangola grass, *Digitaria decumbens*, to be an effective rotation crop for pineapple, because of its nematode resistance and inhibitory effects on some of the most common pineapple nematodes. Epps and Golden (1967) pointed out that Kobe lespedeza should not be grown in rotation with soybean because this caused a build-up of *H. glycines*. Corn and soybeans are highly resistant to *H. trifolii*, and if rotated with red clover this nematode problem may be eliminated (Norton, 1967). Several authors have emphasized the importance of choosing only those varieties or species of legumes and grass for cover crops that do not enhance existing nematode problems (Good et al., 1965; Minton et al., 1967; Brodie et al., 1970). Thomas (1969) drew attention to the fact that grasses and clovers were good hosts for *Longidorus elongatus*, a vector of some viruses severely damaging strawberry and raspberry, and advised against using grass and clover leys in alternative cropping with these berry crops if viruliferous *L. elongatus* were present in the soil.

B. Resistant Varieties

Nematode resistant varieties normally offer the cheapest and best means of control in pasture crops (Bingefors, 1969). Stem nematode resistance in lucerne has interested plant breeders in many countries, resulting in several highly resistant varieties, "Lahontan" being one of the best. The stem nematode resistance of this variety is derived from the old variety "Turkestan", which originates from central Asia. "Lahontan" is highly resistant to stem nematodes in America as well as in Europe, but is not adapted to the climate of northern Europe. The development of varieties with resistance against several pests and diseases is considered important. Lucerne varieties with combined resistance against stem nematodes and *Verticillium* wilt or bacterial wilt, have been released recently or are being developed (Bingefors, 1969). Lucerne varieties differ in their resistance to root-knot nematodes (Allison, 1956) and Reynolds (1955) found northern (hardy) varieties were generally more susceptible to *M. javanica* and *M. incognita acrita* than southern (non-hardy) selections. Resistance in lucerne to *M. hapla* appears to be relatively rare (Stanford et al., 1958); however, a few plants within certain varieties were found highly resistant. Hunt et al. (1969) confirmed that this resistance is conditioned by one dominant gene with tetrasomic inheritance. Combination of resistance against all the important root-knot nematodes and against stem nematode in the same variety is considered an urgent task (Bingefors, 1969).

Red clover has great varietal and intravarietal differences in resistance to the stem nematode. This makes it relatively easy to improve the level of resistance in adapted varieties by mere selection (Fig. 3). The first variety with high resistance was developed in the 1930s in Sweden from plants surviving on soil heavily infested with stem nematodes. Since then intensive breeding work in several European countries has resulted in a number of highly resistant red clover varieties (Bingefors, 1957, 1971). The efficiency in breeding for stem nematode resistance in red clover and lucerne was greatly improved by the development of a carefully controlled laboratory technique, where seedling plants are

D

selected by their response to inoculation with nematodes (Bingefors, 1971), which are themselves produced in callus tissue cultures (Bingefors and Eriksson, 1968).

FIG. 3. Field trial (second-year ley) in nematode-infested soil of red clover varieties with different levels of stem nematode (*Ditylenchus dipsaci*) resistance. (By courtesy of S. Bingefors.)

Breeding for root-knot resistance in red and white clovers has been carried out in Mississippi (Bain, 1959). Inbred red clover lines with combined resistance to five *Meloidogyne* spp. and powdery mildew have been developed, and work is continuing to increase the yield of these lines (Bain, personal communication, 1970). It would be of value to have resistance against *Heterodera trifolii* incorporated in white clover when regarding its great importance as a pasture legume (Norton, 1967). This possibility is shown by investigations in the Netherlands, the U.S.A. and New Zealand (Dijkstra, 1971). This author did not

detect differences in biotypes among several Dutch *H. trifolii* populations, nor did he confirm the correlation of nematode resistance in
white clover with a high glucoside content (Grandison, 1963). Stem
nematode resistance in white clover is of interest to plant breeders in
Denmark (Bingefors, 1969). *Lespedeza* species and strains vary in their
resistance to root-knot nematodes. Breeding lines with resistance to
one or more *Meloidogyne* spp. have been found (Bingefors, 1969).
Offutt and Riggs (1970) showed that irradiation could be used to
improve the resistance of *L. stipulacea* to *M. incognita*.

Breeding nematode-resistant grass varieties is an almost virgin area
of research. However, grass species and varieties differ in their resistance
to various nematodes. Gaskin (1965) found that a few varieties of
Poa pratensis supported reproduction of root-knot (*Meloidogyne naasi*)
while most others did not, and different strains of St Augustine grass
vary in their susceptibility to *M. graminis* (Heald and Perry, 1969).
Varietal differences in tolerance among Bermuda grasses towards root-
knot and some ectoparasitic nematodes were reported by Riggs *et al.*
(1962) and Johnson (1970).

C. Chemicals

Until less phytotoxic, less mammalian-toxic, more effective and
cheaper nematicides are found, chemical control of nematodes in pasture
crops is not advisable or even, in many places, permissible. The residue
problem is of special significance in crops that are used for grazing or
hay. Workers in the Netherlands have found that white clover growth
is considerably improved after treatments with chemicals, although
whether it is due to nematode mortality or improved nutritional
conditions for the crop is debatable. Ennik and Kort (1966) concluded
that a possible use of compounds with systemic effects might open new
avenues for improvement of grass–clover swards. Wagner (1967) reported successful control of stem nematodes in red clover. However,
the chemicals were too poisonous for general use. Seed-borne lucerne
and clover stem nematodes can be controlled by fumigation of the seed
with methyl bromide (Hague and Clark, 1959; Anon., 1968).

Many investigations concern the possible use of nematicides in
established turf (Heald and Perry, 1969). Tarjan (1964) found that
various chemicals tested were effective nematicides in centipede grass
turf. However, nematode populations built up again some months after
treatment. Fumigants are unsatisfactory, owing to their phytotoxic
effects but are useful if applied prior to planting turf. Brodie and
Burton (1967) reported a renewed interest in nematode control in
established turf due to the development of nonvolatile organic

TABLE VII. Figures of crop losses caused by plant nematodes in pasture legumes and turf grasses

Crop and variety	Nematode species	Type and amount of damage	References
Lucerne, "Atlantic"	*Meloidogyne hapla*	growth reduced by 42–58%	Chapman (1960)
	Meloidogyne incognita	growth reduced by 2–7%	Chapman (1960)
"Vernal"	*Pratylenchus penetrans*	mean foliage yield reduced to 83% of control	Willis and Thompson (1969)
Red clover, "Kenland"	*Meloidogyne hapla*	growth reduced by 40–55%	Chapman (1960)
	Meloidogyne incognita	growth reduced by 77–83%	Chapman (1960)
"LaSalle"	*Pratylenchus penetrans*	mean foliage yield reduced to 73% of control	Willis and Thompson (1969)
White clover	*Heterodera trifolii*	leaf production reduced to 30% of control	Wardojo *et al.* (1963)
	Meloidogyne hapla	leaf production reduced to 30% of control	Wardojo *et al.* (1963)
	Pratylenchus penetrans	leaf production reduced to 45% of control	Wardojo *et al.* (1963)
"Ladino"	*Meloidogyne hapla*	dry matter yield reduced by about 60%	Colman (1964)
	Pratylenchus penetrans	mean foliage yield reduced to 93% of control	Willis and Thompson (1969)
Birdsfoot trefoil, "Empire"	*Pratylenchus penetrans*	mean foliage yield reduced to 50% of control	Willis and Thompson (1969)
Agrostis tenuis and *A. palustris*	*Anguina agrostis*	75% reduction in seed yield	Apt *et al.* (1960)

Plant	Nematode	Effect	Reference
Bermuda grass, "Tifdwarf"	*Meloidogyne graminis*	root and top weights reduced by 28%, clipping weight by 69%	Heald (1969)
"Tifton 328"	*Meloidogyne graminis*	average fresh weight of roots 36% less than control	Maur and Perry (1969)
	Belonolaimus longicaudatus and *Hoplolaimus coronatus*	50% reduction in total plant weights	Di Edwardo (1963)
St Augustine grass	*Meloidogyne graminis*	average fresh weight of roots 42% less than control	Maur and Perry (1969)
	Belonolaimus longicaudatus *Trichodorus christiei*	root weight reduced up to 69% growth reduced by 25% and root weight by 49%	Rhoades (1962) Rhoades (1962)
Italian ryegrass	*Hoplolaimus tylenchiformis*, *Tylenchorhynchus claytoni*	top weights reduced by 30–35% in later harvests	McGlohon *et al.* (1961)
Cocksfoot	*Hoplolaimus tylenchiformis* *Trichodorus christiei*	dry foliage reduced by 40–50% dry foliage reduced by 30–40%	McGlohon *et al.* (1961) McGlohon *et al.* (1961)

phosphates with both insecticidal and nematicidal properties. These new nematicides give excellent control of nematodes on turf grasses and their use is gradually becoming standard practice in Florida where high quality turf is desired (Heald and Perry, 1969). Some of these nonvolatile compounds are taken up by the roots to become systemic in the plant.

VI. Economics

Plant-parasitic nematodes considerably decrease pasture yields and increase the costs of turf establishment and maintenance. It is exceedingly difficult, however, to estimate the value of the total losses due to nematodes in these crops. There are not only the direct losses in forage yield and the probable increase of nematode numbers on these perennial crops, but also the value of pastures and leys as soil improvers in the rotation has to be taken into account. The value of turf is largely implicated with the well-being of man; turf has an aesthetic value, and there is an increasing demand for high quality recreational grounds and sports fields.

In order to get a relevant impression of the incidence and economics of nematodes in pasture and turf in various parts of the world, the opinions of some people were requested. According to M. R. Hanna (Lethbridge, Canada) *Ditylenchus dipsaci* in lucerne is the only nematode of any economic significance in these crops in Canada, especially in the irrigated districts of southern Alberta. However, nematodes associated with other disease-producing organisms such as fungi, bacteria and viruses, may be of greater economic concern than is generally recognized. Stem nematodes appear to be a normal hazard to lucerne growers in the irrigated areas of the western U.S.A. and a potential threat to lucerne elsewhere in the world. In the southeastern U.S.A. stem nematode research is of secondary importance compared to the more urgent and very serious "alfalfa weevil" problem (R. T. Sherwood, North Carolina). Root-knot nematodes also cause economic loss in the lucerne crop in the western states, and D. C. Bain (Mississippi) states that nematodes are a problem in pastures on sandy soils in Mississippi. D. C. Norton (Iowa) suspects that nematodes, including *Heterodera trifolii* on the clovers, are responsible for considerable damage in hay and pasture fields, especially during dry periods, and N. A. Minton (Georgia) assumes that nematodes can be blamed for decreased yields and the lack of persistence of certain forage crops in the southeastern U.S.A. According to G. S. Grandison (Glen Osmond, South Australia) *D. dipsaci* is a limiting factor in lucerne and red clover growth in localized areas in New Zealand, and in white clover throughout the country.

In European countries stem nematodes in red clover and lucerne appear to be a major problem. *D. dipsaci* in red clover is widespread throughout Scandinavia, causing great damage in some areas. The use of resistant varieties and changes in cultural practices, however, have effectively controlled the nematode. The nematode problem in lucerne is now overshadowed by the *Verticillium* wilt disease in both Scandinavia and in England (J. F. Southey, Harpenden). According to M. Oostenbrink (Wageningen) lucerne and red clover are damaged by *D. dipsaci* locally in the Netherlands, and various other nematodes seem to be economically important as parasites of meadow grasses and white clover.

Numerous data showing percentage yield reduction appear in the literature (Table VII) and though mainly derived from greenhouse and pot experiments illustrate the detrimental effects on forage production caused by nematodes in susceptible crops. Norton and Wilsie (1967) concluded that nematodes contribute considerably to low forage yields, often reducing yields by as much as 10 or 15%. With the very widespread occurrence of some nematodes the overall effect of even a low yield reduction per hectare can amount to very high values.

A few attempts have been made to estimate in monetary terms the losses due to nematodes. According to "Losses in Agriculture" (Anon., 1965) nematodes caused a 3% reduction in the potential yield of lucerne in the U.S.A. This was equal to an average annual loss during the years 1951–1960 of more than 2 million metric tons or U.S. $48·6 million. Losses in lespedeza, caused by nematodes, were calculated to be 3%, or slightly more than 1800 metric tons valued at $457,000. Total annual losses for eight field and forage crops, including the two mentioned, were valued at about $300 million. Good (1968) assessed the probable loss in lucerne due to nematodes at 4% or nearly 2200 metric tons. Wehunt (1958) estimated the value of nematode damage in white clover in Louisiana to be about $1 million per year, by decreasing the yield of forage alone. Recently (Anon., 1971) losses of several million dollars per year were estimated to occur in the U.S.A. in pasture legumes. Research on the amount and value of nematode damage under field conditions in pasture and turf has lagged behind other crops with higher cash return per hectare, and a better knowledge is greatly needed.

References

Allison, J. L. (1956). *Phytopathology* **46**, 6.
Andersen, K. and Andersen, S. (1970). *Tidsskr. PlAvl* **74**, 559–565.
Anon. (1965). *In* "Losses in Agriculture." United States Department of Agriculture, Washington, D.C. *Agriculture Handbook* No. 291. 120 pp.
Anon. (1968). Advis. Leafl. Minist. Agric. Fish. Fd No. 409. H.M.S.O., London.

Anon. (1971). Estimated crop losses due to plant-parasitic nematodes in the United States. Special Publication No. 1, 7pp. Suppl. *J. Nematol.*

Apt, W. J., Austenson, H. M. and Courtney, W. D. (1960). *Pl. Dis. Reptr* **44**, 524–526.

Ayala, A., Román, J. and González Tejera, E. (1967). *J. Agric. Univ. P. Rico* **51**, 94–96.

Bain, D. C. (1959). *Pl. Dis. Reptr* **43**, 318–322.

Bingefors, S. (1957). *Växtodling* **8**, 1–123.

Bingefors, S. (1960). *K. LantbrHögsk. Annlr* **26**, 317–322.

Bingefors, S. (1967). *LantbrHögsk. Meddn* A79. 63 pp.

Bingefors, S. (1969). *Herb. Abstr.* **39**, 107–111.

Bingefors, S. (1971). *In* "Mutation breeding for disease resistance", Proc. I.A.E.A., Vienna, pp 209-235.

Bingefors, S. and Eriksson, K. B. (1968). *Z. PflZücht.* **59**, 359–375.

Böning, K. (1964). *Bayer. landw. Jb.* **41**, 649–670.

Bovien, P. (1953). *Tidsskr. PlAvl* **56**, 581–591.

Brodie, B. B. and Burton, G. W. (1967). *Pl. Dis. Reptr* **51**, 562–566.

Brodie, B. B., Good, J. M. and Adams, W. E. (1969). *J. Nematol.* **1**, 309–312.

Brodie, B. B., Good, J. M. and Jaworski, C. A. (1970). *J. Nematol.* **2**, 147–151.

Brown, E. B. (1957). *Nematologica* **2**, Suppl., 369–375S.

Chapman, R. A. (1959). *Phytopathology* **49**, 357–358.

Chapman, R. A. (1960). *Phytopathology* **50**, 181–182.

Chapman, R. A. (1964). *Phytopathology* **54**, 417–418.

Colman, R. L. (1964). *Agric. Gaz. N.S.W.* **75**, 1367–1368.

Coursen, B. W., Rohde, R. A. and Jenkins, W. R. (1958). *Pl. Dis. Reptr* **42**, 456–460.

Decker, H. (1969). "Phytonematologie" VEB Deutscher Landwirtschaftsverlag, Berlin.

Di Edwardo, A. A. (1963). *Rep. Fla agric. Exp. Stns* 1963, 109.

Di Edwardo, A. A. and Perry, V. G. (1964). *Bull. Univ. Fla agric. Exp. Stn* **687**, 35 pp.

Dijkstra, J. (1971). *Euphytica* **20**, 36–46.

Ennik, G. C. and Kort, J. (1966). *Neth. J. Pl. Path.* **72**, 184–195.

Ennik, G. C., Kort, J. and v.d. Bund, C. F. (1965). *J. Br. Grassld Soc.* **20**, 258–262.

Epps, J. M. and Golden, A. M. (1967). *Pl. Dis. Reptr* **51**, 775–776.

Faulkner, L. R. and Bolander, W. J. (1967). *Nematologica* **12**, 591–600.

Fritzsche, R. (1970). Proc. IX Int. Nem. Symposium (Warsaw, 1967). *Zesz. Problemowe PNR* nr 92, 293–300.

Fritzsche, R. and Hofferek, H. (1969a). *Arch. Pflanzensch.* **5**, 111–118.

Fritzsche, R. and Hofferek, H. (1969b). *Arch. Pflanzensch.* **5**, 423–429.

Gair, R. (1968). *Pl. Path.* **17**, 145–147.

Gaskin, T. A. (1965). *Pl. Dis. Reptr* **49**, 89–90.

Gerdemann, J. W. and Linford, M. B. (1953). *Phytopathology* **43**, 603–608.

Gibbs, A. J. and Harrison, B. D. (1964). *Ann. appl. Biol.* **54**, 1–11.

Goffart, H. (1962). *SchrReihe Karl-Marx-Univ. Leipzig zu Fragen sozialist. Landw. H.* **8**, 71–78.

Good, J. M. (1968). *Pl. Prot. Bull. F.A.O.* **16**, 37–40.

Good, J. M. and Thornton, G. D. (1956). *Pl. Dis. Reptr* **40**, 1050–1053.

Good, J. M., Minton, N. A. and Jaworski, C. A. (1965). *Phytopathology* **55**, 1026–1030.

Good, J. M., Steele, A. E. and Ratcliffe, T. J. (1959). *Pl. Dis. Reptr* **43**, 236–238.

Goodey, J. B. (1959). *J. Sports Turf Res. Inst.* **10**, 54–60.

Gooris, J. (1968). *Meded. Rijksfakulteit LandbWet. Gent* **33**, 85–100.

Grandison, G. S. (1963). *N.Z. Jl agric. Res.* **6**, 460–462.

Grandison, G. S. (1965). *N.Z. Jl Agric.* **111**, 65.

Guine, S. A. (1965). *Agricultura téc.* **25**, 127–128.

Hague, N. G. M. and Clark, W. C. (1959). *Meded. LandbHogesch. Opzoek-Stns Gent* **24**, 628–636.

Harrison, B. D. and Winslow, R. D. (1961). *Ann. appl. Biol.* **49**, 621–633.

Hawn, E. J. (1969). *J. Nematol.* **1**, 190–191.

Hawn, E. J. and Hanna, M. R. (1967). *Can. J. Pl. Sci.* **47**, 203–208.

Heald, C. M. (1969). *J. Nematol.* **1**, 31–34.

Heald, C. M. and Perry, V. G. (1969). *Agronomy* (American Society of Agronomy) No. 14, 358–369.

Hijink, M. J. and Kuiper, K. (1964). *Nematologica* **10**, 64.

Holdeman, Q. L. and Graham, T. W. (1953). *Pl. Dis. Reptr* **37**, 497–500.

Hunt, O. J., Peaden, R. N., Faulkner, L. R., Griffin, G. D. and Jensen, H. J. (1969). *Crop Sci.* **9**, 624–627.

Hunt, O. J., Griffin, G. D., Murray, J. J., Pedersen, M. W. and Peaden, R. N. (1971). *Phytopathology* **61**, 256-259.

Hutton, E. M. (1969). *N.Z. agric. Sci.* **3**, 142–146.

Johnson, A. W. (1970). *J. Nematol.* **2**, 36–41.

Johnson, A. W. and Powell, W. M. (1968). *Pl. Dis. Reptr* **52**, 109–113.

Krall, E. (1965). *Kratkye itogi nauchn, issl. zash. rast. v Pribalt. zone S.S.S.R.* **6**, 67–68.

Krusberg, L. R. (1959). *Nematologica* **4**, 187–197.

Krusberg, L. R. (1961). *Nematologica* **6**, 181–200.

Kuiper, J. (1953). *Tijdschr. PlZiekt.* **59**, 143–148.

Lewis, S. and Webley, D. (1966). *Pl. Path.* **15**, 184–186.

Lindhardt, K. (1963). *Tidsskr. PlAvl* **67**, 679–687.

McGlohon, N. E. and Baxter, L. W. (1958). *Pl. Dis. Reptr* **42**, 1167–1168.

McGlohon, N. E., Sasser, J. N. and Sherwood, R. T. (1961). *Tech. Bull. N. Carol. agric. Exp. Stn* **148**, 39 pp.

McGuire, J. M., Walters, H. J. and Slack, D. A. (1958). *Phytopathology* **48**, 344.

Mankau, R. and Linford, M. B. (1960). *Bull. Ill. agric. Exp. Stn* **667**, 50 pp.

Martin, G. C. (1961). *Rhodesia agric. J.* **58**, 62–65.

Maur, K. M. and Perry V. G. (1969). *J. Nematol.* **1**, 16.

Minton, N. A. (1965). *Pl. Dis. Reptr* **49**, 856–859.

Minton, N. A., Forbes, I. and Wells, H. D. (1967). *Pl. Dis. Reptr* **51**, 1001–1004.

D*

Moore, J. F. (1971). *Ir. J. agric. Res.* **10**, 207–211.

Norton, D. C. (1965). *Phytopathology* **55**, 615–619.

Norton, D. C. (1967). *Phytopathology* **57**, 1305–1308.

Norton, D. C. (1969). *Phytopathology* **59**, 1824–1828.

Norton, D. C. and Isely, D. (1967). *Pl. Dis. Reptr* **51**, 1017–1020.

Norton, D. C. and Wilsie, C. P. (1967). *Crops and Soils* **19**, No. 8, 17.

Offutt, M. S. and Riggs, R. D. (1970). *Crop Sci.* **10**, 49–50.

Oostenbrink, M. (1953). *Tijdschr. PlZiekt.* **59**, 149–152.

Perry, V. G., Darling, H. M. and Thorne, G. (1959). *Res. Bull. agric. Exp. Stn Univ. Wis.* **207**, 24 pp.

Potter, J. W., Townshend, J. L. and Davidson, T. R. (1969). *Nematologica* **15**, 29–34.

Radewald, J. D., Pyeatt, L., Shibuya, F. and Humphrey, W. (1970). *Pl. Dis. Reptr* **54**, 940–942.

Reynolds, H. W. (1955). *Phytopathology* **45**, 70–72.

Rhoades, H. L. (1962). *Pl. Dis. Reptr* **46**, 424–427.

Riggs, R. D., Dale, J. L. and Hamblen, M. L. (1962). *Phytopathology* **52**, 587–588.

Riispere, U. and Krall, E. (1970). Proc. IX Int. Nem. Symposium (Warsaw, 1967). *Zesz. Problemowe PNR* nr 92, 267–272.

Rohde, R. A. and Jenkins, W. R. (1957). *Phytopathology* **47**, 295–298.

Seinhorst, J. W. and Sen, A. K. (1966). *Neth. J. Pl. Path.* **72**, 169–183.

Singh, N. D. and Norton, D. C. (1970). *Phytopathology* **60**, 1834–1837.

Southey, J. F. (1969). *Pl. Path.* **18**, 164–166.

Stanford, E. H., Goplen, B. P. and Allen, M. W. (1958). *Phytopathology* **48**, 347–349.

Stehman, Jr., E. V. and Norton, D. C. (1969). *Pl. Dis. Reptr* **53**, 121–123.

Sturhan, D. (1963). *Z. angew. Zool.* **50**, 129–193.

Taha, A. H. Y. and Raski, D. J. (1969). *J. Nematol.* **1**, 201–211.

Tarjan, A. C. (1964). *Proc. Fla St. hort. Soc.* **77**, 456–461.

Taylor, C. E. (1967). *Ann. appl. Biol.* **59**, 275–281.

Thomas, P. R. (1969). *Pl. Path.* **18**, 23–28.

Thomas, P. R. (1970). *Ann. appl. Biol.* **65**, 169–178.

Thorne, G. (1961). *In* "Principles of Nematology." McGraw-Hill, New York.

Wagner, F. (1967). *Mitt. biol. BundAnst. Ld- u. Forstw.* H. 121, 58–61.

Wagner, F. (1969). *Mitt. biol. BundAnst. Ld- u. Forstw.* H. 132, 113–115.

Ward, C. H. (1960). *Phytopathology* **50**, 658.

Wardojo, S., Hijink, M. J. and Oostenbrink, M. (1963). *Meded. Landb-Hogesch. OpzoekStns Gent* **28**, 672–678.

Wehunt, E. J. (1958). *Diss. Abstr.* **19**, 426–427.

Wetzel, T. (1969). *Nematologica* **15**, 193–200.

Whitehead, A. G. (1968). *Trans. zool. Soc. Lond.* **31**, 263–401.

Williams, A. S. and Laughlin, C. W. (1968). *Pl. Dis. Reptr* **52**, 162–163.

Willis, C. B. and Thompson, L. S. (1969). *Can. J. Pl. Sci.* **49**, 505–509.

5

Nematode Diseases of Cereals of Temperate Climates

John Kort*

Plantenziektenkundige Dienst
Geertjesweg 15
Wageningen
The Netherlands

I. Introduction to the Crop

Cereals, especially barley and wheat, are some of the oldest domestic crops. Archaeological finds from the Neolithic (the earliest period of the Stone Age, starting ± 3000 B.C.) frequently include seeds of barley, and those from ancient Egypt commonly include wheat. In many parts of the world cereals are the principal crop, both in respect to acreage and to their relative importance in the diet. Cereals and cereal products are the principal resources in most famine international relief actions.

Data on the acreage and yield in cereals in the temperate zones of the world are given in Table I. Two main cereal growing countries in the African continent which border on to the Mediterranean have been included, because most of the nematode species discussed in this chapter occur or are suspected to occur in these countries. The figures for the

* All photographs in this chapter were taken by Plant Protection Service, Wageningen.

TABLE I. Total area (in 1000 ha), and volume (in 1000 metric tons) of cereals produced and their distribution and use in the temperate zones (calculated from F.A.O. Production Yearbook 1968)

Crop	Europe	U.S.S.R.	Canada, U.S.A.	Algeria, Morocco, S. Africa	Australia, N. Zealand	Total	Distribution	Use
WHEAT (*Triticum vulgare*)							Worldwide except tropical regions	Bread grain
Area	27841	67026	25972	5008	9198	145045		
Production	72391	77419	58069	3445	7922	219246		
BARLEY (*Hordeum vulgare* and *H. distichon*)							As wheat	Malting and feed grain
Area	15011	19125	6998	2371	1049	44554		
Production	45592	24662	13533	1440	933	86160		
OATS (*Avena sativa*)							Worldwide except where summers are too cool and too short	Breakfast cereal and feed grain
Area	7822	8688	9491	337	1264	27602		
Production	18454	11581	16146	213	727	47121		

RYE (*Secale cereale*)							
Area	8117	12418	710	31	32	21308	Northern hemisphere, especially Europe / Bread and feed grain
Production	16272	12986	951	14	14	30237	
MIXED GRAIN (Oats/barley)							
Area	1649	88	675	0	0	2412	Mainly Canada, Europe and U.S.A. / Feed grain
Production	4305	142	1387	0	0	5834	
MAIZE (*Zea mays*)							
Area	11039	3485	24860	6043	88	45515	Worldwide, all climates / Feed grain and in the food industry
Production	30830	9163	122793	9897	210	172892	
Total Area	71479	110830	78706	13790	11631	286436	
Total Volume	187844	135953	212879	15009	9805	561490	
Average yield/ha in metric tons	2·62	1·72	2·70	1·09	0·84	1·96	

Soviet Union include the Asian part of the country, whereas the Chinese Republic had to be omitted owing to lack of information.

The importance of cereals in the temperate zones indicated in Fig. 1 probably is demonstrated best by the fact that 62% of the world's total growing area is found in these regions which produce 73% of the world's total cereal production. Information on the distribution and use of different cereals in different parts of the world is shown in Table I.

II. Nematode Diseases

A. Cereal Cyst Nematode, *Heterodera avenae*

1. *World Distribution and Incidence*

For a long time cyst-forming nematodes were considered to be restricted to the temperate zones and, in fact, the first attack by *H. avenae* was reported from Germany (Kühn, 1874). Under different names *H. avenae* has been reported since, from most European countries, Australia (S. Australia, Victoria, New South Wales), Canada (Ontario), Israel, Morocco, South Africa, Japan and India. In India the disease is known as "Molya". The nematode obviously can tolerate much wider temperature ranges than was at first assumed.

The relationship between the economic importance of *H. avenae* and soil type, as has been referred to in many previous publications, has led to the conclusion that *H. avenae* occurs exclusively in light soils. However, this parasite, which is indigenous in many countries, will generally cause economic damage irrespective of soil type when the intensity of cereal cropping exceeds a certain limit; under conditions in the Netherlands, this limit is 60% annually (Kort, 1957). Nevertheless, outbreaks of damage in cereals are more common on light soils than on heavier soils, as will be discussed later (Figs. 2, 3 and 4).

In Britain, Gair (1965) found *H. avenae* in heavy clay, fine sand, and peat soil, but generally regarded it as a pest of light soils. In Germany and Australia also *H. avenae* has been reported from medium and heavy-textured soils (Lücke, 1969; Neubert and Decker, 1970). Meagher (1968) found, under Australian glasshouse conditions, that solonized brown soil and grey soil of heavy structure were more suitable for nematode multiplication than was red-brown earth. Thus, it can be said that *H. avenae* is not restricted to a particular soil type.

The distribution of *H. avenae* within a field is frequently very uneven, probably as a result of local initial infections which are spread mechanically. The patchiness of the field infection depends on the rate of multiplication of the nematode in the crop which in turn depends on the initial infestation level.

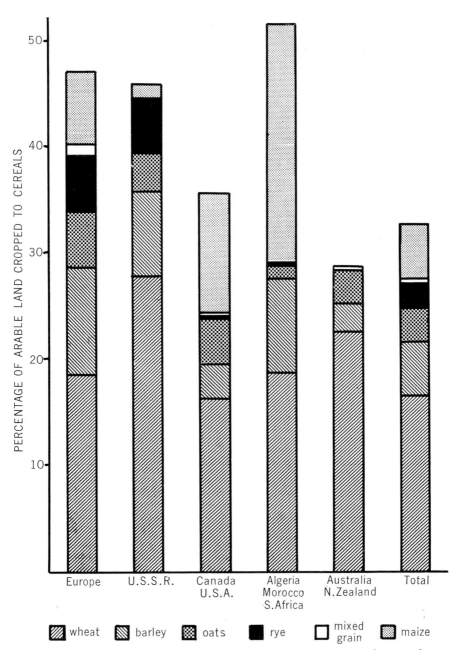

FIG. 1. Percentage of arable land in temperate zones which is cropped to cereals.

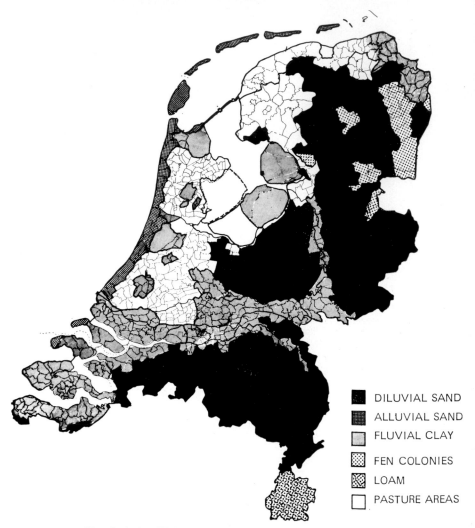

DILUVIAL SAND

ALLUVIAL SAND

FLUVIAL CLAY

FEN COLONIES

LOAM

PASTURE AREAS

FIG. 2. A simplified map of the soil types in the Netherlands.

2. Biology and Disease Symptoms

The host range of *H. avenae* is restricted to graminaceous plants. There is sexual dimorphism; the male remains filiform, whereas the female becomes lemon-shaped and spends its life inside or attached to the root. In the white cyst stage it is clearly visible on infested roots, the swollen body, about 1 mm across, protruding from the root surface. After the female has died, the body wall hardens to a tough, resistant brown cyst, which protects the eggs and larvae. The eggs within the cysts remain viable in the soil for several years.

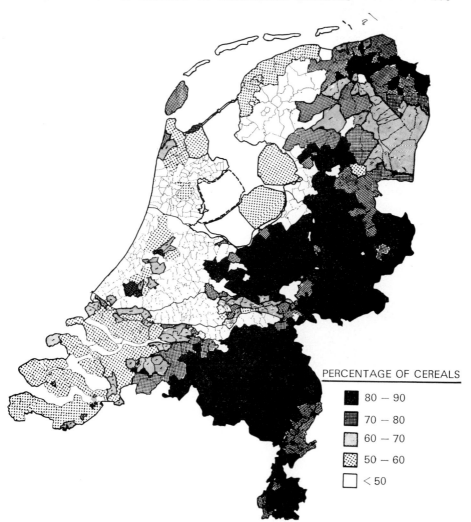

FIG. 3. The intensity of cereal growing in the Netherlands in 1968.

The feeding females induce plant cells in the head region to become enlarged, forming giant cells which restrict both water and food transport through the root. This phenomenon is a major characteristic of cyst-forming nematodes and is essential for female development.

Larval emergence from the cysts is not stimulated by root diffusates of either hosts or non-hosts, but the hatching shows a fixed seasonal rhythm irrespective of the crop. In Europe larval emergence from cysts occurs between mid-March and mid-July. The young females can be found from mid-June on the roots of infected plants. Autumn-sown

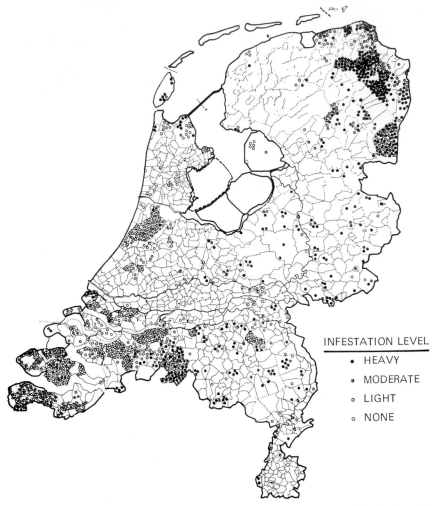

INFESTATION LEVEL

• HEAVY

◉ MODERATE

◦ LIGHT

○ NONE

FIG. 4. The distribution of occurrence of *Heterodera avenae* in the Netherlands. (Data by courtesy of Ir. P. Kleijburg.)

cereals are invaded by second-stage larvae in the autumn but the larvae do not continue their development until the next spring. Under the different climatic and cropping conditions of Australia (Meagher, 1968) *H. avenae* cysts usually appear in greatest numbers in late August and September and in later sowings even in October. Under conditions in Rajasthan (India) large numbers of cysts show on the roots in February (Singh and Swarup, 1964). In both countries these observations were made $3\frac{1}{2}$ months after sowing, so that this is similar to the situation in Europe. *H. avenae* produces only one generation per year and completes its life-cycle within 9–14 weeks of larval invasion (Duggan, 1961) except in autumn sown cereals. Even in India only one generation per crop season has been observed under field conditions.

Andersen (1959) showed conclusively the existence of two pathotypes of *H. avenae* in Denmark and pathotypes have been found since in the Netherlands, Britain, Germany and Sweden. There are indications that they also occur in India (Mathur, personal communication). The pathotypes cannot be distinguished morphologically but their identity can be established by using a series of test plants on which the pathotypes do or do not reproduce (Kort *et al.*, 1964).

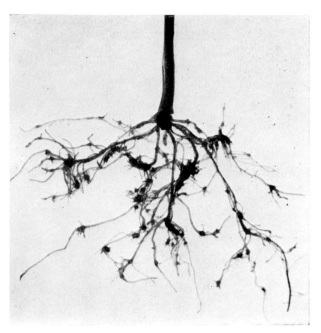

FIG. 5. Young oat plant showing knot-formation at the points of invasion by *Heterodera avenae*.

In plants seriously damaged by *H. avenae*, infested roots divide near the invasion point and side roots similary bifurcate. Affected plants, therefore, have a stunted, knotted root system which may be caused also by water deficiency (Fig. 5). The occurrence of white cysts on the roots from mid-June onwards is the only safe indication of any *H. avenae* attack.

The effect on the aerial parts of an *H. avenae* attack on the roots is also non-specific. Apart from a growth reduction, leaves show discolorations which are identical to severe nitrogen and phosphorus deficiency.

In the field a severe infestation of *H. avenae* is usually apparent by the development of patches of stunted plants which fail to recover (Fig. 6) and in which weeds develop more readily.

Fɪɢ. 6. An infected oat crop, showing patches of poor growth caused by *Heterodera avenae*.

3. Cultural and Environmental Influences

In many European countries there has been an increase since World War II in the incidence of damage in cereals caused by *H. avenae*. During the war, transport difficulties and an increased demand for home-produced food encouraged intensive cereal production. *H. avenae* causes crop failures in those areas where continuous or almost continuous cereal cropping is practised. The present trend of growing a smaller variety of crops on bigger fields decreases the possibilities of an adequate crop rotation system.

Although all cereals are hosts to *H. avenae*, differences in the degree of host susceptibility (crop damage) and in host efficiency (level of infection) occur. Cereals can be arranged in the following order of decreasing susceptibility: oats, spring wheat, spring barley, spring rye and maize. Oats and barley carry the highest infection levels and are, therefore, more efficient host plants than either wheat or rye. Host efficiency is greater in autumn-sown oats than in spring-sown oats (Gair, 1965). Autumn-sown cereals suffer less from *H. avenae* attack than do spring-sown cereals, because at the time of larval invasion of the roots, the root development of autumn-sown cereals is much more advanced than is that of spring-sown cereals which then have just formed seminal roots. In Australia wheat seems to be the most important host crop for *H. avenae*, which is probably explained by the fact that more wheat is grown than any other cereal, rather than by assuming the existence of a special "wheat strain" of the nematode (see McLeod, 1968). Varietal differences in susceptibility have been reported (Goffart, 1941; Jones and Moriarty, 1956; Dieter, 1958; Andersen, 1961; Gair *et al.*, 1962; Gair, 1965; Neubert, 1968).

H. avenae completes its life-cycle in maize although many of the larvae die after entry as a result of root necrosis. Mature females, however, do not break through the root cortex of maize as they do in other cereals and are therefore not fertilized (Johnson and Fushtey, 1966). For this reason maize is an inefficient host for *H. avenae*.

Weather conditions severely modify crop damage initiated by *H. avenae*. Most damage occurs in those years with a moist and cool period at the time of larval hatching, because the soil pores, especially in sandy soils, are filled with sufficient water for the larvae to hatch and to move towards the plant roots (Fidler and Bevan, 1963). If such a favourable period is followed by a spell of drought during the growing season of the crop, plants will suffer even more, because the damaged, deformed root system has a reduced ability to take up water. However, if the rainfall is adequate during the crop growing period and the water holding capacity of the soil is high, almost normal crop development may occur in spite of nematode attack. It is possible to predict so-called "eelworm years" by reference to rainfall records at sowing time or shortly thereafter. Environmental factors such as pH, fertility and humus content of the soil have been shown to be associated with both the severity of crop damage and the infestation level of the soil. Even in instances where an appreciable positive correlation between severity of infestation and pH was demonstrated (Duggan, 1963), reconstruction of the figures could easily lead to an even closer correlation between soil infestation and host crop density. In many other attempts to demonstrate the effect of cultural measures on population dynamics of *H. avenae*, effects were measured by crop yields, and usually the results did not provide a distinction between the role played by these factors and the role of the nematode on crop development. Thus, any cultural measure that favours the crop will be unfavourable to the nematode, and cause less crop damage. This, however, is only true in cases of low or moderate infestation levels. At high levels, the nematode causes severe damage and the farmer suffers a correspondingly severe crop loss.

In severe infestations, overhead sprinkling prevents the crop from being a total failure. Although a susceptible host crop may be safeguarded in this way, the nematode population will increase rapidly in the meantime. The stimulating effect of fertilizer treatment on cyst production in oats, barley, wheat and rye was clearly demonstrated by Hesling (1959) and supported the statement of Peters and Fenwick (1949): "The better the plant the more the cysts."

A good example of this statement was given by Williams (1969) who found that the population of *H. avenae* decreased where the "take-all" fungus (*Ophiobolus graminis*) was predominant, whereas the population maintained itself or increased in number in cases of slight "take-all infections". Cook (1969) demonstrated a distinct correlation between low levels of *H. avenae* and high levels of "take-all" on barley in field and pot experiments.

Slope (1967) and Williams (1969) established that in soil infested with "take-all" partial sterilization with a formaldehyde solution led to a decreased percentage of plants attacked by the fungus but, at the same time, *H. avenae* was increased.

B. Stem Nematode, *Ditylenchus dipsaci*

1. *World Distribution and Incidence*

D. dipsaci has been known for almost 150 years and it is now widespread throughout western and central Europe, U.S.A., Canada, Australia, Brazil, Argentina, and North and South Africa, although it is of greatest economic importance in temperate zones. In some countries it has become less important, because rye has been replaced by summer barley and summer wheat. In the last few years the area cropped to maize has increased and records of damage by *D. dipsaci* in maize have become correspondingly more frequent. Dewez (1940) noted that in a heavily infested area where the density of rye plants was constant, rye was not attacked by *D. dipsaci* in loam brook soil, heavy loam and fluvial clay soils, whereas it was commonly attacked on clayish sand, alluvial sand, light loam and tertiar sandy soils where water conditions were more favourable for nematode activity. In light sandy soils *D. dipsaci* is rarely a problem.

2. *Biology and Disease Symptoms*

D. dipsaci invades the foliage and the base of the stem of cereals, and causes a breakdown of the middle lamellae between cells by secreting pectinase resulting in large inter-cellular spaces in which the nematodes live. They migrate through the tissues and feed on adjacent cells. Reproduction continues inside the plant almost all year round but is minimal at low temperatures and this factor is decisive for the number of generations during the reproductive season.

When an infected plant dies the nematodes return to the soil and infect neighbouring plants. The nematodes are highly mobile in soil and can cover a distance of 10 cm within 2 h; this accounts, in part, for the rapid spread from one plant to another of this nematode.

Within *D. dipsaci* a number of biological races (strains) have been recognized, which are morphologically indistinguishable but differ in host range. The races from rye and oat will be discussed more closely in this chapter. The rye strain is most commonly found on the European mainland, whereas the oat strain seems most prevalent in Britain (Robertson, 1955). The rye strain attacks also oat, maize, mangold, bean, pea, tobacco, onion, flax, clovers (*Trifolium* spp.) and cocksfoot grass (*Dactylis glomerata*). A great number of farm weeds are good hosts for the rye strain and those commonly found in cereal crops are: loose silky-bent (*Apera spica-venti*), couch grass (*Agropyron repens*), common chickweed (*Stellaria media*), cleavers (*Galium aparine*), bindweed (*Polygonum* spp.), spurrey (*Spergula arvensis*), shepherd's purse (*Capsella bursa-pastoris*), pennycress (*Thlaspi arvense*), dead nettle (*Lamium purpureum*), plantain (*Plantago major*) and cornflower (*Centaurea cyanus*). The oat strain attacks oats, onion, pea, bean, vetches and occasionally potatoes, but infection of rye is uncommon. Weed hosts are teasel (*Dipsacus fullonum*), cleavers, common chickweed, thyme-leaved sandwort (*Arenaria serpyllifolia*), mouse-ear chickweed (*Cerastium vulgatum*) and wild oat (*Avena fatua*), of which cleavers presents a difficult problem as the plant is readily transported clinging to animals and men (Staniland, 1945).

Local cell hypertrophy and hyperplasia produce the typical symptoms of *D. dipsaci* attack such as basal swellings, dwarfing and twisting of the stalks and leaves (Fig. 7). Internodes are shortened and many axillary buds produce an abnormal number of halms (15–20 is not exceptional), giving the plant a bushy appearance. Later the nematodes move upwards into the stalks and leaves. These symptoms, which are identical for rye, oats and maize, can often be found in autumn, but are most common in spring (under European conditions).

Heavily infected plants may not survive the seedling stage and their death causes bare patches in the field (Fig. 8). Other attacked plants fail to produce panicles or spikes and the crop assumes a thin appearance. In both instances crop losses are evident. Root development is often much reduced because of the destruction of the cell tissue in the stem base, and so they fail to hold the plant firmly in position. Hence, in heavy winds the plants are uprooted and shoots are easily broken, especially in maize (Fig. 9). In *D. dipsaci* infected crops, wind often contributes to final crop losses.

Fig. 7. A rye seedling attacked by *Ditylenchus dipsaci* showing multi-tillering and crinkled foliage symptoms.

Fig. 8. An infected rye crop showing patches of poor growth caused by *Ditylenchus dipsaci*.

3. Cultural and Environmental Influences

To what extent *D. dipsaci* causes crop losses depends on a combination of factors such as host plant susceptibility, infestation level of the soil, soil type and weather conditions, of which rainfall is very important. Soil type and moisture content greatly influence the activity of nematodes while they migrate through the soil in search of plant hosts and

FIG. 9. The stem base of maize showing deformation as a result of *Ditylenchus dipsaci* infection.

during their penetration of the plant. Soil type influences soil population in the absence of host plants (Seinhorst, 1950). During winter under poor hosts, non-hosts, or during fallow in the growing season, greater nematode populations persist in heavy soils than in lighter soils.

The influence of soil moisture in lighter soils is most important during nematode invasion of the host plant. Saturated soils are unfavourable for both plants and *D. dipsaci*. During cold and moist winters autumn-sown cereals have little chance of escaping attack, whereas spring-sown cereals, generally sown under drier conditions, often suffer less from *D. dipsaci.*

Cereals usually are infected by *D. dipsaci* in the surface layer of the soil and the longer the soil moisture content in this layer is optimum for nematode activity, the greater the chance of a heavy attack. Nematode activity decreases rapidly in well-drained fields.

C. Cereal Root-knot Nematode, *Meloidogyne naasi*

1. *World Distribution and Incidence*

M. naasi is reported from Britain, Belgium, the Netherlands, France, Germany, Yugoslavia, Iran and the U.S.A. The number of records in the last few years on the occurrence of this new species from widely separated countries and the spread within Belgium (Gooris and D'Herde, 1967), England and Wales (Franklin, 1965), suggest that this nematode is probably more widespread than is realized.

M. naasi is one of the few *Meloidogyne* species in temperate climatic areas which occurs in the field. Although it was recorded as a pest from grasses and cereals as early as 1952, it was several years later that it was described by Franklin (1965) as a new species. It is a polyphagous nematode which attacks barley, wheat, rye, sugar-beet, mangold and onion and several monocotyledonous as well as dicotyledonous weeds such as: couch grass (*Agropyron repens*), annual poa (*Poa annua*), slender foxtail (*Alopecurus myosuroides*), reed (*Phragmites communis*), slough grass (*Beckmannia syrigachne*), blue grass (*Merion* sp.), pasture grasses (*Lolium perenne, L. multiflorum, Poa* spp., *Dactylis glomerata, Festuca pratensis*), chickweed (*Stellaria media*), dock (*Rumex* sp.), plantain (*Plantago major*), persicaria (*Polygonum persicaria*) and cat's ear (*Hypochaeris radicata*). In Europe oats is a very poor host as compared with other cereals, whereas in the U.S.A. *M. naasi* readily attacks oats.

2. *Biology and Disease Symptoms*

The young larvae invade the roots of early spring-sown cereals within 1–1½ months of germination (Franklin and Clarke, 1968). Late-sown spring cereals and early-sown autumn cereals are invaded within one month or even earlier. In this early stage of attack small galls on the roots can be observed. The larvae develop and the females become almost spherical in shape, the neck sharply offset from the body. Females with eggs in an egg-sac occur 8–10 weeks after sowing and are usually embedded in the gall tissue.

In studying the biology of *M. naasi* Franklin and Clarke (1968) and Schneider (1967) came to different conclusions concerning the number of complete generations the nematode produces in one year. In England on barley only one generation occurs, whereas in France at least three generations develop on rye grass, possibly because of the year-round growing period of this crop. In the Netherlands (Kuiper, personal communication) embryonated eggs produced on barley gave rise to a complete second generation in the same year as the original inoculation of the rye grass.

FIG. 10. The root system of a barley plant infected with *Meloidogyne naasi* showing the typical galls.

The symptoms of *M. naasi* attack in cereals closely resemble those caused by *H. avenae*. Patches of poorly growing, yellowing plants develop that vary in size from a few square metres to larger areas. Later in the season earing in these patches is delayed or absent. *M. naasi* causes typical galling of the roots, especially the root tips. The galls are typically curve-, horse-shoe- or spiral-shaped, in which respect they differ from galls caused by other root-knot nematodes (Fig. 10). There is no excessive production of secondary roots as there is in infections by *H. avenae*.

3. Cultural and Environmental Influences

This nematode was recognized only recently and little is known about the optimal conditions for its development. According to Watson and

Lownsbery (1970), exposure to low temperatures followed by exposure to warm temperatures stimulates hatching of eggs, and this may vary with the nematode's biotope. In temperate regions seasonal temperature fluctuations from winter to spring probably provide this requirement.

A high water capacity of the soil to greater depth at sowing time favours dispersal of *M. naasi* larvae. Under these conditions and in the presence of graminaceous weeds, the nematode will maintain numbers, even under non-host crops.

D. Other Nematodes

The saying "Opportunity makes the thief" to some extent also accounts for some parasitic nematode species. In the past various nematode species have been recognized as plant parasites of local importance, sometimes described under names which later had to be revised. Several of these nematodes remained minor parasites. Others, as a result of modifications in agricultural economy, have become more important than they were before. Mechanism of agriculture, reallocation of land and a decreased variety of crops per farm are factors which have increased the possibilities for many nematode species to be of economic importance. Other nematodes cause the type of damage to crops that is merely an additional plant growth-limiting factor.

The following nematode species in one way or another represent one of the above-mentioned groups.

1. "Ear cockle" of Wheat, *Anguina tritici*

(a) *World Distribution and Incidence.* The symptoms of the disease were already known in the Middle Ages and William Shakespeare in 1594 in his play "Love's Labour's Lost" wrote: "Sowed cockle, reap'd no corn". In 1743 Turbevill Needham, a Catholic clergyman, discovered the causal agent when examining a crushed gall in a drop of water under a microscope and he reported his discovery of what he called a species of aquatic animal to be denominated worm, eel or serpent to the President of the Royal Society in London.

This seed-borne nematode chiefly attacks wheat, also spelt (*Triticum speltum*) and sometimes rye. *A. tritici* is reported from wherever in the world wheat is grown. In most countries, however, it has become a curiosity since the use of fresh seed and improved seed cleaning methods have been developed. *A. tritici* is still common in many countries around the Mediterranean, in the Balkan States, the Soviet Union and further eastwards.

(b) *Biology and Disease Symptoms.* The nematode is spread in galled seeds when sowing infected seed. A single gall may contain over 10,000 dormant larvae. Once sown, the galls take up water and the larvae emerge and remain between the leaves of the growing plant. The primary leaves become twisted and distorted and the plant may die from a heavy attack. Attacked plants do not show the bushy foliage appearance of a *D. dipsaci* attack.

In growing seedlings the larvae are carried up towards the growing point of the plant and when the ear is formed, the flower head is invaded by the larvae and galls are formed instead of normal flowers. Inside the galls the nematodes mature and the female lays thousands of eggs, from which the larvae hatch and remain dormant in the seed.

In attacked ears the galls are easily recognized by their smaller size and darker colour as compared with normal seeds, but the galls may be confused with those seeds attacked by bunt (*Tilletia tritici*). Under dry conditions the larvae within galls may survive for decades.

(c) *Cultural and Environmental Influences.* In areas which are infested by *A. tritici* lack of farm hygiene and self supply of seed taken from the previous wheat crop mitigate against the elimination of the pest. The nematode is very much favoured by wet and cool weather conditions.

2. Grass Root-gall Nematode, *Ditylenchus radicicola*

(a) *World Distribution and Incidence.* This nematode has been a well-established species in western Europe for many years. Sowerby (1757-1822), the London draughtsman of plants and shells, in 1794 unintentionally figured a galled specimen of *Poa annua* in "English Botany", indicating that the nematode was already common at that time.

D. radicicola exclusively attacks grasses and cereals and it is generally distributed throughout the Scandinavian countries, Britain, the Netherlands, Germany, Poland and the U.S.S.R. It has also been reported from Canada and the U.S.A. The nematode occurs on many grasses of economic importance. Oostenbrink (1953), in examining 700 grass root samples from permanent pastures, found 75% of them infested with *D. radicicola*. Also grasses grown for seed have been found attacked.

(b) *Biology and Disease Symptoms.* Root tips invaded by *D. radicicola* larvae show local swellings which are characteristically spiral-shaped (Fig. 11) and so may be confused with the galls caused by *M. naasi*; mixed populations of these two nematode species have been recorded. A secondary infection occurs when young larvae move from the galls and attack the adjacent healthy roots. The life-cycle of the nematode is completed within 1–2 months, depending on soil temperature.

In the Scandinavian countries *D. radicicola*, under field conditions, causes poor growth in barley and the disease is locally known as "krok". In Sweden rye also is attacked. In Canada galls are found on the roots of wheat, whereas oats and rye are not affected. Outside Scandinavia *D. radicicola* has been shown to attack oats, barley, wheat and rye under laboratory conditions but this has not been confirmed in the field. In an attempt to clarify the host–parasite relationship of *D. radicicola*, s'Jacob (1962) inoculated a number of graminaceous plants

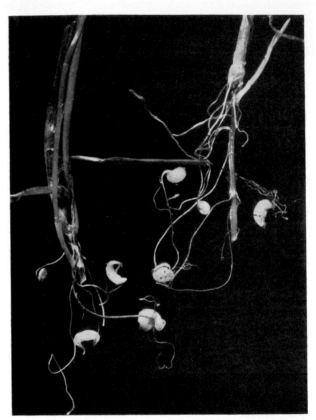

FIG. 11. Roots of *Poa annua* galled by *Ditylenchus radicicola*.

with a Scandinavian and a Dutch population. He found that the Dutch population heavily attacked *Poa annua* but caused only a few galls on the roots of barley, whereas the Scandinavian population behaved conversely, thus indicating the existence of biological races.

(c) *Cultural and Environmental Influences*. Although pot experiments by s'Jacob (1962) showed a close correlation between infestation level of the soil and growth of *Poa annua*, findings of damage in *Poa* grasses in pastures by Lewis and Webly (1966) suggest that *D. radicicola* could be more important in association with soil structure defects such as a pan below the soil surface.

3. Heterodera latipons

This newly described cyst-forming nematode for which no common name is available was earlier reported from the Mediterranean region on the roots of wheat. So far it is known to occur in Libya, Israel and Bulgaria where it is found on barley, wheat, oats and rye.

Little is known of the economic importance of *H. latipons*. However, when compared with other western European cyst-forming nematodes *H. latipons* is very similar to *H. avenae* in cyst shape and colour and in host range (Franklin, 1969). It is possible, therefore, that previous findings of this new nematode species have erroneously been attributed to the economically important *H. avenae*. The description of the symptoms caused by *H. latipons* in the type host—namely stunted, yellowing wheat plants—show them to be very similar to those caused by *H. avenae*.

4. Migratory Nematodes, *Pratylenchus crenatus, P. neglectus* and *P. thornei*

(a) *World Distribution and Incidence.* All three species are commonly found in many countries throughout the temperate zones: *P. crenatus* preferably in light soils, *P. neglectus* in loamy soils and *P. thornei* in heavier soil types. Damage in cereals has been ascribed to all three, particularly to *P. crenatus* in oats and barley, sometimes also in maize and rye, but rarely in wheat. Oats and maize and especially wheat are good hosts to *P. thornei*. *P. neglectus* attacks wheat, barley, rye and maize, in this order of importance, but it occurs sporadically with damaging effects.

(b) *Biology and Disease Symptoms.* *Pratylenchus* species are migratory root-infesting nematodes which are not confined to fixed places for their development and reproduction. Eggs are laid in the soil or inside invaded plant roots. The nematode penetrates the root by means of the stylet, feeds upon the root tissue, lays its eggs between the root cells and may return to the soil. The life-cycle is completed within 20–40 days, depending on the species and on environmental influences.

As a result of nematode attack an affected crop shows patches of poorly growing plants with a poorly developed or stubby root system. A specific symptom on the roots are the so-called "lesions", consisting of oblong, necrotic or rotting areas, caused by the nematode. At these lesions secondary attack by fungi frequently occurs and thereafter the nematodes leave these lesions and return to the soil to penetrate healthy roots elsewhere.

(c) *Cultural and Environmental Influences.* The recognition of high populations of *Pratylenchus* species in poorly growing cereal crops may indicate that these nematodes are responsible for the damage, but only critical examination will prove whether or not they are the causative agent. Many root diseases are caused by a complex of organisms, one of which may be a nematode. In some instances, the nematodes are associated with fungi in a disease complex but may remain unnoticed. In Canada *Rhizoctonia solani* and *P. neglectus* were found to be closely and consistently associated with a root rot of winter wheat. Although the combined effect of the two pathogens upon the growth of wheat was almost twice as great as the effect produced by either pathogen alone, no direct proof could be obtained for an interrelation of the two organisms in the infection court (Mountain and Benedict, 1956).

In other instances, the nematode is itself the pathogen but is harmful only when an environmental factor encourages nematode development. Here we could modify Peters and Fenwick's statement by saying: "The worse the plant the more damaging the nematode".

In his report on the spread of *Pratylenchus* species and the damage they caused in cereals, Kemper (1966, 1967) found that barley could be damaged by 8 *Pratylenchus* individuals/g of soil at favourable pH values. In most instances, however, heavy damage was observed at pH (KCl) 4–4·5. At these levels a lime-requiring crop like barley is predisposed to any other growth-limiting factor such as nematodes. He further stated that the tolerance to nematode attack is very variable and depends on many factors including that of soil acidity.

III. Control

A. Crop Rotation

The principle of rotation for nematode control is to grow host and non-host crops in such a balance as to prevent low nematode populations from building up to damaging levels and also to decrease high nematode populations to a level that is satisfactory for good crop growth. In countries where a soil sampling service is operating, field sampling supplies the necessary information on which species of nematodes are present and, therefore, contributes significantly in setting up an efficient rotation scheme.

Of all the control measures commonly used, crop rotation that includes non-host crops is the most frequently recommended method of controlling nematodes. Contrary to that which occurs with many other cyst-forming nematodes, a field population of *H. avenae* is greatly decreased by non-host crops or by fallow. Duggan (1958) found the following average percentage population decrease after lucerne (53), roots (46), rye grass–cocksfoot mixture (48) and fallow (50). Hesling

(1958) and Kort (1959) also found a decrease varying between 45 and 65% per year. In Australia (Meagher and Rooney, 1966) a single year's fallow decreased the nematode population by 56–75%. On the basis of these figures a non-host crop grown for two successive years, even on moderately infested fields, will enable farmers to return to susceptible crops thereafter without much risk. The existence of pathotypes decreases the opportunities for rotation and also requires that pathotypes are tested for their host range. Frequently only one pathotype is involved, or at least it is predominant, in the particular area. However, Anderson (1961) and Kort *et al.* (1964) have shown that a mixture of pathotypes in the same field is relatively common. Grasses, as a rule, are poor hosts for *H. avenae* although this varies with the pathotype (Kort, 1964; Neubert, 1967).

It is very difficult to forecast the degree of crop losses caused by *D. dipsaci*. Severe symptoms may occur in fields where no disease symptoms were observed in the previous years. On the other hand a susceptible crop, sown in infested land, does not always develop symptoms. Such observations made in different years on the same field, caused Seinhorst (1950) to conclude that weather conditions must be the factor determining the presence or absence of symptoms. However, it is not known in what way the population dynamics of *D. dipsaci* is controlled by weather.

The occurrence of different races of *D. dipsaci* complicates the problem of control. Eriksson (1965) and Sturhan (1966) successfully crossed different races, of which the progeny frequently had a different host range from each of the parents. Such results make it difficult to make suitable recommendations for crops in a rotation. However, something can be done to minimize the risk of heavy crop losses. Rye, oats and maize should not be grown following any crop (except clovers and lucerne) that was attacked by *D. dipsaci* in the previous year. However, the differences in host range between races may fluctuate from field to field owing to the occurrence of mixtures and hybridization of races. Seinhorst (1963) concluded that more than 10 nematodes/500 g of soil in loamy-sand soils would lead to crop damage in rye and that this level of infestation could be maintained by frequent rye and oat cropping. Effective non-host crops for both the rye and oat race of *D. dipsaci* are barley, wheat, lucerne, carrot and clover, and the potato is very little affected. In clover and lucerne foliage deformations may occur, mainly in the seedling stage, but the nematode does not (or hardly ever) reproduce in these crops. Flax also is a poor host, but deformations in this crop may diminish the quality of the fibres and therefore should not be grown.

Information on crop damage caused by *M. naasi* is far from complete.

E

There are indications that those crops in which noticeable crop losses due to *M. naasi* attack have been observed are not those hosts in which reproduction occurs most readily, whereas in those which are good hosts, like pasture grasses, *M. naasi* does not influence the breakdown of the ley (Lewis and Webley, 1966). The planting of beet or onion following pasture or cereals, except oats, should be avoided. The combination of cereals and undersown grasses gives rise to a maximum increase of the soil infestation level. Generally, it is risky to follow the cultivation of grasses with any susceptible host crop. Beet, onion and barley are the most susceptible crops and should be omitted from close rotation even in moderately infested land. When cereals are grown on infested land the following crop should be a non-host. Heavily infested land should be cropped exclusively with non-host crops. There are reports of *M. naasi* attacking potatoes, flax, maize, beans, clovers and other leguminous crops. Oats is a poor host for *M. naasi*, but does not decrease the soil infestation level significantly.

Fields infested with *A. tritici* should be planted to any crop except wheat, spelt or rye for one or two years, as the nematodes cannot survive in the soil for that period without a host plant.

The biology of *Pratylenchus* species is still poorly known and so negligence in crop management should be avoided. In fields heavily infested with one of the three migratory nematode species discussed, certain host cereals, grasses and leguminous crops should be excluded from the rotation, and replaced by mangold, sugar-beet, flax, peas or beans. Cruciferous crops and potatoes will help to decrease the infestation level of *P. crenatus* in the soil (Oostenbrink *et al.*, 1956).

B. Cultural Practices

Within the framework of cultural practices there are only a few measures that can be recommended for controlling nematodes in cereal crops. For crop rotation to be most effective, soil sample investigations are valuable in order to determine in which fields it is safe, or not too risky, to grow susceptible cereals. Therefore, crop-damaging nematode levels have to be known for different nematode species under varying conditions of host crop and soil type. Unfortunately, such data are far from complete but those which are available, partly mentioned under "Crop rotation" (III.A) and "Economics" (IV), may serve as a basis for practical application, although circumstances may vary from country to country.

In 1955–1964 in the Netherlands pre-cropping soil samples from three areas on sandy, peat and silt soils were taken, and *H. avenae* cyst and larval counts made. During the following season previously sampled fields cropped to cereals were grouped into four categories of growth: regular, irregular, patchy and poor, and cropped to cereals. It was found that in order to protect oats from noticeable loss in yield 2 eggs/g of soil should not be exceeded on both sandy and peat soils and that 8 eggs/g of soil should not be exceeded on heavier soil types. To minimize loss in yield for spring-sown wheat these values are approximately the same as those for oats, but the value for spring-sown barley should not exceed 3 eggs/g on light soils and 10–12 eggs/g on heavier soils. It should be emphasized that our figures are not comparable to those given by Gair (1965) because of the different conditions of growth and sampling. Laboratories for soil and crop testing which are operating in different countries are valuable to farmers because they analyse field soils for pests such as *H. avenae*, *D. dipsaci* and *Pratylenchus* spp. and give the degree of infestation and advice on subsequent control and cropping.

It can be concluded that in view of the environmental influences on the development of nematodes, crop losses caused by them may be attributed to negligence in crop management. This applies especially to *D. dipsaci* in that the drainage of fields should be good in order to dry the surface layer of the soil so that nematode activity is restricted. *D. dipsaci*-infested straw preferably should be burned.

Weed control is an important fact to consider, especially in those instances where weeds are excellent hosts for nematodes. Wild oats (*Avena fatua*) enables *H. avenae* to maintain a fairly high infestation level of the soil. Control of this weed is recommended in root crops in which it cancels the clean-up effect of the non-host. A number of excellent weed hosts to *D. dipsaci*, mentioned before, have a similar effect in lighter soils in that they quickly build up the nematode population in the absence of host crop plants. The weed grasses Couch Grass and Slender Foxtail should be controlled in *M. naasi*-infested fields.

Although it is difficult to prevent spread of nematodes from one field to another special attention should be given to the spread of *D. dipsaci* which at low infestation levels may easily lead to crop losses. Where onions are applied in the rotation the use of clean onion seed is essential. *D. dipsaci* is found mainly in the seed debris, rather than in the seed itself. Effective seed cleaning helps in preventing spread of not only *D. dipsaci* but also of *A. tritici*. In both instances infested lots of seed are easily cleaned by using a winnowing mill where nematode-infested debris, and the light-weight galls respectively are blown away. *D. dipsaci* can also be controlled by fumigation of the seed with methyl bromide.

C. Resistant Varieties

The best prospect for an effective control of *H. avenae* is offered by growing resistant varieties in which females are unable to reach maturity so that there is a decrease in the field population (Andersen, 1967; Cotten, 1970a). In the European countries in which known pathotypes of *H. avenae* occur, testing for resistance of barley and oat genotypes has started and promising lines have been selected for further work. From results obtained by using Andersen's standard differential genotypes it can be assumed that the pathotypes known as "race 1" in Denmark and Britain and pathotype "A" in the Netherlands are synonymous, and similarly the British and Danish race "2" is synonymous with the Dutch pathotype "C". Resistance to the first group was found in the Danish barley varieties Drost, Pendo, Fero, Brage and Amsel, whereas varieties of LP 14 and LP 191 are resistant to the second group. Resistance to both groups was found in the barley varieties Morocco CI 3902 and Marocaine 079 CI 8334, in varieties of the wild oat, *Avena sterilis*, and in the wheat variety Loros.

Workers in eleven countries are cooperating, using standardized test methods, to find cereal lines with resistance to all known variations in pathogenicity in European and Australian populations of *H. avenae*. Selection of genotypes is still in progress, but meanwhile resistance is being introduced into commercial barley and oat varieties. Attention is being given to the inheritance of the resistance and to genetic analysis of the progeny of back-crosses with different genotypes (Andersen and Andersen, 1968 and 1969; Cotten and Hayes, 1969).

The use of resistant varieties is probably the only economic method to control *D. dipsaci* in cereals. In the Netherlands the breeding of rye resistant to *D. dipsaci* started in 1943, by utilizing a resistant local variety "Ottersumse". This resulted in the resistant winter rye variety "Heertvelder", which on infested land yields about 50% more than existing susceptible rye varieties. However, in fields with no or only a few nematodes this variety is outyielded by all susceptible rye varieties. Spring oat varieties resistant to *D. dipsaci*, like "Manod" (S.235) and "Early Miller" and resistant winter oats, like "Peniarth" (S.238) and "Maris Quest", have been developed at the Welsh and Scottish Plant Breeding Stations. Cotten (1969) reports successful control of *D. dipsaci* by using these varieties. The variety "Early Miller" is classed as tolerant to *D. dipsaci* rather than strictly resistant. A successful Belgian resistant oat variety is "Greta".

D. Nematicides

The use of chemicals to control nematodes in cereals, which usually are grown over large areas, holds little promise of ever being an economic possibility (Williams, 1969; Williams and Salt, 1970). Systemic nematicides in current use have had less affect on cyst-forming than on free-living nematodes. In the Netherlands nematicides commonly are used in controlling the potato cyst nematode and it has been observed that *H. avenae* is noticeably decreased by such fumigation but probably not sufficiently so.

IV. Economics

Although it is impossible to be precise about the annual crop losses caused by any pathogen, estimates have been made for different nematode species. Crop losses very much depend on the nematode species, the infestation level in the soil, the crop variety and the environmental influences. Generally, crop losses caused by nematodes are most severe in certain seasons after a spell of moist and cool weather in the post-sowing period.

An estimate of crop losses can be obtained by comparing crop yields with the level of plant-parasitic nematode populations in the field. Improved yields as a result of soil fumigation are not a reliable guide to the effect of nematodes on crop losses because the fumigants not only kill nematodes but, under certain conditions, also affect other pathogens of the same host and influence plant growth directly by changing the availability of soil nutrients. Annual crop losses in cereals caused by *H. avenae* have been estimated at DKr 50 million in Denmark, £2 million in Britain, SKr 50 million in Sweden and several million dollars in Australia. The influence of the level of the *H. avenae* infestation in the soil on crop yield for varieties of oats and barley are shown (Table II) (Andersen, 1962). Gair (1965) showed that 2 eggs/g of soil caused neither obvious stunting of plants nor patchiness in oats, but at 16 eggs/g of soil the crop is reduced by as much as 35% in some instances. At the latter initial level of infestation the yield of wheat is reduced by 20%, whereas barley suffers 25–35% loss at 32 eggs/g of soil. Similarly, Dixon (1968) found that after a certain infestation level every increase of 10 eggs/g of soil results in a loss of 380 kg/ha in spring oats, 190 kg/ha in spring wheat and 76 kg/ha in spring barley, which confirms the earlier mentioned order of crop susceptibility for *H. avenae*. Recent investigations in Germany (Küthe and Dern, 1970) showed that in susceptible maize varieties where 50–70% of the plants were infested by *D. dipsaci* there was a decrease in yield ranging from 18 to 31%.

After an extensive survey in China, Chu (1945) estimated that the annual decrease in yield of wheat caused by *A. tritici* was not greater than 0·25% which was equal to 5448 metric tons of wheat. More interesting are his figures for some individual fields in which the sowing seed was contaminated with galls at the rate of 2·5, 6·5 and 8·5% of total weight and which resulted in yield decreases of 30, 54 and 69%

respectively. Situations like these still occur in some countries where crop losses of 30–50% are not exceptional.

In spite of the long list of host plants for *D. radicicola*, very little is known about the economic damage it causes. It is usually very difficult to prove such a loss in pastures as weakened plants are easily overgrown by neighbouring grasses.

TABLE II. Influence of the infestation level of *Heterodera avenae* in the soil on the yield (expressed as a percentage of the control) of oats (var. Sun II) and barley (var. Herta) in pot experiments and of barley (var. Bonus) in field experiments in 1959–1961. Courtesy of Sigurd Andersen (1962)

Infestation level of the soil in eggs/g		Relative yield		
Sun II Herta	Bonus	Sun II*	Herta*	Bonus†
0	0–0·9	100	100	100
1	1·0–2·4	73	91	97
2·5	2·5–4·9	66	88	93
5	> 5·0	54	84	91

* Average from 10 replicates.
† Statistical analysis of 550 field experiments.

Experiments with *Pratylenchus* species have failed to prove convincingly the primary role of these nematodes in cereal damage. In instances where nematode numbers in the field are associated with poor growth in cereals, but where inoculation experiments do not confirm a relationship, other limiting factors must be involved.

Obviously, records of exact figures on crop losses due to nematode attack in cereals are scarce, but as indicated by Cotten (1970b), "Present day nematode problems in the temperate cereal growing areas of the world are less obvious to the casual observer and tend to be overlooked because of their insidious nature."

References

Andersen, S. (1959). *Nematologica* 4, 91–98.
Andersen, S. (1961). *Meddr K. Vet. -og Landbohøjsk. afd. Landbr. PlKult.* 68, 179 pp.
Andersen, S. (1962). *Ugeskr. Landm.* 33, 7 pp.
Andersen, S. (1967). *Tidsskr. Landøkon.* 94–112.

Andersen, S. and Andersen, K. (1968). *Nematologica* **14**, 128–130.

Andersen, S. and Andersen, K. (1969). *E.P.P.O. Public. Ser. A.*, No. 54, 29–36.

Chu, V. M. (1945). *Phytopathology* **35**, 288–295.

Cook, R. (1969). *Proc. 5th Br. Insectic. Fungic. Conf.* Vol. 3, 626–634.

Cotten, J. (1969). *Proc. 5th Br. Insectic., Fungic. Conf.* Vol. 1, 164–168.

Cotten, J. (1970a). *Ann. appl. Biol.* **65**, 163–168.

Cotten, J. (1970b). *Span* **13**, 150–152.

Cotten, J. and Hayes, J. D. (1969). *Heredity, Lond.* **24**, 593–600.

Dewez, W. J. (1940). *Tijdschr. PlZiekt.* **46**, 194–204.

Dieter, A. (1958). *NachrBl. dt. PflSchutzdienst-Berlin* **12**, 155–158.

Dixon, G. M. (1968). *C.r. 8th Int. Symp. Nem. Antibes* 1965: 55.

Duggan, J. J. (1958). *Econ. Proc. R. Dubl. Soc.* **4**, 103–118.

Duggan, J. J. (1961). *Sci. Proc. R. Dubl. Soc. Ser. B* **1**, 21–24.

Duggan, J. J. (1963). *Ir. J. agric. Res.* **2**, 105–109.

Eriksson, K. B. (1965). *Nematologica* **11**, 244–248.

Fidler, J. H. and Bevan, W. J. (1963). *Nematologica* **9**, 412–420.

Franklin, M. T. (1965). *Nematologica* **11**, 79–86.

Franklin, M. T. (1969). *Nematologica* **15**, 535–542.

Franklin, M. T. and Clarke, S. (1968). *Rep. Rothamsted exp. Stn for 1967*, 145–146.

Gair, R. (1965). *In* "Plant Nematology" (J. F. Southey, ed.). *Tech. Bull. Minist. Agric. Fish. Fd* No. 7, 199–211.

Gair, R., Price, T. J. A. and Fiddian, W. E. H. (1962). *Nematologica* **7**, 267–272.

Goffart, H. (1941). *Arb. biol. ReichsAnst. Land-u. Forstw. Berlin* **2**, 141–161.

Gooris, J. and D'Herde, J. (1967). *Landbouwk. Tijdschr., Brussel* **20**, 1055–1065.

Hesling, J. J. (1958). *Nematologica* **3**, 274–282.

Hesling, J. J. (1959). *Ann. appl. Biol.* **47**, 402–409.

s'Jacob, J. J. (1962). *Nematologica* **7**, 231–234.

Johnson, P. W. and Fushtey, S. C. (1966). *Nematologica* **12**, 630–636.

Jones, F. G. W. and Moriarty, F. (1956). *Nematologica* **1**, 326–330.

Kemper, A. (1966). *Mitt. Biol. BundAnst. Berlin* **118**, 107–116.

Kemper, A. (1967). *Mitt. Biol. BundAnst. Berlin* **121**, 88–92.

Kort, J. (1957). *Versl. Meded. plziektenk. Dienst Wageningen* **130**, 143–146.

Kort, J. (1959). *Tijdschr. PlZiekt.* **65**, 1–4.

Kort, J. (1964). *Meded. LandbHogesch. OpzoekStns Gent* **29**, 783–787.

Kort, J., Dantuma, G. and Essen, A.v. (1964). *Neth. J. Pl. Path.* **70**, 9–17.

Kühn, J. (1874). *Landw. Jbr.* **3**, 47–50.

Küthe, K. and Dern, R. (1970). *Gesunde Pfl.* **22**, 101–104.

Lewis, S. and Webley, D. (1966). *Pl. Path.* **15**, 184–186.

Lücke, E. (1969). *Z.PflKrankh. PflPath. PflSchutz* **76**, 269–276.

McLeod, R. W. (1968). *Agric. Gaz. N.S.W.* **79**, 293–295.

Meagher, J. W. (1968). *J. Agric. Vict. Dep. Agric.* **66**, 230–233.

Meagher, J. W. and Rooney, D. R. (1966). *Aust. J. exp. Agric. Anim. Husb.* **6**, 425–431.

Mountain, W. B. and Benedict, W. G. (1956). *Phytopathology* **46**, 241–242.

Neubert, E. (1967). *NachrBl. dt. PflSchutzdienst, Berl. NF* **21**, 66–68.

Neubert, E. (1968). *NachrBl. dt. PflSchutzdienst, Berl.* **22**, 55.

Neubert, E. and Decker, H. (1970). *NachrBl. dt. PflSchutzdienst, Berl.* **24**, 195–198.

Oostenbrink, M. (1953). *Tijdschr. PlZiekt.* **59**, 149–152.

Oostenbrink, M., s'Jacob, J. J. and Kuiper, K. (1956). *Nematologica* **1**, 202–215.

Peters, B. G. and Fenwick, D. W. (1949). *Ann. appl. Biol.* **36**, 364.

Robertson, D. (1955). *Scott. Agric.* **24**, 209–212.

Schneider, J. (1967). *Phytoma* **185**, 21–25.

Seinhorst, J. W. (1950). *Tijdschr. PlZiekt.* **56**, 289–348.

Seinhorst, J. W. (1963). *Meded. Dir. Tuinb.* **26**, 349–358, 391.

Singh, K. and Swarup, G. (1964). *Indian Phytopath.* **17**, 212–215.

Slope, D. B. (1967). *Ann. appl. Biol.* **59**, 317–319.

Staniland, L. N. (1945). *Ann. appl. Biol.* **32**, 171–173.

Sturhan, D. (1966). *Z.PflKrankh. PflPath. PflSchutz* **73**, 168–174.

Watson, T. R. and Lownsbery, B. F. (1970). *Phytopathology* **60**, 457–460.

Williams, T. D. (1969). *Ann. appl. Biol.* **64**, 325–334.

Williams, T. D. and Salt, G. A. (1970). *Ann. appl. Biol.* **66**, 329–338.

6
Nematode Diseases of Rice

M. Ichinohe

National Institute of Agricultural Sciences
Nishigahara, Kita-Ku
Tokyo, Japan

I. The Rice Crop

The rice plant, *Oryza sativa* L., is an annual grass that originated in the Orient and has been cultivated for thousands of years. Rice is planted either by broadcasting seed, or by transplanting seedlings that have been raised in a seedbed. The plant requires a large amount of sunlight, especially during the period between heading and harvesting. In general, adequate irrigation during growth is essential in order to obtain a high yield as well as high quality. There are two groups of rice varieties which are adapted to either upland (non-irrigated) or lowland (paddy) cultivation.

Since an increase in rice production is of economic importance in the Asian countries, considerable research has been placed on breeding, plant physiology, fertilization, and on the diseases and insect pests in relation to rice production. However, relatively little work has been done on the nematode pests.

There has undoubtedly been much confusion in terminology of the various parts of rice plants, and so the following definitions will help

E*

to interpret the data better. The culm (or stem) is made up of a series of nodes, alternating with internodes, and each node bears a leaf and a bud. The bud may grow into a tiller (or shoot). Tillers grow out of the main culm in alternate order, and the primary tillers grow from the lowermost nodes. A panicle (or ear) is a group of spikelets borne on the uppermost node of the culm. An unhulled grain (or seed) constitutes the kernel and the paleae (hull, or husk).

II. Nematode Parasites of Rice Plants

The nematode species known to be associated with rice are listed below (Table I) based mainly on Timm (1965) and Sher (1968). Investigations into the biology of nematode diseases of rice have concentrated upon the four genera *Aphelenchoides, Ditylenchus, Heterodera* and *Hirschmanniella* (Table II).

TABLE I. Nematodes associated with rice

Aphelenchoides besseyi	*Hirschmanniella spinicaudata*
*Criconemoides curvatum	*Hirschmanniella thornei*
*Criconemoides komabaensis	*Hoplolaimus galeatus*
*Criconemoides onoensis	*Hoplolaimus indicus*
*Criconemoides rusticum	*Meloidogyne graminicola*
Ditylenchus angustus	*Meloidogyne incognita*
Helicotylenchus crenacauda	*Meloidogyne javanica*
Helicotylenchus multicinctus	*Meloidogyne thamesi*
Helicotylenchus retusus	*Paralongidorus beryllus*
Hemicriconemoides cocophillus	*Paralongidorus citri*
Heterodera oryzae	*Pratylenchus brachyurus*
Hirschmanniella belli	*Tylenchorhynchus clavicaudatus*
Hirschmanniella caudacrena	*Tylenchorhynchus martini*
Hirschmanniella imamuri	*Tylenchorhynchus palustris*
Hirschmanniella mucronata	*Xiphinema indicum*
Hirschmanniella oryzae	*Xiphinema orbum*

* *Criconemoides* = *Macroposthonia*.

The rice root-knot nematode, *Meloidogyne graminicola*, is common in the seedbeds in Thailand, Laos, and India, and many rice varieties have been known to be hosts to this nematode in the U.S.A. (Golden and Birchfield, 1968). The rice stunt nematode, *Tylenchorhynchus martini*, is widespread in rice paddies in the U.S.A., Japan, Philippines, Thailand, India, East Pakistan, and some countries in Africa. Four closely related species of *Criconemoides* (= *Macroposthonia*) have been recorded around rice roots, and some are distributed in the U.S.A., Japan, Thailand, India, and East Pakistan.

TABLE II. World distribution of the four major nematode pests of rice

Nematode species	Type of injury	Country in which the species occurs	References
Aphelenchoides besseyi	Panicles and seeds infected	Brasil, Cameroon, Central African Republic, Ceylon, Chad, Comoro Islands, Congo, Dahomey, East Pakistan, El Salvador, Gabon, Ghana, India, Indonesia, Italy, Ivory Coast, Japan, Madagascar, Nigeria, Philippines, Senegal, Sierra Leone, Taiwan, Thailand, Togo, U.S.A. (Hawaii and mainland), U.S.S.R., Uzbek S.S.R.	Peachey *et al.* (1966) Barat *et al.* (1969)
Ditylenchus angustus	Culms of lowland rice infected	Burma, East Pakistan, India, Madagascar, Malaya, Philippines, Thailand, United Arab Republic	Vuong Huu-Hai (1969)
Hirschmanniella spp.	Roots of lowland rice infected	Formosa, India, Indonesia, Japan, Malaysia, Nigeria, El Salvador, Sierra Leone, U.S.A., Venezuela	Sher (1968)
Heterodera oryzae	Roots in seedbed and/or upland rice infected	India, Ivory Coast, Japan	Luc and Berdon Brizuela (1961) Rao (1970)

A. The White Tip Nematode, *Aphelenchoides besseyi*

The white tip disease of rice caused by *Aphelenchoides besseyi* was first discovered in Japan around 1940, and more work has been done on this than on any other nematode disease of rice.

1. Plant Symptoms

The upper 2–5 cm tip of the leaf turns pale yellow to white in the tillering stage, then brown, necrotic, and frayed (Fig. 1). Flag leaves

FIG. 1. Rice leaves with "white tip", the typical symptom of infection with *Aphelenchoides besseyi*. (By courtesy of N. Fukazawa.)

are characteristically shortened and twisted at their apical portions and symptoms are most conspicuous at the booting stage (Fig. 2). The disease occurs both in upland and lowland fields but more severely in the latter. The white tips of leaves of infected plants are not always manifest in certain varieties and/or under particular growing conditions.

In Japan the disease was once, and in some localities still is, called "fire-fly blast", because at the booting stage the trembling of numerous bleached leaf tips in the field looks like fire-flies, and "blast" is a pronoun of rice disease in Japan.

2. Nematode Bionomics

A. besseyi causes "ear-blight" of Italian millet, *Setaria italica* (Yoshii and Yamamoto, 1950a). Foxtail, *Setaria viridis*, is also a host, but not so good as rice; crab grass, *Panicum sanguinale*, and *Cyperus iria* are slightly infected, but *Panicum crus-galli* is not (Yoshii and Yamamoto, 1950b).

FIG. 2. Rice flag leaves, shortened and twisted as a result of *Aphelenchoides besseyi* infection. The right two are uninfected. Reproduced with permission from Yoshii and Yamamoto (1950a).

The nematodes are located first inside the leaf sheath of rice seedlings, although the number is always small. They increase in number rapidly on the young panicle at the booting stage, and then proceed to the exterior of the glumes, and later inside the paleae (Goto and Fukatsu, 1952). At harvest they coil up and become quiescent in the seeds. The number of viable nematodes infesting seeds varies, ranging from 0 to 64 per seed; the numbers in 100 seeds from severely and slightly affected paddies were 1241 and 132, respectively, of which nearly 90% inhabited the inner surface of the husks and the rest were on the kernel (Fukano, 1962).

The largest number of nematodes swim away from infected seeds at 25–30°C within 72–88 h of commencement of water-soaking (Tamura and Kegasawa, 1957). The nematode development is initiated at 13°C, the optimum being at 21–23°C; the life-cycle takes 10 days at 21°C and 8 days at 23°C (Sudakova, 1968).

The nematode is attracted to the young, growing parts of rice plants and to the aqueous extract of the germinating seed, but not to the ungerminated seed, husks, or other old plant parts. The attractiveness of seedlings of different varieties is strongly correlated with susceptibility in the field (Goto and Fukatsu, 1956).

A. besseyi can be cultured on each of three *Alternaria* spp. but not on *Hypochunus sasakii* or *Fusarium bulbigenum* (Iyatomi and Nishizawa, 1954). The fact that this nematode can be reared on fungi implies that this species is capable of surviving and possibly reproducing in the field in the absence of a host plant. Records show that nematodes retain their viability in rice grains stored dry for at least 3 years (1945–1948), although survival decreased from 62·9% for the newly harvested to 46·9% for the stored rice grain (Yoshii and Yamamoto, 1950b).

The incidence of the stem-rot disease caused by *Helminthosporium sigmoideum* on potted rice was decreased when the plants had been previously infected with *A. besseyi*, suggesting that the nematodes affected the physiology of the rice plant (Nishizawa, 1953b). Nonaka (1959) considered that this interaction was related to a higher respiration rate of the plant resulting from the greater activity of respiratory enzymes induced by nematode invasion.

B. The Rice Stem Nematode, *Ditylenchus angustus*

The disease caused by the rice stem nematode, *Ditylenchus angustus*, has long been known as "ufra", a name which originated from local use at the head of the Bay of Bengal (East Pakistan) where this disease was discovered in 1912.

1. Plant Symptoms

The earliest sign in the field is chlorosis or streaks of the upper leaves of rice plants that are more than 2 months old. There are two distinct types of symptom (Padwick, 1950): one is the "swollen ufra" where the panicle remains enclosed within the leaf sheath and there is a strong tendency towards branching of the stem in the infected portions; and the other type, called the "ripe ufra", where the panicle emerges and produces normal grain only near the tip. The peduncle turns dark brown, and flowers on the lower parts remain unfertilized.

2. Nematode Bionomics

D. angustus is an obligate ectoparasite of eight species of *Oryza* (Hashioka, 1963), and is found abundantly in the field of harvest, mainly at the base of the peduncle, the stem just above the upper nodes, or inside the glumes of the panicle where the typical brown discoloration occurs (Padwick, 1950).

The nematode coils up and becomes inactive when the plants ripen during the dry season, but when moistened, the nematode uncoils and rapidly regains motility. Alternate moribund and motile behaviour can be repeated a number of times by alternately drying and moistening, and the nematode is known to survive at least 6 months of dry conditions (Padwick, 1950). According to Taylor (1969), cottony masses are formed by these quiescent nematodes, and remain on the stubble when the crop is harvested.

C. The Rice Root Nematodes, *Hirschmanniella* spp.

The rice root nematodes, *Hirschmanniella* spp., are one of the most common nematode species inhabiting rice paddies throughout the world. Seven species attack rice roots (Sher, 1968), although they were earlier considered to be the single species, *Hirschmanniella oryzae*.

1. Plant Symptoms

The roots invaded by the nematodes are characterized by yellowish to rusty brown discoloration of the cortical tissue with the darkest colour around the stele and the base of the root hairs where the nematodes are found most frequently (van der Vecht and Bergman, 1952). This discoloration becomes darker on 2–3 month old plants.

There has been discussion on the relationship between *Hirschmanniella* infestation and the occurrence of "omo mentek" in Indonesia, one of the most complicated rice diseases causing reddish discoloration of leaves and stagnation of plant growth. The possible relationship has not yet been proven despite the investigations in Indonesia (van der Vecht and Bergman, 1952). According to Ou (1965), however, "omo mentek" is caused not by a nematode but by a virus, possibly similar to that causing "penyakit merah" in Malaysia and tungro virus disease in the Philippines: he uses as evidence (a) similarity in plant symptoms, (b) the wide occurrence of the virus in the region where "omo mentek" has been seen, and (c) similarity in varietal reaction to the diseases.

2. Nematode Bionomics

Although it is likely that *Hirschmanniella oryzae* is a mixture of two species, van der Vecht and Bergman (1952) revealed the following in

Indonesia (Java). The adults of both sexes invade young roots some distance from the tip, and often use the opening produced by other individuals. The nematode moves through the air channels between the lamellae of the radial parenchyma of the root (Fig. 3). The thin laterals without air channels are not infected. Large numbers of nematodes can be found in the base of the seedling coleoptile, but never in the lower part of the leaf sheaths. The female commences oviposition a few days

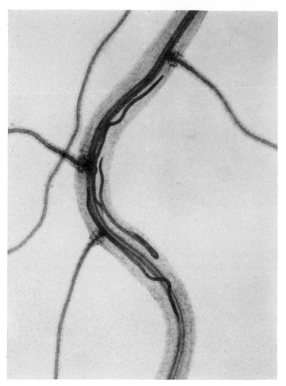

FIG. 3. Three adults of *Hirschmanniella oryzae* in rice roots. Magnification— × 40.
(By courtesy of T. Nishizawa.)

after invasion. The minimum time for development from egg to adult is at least one month, and the multiplication factor per generation may be as high as thirteen.

Different combinations of two *Hirschmanniella* spp. inhabit most rice-growing countries, and each combination consists of the common species *Hirschmanniella oryzae*, and a second species that is specific for each country except for the U.S.A. (Table III).

Hirschmanniella spinicaudata parasitizes lowland rice, but the irrigated soil conditions are often unsuitable for its reproduction (Luc, 1968). It is resistant to drying of the soil. The bionomics of *Hirschmanniella oryzae* and *Hirschmanniella imamuri* have been compared extensively in Japan, in which almost all rice paddies are infested with both species except for small, limited locations where only one of them occurs. Kuwahara and Iyatomi (1970) consider that there is only one generation a year of *Hirschmanniella imamuri* and two generations of *Hirschmanniella oryzae*.

TABLE III. The important *Hirschmanniella* spp. on rice in different countries.
(Based on Sher, 1968)

Country	First species	Second species
India	*H. oryzae*	*H. mucronata*
Indonesia (Java)	*H. oryzae*	*H. thornei*
Japan	*H. oryzae*	*H. imamuri*
Nigeria	*H. oryzae*	*H. spinicaudata*
U.S.A. (California)	*H. belli*	*H. spinicaudata*
U.S.A. (Louisiana)	*H. caudacrena*	—

D. The Rice Cyst Nematode, *Heterodera oryzae*

The rice cyst nematode, *Heterodera oryzae*, was first described on "lowland temperate rice" in the Ivory Coast (Luc and Berdon Brizuela, 1961). Prior to this report, however, this nematode had been known in Japan on upland rice roots.

1. Plant Symptoms

The nematode retards growth and causes severe leaf chlorosis and a reduction in the number of tillers of upland rice seedlings 2 weeks after germination. Roots are characterized by fewer root hairs and by a brownish discoloration. There is a tendency toward an uneven display of the disease symptoms within the field and a remarkable fluctuation from year to year. It is not rare to have no symptoms appear in a field which had been heavily infected during the preceding year (Watanabe *et al.*, 1963).

2. Nematode Bionomics

The life-cycle of *Heterodera oryzae* is short; the second-stage larvae emerge from the cysts between the 26th and 30th day after inoculation of the root with larvae. There is tylose formation around the female body but no true giant cell is formed; this reaction may be peculiar to some monocotyledonous roots. Single larva inoculations indicate that a male is essential for reproduction. Two, or probably three, generations may take place during 4 months of a rice growing season in the Ivory Coast (Berdon Brizuela and Merny, 1964; Merny, 1968).

III. Cultural and Environmental Influences on the Nematode Diseases of Rice

Aphelenchoides besseyi is disseminated primarily through nematode-infested grain, husks, or plant remains, and secondarily through irrigation water into which nematodes are released from the primary sources (Fukano, 1962). *Ditylenchus angustus* is spread mainly through irrigation water from diseased ratoons or plant remains, but not through seeds (Hashioka, 1963).

In eastern Bengal and adjacent districts of Assam, the main type of paddy, known as "aman" in which rice is sown between March and May and harvested during November and December, induces "ufra" disease (*D. angustus*). This is often severe on the broadcast paddy in the low-lying areas which remain submerged for a long period of the year, but not usually on those which are transplanted on to higher land (Padwick, 1950). *Heterodera oryzae*, in Japan, never occurs on the transplanted paddy rice, but on only the broadcast upland rice and on the lowland rice variety which is broadcast under well-drained conditions and later followed by irrigation (Kawashima, 1968).

On the basis of the relationship between temperature of the irrigation water and elevation of the growing point of rice where *A. besseyi* attacks, Fukano (1962) suggests that deep-water irrigation decreases the disease incidence when the temperature exceeds 19–23°C. This is the optimum temperature for nematode motility from tillering to the young panicle formation stage.

Excessive nitrogenous fertilizer induces a high infection rate of *A. besseyi* (Tamura and Kegasawa, 1959). *Criconemoides onoensis* causes characteristic malformation of both primary and secondary rice roots in the oxidized soil layers below the biochemically reduced layer, and *Hirschmanniella* sp. causes decay of primary root tips in the oxidized soil layer (Hollis, 1967). Laboratory tests have proved that the rice stunt nematode, *Tylenchorhynchus martini*, is killed by hydrogen sulphide within 5–10 days at concentrations similar to those found in flooded fields (Rodriguez-Kabana *et al.*, 1965), or by the concentrations of *n*-butyric acid and propionic acid produced 4 days after flooding from a corn-meal supplement applied to rice soil (Hollis and Rodriguez-Kabana, 1966). *T. martini* has a comparatively low optimum soil moisture level of 40–60% field capacity (Johnston, 1958).

Rice variety "Tsurugiba" never exhibits white tip symptoms even when heavily infested with *A. besseyi*, but infection can reduce grain yield by more than 10%. It is likely that these symptomless varieties may act as nematode carriers (Fukano, 1962).

Most adult *Hirschmanniella imamuri* which have hibernated in old rice stubble roots are released into the paddy soil by the time of rice transplanting in June. A high population level as well as a high adult ratio (68·5%) of *H. imamuri* have been observed at transplanting time in the roots of barnyard grass (*Panicum crus-galli* var. *frumentaceum*), one of the common rice paddy weeds in Japan (Goto, 1969). Paddy weeds sometimes may play a significant role as reservoirs for further nematode infection of rice.

Since desiccation as well as high humidity are needed for *D. angustus* to survive until the next planting and to migrate to the host respectively, it appears that this species is highly adapted to the local environment, and this may explain the limited spread of the "ufra" disease to eastern Bengal, where humidity is high during the growing and ripening seasons and rainfall is almost negligible during the winter and spring months (Padwick, 1950).

IV. Control Measures

A. Physical Control

The effective warm-water treatment of rice seeds to control *Aphelenchoides besseyi* consists of pre-soaking below 20°C for 16–20 h and subsequently soaking at 51 (50–52)°C for 7 (5–10) min (Yoshii and Yamamoto, 1950c). Yoshii and Yamamoto (1951) revised this treatment to soaking in warm water at 56–57°C for 10–15 min for seeds without pre-soaking. This treatment can be done up to 60 days prior to sowing, provided that the seeds are dried sufficiently after treatment.

The warm-water treatment involves some risk of injury to seed germination unless the temperature is precisely maintained. Several authors have modified the water temperature, period of time for soaking, with or without pre-soaking in order to improve the nematicidal effect and to minimize seed damage by the treatment.

B. Chemical Control

Attempts have been made to find non-phytotoxic chemicals which efficiently rid rice seed of *A. besseyi*. The first promising chemicals are three rhodanine compounds: 3-methyl-5-ethyl rhodanine (REE or "N-168"), 3-*p*-chlorophenyl-5-methyl rhodanine ("N-244"), and 3-*p*-chlorophenyl-5-ethyl rhodanine ("N-245"), any one of which has been found effective at 112 g of 10% dust per bushel of seed (Todd and Atkins, 1959). Seeds soaked in an aqueous solution of 20% emulsifiable concentrate of REE at a concentration of 1 : 100–300 for 24 h or 1 : 400–500 for 48 h, at 15°C, resulted in more than 95% nematode kill (Fukano, 1962).

Many types of chemical application control *A. besseyi* in seeds with varying degrees of efficiency. They include: soaking seeds in an aqueous solution of mercuric chloride or silver nitrate, fumigation of seeds with methyl bromide and treatment of seeds with phosphorous insecticide dusts such as parathion, systox, malathion, dipterex, or diazinon. Some of these treatments are not now recommended because of possible mammalian toxicity.

Experiments have proved that soil fumigation with either one of methyl bromide, dichloropropane-dichloropropene (DD mixture), ethylene dibromide (EDB), or dibromochloropropane (DBCP) controls *Hirschmanniella* spp., *Heterodera oryzae*, *Tylenchorhynchus martini*, and other root-infesting nematodes.

C. Cultural Control

There are three degrees of resistance of rice varieties to white tip disease: (*a*) susceptible and manifesting white tip symptoms, (*b*) susceptible but symptomless and (*c*) immune or no indication of nematode invasion. Most varieties so far known in Japan belong to the first group, while very few or no varieties belong to the second or third group (Fukano, 1962). From a 4-year test on the varietal resistance to white tip disease in Japan, the following nine symptomless varieties were found: Nôrin 8, Tôsan 36, Tôsan 37, Tôsan 38, Tôsan 58, Nankai 3, Asa Hi, Asahi 1 (Fukuoka), and Kagoshima-Asahi 1 (Nishizawa, 1953a). In the U.S.A., resistant rice varieties are Arkansas Fortuna, Nira 43, Bluebonnet, Improved Bluebonnet, Century 231, and Century 52 (Cralley, 1952). The Italian rice varieties of Rinaldo Bersani, Carnaroli, and Pierrot, show some resistance to the white tip disease (Orsenigo, 1955).

In Thailand, "Khao Tah Oo" resistant to the leaf blast disease was most resistant to *D. angustus* (42·9% infected) whereas "Khao Tah Haeng 17" was susceptible both to the leaf blast disease and to the nematode (80·9% infected) (Hashioka, 1963).

Since *D. angustus* is spread mainly through irrigation water, manual removal or burning of the straw is the practice in Burma, although this is somewhat difficult because the land is under water for much of the time (Padwick, 1950). In southern Thailand where rice is traditionally harvested by ear cutting, it is recommended that the rice be cut low, all stubble buried, crop remains burnt, and all ratoons eliminated to control *D. angustus* (Hashioka, 1963).

V. Economics of Yield Loss Caused by Nematodes and Paddy Soil Treatment

According to the Food and Agriculture Organization (1968), the world total rice production in 1967 was estimated to be 275,942,000 metric tons, of which 91·3% was produced by countries in the Orient (Far East), i.e. China Mainland 92,000, India 56,787, Pakistan 19,005, Japan 18,770, Indonesia 13,932, respectively in 1000 ton units. The highest average yield per hectare was recorded in Australia, followed by Spain, Morocco, Japan, and Italy.

Only a limited amount of data is available on the effect of nematode diseases on the yield of rice. Table IV indicates that the yield loss caused by *A. besseyi* amounts to as much as 30–50%, although this is variable depending on conditions such as variety, the initial nematode population level, or the cultural practices.

The normal rice yield in southern Thailand was about 2000 kg per hectare, but this was seldom reached because *D. angustus* caused a 20–90% loss in yield (Hashioka, 1963). Two-thirds of 505 hectares at Nagori was affected by *D. angustus* with a loss of Rs. 62 per hectare (4·60 Pakistan Rs. = 1 U.S. dollar), 650 hectares in the Dacca district at Rs. 86 per hectare, and in one small portion of Noakhali district, a loss of 200,000 maunds (7500 metric tons) was estimated in 1910 (Padwick, 1950).

The estimated reduction in rice yield caused by *Hirschmanniella* spp. is still controversial because of considerable variability in the conditions of experiments. In field experiments in Thailand, Taylor (1968) applied two dosages of emulsified DD mixture or DBCP to rice paddy plots and to seedbeds, resulting in a 24–36% increase in grain yield in the treated plots. Similar results were obtained by Yamsonrat (1967) through pot experiments in Thailand.

Several experiments in rice paddies have been conducted in Japan, mostly since 1963, and the results have been reviewed (Ichinohe, 1968). DD mixture, 30% EDB, or DBCP (80% E.C.) at rates of 300–500 litres/ha for the first two, and approximately one-tenth the DD dosage for DBCP (active ingredient), were used for treated paddy plots. The DD treatment resulted in 10–38% yield increases in the well-drained paddy, and less in the moderately and ill-drained paddies. DBCP treatment increased the average yield by about 10%, but EDB induced plant abnormalities involving a short culm, unfolded leaves, dense to dark green in colour, or exceptionally active tillering. In addition,

TABLE IV. Estimates of the reduction in rice yield caused by *Aphelenchoides besseyi*

Variety	Grain yield Treated	Grain yield Untreated	Yield loss (%)	Procedures for estimation	References
Asa Hi	5·11*	3·67	28·2	Random 100 hills (= plants) possessing white tip on every culm ("untreated") compared with another 100 without it ("treated").	Yamada and Shiomi (1950)
Early Prolific (susceptible)	3941†	1810	54·1	Seed previously treated with N-244 for "treated" plots, and nematode inoculation made for "untreated" plots.	Atkins and Todd (1959)
Magnolia (susceptible)	2891	2389	17·4		
Sunbonnet (resistant)	4140	3133	24·3		
Fortuna (resistant)	3822	3843	None		
Jikkoku	1473‡	1126	23·6	Seed previously treated with 1 : 100 liquid of REE 20% E.C. for "treated" plots.	Fukano (1962)
Saikai 29	1·77§	0·95	46·3	Random 25 hills with white tip in various degrees sampled, from which 2 culms per hill, 1 with ("untreated") and 1 without white tip ("treated"), compared.	Fukano (1962)
Nôrin 18	2·21	1·33	39·8		

* litres/100 hills (= plants).   † lb/acre.   ‡ g/50 hills.   § g/panicle.

experiments indicated that soil treatment should be accompanied by a reduction in the amount of nitrogenous fertilizer by at least 20–40%, and in heavy clay soil by 50–60%, otherwise the rice plant grows too rapidly and heavy fungus infection and lodging occurs.

Virtually every rice paddy in Japan is infested with *Hirschmanniella*, which may suggest that rice varieties now grown best in paddies are those relatively resistant to this nematode, because they have been developed for hundreds of years on the basis of rice paddies inhabited by the nematodes. By the same token, most varieties respond appropriately to the normal application of fertilizer, and its amount should be decreased when the paddy is treated with nematicides.

The annual decrease in yield of upland rice (Table V) caused by *Heterodera oryzae* have been examined in Japan by a 4-year continual cropping in the same field plot (Watanabe *et al.*, 1963). Despite this loss, however, the application of nematicides to upland rice is still uneconomical due to its smaller cash value.

TABLE V. Annual decrease in upland rice yield of "Nôrin 21" caused by *Heterodera oryzae*. (Watanabe *et al.*, 1963)

Year	Grain weight*
1951	2870
1952	3030
1953	1030
1954	950

* kg/ha converted from 20 m² plot × 2 replications.

According to agricultural statistics of the Japanese government, the nation's average rice yield per hectare in 1969 was 4330 kg of lowland rice and 2030 kg of upland rice. On the basis of rice price of Y.269,672 per 150 kg (360 Japanese Y. = 1 U.S. dollar), the gross receipts of the lowland rice grower are calculated as Y.583,828 per hectare. Since the cost to produce the average lowland rice yield per hectare amounts to Y.314,156, the net returns of the growers was estimated at Y.269,672 per hectare. The price of 200 litres of either DD mixture or 30% EDB is approximately Y.26,000, and the labour expenses for application is much more than this price. This makes chemical control of nematodes economical only in special circumstances.

Serious damage caused by *Meloidogyne graminicola* to rice seedlings has been reported from Thailand. The seedbed area is usually only about one-twentieth of the paddy area, and the expense of using nematicides on it would not be great. Calculations indicate that the cost of treating enough seedbeds to plant one hectare of rice would be about U.S. $8·00 (Taylor, 1968).

Since the environmental conditions under which a rice plant grows are quite unique, the nematode species associated with rice and the recommended control measures are somewhat different from those of other crops. More data as to the role these nematodes play in rice culture should be accumulated, especially the roles of *Hirschmanniella* spp. and *Heterodera oryzae*. Control measures should be considered in relation to the nematode behaviour under standing water and to the cultural practices peculiar to rice. Rice nematodes are of considerable economic importance because rice is one of the world's major sources of carbohydrate, especially for the people in Asian countries.

References

Atkins, J. G. and Todd, E. H. (1959). *Phytopathology* **49**, 189–191.

Barat, H., Delassus, M. and Vuong, Huu-Hai (1969). *In* "Nematodes of Tropical Crops" (J. E. Peachey, ed.), Tech. Commun., No. 40, Commonw. Bur. Helminth. pp. 269–273.

Berdon Brizuela, R. and Merny, G. (1964). *Revue Path. vég. Ent. agric. Fr.* **43**, 43–53.

Cralley, E. M. (1952). *Arkans. Fm Res.* **1**, 6.

Fukano, H. (1962). *Res. Bull. Fukuoka pref. agric. Exp. Stn* **18**, 1–108.

Golden, A. M. and Birchfield, W. (1968). *Pl. Dis. Reptr* **52**, 423.

Goto, K. and Fukatsu, R. (1952). *Ann. phytopath. Soc. Japan* **16**, 57–60.

Goto, K. and Fukatsu, R. (1956). *Bull. natn. Inst. agric. Sci., Tokyo* **6**, 123–149.

Goto, M. (1969). *Yamagata Nôrin Gakkaiho* No. 26, 43–51.

Hashioka, Y. (1963). *Pl. Prot. Bull., F.A.O.* **11**, 97–102.

Hollis, J. P. (1967). *Pl. Dis. Reptr* **51**, 167–169.

Hollis, J. P. and Rodriguez-Kabana, R. (1966). *Phytopathology* **56**, 1015–1019.

Ichinohe, M. (1968). *Rev. Pl. Prot. Res., Tokyo* **1**, 26–38.

Iyatomi, K. and Nishizawa, T. (1954). *Jap. J. appl. Ent. Zool.* **19**, 8–15.

Johnston, T. (1958). *Proc. La Acad. Sci.* Yr 1957, **20**, 52–55.

Kawashima, K. (1968). *Res. Bull. Fukushima pref. agric. Exp. Stn* **4**, 17–31.

Kuwahara, M. and Iyatomi, K. (1970). *Jap. J. appl. Ent. Zool.* **14**, 117–121.

Luc, M. (1968). *In* "Tropical Nematology" (G. C. Smart, Jr., and V. G. Perry, eds), pp. 93–112. University of Florida Press, Gainesville.

Luc, M. and Berdon Brizuela, R. (1961). *Nematologica* **6**, 272–279.

Merny, G. (1968). International Symposium of Nematology (8th), Antibes, 1965. Reports, p. 53.

Nishizawa, T. (1953a). *Bull. Kyushu agric. Exp. Stn* **1**, 339–349.

Nishizawa, T. (1953b). *Ann. phytopath. Soc. Japan* **17**, 137–140.

Nonaka, F. (1959). *Sci. Bull. Fac. Agric. Kyushu Univ.* **17**, 1–8.

Orsenigo, M. (1955). *Riso* **4**, 15–17.

Ou, S. H. (1965). IRC News Letter **14**(2), 4–10.

Padwick, G. W. (1950). *In* "Manual of Rice Diseases" pp. 104–115. *Commonwealth Mycological Institute*, London.

Peachey, J. E., Larbey, D. W. and Cain, S. C. (1966). *Helminth. Abstr.* **35**, 337–339.

Rao, Y. S. (1970). Study of plant parasitic nematodes affecting rice production in the vicinity of Cuttack (Orissa) India. Final Technical Report (Mimeog.) Indian Council of Agricultural Research 1–115.

Rodriguez-Kabana, R., Jordan, J. W. and Hollis, J. P. (1965). *Science, N.Y.* **148**, 524–526.

Sher, S. A. (1968). *Nematologica* **14**, 243–275.

Sudakova, M. I. (1968). *Parazitologiya* **2**, 71–74.

Tamura, I. and Kegasawa, K. (1957). *Jap. J. Ecol.* **7**, 111–114.

Tamura, I. and Kegasawa, K. (1959). *Jap. J. Ecol.* **9**, 65–68.

Taylor, A. L. (1968). *In* "Tropical Nematology" (G. C. Smart, Jr., and V. G. Perry, eds), pp. 68–80. University of Florida Press, Gainesville.

Taylor, A. L. (1969). *In* "Nematodes of Tropical Crops" (J. E. Peachey, ed.) Tech. Commun., No. 40, Commonw. Bur. Helminth. pp. 264–268.

Timm, R. W. (1965). *In* "A preliminary survey of the plant parasitic nematodes of Thailand and the Philippines" p. 18. South-East Asia Treaty Organization, Bangkok.

Todd, E. H. and Atkins, J. G. (1959). *Phytopathology* **49**, 184–188.

Vecht, J. van der and Bergman, B. H. H. (1952). *Pember. Balai Besar Penjel. Pertan.* **131**, 1–82

Vuong, Huu-Hai (1969). *In* "Nematodes of Tropical Crops" (J. E. Peachey, ed.) Tech. Commun., No. 40, Commonw. Bur. Helminth. pp. 274–287.

Watanabe, T., Yasuo, M., Ishii, K., Nagai, M. and Ichiki, K. (1963). *J. centr. agric. Exp. Stn* **5**, 1–44.

Yamada, W. and Shiomi, T. (1950). *Res. Bull. Okayama pref. agric. Exp. Stn* **46**, 15–28.

Yamsonrat, S. (1967). *Pl. Dis. Reptr* **51**, 960–963.

Yoshii, H. and Yamamoto, S. (1950a). *J. Fac. Agric. Kyushu Univ.* **9**, 209–222.

Yoshii, H. and Yamamoto, S. (1950b). *J. Fac. Agric. Kyushu Univ.* **9**, 223–233.

Yoshii, H. and Yamamoto, S. (1950c). *J. Fac. Agric. Kyushu Univ.* **9**, 293–310.

Yoshii, H. and Yamamoto, S. (1951). *Sci. Bull. Fac. Agric. Kyushu Univ.* **12**, 123–131.

7
Nematode Diseases of Sugar-cane

S. K. Prasad

Indian Agricultural Research Institute
New Delhi-12, India

I. Crop Production

Sugar-cane (*Saccharum officinarum* L.) is grown anywhere within the tropics except at high altitudes with freezing temperatures, or in deserts or other areas unsuited to active plant life. The areas of intensive plantation growth are shown in Fig. 1, and the northern and southern limits are approximately 31°N and 31°S respectively. The more important sugar-cane growing countries in the world are: Australia, Barbados, Cuba, Fiji, Formosa, India, Indonesia, Mauritius, the Philippines, South Africa, South America, and the U.S.A. (mainland and Hawaii).

In the true tropics, where sugar-cane grows best, it grows more or

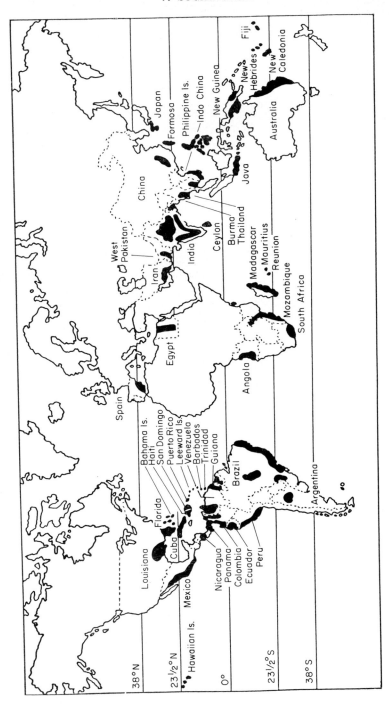

Fig. 1. Sugar-cane growing areas of the world.

less continuously and requires from 12 to 24 months to reach maturity, which makes it very vulnerable to pest attack. There is practically no area in which sugar-cane is grown where it is not associated with one or more plant-parasitic nematodes.

II. Nematodes Associated with Sugar-cane

The first report of a nematode parasite of sugar-cane was by Treub (1885), from Java. Later Soltwedel (1888), Cobb (1893, 1906, 1909), Stewart (1925), Muir and Henderson (1926), Stewart and Hansson (1928), Stewart *et al.* (1928), Cassidy (1930) and Van Zwaluwenburg (1930, 1931, 1932) reported nematodes associated with sugar-cane. Cobb described many species of nematodes from the rhizosphere of sugar-cane and postulated that the widespread unknown cause of reduction in sugar-cane growth was due to the combined action of these nematodes.

A considerable amount of literature on nematodes associated with sugar-cane has accumulated since then, and the reviews on nematode pests of sugar-cane by Roman (1968) and Williams (1969) need special mention. Over 15 genera and 50 species of plant-parasitic nematodes have been reported from sugar-cane but most records are of nematodes extracted from soil, or from mixed root and soil samples and therefore evidence of their feeding and reproducing on sugar-cane is only circumstantial. At least 20 new species have been described whose type host is sugar-cane. These are: *Belonolaimus lineatus* (Ramirez, 1964), *Criconemoides mauritiensis* (Williams, 1960a), *Helicotylenchus apiculus* (Roman, 1965), *H. caribensis* (Roman, 1965), *H. concavus* (Roman, 1961), *H. curvatus* (Roman, 1965), *H. elegans* (Roman, 1965), *H. parvus* (Williams, 1960), *H. truncatus* (Roman, 1965), *Hemicriconemoides sacchari* (Sood, 1967), *Heterodera sacchari* (Luc and Merny, 1963), *Hoplolaimus puertoricensis* (Roman, 1964a), *Longidorus laevicapitatus* (Williams, 1959), *Meloidogyne javanica* (Treub, 1885), *Radopholus similis* (Cobb, 1893), *Trichodorus acaudatus* (Siddiqi, 1960), *T. minor* (Colbran, 1956), *Tylenchorhynchus brevilineatus* (Williams, 1960a), *T. crassicaudatus* (Williams, 1960b), *T. curvus* (Williams, 1960b) and *T. martini* (Fielding, 1956).

Amongst the Tylenchoidea, the nematodes reported to be associated with sugar-cane are species of *Meloidogyne*, *Pratylenchus*, *Heterodera*, *Rotylenchulus*, *Radopholus*, *Hoplolaimus*, *Rotylenchus*, *Helicotylenchus*, *Scutellonema*, *Belonolaimus*, *Tylenchorhynchus*, *Criconema*, *Criconemoides*, *Paratylenchus*, *Cacopaurus*, *Hemicycliophora*, *Caloosia*, *Tylenchus*, *Ditylenchus*, *Trophurus*, and *Dolichodorus*.

The members of Aphelenchoidea recovered from sugar-cane roots are *Aphelenchoides* spp. and *Aphelenchus* spp.

The most important members of the Dorylaimida group associated with sugar-cane are species of *Xiphinema*, *Longidorus*, *Trichodorus*, and *Paralongidorus*. A more detailed commentary on the most important nematode pests of sugar-cane now follows.

A. Root-knot Nematodes, *Meloidogyne* spp.

The root-knot nematodes, *Meloidogyne* spp., are very common in sugar-cane fields and *M. javanica*, *M. incognita*, *M. arenaria* and *M. thamesi* have been reported from several cane growing areas of the world (Williams, 1969). They cause galls or knots of varying sizes along the root, usually near the root tips, resulting in chlorosis and stunting. As in other grasses, root galls on sugar-cane are small and may take the form of nodules and elongated curled thickenings at or near the root tips which can easily be overlooked. In heavily infested plants the leaves are characterized by waxy golden-yellow bands extending from the tips to the sheath (Matz, 1925). The larvae enter the root through the region of the root tip. Heavy larval infections stop root growth by affecting the meristematic zone. The pressure from the expanding giant cells, maturing nematodes and egg masses causes mechanical damage to the root tissues such as blockage or malformation of the xylem vessels. Lateral root proliferation is sometimes but not always associated with gall formation but pronounced root curvature is associated with nematode infection. The external manifestation of injury varies considerably. Rao (1961) observed stray attack of *M. javanica* in sugar-cane fields in Coimbatore, Udumalpat and Nellikuppam areas (India). In Udumalpat area, the patches of affected crop showed signs of yellowing and stunting. These patches of soil were known to be poor and infertile and cane roots had numerous galls of varying sizes. In the Nellikuppam area, however, symptoms of poor growth were generally less conspicuous; this may have been due to the practice of heavy manuring in this area, which apparently compensated to some extent for the damage done by the nematode.

B. Lesion Nematodes, *Pratylenchus* spp.

The species of root-lesion nematodes reported from the roots and the rhizosphere of sugar-cane are: *Pratylenchus zeae*, *P. brachyurus*, *P. delattrei*, *P. scribneri*, *P. coffeae*, *P. pratensis*, *P. sacchari* and *P. thornei*. In sugar-cane fields *P. zeae* and *P. brachyurus* are widespread. The nature of root damage to sugar-cane is not well known but they seem to invade only the cortical parenchyma. The cells adjacent to the nematodes become brownish and often collapse. The cane roots attacked by *P. zeae* are thickened with a few fine roots and show dark, round or elongated lesions. The reduction in yield of sugar-cane increases as the nematode population densities increase up to 48/pint of soil and 255/g of root tissue. *P. zeae* multiplies more on sugar-cane when the sets are also inoculated with *Phytophthora* spp. (Khan, 1963). A pathogenicity trial with *P. zeae* and *Phytophthora megasperma* alone and in combination indicated that both the organisms affected sugar-cane growth independently. In fallow soil *P. thornei* was found at soil depths of around 40 cm, and when sugar-cane was planted the nematodes moved up to the developing roots (Oteifa *et al.*, 1963).

C. Burrowing Nematodes, *Radopholus* spp.

All stages of the burrowing nematodes, *Radopholus similis* and *R. williamsi*, have been found in the roots of sugar-cane. *R. similis* has been reported from Hawaii, Java, the Philippines, India, Louisiana, Florida, Cuba, Australia and Mauritius. It enters young sugar-cane roots and feeds on the cortex. It causes lesions and cavities and the infected root tissues lose their whitish colour, become reddish and finally turn purplish black. The vascular tissues may be affected also.

D. Cyst Nematode, *Heterodera* spp.

The cyst nematode, *Heterodera sacchari*, has been reported from Congo, India, Ivory Coast and Nigeria. Cysts were collected from soil around sugar-cane roots in the brownish-grey sandy soil. Cysts were not present in heavy clay soils. Little is known about the damage caused by *H. sacchari* to sugar-cane.

E. Reniform Nematodes, *Rotylenchulus* spp.

The reniform nematodes *Rotylenchulus parvus* and *R. reniformis* have been found in sugar-cane fields. Immature stages have been reported from Congo, Mauritius, Natal, Puerto Rico, Rhodesia and Venezuela. *R. parvus* infests roots of both sugar-cane and the weed, *Bidens pilosa*. Roman (1964) suggested that sugar-cane can be rotated with pineapple as a control measure against *R. reniformis* on pineapple. Birchfield and Brister (1962) concluded from pot trials that sugar-cane is immune to *R. reniformis*, but suggested that different strains might exist within the species because a morphologically similar form in the Dominican Republic was associated with sugar-cane.

F. Lance Nematodes, *Hoplolaimus* spp.

The lance nematodes, *Hoplolaimus galeatus*, *H. puertoricensis* and *H. seinhorsti*, have been reported from cane roots, but the action of these nematodes on sugar-cane has not been studied.

G. Spiral Nematodes, *Helicotylenchus* spp.

Helicotylenchus dihystera, *H. erythrinae*, *H. multicinctus*, *H. concavus*, *H. retusus*, *H. nannus*, *H. parvus*, *H. egyptiensis*, *H. crenacauda*, *H. apiculus*, *H. borinquensis*, *H. caribensis*, *H. curvatus*, *H. elegans* and *H. truncatus* have been reported from sugar-cane fields. *H. dihystera* was reported for the first time by Cobb (1893) from sugar-cane.

Infected roots show disorganization and collapse of the cells of the cortical tissues. Sloughing off of the epidermal cells is also observed. After penetration the nematode makes a pathway through the root tissues by destroying the cells in its path. The necrotic damage to cells in the area of the feeding site also extends to cells some distance away from the feeding site. Sometimes the nematodes penetrate deep into the cortex and approach the stele. All such infected roots show disorganization of cortical cells and brown lesions. Jensen *et al.* (1959) observed *H. dihystera* with the head end embedded in cane roots and remarked that secondary infection generally seems to occur around the feeding sites. Apt and Koike (1962a) observed blunt, malformed roots and a reduction in the number of small branch rootlets. David and Thirvengadam (1961) reported that in the Pangalur factory zone (India), a field of var. Co_{499} generally infected with *Helicotylenchus erythrinae* yielded 53 metric tons/hectare whereas the average yield in the area was 100 metric tons/hectare. The jaggery (gur) prepared from the juices of these canes was also of very poor quality.

Birchfield (1965) found that *H. nannus* depressed growth and reduced dry weight of cane particularly in combination with the fungus *Pythium graminicola*.

H. *Rotylenchus* spp., *Scutellonema* spp. (Spiral Nematodes)

R. robustus, *R. buxophilus*, *R. brevicaudatus* and *S. magniphasmum*, *S. unum*, *S. brachyurum* and *S. clathricaudatum* have been recorded from sugar-cane fields. *S. magniphasmum* and *S. unum* have been described from sugar-cane roots and adjacent soil in Southern Rhodesia and Kenya respectively. These spiral nematodes generally feed externally on roots but sometimes some of them may embed the anterior end of the body in the root tissues or may be even completely embedded. No pronounced symptom is associated with the feeding of these nematodes.

I. Sting Nematodes, *Belonolaimus* spp.

The sting nematodes, *Belonolaimus gracilis* and *B. lineatus*, have been found in soil around the roots of declining sugar-cane. They stop apical growth and may cause necrosis of cortical tissue. *B. gracilis* lives in and reproduces on cane roots, although no host symptoms or growth reduction is evident.

J. Stylet Nematodes, *Tylenchorhynchus* spp.

The stylet nematodes, *Tylenchorhynchus martini*, *T. claytoni*, *T. acutus*, *T. nudus*, *T. dactylurus*, *T. curvus*, *T. crassicaudatus*, *T. brevilineatus* and *T. elegans*, have been recovered from sugar-cane fields. Birchfield and Martin (1956) found sugar-cane roots attacked by *T. martini* to be abnormally blunt, sparse and irregular. *Tylenchorhynchus* spp. are free-living and their feeding on root hairs and epidermal cells at root tips may result in poor growth.

K. Dagger, *Xiphinema* spp.; Needle, *Longidorus* spp.; Stubby-root, *Trichodorus* spp. Nematodes

The species recovered from sugar-cane fields are *Xiphinema americanum*, *X. elongatum*, *X. insigne*, *X. eusiculiferum*, *X. truncatum*, *X. vuittenezi*, *X. pratense*, *Longidorus laevicapitatus*, *L. africana*, *Trichodorus christiei*, *T. minor*, *T. porosus*, *T. rhodesiensis*, *T. acaudatus* and *T. mirzai*. Noting is known about the action of *Xiphinema* species on sugar-cane but their distribution and abundance in Mauritius was found to be correlated with the soil type, climate and altitude. In Hawaii, *X. americanum* was not found in cane fields above 700 ft. *L. africana* was described from cane roots. Apt and Koike (1962a, 1962b) and Jensen *et al.* (1959) demonstrated the pathogenicity of *T. christiei*. This nematode caused severe root stunting and prevented development of fine feeder roots of seedling canes grown in pots. The tap root growth decreased in direct proportion to the number of nematodes present.

L. Other Nematodes

With the exception of *Caloosia*, all the following species in this group are free-living ectoparasites of sugar-cane roots and some of them have been associated with poor growth of sugar-cane: *Criconema brevicaudatum*, *Criconemoides rusticum*, *C. curvatum*, *C. sphaerocephalus*, *C. mutabilis*, *C. onoensis*, *C. ornata*, *C. mauritiensis*, *Hemicriconemoides cocophillus*, *H. obtusus*, *H. sacchari*, *H. wessoni*, *Paratylenchus minutus*, *Hemicycliophora membranifer*, *H. parvana*, *H. penetrans*, *Dolichodorus heterocephalus*, *Caloosia longicaudata* and *Cacopaurus* spp.

Hemicriconemoides sacchari is the most common parasitic nematode in the sugar-cane fields in the Indian union. All stages are encountered in the soil and in the sugar-cane roots. The result is browning of the roots and sloughing off of the epidermal cells. The infected plants have poorly developed root systems (Sood, 1967).

III. Nematode Disease Complexes

Sometimes associations occur between a nematode species and a species of fungus, bacterium and/or virus. Disease complexes in which a nematode and another organism are both required to cause the disease syndrome have not yet been demonstrated for sugar-cane as for many other crops. Tests to determine if ratoon stunting and chlorotic streak diseases could be transmitted by nematodes have yielded negative results, and the possibility of nematodes transmitting virus diseases of sugar-cane has not yet been demonstrated, although the virus vectors *Xiphinema*, *Longidorus* and *Trichodorus* species occur in sugar-cane fields. *Meloidogyne*, *Helicotylenchus* and *Pythium arrhenomanes* each reduce sugar-cane growth and there is only a slight additional effect when *Meloidogyne* and *P. arrhenomanes* or *Helicotylenchus* and *P. arrhenomanes* occur together. Similarly *Pratylenchus zeae* and *Phytophthora megasperma* are found to affect sugar-cane growth independently.

IV. Environmental and Cultural Influences

Nematode populations differ qualitatively and quantitatively with soil type, climate, cropping pattern and cultural practices. Some idea of their prevalence in cane soils and roots can be obtained from the observations made by Martin and by Innes and Chinloy. Martin (1962) in Rhodesia recovered 1000–25,000 plant-parasitic nematodes from sugar-cane roots while Innes and Chinloy (1960) in Jamaica recovered 780–17,500 tylenchid nematodes from 568 ml of soil and roots. The population of plant-parasitic nematodes generally is larger in sandy or light soils (Van Zwaluwenburg, 1932; Rands and Abbott, 1939; Dick, 1958; Williams, 1963; Hu and Chu, 1964). The damage to sugar-cane caused by *Meloidogyne* is most pronounced in such light soils and the injury is accentuated by lack of nutrients. In Hawaii, the population of *Radopholus similis* decreases if rainfall is greater than 75 in. or less than 30 in. over a 3 month period. *Meloidogyne* populations in Hawaii increase with altitude while the opposite holds for *Pratylenchus zeae*, *Radopholus similis* and *Xiphinema americanum*. *X. americanum* is not found in cane fields above 700 ft. and *R. similis* and *P. zeae* are rarely found above 2000 ft (Van Zwaluwenburg, 1932; Anon., 1961). Weeds

F

assist in maintaining the numbers of *Meloidogyne* in cane fields. The population of plant-parasitic nematodes usually is very high in fields where sugar-cane is grown continuously year after year (Prasad *et al.*, 1965).

V. Control Methods

A. General

Although nematodes were shown to cause injury to sugar-cane as early as 1887, the first report of their control was not published until 1926.

In Hawaii, the yield of sugar-cane grown in rotation with jack beans, a nematode trap crop, was found to be greater than the yield following a year of clean fallow (Muir and Henderson, 1926). Winchester (1964) reported that in Florida one year's growth of Pangola grass (*Digitaria decumbens*) controls *Meloidogyne* spp. sufficiently well to provide 3 years of increased sugar-cane yields.

The decline of economically acceptable varieties and the never-ending demand for new ones is a well-recognized problem in the case of sugar-cane. Some sugar-cane varieties show different degrees of susceptibility to different nematode species (Rao, 1961). Co_{527}, Co_{290} and a few others were relatively less affected by *Meloidogyne* spp. Sugar-cane varieties Co_{413}, N. Co_{310} and $48D_{12}$ in the United Arab Republic are good hosts of *Pratylenchus thornei* and *Longidorus elongatus* but poor hosts of *Tylenchorhynchus martini* while var. $48D_{12}$ is a good host of *Helicotylenchus dihystera* and *Rotylenchulus reniformis* (Oteifa *et al.*, 1963). Co_{1104} and Co_{1234} varieties are less favourable for the multiplication of *Hemicriconemoides sacchari* and *Helicotylenchus dihystera* (Rao, 1966; Sood, 1967).

The natural enemies like sporozoan parasites, nematode-trapping fungi, predatory nematodes and nematophagous insects and mites, etc., may be important in regulating nematode populations in the soil. Williams (1960b) recorded a sporozoan infecting 34% of *Meloidogyne* females in cane roots in Mauritius. A similar parasite was reported to decrease *Xiphinema* populations in Mauritius (Williams, 1967).

Five species of nematode-trapping fungi, viz. *Arthrobotrys cladodes*, *A. conoides*, *A. oligosperma*, *Dactylella ellipsospora* and *Dactylella* sp., were isolated by Chu and Hsu (1965) from Taiwan sugar-cane soils. In pot experiments *A. oligosperma* was found to be beneficial in trapping *Meloidogyne* spp., and the greatest effect was at 27°C whereas at the extremes of 12°C and 32°C growth of the mycelium was checked and its effect was less. The damage from nematodes was greatly reduced when the fungus developed well.

The topsoil of a field in Nigeria infested with the cysts of *Heterodera sacchari* was scraped to a depth of 5–10 cm, molasses was spread over it, and the topsoil replaced. The untreated field showed an average of 48 cysts per soil sample, whereas 13 cysts per sample were recovered from the treated fields, suggesting that molasses acted as a limiting factor in the increase of *H. sacchari*. Stewart *et al.* (1928) also found fewer nematodes of *Meloidogyne* spp. in roots of sugar-cane grown in soil treated with molasses ("mud press") or compost than in check soil or soil that had received only inorganic additions. The addition of molasses was followed by an increased population of predaceous *Mononchus* spp.

B. Chemical

A number of reports on the success of nematicides against some of the nematodes associated with sugar-cane have been published. Since sugar-cane is a long duration crop the nematicide can be applied as: pre-plant treatment, treatment at planting, post-plant treatment and stubble treatment after harvest.

The earliest published use of nematicides on sugar-cane was by Chu and Tsai (1957) in Taiwan. The number and length of millable canes and their weights were improved and the sugar yield increased from 2 to $3\frac{1}{2}$ times by pre-planting application of nematicides like EDB (ethylene dibromide) and MBr (methyl bromide). Bates (1957) in Guyana and Williams (1959) in Mauritius obtained similar gains. Williams reported a 15–34% increase in yield. Hawaiian workers, who had earlier reported that nematodes were not important in sugar-cane, obtained favourable yield responses to the application of DD, EDB, MBr and DBCP (1, 2-dibromo-3-chloropropane) (Anon. 1960). When the nematicide was applied under ideal situations (good soil preparation, favourable weather and moisture conditions, a waiting period between fumigation and planting) there was an average increase in cane yields of about 34% over a 5 year period in tests in Florida (Winchester, 1968a). Singh (1960) obtained an increased yield of 30–78% of stripped canes when either DD or Vapam was applied 3 weeks before planting. Dick (1968) obtained yield increases from nil to as much as 80 tons of cane per hectare.

Average yield responses of sugar-cane to furrow application of DD, Fensulfothion, Diazinon, Carbofuran, Methomyl and Mocap at planting in Florida during 1965-1969 resulted in 21–33% yield increases. Birchfield (1969) reported that 10% granular form in Aldicarb, B-25141 and VC 9-104, incorporated in the furrow at planting time were effective in low dosages. These have increased yields by tons of cane and many kilograms of sugar per hectare, and have performed consistently well under different seasonal influences.

The first published data on the application of DD, EDB and organophosphates to growing cane appears to be that of Winchester (1964). Either the liquid or granular formulations of the organo-phosphates may be applied when the new plants are about 30 cm in height, in a 37·5 cm band centred over the cane row but directed towards the ground. The treated band is then covered by a shallow 2·5–5 cm layer of soil. This offers protection from the sun and the soil moisture helps to move the nematicide into the root zone. In some cases Winchester (1968b) obtained yield responses of more than 100% and the numbers of nematodes, in this case *Hemicycliophora parvus*, were considerably decreased.

Vapam gave maximum kill of *Hemicriconemoides sacchari*, and the sugar content in canes grown in Vapam treated soil was reported to be 16% compared with the 10–11% sugar content of canes grown in untreated soil or in soil treated with other nematicides like DD, Nemagon, etc. (Sood, 1967). In Puerto Rico, Roman (1968) reported that the control of spiral nematodes with Diazinon and Dasanit increased sugar yield by 55%.

The attainment of ideal conditions for successful control of plant-parasitic nematodes is less necessary when emulsifiable soil fumigants like DBCP are used. An application of 17 litres per hectare of concentrated Nemagon incorporated into the irrigation water increased the germination of sugar-cane by 50% and there was an increase in the overall plant height by 10 cm. DBCP applied in irrigation water at the rate of 5·6 and 11·2 litres per hectare resulted in lower nematode populations. Birchfield (1969) observed that the use of DBCP increases the sugar content of cane and has eth advantage that it is non-phytotoxic and so can be used at planting time. Roman (1965) always obtained an increase of cane and sugar by the application of DBCP. There is a possibility that two applications a year would be more effective than one. The second application, about 5–6 months before harvest, would tend to retard the build-up of the nematode population.

VI. Economic Importance

Different types of plant-parasitic nematodes, e.g. endoparasites like *Meloidogyne, Heterodera, Pratylenchus, Radopholus*, semiendoparasites like *Hoplolaimus, Helicotylenchus, Rotylenchus, Tylenchorhynchus, Rotylenchus* and ectoparasites like *Belonolaimus, Xiphinema, Dolichodorus, Longidorus, Trichodorus, Criconema, Criconemoides* and *Hemicycliophora* are found in soils of sugar-cane fields. Many species and genera may exist together in a state of mutual tolerance in which no serious competition can be observed as is evident from the occurrence and distribution of the nematode species associated with sugar-cane at Daurala and

Mavana farms, India (Prasad *et al.*, 1965). The nematode fauna of cane fields in different countries are similar in that the population characteristics are determined by the soil type, climate, cultural practices, etc. Therefore it is difficult to assess accurately the population of nematodes in sugar-cane and even more difficult to assess the losses caused by them. However, the nematode problem in sugar-cane appears to be becoming more acute every year. The main cause of this increase in seriousness seems to be the continuity of the standing crop, in some stage or other, throughout the year (from one year in some countries to an average of 7 years in Mauritius) which facilitates uninterrupted breeding and multiplication of nematodes. Prasad *et al.* (1965) observed that the population of *Hoplolaimus* and *Helicotylenchus* was very high in fields where sugar-cane was grown continuously for 20–25 years. In spite of the fact that the cultivators were using many improved agricultural practices, the crop was showing stunting, drying and discoloration of leaves.

There are several methods of assessing crop losses. Since nematodes do not kill the entire plant, the percentage of injured or killed plants does not indicate the loss in yield. Three methods of assessing crop losses as a result of nematode damage are described below:

1. Comparing yields of plants artificially infected with that of uninfected plants when all the other conditions are similar.

Some progress has been made in correlating damage with the number of nematodes present. At 1000 and 10,000 inoculation levels of *Helicotylenchus dihystera*, the canes are smaller than those grown in steam-sterilized soil. In respect to *Hemicriconemoides sacchari* a population of 10,000 nematodes per 500 g of soil is required to affect the plant growth significantly.

Experiments with *Helicotylenchus dihystera, Hemicriconemoides sacchari, Trichodorus christiei, Tylenchorhynchus martini, Pratylenchus zeae* and *Meloidogyne incognita* have shown that each one of these species of nematode above a certain population density is pathological to sugar-cane. To get a clear picture of the crop losses it is necessary to know the harmful population levels of different species under different conditions. There is a further difficulty in that under natural conditions there is almost always a mixture of populations of different nematode species as well as populations of predatory nematodes, nematode-trapping fungi, etc., that also affect the crop losses. Therefore, only a rough estimate of the role of nematodes in decreasing sugar-cane yields can be obtained by this method.

2. Comparing yields of naturally infected plants with those of uninfected plants in the same field when all the other conditions are similar.

To determine the losses in yield of sugar-cane by a specific nematode species a comparison of uninfected and infected plants in the same field with all other conditions as similar as possible must be made. Unfortunately, quantitative estimates based on such actual measurements made under uncontrolled field conditions are not yet available for sugar-cane.

3. Comparing the yields of naturally infected plants with those of plants artificially protected from infection by chemical treatment.

The increase in yield caused by nematicide treatment of nematode-infested soils is not identical with the mere absence of nematodes as the nematicide can have effects other than killing the nematodes. Thus the nematicide may also control other damaging organisms and so increase the yield; the nematicide may also control the insect predators, predacious nematodes, and nematode-trapping fungi in such a way as to increase the rapid build-up of the nematode population and so reduce the yield. Also there may be various effects of the nematicide directly on the plant or on the flora and fauna of the soil, which may in turn affect yield. The elimination of the nematode by a nematicide does not indicate exactly how much damage the nematode would have done. However, it does give an indication of the overall benefits to be gained by the use of nematicides.

The main criteria which are considered in assessing the effectiveness of a nematicide are:

(i) Technical effectiveness: basically the nematode kill, but also interactions with other soil organisms, (ii) crop effectiveness: considering the yield, quality of crop and value of yield and quality, and finally, (iii) economic effectiveness, which accounts for the cost : potential benefit ratio or percentage return on investments, this being the difference between the combined value of yield and quality resulting from treatment, and the costs of treatment. There is very little of such information available for the control of nematodes in sugar-cane. Workers from several countries including Hawaii obtained significantly greater economic returns by using soil fumigants to control nematodes. This is particularly true for those areas where the agricultural economy is based mainly on the production of sugar-cane, the major part of the land available for cultivation is devoted to this purpose year after year and the commercial varieties are susceptible to nematode attack.

There are some reports of percentage kill of nematodes or increase in yield of sugar-cane as well as an increased brix and sugar content as a result of soil fumigation, but there is none which indicates the cost : potential benefit ratio or percentage return on investments.

The question of felt loss due to nematodes, although at first glance it looks simple, has proved to be difficult to assess. In the absence of any correct estimates, varous "guesstimates" have been vague. One of the most universally reported guesses is, for example, that nematodes cause about a 10% loss annually and on this basis, the annual loss sustained by sugar-cane as a result of plant-parasitic nematodes, amounts to about 3,649,500 tons raw value of cane sugar. (World cane sugar produced in 1967-1968 was 36,495,000 tons raw value: according to F.A.O. Production Year Book 1968.) This calculation gives only a rough and perhaps a misleading idea of the magnitude of the problem and makes out a *prima facie* case for devoting serious attention to it.

References

Anon. (1960). *Rep. Hawaiian Sug. Plrs' Ass. Exp. Stn* 23.
Anon. (1961). *Rep. Hawaiian Sug. Plrs' Ass. Exp. Stn* **22–23**.
Apt, W. J. and Koike, H. (1962a). *Phytopathology* **52**, 798–802.
Apt, W. J. and Koike, H. (1962b). *Phytopathology* **52**, 963–964.
Bates, J. F. (1957). *Proc. Br. W. Indies Sug. Technol.* 1957, 100–104
Birchfield, W. (1965). *Phytopathology* **55**, 1051–1052.
Birchfield, W. (1969). *Pl. Dis. Reptr* **53**, 530–533.
Birchfield, W. and Brister, L. R. (1962). *Pl. Dis. Reptr* **46**, 683–685.
Birchfield, W. and Martin, W. J. (1956). *Phytopathology* **46**, 277–280.
Cassidy, G. (1930). *Hawaii. Plrs' Rec.* **34**, 379–387.
Chu, H. T. and Hsu, S. C. (1965). *Rep. Taiwan Sug. Exp. Stn* **37**, 81–88.
Chu, H. T. and Tsai, T. K. (1957). *Rep. Taiwan Sug. Exp. Stn* **16**, 73–79.
Cobb, N. A. (1893). *Agric. Gaz. N.S.W.* **4**, 808–833.
Cobb, N. A. (1906). *Bull. Div. Path. Physiol. Hawaiian Sug. Plrs' Ass. Exp. Stn.* **5**, 163–195.
Cobb, N. A. (1909). *Bull. Div. Path. Physiol. Hawaiian Sug. Plrs' Ass. Exp. Stn.* **6**, 51–74.
Colbran, R. C. (1956). *Qd Jl agric. Sci.* **13**, 123–126.
David, H. and Thiruvengadam, C. R. (1961). *Proc. Conf. Sugcane Res. Dev. Wkrs India* **4**, 528–532.
Dick, J. (1958). *Bull. S. Afr. Sug. Ass. Exp. Stn* **8**, 6 pp.
Dick, J. (1968). *In* "Pests of Sugarcane" (J. R. Williams, J. R. Metcalfe, R. W. Montgomery and R. Mathes, eds), pp. 531–539. Elsevier, Amsterdam.
Fielding, M. J. (1956). *Proc. helminth. Soc. Wash.* **23**, 47–48.
Hu, C. H. and Chu, H. T. (1964). *Rep. Taiwan Sug. Exp. Stn* **33**, 63–82.
Innes, R. F. and Chinloy, T. (1960). *Proc. Br. W. Indies Sug. Technol.* 1960, 28–33.
Jensen, H. J., Martin, J. P., Wismer, C. A. and Koike, K. (1959). *Pl. Dis. Reptr* **43**, 253–260.
Khan, S. A. (1963). *Proc. int. Soc. Sug. Cane Technol.* **11**, 711–717.
Luc, M. and Merny, G. (1963). *Nematologica* **9**, 31–37.
Martin, G. C. (1962). *Rhodesia agric. J.* **59**, 28–35.
Matz, J. (1925). *Phytopathology* **15**, 559–563.
Muir, F. and Henderson, G. (1926). *Hawaii. Plrs' Rec.* **30**, 233–250.

Oteifa, B., Rushdi, M. and Salem, A. (1963). *Bull. Sci. Technol. Assiut Univ.* **6**, 271–279.

Prasad, S. K., Mathur, V. K. and Chawla, M. L. (1965). *Labdev J. Sci. Technol.* **3**, 211.

Ramirez, C. T. (1964). *J. Agric. Univ. P. Rico* **48**, 127-130.

Rands, R. D. and Abbott, E. V. (1939). *Proc. int. Soc. Sug. Cane Technol.* **6**, 202–212.

Rao, G. N. (1961). *Sci. Cult.* **27**, 94–96.

Rao, V. N. (1966). Studies on *Helicotylenchus dihystera* (Cobb, 1893) Sher, 1961. Ph.D. Thesis, Indian Agricultural Research Institute, New Delhi.

Roman, J. (1961). *J. Agric. Univ. P. Rico* **45**, 300–303.

Roman, J. (1964a). *J. Agric. Univ. P. Rico* **48**, 127–130.

Roman, J. (1964b). *J. Agric. Univ. P. Rico* **48**, 162–163.

Roman, J. (1965). *Tech. Pap. Univ. P. Rico agric. Exp. Stn* **41**, 23 pp.

Roman, J. (1968). *In* "Tropical Nematology" (G. C. Smart and V. G. Perry, eds), pp. 61–67. University of Florida Press, Gainesville.

Siddiqi, M. R. (1960). *Proc. helminth. Soc. Wash.* **27**, 22–27.

Singh, K. (1960). *Indian Phytopath.* **13**, 181–182.

Soltwedel, F. (1888). *Tijdschr. Land-en Tuinb. Boschkultuur*, 1898: 7–15.

Sood, U. (1967). "Studies on the genus *Hemicriconemoides* Chitwood and Birchfield, 1957" Ph.D. Thesis, University of Agra, Agra.

Stewart, G. R. (1925). *Hawaii Plrs' Rec.* **29**, 400–409.

Stewart, G. R. and Hansson, F. (1928). *Hawaii Plrs' Rec.* **32**, 217–223.

Stewart, G. R., Muir, F., van Zwaluwenburg, R. H., Cassidy, G. H. and Hansson, F. (1928). *Hawaii Plrs' Rec.* **32**, 205–217.

Treub, M. (1885). *Meded. plTuin. Batavia* **2**, 1–39. Also in *Bijlage Archief Java-Suikerindustr.* 1898, 3–6.

Van Zwaluwenburg, R. H. (1930). *Proc. int. Soc. Sug. Cane Technol.* **3**, 216–225.

Van Zwaluwenburg, R. H. (1931). *In* "The Insects and other Vertebrates of Hawaiian Sugarcane Fields" (F. X. Williams, ed.), pp. 352–368. Honolulu, Hawaii.

Van Zwaluwenburg, R. H. (1932). *Proc. int. Soc. Sug. Cane Technol.* **4**, Bull. 5, 4 pp.

Williams, J. R. (1959). *Occ. pap. Maurit. Sug. Ind. Res. Inst.* **3**, 28 pp.

Williams, J. R. (1960a). *Occ. pap. Maurit. Sug. Ind. Res. Inst.* **4**, 30 pp.

Williams, J. R. (1960b). *Nematologica* **5**, 37–42.

Williams, J. R. (1963). *Proc. int. Soc. Sug. Cane Technol.* **11**, 717–722.

Williams, J. R. (1967). *Nematologica* **13**, 336–342.

Williams, J. R. (1969). Nematodes as pests of sugarcane. *In* "Pests of sugarcane" (J. R. Williams, J. R. Metcalfe, R. W. Mongomery and R. Mathes, eds), pp. 503–530. Elsevier, Amsterdam.

Winchester, J. A. (1964). *Proc. Soil Crop Sci. Soc. Fla* **24**, 454–457.

Winchester, J. A. (1968a). *Biokemia* **16**, 15–17.

Winchester, J. A. (1968b). *Nematologica* **14**, 18–19.

8

Nematodes of Tobacco

D. L. Milne

Citrus and Sub-Tropical Fruit Research Institute
Nelspruit
E. Transvaal, South Africa

I. Tobacco Production

When Columbus discovered America, the natives were found to be cultivating and using tobacco, and the same was true of Cuba in 1492, where Columbus found the inhabitants smoking the equivalent of today's cigar. By 1616 tobacco culture had become quite general in the colony of Virginia and soon it was noticed that the sandy soils rapidly became tobacco "tired", thereby leading to the practice of clearing virgin coastal forest for each new planting. In the year 1640 Virginia produced more than 450 metric tons of tobacco, and by the end of the century, smoking had been introduced into practically every country of the world (Garner, 1951). Tobacco is now produced in most parts of the world, though Asia and North America account for a large percentage of the total world production of approximately 4·5 million metric tons per annum. This is made up of the lighter flue-cured types used mainly in cigarette manufacture, and the heavier or more aromatic air- and sun-cured types, which in general receive heavier fertilizer applications and are used as cigar and pipe tobaccos (Garner, 1951).

Tobacco's economic value has in most countries become twofold, income being derived from sale of the commodity and also from its taxation. The susceptibility to nematodes and high economic value of tobacco per hectare have made it one of the crops to show the most spectacular economic returns from soil fumigation and other methods of nematode control.

II. Nematodes Associated with the Crop, and the Damage they Cause

A. Root-knot Nematodes, *Meloidogyne* spp.

1. Distribution, Biology and Symptoms

The root-knot or gall nematodes are economically the most important nematode pests of tobacco. The species most commonly recorded on tobacco are *M. incognita* and *M. javanica* (Lucas, 1965; Milne, 1961), the former occurring as the predominant species in temperate and the latter in subtropical and tropical zones. These species quite often occur together in the same field and are sometimes in association with *M. hapla* or *M. arenaria*.

The life-cycles of these species in tobacco roots are typical of the genus. On emergence from the egg, the second-stage larva is attracted to the growing point of the young roots where it enters just behind the root tip. Entry by one individual often creates conditions attractive to others, so that penetration by one larva is often followed by mass invasion. The larvae feed in the stele, inducing the production of giant cells and a gall or "knot". Such galls become visible within 48 h of larval penetration. Females soon become sedentary within the galls, swelling up to form the characteristic flask or gourd shape. The mature female lays up to 400 eggs in a jelly-like egg sac produced towards the exterior of the root.

The life-cycle on tobacco takes from 21 days in summer up to 56 or more days during cool weather, and approximately 9300 Centigrade hours above the minimum threshold of 7·5°C is required for *M. javanica* to complete its life-cycle on tobacco (Milne and du Plessis, 1964). Races of *M. javanica* vary in their tolerance to temperature extremes. Those from the tropics are more resistant to high temperatures while those from temperate climates are more tolerant to cold (Daulton and Nusbaum, 1961).

Development of the giant cells and growth of the females causes extensive distortion and blocking of the vascular tissue which seriously slows water and nutrient transport. Secondary larval invasion causes the galls to coalesce and eventually the gross root system begins to decay.

The above-ground symptoms are those typical of drought and nutrient deficiency. The plants are dwarfed or stunted in growth, wilt prematurely and the leaves are yellow and thin, scorching easily and eventually turning brown. Where late infestations occur, the leaf lamina may turn white while the veins remain green, and during flue-curing the leaf turns an uneven yellow and green. Under conditions of severe infestation, weed growth becomes competitive for water and nutrients and there may be a total loss of the crop (Fig. 1). Leaves of

FIG. 1. Total loss of the tobacco crop due to a severe attack by *Meloidogyne javanica*. No rotation or fumigation was applied to this land.

root-knot infested plants are more susceptible to attack by brown spot (*Alternaria tenuis*), anthracnose (*Colletotrichum destructivum*), *Cercospora* leaf spot and powdery mildew (*Erisyphe cichoracearum*) (Tsumagari and Tanaka, 1954a). Furthermore, a combination of root-knot eelworm and *Fusarium* has been shown to increase considerably the occurrence of brown spot (Powell and Batten, 1969).

Meloidogyne infestations occur typically as patches in tobacco plantations and can be recognized by the yellowing and wilting of the plants in these areas. This patchiness in distribution makes the sampling of soil prior to planting notoriously unreliable as a basis for recommending fumigation. In general, however, sandy soils are more heavily infested than clay soils because of the more favourable conditions for nematode hatching and movement. Virgin soils should not necessarily be regarded as being free from root-knot nematodes.

In addition to the direct damage they cause, root-knot nematodes have been associated with an increase in various other tobacco root diseases.

2. Relationship of Root-knot Nematodes to Other Tobacco Root Diseases

(a) Black Shank, *Phytophthora parasitica* var. *nicotianae*. In 1949 black shank resistant varieties of flue-cured tobacco such as Dixie Bright 101 were released in the U.S.A. Although it was highly successful in most areas, by 1951 Dixie Bright 101 was suffering from stunting in some fields (Moore *et al.*, 1955). This stunting was associated with the presence of *Meloidogyne* spp., *Pratylenchus* spp., or *Tylenchorhynchus claytoni*, and it was found that application of fumigants helped in these cases (Sasser *et al.*, 1955). The latter authors carried out a series of experiments in which tobacco seedlings were inoculated with *M. incognita* either alone, or in combination with the black shank fungus. When the fungus alone was used, only a very low percentage of the plants developed symptoms after two weeks. However, when the fungus and root-knot nematodes were combined, the plants wilted within a week and a very high percentage developed black lesions within two weeks. The longer the lapse of time between infestation with root-knot nematodes and inoculation with the fungus, the more severe were the symptoms of black shank. Apparently the role of the nematode in this association is not merely one of wounding the roots but rather that of changing the root physiology such that the root is more suitable for fungal development (Sasser *et al.*, 1955). Infestation by the nematode is in fact obligatory for the development of the disease on black shank resistant varieties (Nusbaum and Todd, 1970). Where root-knot resistant varieties are not yet available it would be wise to fumigate the soil if proper use is to be made of black shank resistant varieties (Sasser *et al.*, 1955).

When the root-knot resistant variety NC 95, resistant to *M. incognita*, was released in Florida, it was observed that *M. javanica* infestations were commonly associated with black shank infections of this variety (Miller *et al.*, 1968). Laboratory investigations then showed that addition of the fungus alone to NC 95 plants led to a 22% black shank infection, whereas addition of *M. javanica* together with the fungus, led to a 64% black shank infection (Miller, 1968). Nusbaum and Todd (1970) have also demonstrated that where high-yielding varieties with moderate black shank resistance are used, the optimum results are obtained by pretreating the soil with a combination fungicide-nematicide. There is therefore no doubt that root-knot nematodes play an extremely important role in the black shank disease of tobacco, thereby further increasing the economic importance of this nematode group.

(b) Sore-Shin, *Rhizoctonia solani.* Using both root-knot susceptible and root-knot resistant varieties, Powell and Batten (1967) found that if root-knot eelworm infection occurred as long as 3 weeks before inoculation with the sore-shin organism, *Rhizoctonia solani,* then the plants would be predisposed to severe sore-shin attack. Root-knot resistant NC 95, on the other hand, developed no sore-shin, regardless of the time lapse between inoculation with the two organisms. It was therefore concluded that *M. incognita* was important in increasing the severity of sore-shin in root-knot nematode susceptible varieties. However, Hartill (1968) in Rhodesia, found no consistent relationship between infestation with *M. javanica* and consequent development of sore-shin.

(c) Fusarium Wilt, *Fusarium oxysporum* var. *nicotianae.* Powell and his co-workers have shown that the root-knot nematode species, *M. incognita, M. javanica* and *M. arenaria,* all predispose tobacco root tissue to attack by *Fusarium* wilt, the indication being that the giant cells themselves are highly suitable for development of the fungus (Melendéz and Powell, 1967; Porter and Powell, 1967). In their experiments they found that when plants were infected with nematodes 3–4 weeks before *Fusarium* inoculation, there was an average increase of wilt incidence up to the 50% level in both nematode susceptible and resistant varieties.

It has been shown that root-knot nematodes will increase the incidence of *Fusarium* wilt even when the nematode populations themselves are insufficient to cause direct damage to the tobacco plant (Nusbaum and Todd, 1970). These authors have also shown that in soils infested with both *Fusarium* and root-knot, fumigation with a nematicide will delay the onset of wilt symptoms.

(d) Granville Wilt, *Pseudomonas solanacearum.* In experiments in which the Southern bacterial wilt or Granville wilt organism and root-knot nematode, *M. incognita,* were added to tobacco plants singly or in combination, the infection with nematodes led to injury of the roots and consequent earlier and more severe Granville wilt infection (Lucas *et al.,* 1955; Johnson and Powell, 1969). These authors recommend that where wilt-resistant varieties such as Oxford 25 and Dixie Bright 101 are used, the greatest benefit is derived when the soil is previously treated against root-knot eelworm.

(e) Other root diseases. Neither *Pythium ultimum* nor *Trichoderma harzianum* cause root decay of tobacco in the absence of nematodes, but where infection with *Meloidogyne* occurs 4 weeks before inoculation, both Coker 316 and Dixie Bright 101 suffer severe root decay from a combination of these soil pathogens (Melendéz and Powell, 1969).

Mayol and Bergeson (1970) showed that plants inoculated with *M. incognita* under aseptic conditions suffer only a 37% leaf weight loss, whereas in the presence of ordinary soil bacteria and fungi, this loss may increase to 75%. They also conclude that the most likely requirement for secondary invasion is the modification of host tissues by the nematodes.

B. Root lesion or Meadow Nematodes, *Pratylenchus* spp.

1. Distribution, Biology and Symptoms

In North America *Pratylenchus pratensis, P. neglectus, P. brachyurus* and *P. zeae* have been recorded on tobacco, whereas in South Africa *P. hexincisus, P. thornei, P. vulnus, P. brachyurus, P. minyus* and *P. zeae* have been found in tobacco (Milne, 1961; Honey, 1967). *Pratylenchus* species are generally world-wide in distribution although *P. brachyurus* and *P. zeae* are more common in tropical areas and *P. penetrans, P. thornei* and *P. minyus* are more common in temperate regions (Corbett, 1970).

Lesion nematodes are attracted to tobacco roots where they force aside the epidermal cells just behind the root-hair region. The migratory root lesion nematodes are endoparasitic and feed and lay eggs within the cortex, the life-cycle being completed in 50–60 days. Their movement and feeding activity causes disintegration of the cortical cells which leads to browning of the root tissues. These symptoms are known as "brown root rot" of tobacco (Graham, 1951; Mountain, 1954; Graham and Heggestad, 1959). The disease is characterized by root-pruning (Figs. 2a and b) and water-soaked, lesions on the roots, although Mountain (1954) found that these symptoms were very much less severe when roots were infected under otherwise aseptic conditions. In the field, infested areas occur typically in well-defined patches in which, in addition to the root symptoms, plants are usually stunted, wilt prematurely and may even die. Root lesion nematodes are undoubtedly one of the most important groups of nematodes causing losses to the tobacco crop.

In general tobacco is not regarded as a very favourable host for multiplication of lesion nematodes, although Honey (1967) found that *P. hexincisus* populations increased steadily on tobacco without causing severe symptoms and *P. brachyurus* has been recorded as increasing on cultivar Hicks (Southards, 1966). On the other hand some tobacco varieties are sensitive to attack by lesion nematodes which have built up on a previous more favourable host (Lucas, 1965). In this respect it is important to note that various plants commonly used in rotations are favourable for the build-up of lesion nematodes. Such crops include

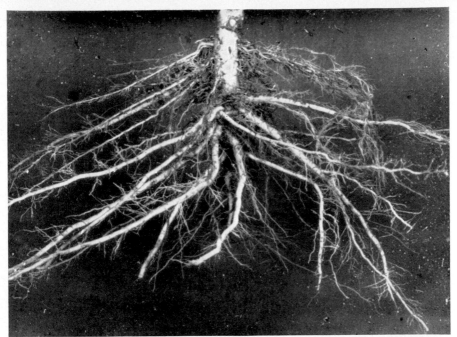

FIG. 2. (a) Roots of tobacco plant grown in soil infested with *Pratylenchus*. (By courtesy of C. J. Nusbaum.)

(b) Roots of tobacco grown in the same *Pratylenchus*-infested soil after fumigation with methyl bromide. (By courtesy of C. J. Nusbaum.)

Crotalaria juncea (sunn-hemp), *Phleum pratense* (timothy), *Secale cereale* (rye) and *Zea mays* (corn). The weed *Schkuria pinnata* carries large numbers of *P. hexincisus* (Honey, 1967).

In general, lesion nematodes are poor competitors when other nematode genera are present, as has been shown to be the case when either root-knot (Graham *et al.*, 1964; Johnson and Nusbaum, 1970) or cyst nematodes are present (Miller, 1970). They have, however, been incriminated in the increase of various other tobacco root diseases.

2. *Relationship of Lesion Nematodes to Other Tobacco Root Diseases*

(a) Black Shank, *Phytophthora parasitica* var. *nicotianae*. Inagaki and Powell (1969) inoculated tobacco plants with the black shank fungus alone and in combination with a population of the lesion nematode *P. brachyurus*. They found that when the nematodes were added either a week before or together with the fungus, there was a more rapid and severe development of black shank symptoms than when the fungus was used alone. They concluded that the main role of the nematodes was as wounding agents. However, neither *Pratylenchus* infection nor mechanical wounding of the roots reduced the inherent resistance of the varieties NC 2326 or NC 95 to black shank.

(b) Black Root Rot, *Thielaviopsis basicola*. In experiments in South Africa *P. hexincisus* slightly increased the development of black root rot on the cultivar Hicks, whereas this did not apply in the case of Delcrest (Honey, 1967). Various races of *P. penetrans* did not break down the resistance of Burley 49 to black root rot, the tobacco varieties susceptible to *T. basicola* being tolerant to *P. penetrans* and *vice versa* (Olthoff, 1968).

C. Other Nematodes

1. *Tylenchorhynchus claytoni* (Stunt Nematode)

The tobacco stunt nematode *Tylenchorhynchus claytoni* is a small ectoparasitic nematode which has been recorded as causing moderate stunting of tobacco in both the U.S.A. and Canada (Lucas, 1965). Wouts (1966) has recorded *T. capitatus* as occurring commonly in association with tobacco in New Zealand, though no indication of pathogenicity is given. Milne (1961) and Shepherd (1968) found a *Tylenchorhynchus* species commonly associated with tobacco in South Africa and Rhodesia. The life-cycle of *T. claytoni* takes approximately one month at favourable temperatures (30–34°C) and the nematode can survive in the soil for 10 months without a host. Krusberg (1959) found that these nematodes feed on the epidermal cells between the root hairs, and that tobacco, corn, small grains, potatoes and sweet potatoes (*Ipomoea batata*) increased the stunt nematode population. Holdeman (1956) found that plants of the *Fusarium*-wilt-susceptible variety Oxford 1-181, inoculated with both stunt nematodes and the *Fusarium* wilt fungus, were more severely affected by wilt disease than those to which the fungus alone was added. However, Lucas and Krusberg (1956) were unable to find any consistent relationship between the presence of stunt nematodes and the occurrence of Granville wilt (*Pseudomonas solanacearum*) of tobacco.

2. *Heterodera tabacum* (Tobacco Cyst Nematode)

H. tabacum, not to be confused with *H. marioni*, is a true cyst nematode (Lownsbery and Lownsbery, 1954), recorded in the U.S.A. on both shade-grown and flue-cured tobacco. The brownish cysts are visible on infested roots (Fig. 3) and an attack by this nematode leads to reduction of the root system with consequent stunting and wilting of the plant (Lucas, 1965). When this nematode occurs in combination

Fig. 3. Cysts of the tobacco cyst nematode, *Heterodera tabacum*, on tobacco root. (By courtesy of W. W. Osborne.)

with the stunt nematode or lesion nematodes, it acts competitively with the other species and decreases their populations (Miller and Wirheim, 1968). This is particularly the case with *Pratylenchus penetrans* which eventually disappears from a mixed population in which cyst nematodes are present (Miller, 1970).

In experiments on the effect of high temperatures on this nematode, Miller (1969a) found that heating soil to 32°C for 6 days killed the larvae of cyst nematodes, and heating at 42°C for 24 h killed the unhatched eggs. Miller (1969b) found also that applications of the systemic fungicides benomyl and thiabendazole suppressed invasion of tobacco roots by this nematode.

An indication of the potential economic importance of this nematode was shown by Osborne (1967) who found that fumigation of tobacco cyst nematode-infested soil led to a 57% increase in yield over the untreated checks.

3. *Trichodorus* spp. (Stubby-root Nematodes)

Although Meagher (1969) has found *T. lobatus* stunting tobacco roots in Australia, members of the genus *Trichodorus* have gained recognition in recent years as vectors of the tobacco rattle virus (TRV). The nematode is ectoparasitic, feeding at the root tips, thereby causing root growth to cease and leading to the "stubby root" appearance of such roots.

Ten species of *Trichodorus* have been recorded as vectors of the tobacco rattle virus. They are *T. pachydermus*, *T. primitivus*, *T. alii*, *T. christiei*, *T. allius*, *T. anemones*, *T. cylindricus*, *T. nanus*, *T. similis* and *T. teres* (Jensen and Allen, 1964; Van Hoof, 1964; Lucas, 1965; Ayala and Allen, 1966; Bjørnstad and Støen, 1967; Van Hoof, 1968).

TRV is known to occur in parts of Europe and in the U.S.A. where it causes also "corky ringspot" of potatoes. The symptoms of the disease on tobacco are similar to those caused by "Kroepoek" or "Kroesblaar", the tomato spotted wilt virus (TSV). The leaf edges turn reddish brown and the leaves are pocked or crinkly in appearance, with small necrotic lesions. Hence, where a source of the virus is present, *Trichodorus* could constitute a serious economic threat to tobacco if left uncontrolled. Owing to the limited number of sites at which both the virus and nematode occur together in tobacco areas, however, it must be regarded as a pest of minor importance to tobacco at present.

4. *Xiphinema americanum* (Dagger Nematode)

These large ectoparasitic nematodes are universally distributed and occur on a wide variety of host plants. Their main importance relates to the fact that they transmit tobacco ringspot virus (TobRV). McGuire (1964) showed that the nematodes acquire the virus within 24 h and that all developmental stages are capable of acting as vectors. They remain infective for a period of 49 weeks (Lucas, 1965). TobRV symptoms are expressed 13–25 days after transmission, typical symptoms consist of concentric necrotic lesions, some of which follow the veins of the leaves. Recently *X. coxi* was also shown to be a vector of the disease (Van Hoof, 1971). TobRV disease is almost world-wide in occurrence and for this reason dagger nematodes are potentially of considerable economic importance to tobacco growers.

5. *Ditylenchus dipsaci:* Cause of "Stem-break"

This nematode is the only one recorded as feeding above ground on tobacco. During wet weather the nematode enters the stem of the tobacco plant and causes the development of small yellow swellings. Later these galls rot, the plant stops growing and eventually the stem breaks and the plant falls over. At present this nematode disease is known only in Holland, France and Germany; the alfalfa strain of *D. dipsaci* apparently does not attack tobacco in the U.S.A. (Lucas, 1965).

6. Other Genera

In 1955, Graham found that both *Rotylenchus brachyurus* and a *Criconemoides* sp. cause stunting of tobacco, when present in sufficient numbers.

The following nematode species have been found associated with tobacco but are not recorded as being of economic importance: *Longidorus martini* in Rhodesia (Merry, 1966), *Paratylenchus* sp. in Canada (Olthoff, 1968), *Tetylenchus nicotianae* in Japan (Yokoo and Tanaka, 1954) and *Helicotylenchus* in the U.S.A. (Lucas, 1965).

III. Relationship of Farming Practices to the Occurrence and Spread of Nematodes

A. Virgin Lands and Fallow

A grower is taking a risk who assumes that he will have no eelworm problems because he is planting on virgin soil. *Meloidogyne* species are so universally distributed on such a wide range of grasses and trees that there are very few lands which are free of these nematodes. Similarly, it is an ineffective control measure to fallow land in order to decrease the numbers of root-knot nematodes, because the volunteer weeds and grasses are often good hosts. In Rhodesia *M. javanica* is still viable in tobacco soils after more than 4 years of *clean* fallow (Martin, 1967).

B. Irrigation

Root-knot nematodes are easily washed down irrigation canals, particularly where the soil is regularly moved in order to open such channels, or where livestock have access to the water. When these channels run through infested soil, they should be fumigated prior to planting. Where rivers are tapped for irrigation purposes it is advisable to use storage dams or reservoirs in which the nematodes can settle out. As *Meloidogyne* larvae can survive under water for three weeks, it is wise to pump irrigation water from the surface of such reservoirs.

C. Date of Planting

J. Shepherd (personal communication) found that in Rhodesia the date of planting can have a considerable effect on the degree of disease development caused by *Meloidogyne*. When planted in late summer, conditions were rather too hot and dry for the tobacco, but were extremely favourable for rapid development of *Meloidogyne*. The consequence was that in these late plantings the yield obtained was equivalent to only 37 kg of tobacco per hectare—a total loss of the crop. Tobacco planted two months earlier, when temperatures were lower and the rate of breeding of *Meloidogyne* greatly reduced, yielded 2743 kg of tobacco per hectare—a 7000% increase in yield. This indicates the importance of selecting the best planting season in areas where *Meloidogyne* is a problem.

D. Ploughing

1. Wet Ploughing

In some of the heavier soils used for tobacco planting, experience has shown that ploughing of these soils when very wet will cause a slight reduction in the eelworm population. No explanation for this phenomenon has yet been found.

2. Normal Ploughing

Ploughing out or pulling out of infested plants in sandy soils immediately after harvest significantly decreases the *M. incognita* population occurring in the soil the following season (Nusbaum, 1959). However, Milne (1963b) working with a loam soil infested with *M. javanica* found that pulling out the plants resulted in an infection equivalent to that in the plots where the plants were ploughed in immediately after harvest. While the beneficial effect of ploughing on sandy soils can be expected to be greater than that on heavy soils, repeated ploughing of fallow soil could be detrimental to soil structure.

IV. Nematode Control

A. Control in Seedbeds

In order to produce a healthy stand of tobacco there is no doubt that it is most important to commence with disease-free seedlings. Caution must be used when planting in virgin soils. The earlier practice of clearing indigenous pine tree lands before making seedbeds was effective, but this was probably because the pine trash was burned on the beds, thereby effectively heat-sterilizing the soil. As a result of the intensive production of plants in the seedbed area, a large expenditure per unit area is justified.

1. Chemical Control in the Seedbeds

In selecting a nematicide for use in tobacco seedbeds, the trend nowadays is to use an "all-rounder" treatment for the eradication of nematodes, weeds and harmful soil micro-organisms. However, if a nematicide only is required, ethylene dibromide (EDB) and 1,3 dichloropropene (Telone) or a mixture of dichloropropene and dichloropropane (DD) are most commonly recommended. These materials are normally injected into the soil with a manual soil injector (Fig. 4) at the rate of 130 ml/m² for DD and 65 ml/m² for 42% EDB.

Various commercial formulations of EDB may be obtained, varying from a 21% solution in white spirits to a 92% water miscible formulation, or even technical EDB. However, if the material is applied accurately, its efficacy is related to the quantity of active material applied and not to the formulation used. In all cases at least 2 weeks should elapse between treatment with EDB or DD and the sowing of tobacco seed. On heavier soils during wet weather a longer aeration period may be required, and it is wise to smell the soil for residues before sowing.

New materials which have proved effective when used in tobacco seedbeds include the highly toxic but effective nematicide "Temik" or "Aldicorb" ((2-methyl-2-methylthio) propionaldehyde O-(methylcarbamoyl)-oxime), the non-phytotoxic granular nematicide "Mocap" (O-ethyl S, S-dipropyl phosphorodithioate) and the new products "Nemacur" or BAY 68138 ((ethyl 4-methylthio)-*m*-tolyl isopropylphosphoramidate), "Lannate" (S-methyl N (methylcarbamoyl oxy) thioacetimidate), "Dasanit" (O, Diethyl O-(P-(methylsulfinyl) phenyl phosphorothioate) and "Penphene" (tetrachlorothiophene).

Methyl bromide is a highly toxic, odourless gas which is extremely suitable for use in tobacco seedbeds. It is applied under a gas-tight sheet which is sealed at the edges and is suspended about 10 mm above the soil surface (Dorsin, 1968). Two or three days prior to fumigation, the seedbed area should be well watered in order to induce weed-seed germination, as dry, dormant seeds are little affected by the gas. The soil tilth should be sufficiently fine to permit even distribution of the gas, and the soil or tarp should remain in place for 24–48 h. Where possible, fumigation on sloping soil should be carried out on the contour, as the heavy gas drains to the lower end of the area to be treated. Methyl bromide at 50 g/m² is highly effective against nematodes and weeds (including "nutgrass" *Cyperus rotundus*), and at 100 g/m² will also give control of black root rot (*Thielaviopsis basicola*), anthracnose (*Colletotrichum destructivum*) and black shank (*Phytophthora parasitica* var. *nicotianae*). Seed may be sown within 48 h of fumigation.

FIG. 4. Injector gun used for subsoil placement of fumigants such as EDB and DD.

Mixtures of DD or EDB with chloropicrin and propargyl bromide also are highly effective for use in seedbeds against nematodes, weeds and the black shank fungus. Chloropicrin is commonly used for seedbeds in the tobacco growing areas of Japan, where it is regarded as effective against nematodes and against damping-off soil organisms, especially in sandy soils (Tanaka and Tsumagari, 1954).

2. Heat Treatment of Seedbeds

Burning of pine wood trash on seedbeds effectively frees soil of nematodes. Similar results have recently been obtained by burning 24 kg of rice husks/m³ (Krishnamurthy and Elias, 1969) on tobacco seedbeds in India. A long period of heat with deep penetration is required, and the surface burning of petroleum products is therefore entirely useless.

Steam sterilization requires a high initial outlay for the boiler system, but if properly applied gives extremely good results against nematodes, weeds and soil fungi. A permanently installed system of perforated pipes laid 15–20 cm below the surface successfully steam-sterilizes seedbeds. A soil temperature of 180°F (81°C) must be maintained for at least 1 h, and the area to be treated is therefore covered with wooden or metal covers to retain the heat.

B. Control in the Lands

1. Chemical Control

The materials available for chemical treatment of tobacco lands are much the same as those used in the seedbeds, except that the more expensive materials do not usually justify field use.

(a) *DD* and *EDB*. DD (Vidden D) is used at the rate of 112–125 litres per hectare as an overall treatment, or 90–169 litres per hectare in row-placement. Similarly EDB is applied at the rate of 56-112 litres of 42% material (or 14–28 litres of active material) per hectare, for total land treatment. Row or planting-site treatment is usually recommended where sprinkle irrigation or natural rainfall are the main sources of water. Where furrow irrigation is practised, with its concomitant movement of soil, overall land treatment is generally justified. These two nematicides generally give effective reduction of the *Meloidogyne* population even though galling may develop on the plants late in the season. *Pratylenchus* is effectively controlled by both these materials although opinions have differed from time to time on their relative efficacy against this nematode group (Owens and Ellis, 1951; Nusbaum, 1955; Graham, 1962; Tappan and Kincaid, 1962; Milne, 1963b). DD is generally regarded as more effective at low temperatures although neither DD nor EDB should be used when soil temperatures are below 55°F (12°C).

A lapse of at least 2 weeks should be allowed between using these materials and the planting of tobacco transplants. Tobacco seedlings are very sensitive to these chemicals and planting on an inadequately aerated soil causes scorching of the roots with consequent yellowing, wilt, and often death of the plants. Smelling the soil is a sure way of establishing that no harmful residues remain.

After 25 years of use there is as yet no indication of *Meloidogyne* building up resistance to EDB or DD. One problem arising from the use of these materials, however, is the increase of halogens (bromine and chlorine) in the leaves of tobacco grown on treated soil. In soils naturally high in chlorine the continued application of DD over a period of 6 years has greatly increased the halogen content of leaves. EDB causes a less severe build-up of halogens than does DD (Milne, 1963a).

Early workers observed that plants grown in fumigated soil were often a darker green than those grown in untreated soil and research has since shown that fumigation with EDB or DD inhibits the nitrification of ammonium nitrogen by nitrifying bacteria, and that this sometimes aggravates chlorine toxicity in brackish soils (Kincaid and Volk, 1952; Elliot and Mountain, 1963; Lucas, 1965). Furthermore the consequent late release of nitrogen may be detrimental to the flue-cured tobacco crop.

Heavier tobacco types, such as dark air-cured tobacco, can be very much more heavily fertilized than flue-cured tobacco, and this often decreases the effects of nematode attack. The detrimental effect of high nitrogen levels on the quality of flue-cured tobacco, however, makes it impractical to use heavy nitrogen applications to reduce the effects of nematode attack.

In hot, dry, sandy soils the depth of placement of the fumigant is of great importance in obtaining effective protection of the crop, as shallow placement leads to rapid loss of the fumigant from the surface. It is necessary under such conditions to place the fumigant at 23–38 cm in order to obtain effective control (Daulton, 1967). Post-treatment rolling of sandy soils, where possible, increases the efficacy of the treatment.

(b) *Other Chemicals.* As mentioned under the section on seedbeds, the trend nowadays is towards multipurpose chemical mixtures which will suppress weeds, nematodes, and other soil pathogens. Materials of this kind which have shown promise for use on tobacco include mixtures of methyl bromide with chloropicrin and a petroleum solvent, and materials such as SMDC (Na, N-methyl dithiocarbamate) and DMTT

(3,5 dimethyl tetrahydro 1,3,5 2H-thiadiazine 2 thione) both of which have a suppressing effect on *Fusarium* wilt (Good, 1964). Others in this category which combine various known materials include "Telone PBC" (a mixture of dichloropropene, chloropicrin and propargyl bromide), "Vorlex" (a mixture of methyl isothiocyanate and chlorinated C_3 hydrocarbons), "Vorlex 201" ("Vorlex" plus chloropicrin), "Terrocide" (EDB plus chloropicrin) and "Experimental Compound 2680" (dichloropropene and chloropicrin). Most of these products are effective not only against nematodes but also against black shank and bacterial wilt (Todd, 1966; Nusbaum and Todd, 1970).

Other general nematicides which have shown promise and some of which are already in use on tobacco are "Mocap", "Penphene", "Dasanit", "TD 1287" (tetrachlorothiophene plus 1,4 dichloro-2-butyne) and Bay 68138 or "Nemacur".

Two systemic materials have recently been introduced; these include the very toxic material "Aldicarb" or "Temik" which is effective not only against root-knot and root lesion nematodes but also against leaf miner and aphids on tobacco (Legge and Shepherd, 1967). It sometimes causes a slight delay in the maturation of tobacco (Brodie and Dukes, 1966). The other new systemic nematicide is "D 1410" (*S*-methyl 1-(dimethyl-carbamoyl)-N (methylcarbamoyl) oxy thioformimidate). This material can be applied as a foliar spray or as a soil drench, and is translocated downwards from the leaves to the roots where it prevents invasion by *Meloidogyne* and *Pratylenchus*, though there has been some marginal leafburn associated with foliar applications (Radewald *et al.*, 1970).

It is interesting to note that the systemic fungicides benomyl and thiabendazole have a suppressing effect on the invasion of roots by larvae of *Heterodera tabacum* (Miller, 1969b). Another material with interesting systemic side-effects is maleic hydrazide (MH 30), the desuckering chemical sometimes used on tobacco. Nusbaum (1958) first noticed that foliar applications of this material reduced *Meloidogyne* multiplication in the roots and this was later confirmed by Peacock (1963) and Davide and Triantaphyllou (1968), who found that maleic hydrazide reduced the rate of development and sex ratio of both *M. incognita* and *M. javanica*.

A group of poisons not yet mentioned includes the toxins produced by many *Pseudomonas* strains, and which have been found to be highly effective in suppressing saprophytic nematodes (Iizuka *et al.*, 1962) and which might therefore lead to new developments in this field.

In conclusion it may be stated that the high level of efficacy and wide range of organisms controlled, make the future use of chemicals in tobacco a highly economic proposition (see Section V; Fig. 5).

2. Rotation

The selection of a suitable rotation for tobacco is greatly complicated by the fact that there are so many species of nematodes which attack the crop. The difference in host ranges of the *Meloidogyne* species and the various races of these species further complicates the issue. What is more, tobacco is susceptible to a wide range of fungus and bacterial

FIG. 5. Fumigation makes the difference! The row of tobacco plants on the left was fumigated with EDB. The row on the right was untreated and was completely ruined by *Meloidogyne* attack.

soil diseases which may also be increased by alternative crops. The high value of the tobacco crop and ever decreasing availability of arable soil therefore make it imperative that one should give careful consideration to the economics of using crop rotations with tobacco. It is now seldom feasible to move on to virgin lands each year, or to fallow lands for long periods. Also, the availability of suitable chemicals makes the continued production of tobacco on one piece of land an economic proposition.

The question of clean fallow or weed fallow has already been considered, and in general these systems cannot be recommended unless the volunteer weeds are known to be resistant to nematodes and there is almost limitless availability of arable land.

Crops to be avoided in tobacco rotations include the majority of vegetables, especially tomatoes, other solanaceous plants and the cucurbits, all of which are highly susceptible to *Meloidogyne*. Where flue-cured tobacco is to be grown, leguminous plants which add excess nitrogen to the soil should be avoided. Timothy, rye and, to a certain extent, corn cause a build-up of large populations of *Pratylenchus* and release toxins in the soil when decomposing, thereby leading to two causes of "brown root rot" of tobacco (Lucas, 1965).

In South Carolina, tobacco grown after cotton (*Gossypium hirsutum*) is injured by *Pratylenchus* (Good, 1968) and in North Carolina cotton is generally regarded as fairly resistant to *M. incognita*. However, Sasser and Nusbaum (1955) have shown that there is more than one race of *M. incognita*, one race attacking cotton only, and another race attacking both cotton and tobacco. In South Africa inoculation experiments have shown that both *M. incognita* and *M. javanica* populations will breed very slowly on cotton and that these populations will attack tobacco. In India, however, *Gossypium hirsutum* is listed as resistant to *M. javanica* (Chadwani and Reddy, 1967) and *G. barbadense* is resistant to *M. incognita* (Krishnamurthy and Elias, 1967).

Corn (*Zea mays*) is often used successfully in rotations with tobacco in spite of the fact that it will generally maintain *Meloidogyne* populations and cause a build-up of both *Pratylenchus* and *Tylenchorhynchus*. The possibility of developing nematode-resistant corn varieties for use in tobacco rotations is worthy of further investigation.

Although groundnuts are susceptible to attack by *M. hapla* and *M. arenaria*, they are generally regarded as resistant to *M. incognita* and *M. javanica* and are therefore often recommended in rotations where these two species predominate. Recently, however, *M. javanica* races have been found which attack groundnuts in both Rhodesia and Georgia (Minton *et al.*, 1969). Groundnuts suppress populations of both *Pratylenchus brachyurus* and *Xiphinema americanum* (Good, 1968) and should therefore still serve a useful function in tobacco rotations.

Fescue (*Festuca elatior*) is very useful in tobacco rotations, causing a general reduction in the *Meloidogyne* problem, reducing the incidence of bacterial wilt considerably and also reducing black shank damage (Lucas, 1965; Nusbaum and Todd, 1970). When fescue rotations are used in conjunction with broad spectrum fumigant mixtures such as "Vorlex 201" and nematode resistant varieties such as NC 95, the reduction in bacterial wilt, black shank and nematodes is such that crop yields are increased considerably (Nusbaum and Todd, 1970).

The use of pasture grasses in rotation with tobacco has been investigated extensively by van der Linde *et al.* (1959) in South Africa and by Daulton (1963) and Shepherd (1968) in Rhodesia. In general they found that the Ermelo strain of *Eragrostis curvula* was highly resistant to *M. incognita* and *M. javanica* and was effective in reducing root-knot populations when grown for 3–4 years. Milne (1963b) found a significant reduction of *M. javanica* even after only one year of well-established *E. curvula*. The spiral nematode, *Scutellonema brachyurum*, is capable of building up considerable populations on *E. curvula* and these populations are maintained by the subsequent tobacco crop without causing obvious damage (Shepherd, 1968). *Chloris gayana* var. Katambora and *Panicum maximum* var. Sabi cause effective reduction of the *Meloidogyne* population when grown continuously for 4 years (Daulton, 1963; Shepherd, 1968), whereas *Chloris gayana* var. Giant and *Setaria sphacelata* var. Kazangula were ineffective. *Chloris gayana* var. Giant was also found to support a large population of *Rotylenchulus parvus* which slowly declined under tobacco. *Cenchrus ciliaris* has been shown to be highly resistant to *M. incognita*, though very lightly attacked by *M. javanica* (van der Linde *et al.*, 1959). The considerable susceptibility of many grasses to *Meloidogyne* species makes their use in rotations one which should be applied with caution.

The French Marigolds *Tagetes patula* and *T. erecta* exude toxins into the soil which suppress *M. javanica* and *Pratylenchus penetrans* populations (Daulton and Curtis, 1964; Krishnamurthy and Elias, 1967; Miller and Ahrens, 1969), although Miller and Ahrens (1969) found that *T. patula* will maintain a high population of *Tylenchorhynchus claytoni*. *Tagetes minuta*, the "Mexican Marigold" or "Khakibos" weed suppresses *Meloidogyne* less effectively than does *T. erecta*.

Various *Crotalaria* or sun-hemp species are used in tobacco rotations, although usually they are not used directly before a flue-cured crop because of the accumulation of nitrogen in the soil (Fig. 6). In South Africa, *C. juncea* is superseded by a winter oat crop before tobacco. Both *C. juncea* and *C. spectabilis* are effective in reducing *Meloidogyne* populations but a *Pratylenchus* species builds up on *C. juncea* and *C. spectabilis* is susceptible to *Fusarium* wilt. Good (1968) found that *C. spectabilis* built up a high population of *Pratylenchus brachyurus*, but at the same time reduced the *Xiphinema americanum* population. In Madagascar, de Guiran (1965) found both the annual *C. fulva* and the perennial *C. grahamiana* effective in reducing *Meloidogyne* in tobacco soils.

It will be realized that there is no single solution to the problem of finding a suitable rotation for use with tobacco, as species, crops and climate vary from area to area. It is, however, important that where suitable knowledge is available, the best rotation system should be selected on the basis of long term disease reduction, soil improvement and maximum economic production per unit area. As plant breeders produce more and more disease- and nematode-resistant varieties, so the scope of including other high value crops in rotation with tobacco will be extended.

FIG. 6. Ploughing in of sunn-hemp (*Crotalaria juncea*). This crop is used in rotation with tobacco in order to reduce the *Meloidogyne* population in the soil.

3. Breeding

The search for a commercially acceptable tobacco variety resistant to *M. incognita* ended in 1962 with release of the flue-cured variety NC 95 (Moore *et al.*, 1962). This was the culmination of more than 25 years of research on the original TI 706 introduced from Honduras, and was the result of the continued efforts of many workers (Clayton *et al.*, 1958; Drolsom and Moore, 1958; Graham, 1961; Graham *et al.*, 1961). The major breakthrough in this search for resistance was that of Clayton and his colleagues who succeeded in breaking the linkage of small leaves with resistance in TI 706 by crossing it with the amphi-diploid 4n (*N. sylvestris* × *N. tomentosiformis*).

NC 95 has the added advantage that it carries resistance to black shank, *Fusarium* wilt, and to a certain extent, bacterial wilt and brown spot. The result is that when used in rotation with fescue and after soil treatment with "Vorlex", dramatic reductions in nematode and bacterial wilt damage occur. For example, when the wilt-resistant variety SG 36 was planted after tobacco, without chemical treatment, the *Meloidogyne* index was 57% and the bacterial wilt index was 69%. However, when NC 95 was used, after fescue and including a "Vorlex" treatment, the *Meloidogyne* index was 0%, the bacterial wilt index 3%, and the crop value was nearly trebled (Nusbaum and Todd, 1970) (Table I).

The mechanism of resistance of NC 95 to *M. incognita* apparently depends on a hypersensitive host root reaction (Powell, 1962; Milne, 1966).

Unfortunately NC 95 is not resistant to *M. javanica* or *M. arenaria* (Lea and Willetts, 1963; Graham, 1964b; Milne, 1966; Miller, 1968), and where *M. javanica* is present, black shank is a problem on this variety (Miller, 1968). Although it is susceptible to *Pratylenchus* species, Graham *et al.* (1964) found that *Pratylenchus* breeds more slowly on NC 95 than it does on Hicks. Recently in North Carolina a new race of *M. incognita* was found which attacks and breeds on NC 95 (Graham, 1969). This makes it all the more imperative that rotation and fumigation should still be practised when introducing a new root-knot resistant variety, in order to reduce the chances of early development of pathogenic races of the nematode.

Other tobacco varieties resistant to *M. incognita* include Florida 22 which is, however, susceptible to black shank and bacterial wilt (Clark, 1961), and the Japanese variety RK 70 on which *M. incognita* breeds very slowly (Tsumagari and Tanaka, 1954b).

There is no commercial variety of tobacco which is resistant to *M. javanica*. Various workers have carried out tests on a number of *Nicotiana* species in a search for *M. javanica* resistance (Chapman, 1957; Burk and Dropkin, 1961; Calitz and Milne, 1962; Schweppenhauser *et al.*, 1963). In all instances the most significant resistance was found in the species *N. repanda* and the amphidiploid hybrid 4*n* (*N. repanda* × *N. sylvestris*), and in *N. longiflora*. Investigations by Milne *et al.* (1965) showed that *N. repanda* resistance to *M. javanica* was also dependent on hypersensitive reactions within the root, this reaction apparently being related to the greater content of chlorogenic acid, and its oxidase, together with a quinone-inactivating compound in the roots of *N. repanda*. Although crosses of *N. repanda* with tobacco are frequently sterile, and this has slowed down the breeding programme with this species, in South Africa C. J. Nuss (personal communication) is close to successful incorporation of *N. repanda* resistance in tobacco.

TABLE I. Increases in tobacco production due to various treatments

Reference	"Treatment"	Yield in kg per hectare "untreated"	"treated"	Percentage increase in yield
Brodie and Dukes (1966)	Fumigation	749	1576	110%
Brodie and Dukes (1967)	Fumigation	614	2555	300%
Daulton (1965)	Fumigation	487	1573	223%
	Rotation	487	1598	229%
	Rotation and fumigation	487	2103	330%
Graham (1964a)	Fumigation	1430	1671	17%
Milne (1962)	Fumigation	555	1445	160%
	Fumigation	566	907	60%
Tappan and Kincaid (1966)	Fumigation (*Pratylenchus* only)	1607	1915	27%

Reference	"Treatment"	Crop value per hectare (dollars)		Percentage increase in value
Nusbaum and Todd (1970)	Fumigation	2103	3227	53%
	Fumigation	1522	2513	65%
	Rotation	1522	3420	123%
	Rotation and fumigation	1522	3645	139%
	Rotation and fumigation + resistant cultiver	1522	4070	167%

Another source of resistance to *M. javanica* comes from *N. tabacum* selections made in Rhodesia. Varieties such as R83 and R388 show the hypersensitive reaction to attack by *M. javanica* (Milne, 1966) and although these are far from being commercial varieties, there are considerable possibilities of using this source of resistance in future breeding programmes (Schweppenhauser, 1968).

As has been stated previously, NC 95 is not resistant to *Pratylenchus* species, but the shade-grown variety "Florida 15" is tolerant to *P. penetrans* and *P. brachyurus* and could therefore play a useful role in breeding programmes (Tappan *et al.*, 1967). Southards (1966) has shown that lines of flue-cured tobacco can be selected which are tolerant to attack by *P. brachyurus*.

It seems probable that there is a chance of finding other sources of nematode resistance in *N. tabacum* which are as yet untapped, particularly in large areas of commercially grown tobacco. These sources, together with better utilization of wild species, could give growers great hopes for the future, and undoubtedly the investment of funds in a breeding programme is an economically sound one.

V. The Role of Nematodes in the Economics of Tobacco Production

The total world tobacco production is now in the vicinity of 4·5 million metric tons per annum, and even when valued at a conservative 55 cents per kg, this is equivalent to a total value of $4 billion. In addition, tobacco is one of the most heavily taxed agricultural products in the world, so that for example in the U.S.A. in 1965 the tobacco crop was worth $1 billion to the farmer and $3 billion in taxes (Lucas, 1965). The total world value of the crop, including taxes, must therefore be well above $10 billion per annum.

What inroads are made by nematodes on this figure? Analysis of many fumigation experiments has shown that if one commences with a yield of 2242 kg per hectare, there will be a loss of approximately 45 kg in yield for each 10% increase in the root-knot nematode infestation. In Rhodesia in 1963, total losses due to nematode attack were estimated at $10 million, yields being reduced in some areas by as much as 1121 kg/ha (Daulton, 1963). In North Carolina, losses due to nematode attack have been gradually reduced from a fantastic $24 million in 1955 to $6 million in 1964 and $4·6 million in 1965 (Lucas, 1965; Todd, 1966). The major part of these losses due to nematode attack can be ascribed to *Meloidogyne* species, although *Pratylenchus* species were held responsible for losses of $5 million in 1956 in North Carolina alone (Lucas, 1965).

The Food and Agriculture Organization (FAO) of the United Nations has estimated that in spite of intensive research and extension, tobacco crop losses due to nematodes in the U.S.A. average 4% per annum (Anon., 1968); the world average loss can be expected to be nearer 10%, and must therefore total approximately $400 million per annum. But this is only the *direct* loss; what of the *indirect* losses?

Taxes on tobacco are unusually high, approaching or exceeding the actual value of the crop itself. It therefore follows that for every dollar's worth of crop loss caused by nematodes there is at least a dollar's worth of indirect loss in taxes. Other indirect losses of importance resulting from nematode attack include the following: losses due to secondary attack of tobacco by both root and leaf pathogens; losses due to the increase of weed control problems because of lack of competition by tobacco roots; loss of usefulness of varieties resistant to other diseases in the absence of nematodes; losses due to non-utilization of fertilizer and water applied; losses due to the expense involved in fumigation; reduced income while using certain rotation crops; and loss of export markets due to the resulting higher costs of production.

The justification for trying to eradicate nematodes is clear enough from the above facts, but how economical are the control measures at present available? The effects of various recommended treatments on the increase in yield and value of the crop are shown in Table I. Yield increases will depend on the nematode population prior to fumigation, but, as can be seen in the table, combined treatments can lead to increases of up to 330%. Not only have there been these tremendous yield increases, but by better utilization of soil nutrients, many farmers have been able to reduce fertilizer use by as much as 20% when nematodes were controlled by fumigation (Good, 1968).

The cost of chemical treatment is as low as $45 per hectare (Todd, 1966) though the new multi-purpose chemicals may cost somewhat more. Daulton (1962) estimates that in Rhodesia, fumigation costs are covered by a yield increase of 39 kg of cured leaf per hectare, whereas the increases expected from fumigation vary from 225 to 1120 kg/ha— ample return for the small investment.

From the above evidence it is clear that if fumigation, rotation and the use of nematode-resistant varieties are applied effectively, nematode control on tobacco will continue to be a highly economic proposition.

References

Anon. (1968). *F.A.O. Pl. Prot. Bull.* **16**, 37–40.
Ayala, A. and Allen, M. W. (1966). *Nematologica* **12**, 87.
Bjørnstad, A. and Støen, M. (1967). *Norsk Landbr.* **8**, 12–14.

G

Brodie, B. B. and Dukes, P. D. (1966). *Fungicide–Nematocide Tests* **22**, 118.

Brodie, B. B. and Dukes, P. D. (1967). *Fungicide–Nematocide Tests* **23**, 135.

Burk, L. G. and Dropkin, V. H. (1961). *Pl. Dis. Reptr* **45**, 734–735.

Calitz, P. C. and Milne, D. L. (1962). *S. Afr. J. agric. Sci.* **5**, 123–126.

Chadwani, G. H. and Reddy, T. S. N. (1967). *Indian Phytopath.* **20**, 383–384.

Chapman, R. A. (1957). *Phytopathology* **47**, 5.

Clark, F. (1961). *Circ. Fla. agric. Exp. Stn* **S-134**, 12 pp.

Clayton, E. E., Graham, T. W., Todd, F. A., Gaines, J. G. and Clark, F. A. (1958). *Tobacco Science* **2**, 53–63.

Corbett, D. C. M. (1970). *Nematologica* **15**, 550–556.

Daulton, R. A. C. (1962). *Rhod. J. agric. Res.* **59**, 216–217.

Daulton, R. A. C. (1963). *Rhod. J. agric. Res.* **60**, 150–152.

Daulton, R. A. C. (1965). *Down to Earth* **20**, 2–32.

Daulton, R. A. C. (1967). *Down to Earth* **22**, 20–21.

Daulton, R. A. C. and Curtis, R. F. (1964). *Nematologica* **9**, 357–362.

Daulton, R. A. C. and Nusbaum, C. J. (1961). *Nematologica* **6**, 280–294.

Davide, R. G. and Triantaphyllou, A. C. (1968). *Nematologica* **14**, 37–46.

de Guiran, G. (1965). *Trav. Congr. Protect. Cult. Trop. Marseilles* 681–685.

Dorsin, U. G. (1968). *Z. Pflanzenkr.* **75**, 665–673.

Drolsom, P. N. and Moore, E. L. (1958). *Pl. Dis. Reptr* **42**, 596–598.

Elliot, J. M. and Mountain, W. B. (1963). *Can. J. Soil Sci.* **43**, 18–26.

Garner, W. W. (1951). *In* "The Production of Tobacco". Blakiston, New York.

Good, J. M. (1964). *Pl. Dis. Reptr* **48**, 199–203.

Good, J. M. (1968). *In* "Tropical Nematology" (G. C. Smart and V. G. Perry, eds), pp. 113–138. University of Florida Press, Gainesville, U.S.A.

Graham, T. W. (1951). *S. Carolina agric. Exp. Stn Bull.* **390**, 25 pp.

Graham, T. W. (1955). *Phytopathology* **45**, 347.

Graham, T. W. (1961). *Pl. Dis. Reptr.* **45**, 692–695.

Graham, T. W. (1962). *Fungicide–Nematocide Tests* **18**, 84–85.

Graham, T. W. (1964a). *Fungicide–Nematocide Tests* **20**, 111.

Graham, T. W. (1964b). *Phytopathology* **54**, 623.

Graham, T. W. (1969). *Phytopathology* **59**, 114.

Graham, T. W. and Heggestad, H. E. (1959). *Tobacco Sci.* **5**, 172–178.

Graham, T. W., Chaplin, J. F. and Ford, Z. T. (1961). *Tobacco Sci.* **7**, 107–111.

Graham, T. W., Ford, Z. T. and Currin, R. E. (1964). *Phytopathology* **54**, 205–210.

Hartill, W. F. T. (1968). *Rhod. J. agric. Res.* **6**, 77–79.

Holdeman, Q. L. (1956). *Phytopathology* **46**, 129.

Honey, A. de S. (1967). Studies on the genus *Pratylenchus* and a preliminary assessment of its importance to tobacco in South Africa. M.Sc. thesis (unsubmitted).

Hoof, H. A. Van (1964). *Nematologica* **10**, 141–144.

Hoof, H. A. Van (1968). *Nematologica* **14**, 15–16.

Hoof, H. A. Van (1971). *Neth. J. Pl. Path.* **77**, 30–31.

Inagaki, H. and Powell, N. T. (1969). *Phytopathology* **59**, 1350–1355.

Iizuka, H., Komagata, K., Kawamura, T., Kunii, Y. and Shibuya, M. (1962). *Agric. Biol. Chem.* **26**, 199–200.

Jensen, H. J. and Allen, Jr. T. C. (1964). *Pl. Dis. Reptr* **48**, 333–334.

Johnson, A. W. and Nusbaum, C. J. (1970). *J. Nematol.* **2**, 334–340.

Johnson, H. A. and Powell, N. T. (1969). *Phytopathology* **59**, 486–491.

Kincaid, R. R. and Volk, G. M. (1952). *Univ. Fla. agric. Exp. Stn Bull.* **490**, 24 pp.

Krishnamurthy, G. V. G. and Elias, N. A. (1967). *Indian Phytopath.* **20**, 374–377.

Krishnamurthy, G. V. G. and Elias, N. A. (1969). *Indian J. agric. Sci.* **39**, 263–265, in *Trop. Abstr.* **25**, 201 (1960).

Krusberg, L. R. (1959). *Nematologica* **4**, 187–197.

Lea, H. W. and Willetts, H. J. (1963). The development of a nematode resistant tobacco variety for Australian conditions. I. An evaluation of the T.I. 706 source of root-knot resistance in New South Wales. *Proc. 3rd World Tob. Sci. Congr.* Salisbury, 230.

Legge, J. B. B. and Shepherd, J. A. (1967). *Rhodesia Zambia Malawi J. agric. Res.* **5**.

Lownsbery, B. F. and Lownsbery, J. W. (1954). *Proc. helminth. Soc. Wash.* **21**, 42–47.

Lucas, G. B. (1965). "Diseases of Tobacco." Scarecrow Press, New York.

Lucas, G. B. and Krusberg, L. T. (1956). *Pl. Dis. Reptr* **40**, 150–152.

Lucas, G. B., Sasser, J. N. and Kelman, A. (1955). *Phytopathology* **55**, 537–540.

Martin, G. C. (1967). *Rhod. J. Agric. Res.* **64**, 112–114.

Mayol, P. S. and Bergeson, G. B. (1970). *J. Nematol.* **2**, 80–83.

McGuire, J. M. (1964). *Phytopathology* **54**, 799–801.

Meagher, J. W. (1969). *Proc. First Int. Citrus Symp.* **2**, 999–1006.

Melendéz, P. L. and Powell, N. T. (1967). *Phytopathology* **57**, 286–292.

Melendéz, P. L. and Powell, N. T. (1969). *Phytopathology* **59**, 1348.

Merry, G. (1966). *Nematologica* **12**, 385–395.

Miller, C. R. (1968). *Phytopathology* **58**, 553.

Miller, P. M. (1969a). *Pl. Dis. Reptr* **53**, 191–193.

Miller, P. M. (1969b). *Phytopathology* **59**, 1040–1041.

Miller, P. M. (1970). *Pl. Dis. Reptr* **54**, 25–26.

Miller, P. M. and Ahrens, J. F. (1969). *Pl. Dis. Reptr* **53**, 642–646.

Miller, P. M. and Wirheim, S. E. (1968). *Pl. Dis. Reptr* **52**, 57–58.

Miller, C. R., Mullin, R. S. and Whitty, E. B. (1968). *Pl. Dis. Reptr* **52**, 984.

Milne, D. L. (1961). *S. Afr. J. agric. Sci.* **4**, 217–223.

Milne, D. L. (1962). *S. Afr. J. agric. Sci.* **5**, 305–313.

Milne, D. L. (1963a). *Fmg S. Afr.* **38**, 36–37.

Milne, D. L. (1963b) *Fmg S. Afr.* **39**, 59-60.

Milne, D. L. (1966). *S. Afr. J. agric. Sci.* **9**, 422–435.

Milne, D. L. and du Plessis, D. P. (1964). *S. Afr. J. agric. Sci.* **7**, 673–680.

Milne, D. L., Boshoff, D. N. and Buchan, P. W. W. (1965). *S. Afr. J. agric. Sci.* **8**, 557–570.

Minton, N. A., McGill, J. F. and Golden, M. (1969). *Pl. Dis. Reptr* **53**, 668.
Moore, E. L., Drolsom, P. N. and Clayton, E. E. (1955). *Phytopathology* **45**, 349.
Moore, E. L., Gwynn, G. T., Powell, N. T. and Jones, G. L. (1962). *Research and Farming* **21**, 9.
Mountain, W. B. (1954). *Can. J. Bot.* **32**, 737–759.
Nusbaum, C. J. (1955). *Phytopathology* **45**, 349.
Nusbaum, C. J. (1958). *Phytopathology* **48**, 344.
Nusbaum, C. J. (1959). *Phytopathology* **49**, 547–548.
Nusbaum, C. J. and Todd, F. A. (1970). *Phytopathology* **60**, 7–12.
Olthof, T. H. A. (1968). *Nematologica* **14**, 482–488.
Osborne, W. W. (1967). *Fungicide–Nematocide Tests* **23**, 138.
Owens, R. G. and Ellis, D. E. (1951). *Phytopathology* **41**, 123–126.
Peacock, F. C. (1963). *Nematologica* **9**, 581–583.
Porter, D. M. and Powell, N. T. (1967). *Phytopathology* **57**, 282–285.
Powell, N. T. (1962). *Phytopathology* **52**, 25.
Powell, N. T. and Batten, C. K. (1967). *Phytopathology* **57**, 826.
Powell, N. T. and Batten, C. K. (1969). *Phytopathology* **59**, 1044.
Radewald, J. D., Shibuya, F., Nelson, J. and Bivens, J. (1970). *Pl. Dis. Reptr* **54**, 187–190.
Sasser, J. N. and Nusbaum, C. J. (1955). *Phytopathology* **45**, 540–545.
Sasser, J. N., Lucas, G. B. and Powers, H. R. Jr. (1955). *Phytopathology* **45**, 459–461.
Schweppenhauser, M. A. (1968). *CORESTA Info. Bull.* **1**, 9–20.
Schweppenhauser, M. A., Raeber, J. G. and Daulton, R. A. C. (1963). *Proc. 3rd World Tob. Sci. Congr. Salisbury*, 222–229.
Shepherd, J. A. (1968). *Rhod. J. agric. Res.* **6**, 19–26.
Southards, C. J. (1966). *Diss. Abstr.* **26**, 4164–4165.
Tanaka, I. and Tsumagari, H. (1954). *Bull. Kagoshima Tob. Exp. Stn* **9**, 1–16.
Tappan, W. B. and Kincaid, R. R. (1962). *Fungicide–Nematocide Tests* **18**, 85–86.
Tappan, W. B. and Kincaid, R. R. (1966). *Fungicide–Nematocide Tests* **22**, 119.
Tappan, W. B., Kincaid, R. R. and Smart, G. C. (1967). *Fungicide–Nematocide Tests* **23**, 142–143.
Todd, F. A. (1966). Where to with nematocides in tobacco? *Mimeo report, N.C. State University*, 3 pp.
Tsumagari, H. and Tanaka, I. (1954a). *Bull. Kagoshima Tob. Exp. Stn* **9**, 26–40.
Tsumagari, H. and Tanaka, I. (1954b). *Bull. Kagoshima Tob. Exp. Stn* **9**, 40–57.
van der Linde, W. J., Clemitson, Jean G. and Crous, Martha E. (1959). *S. Afr. Agric. T.S. Sci. Bull.* **385**.
Wouts, W. M. (1966). *N.Z. Jl Sci.* **9**, 878–881.
Yokoo, T. and Tanaka, I. (1954). *Bull. Kagoshima Tob. Exp. Stn* **9**, 59.

9
Nematode Diseases of Cotton

J. N. Sasser

Department of Plant Pathology
North Carolina State University
Raleigh, North Carolina, U.S.A.

I. Production and Importance of Cotton as a World Fiber Crop

Cotton, *Gossypium* spp., is a major crop in world agriculture. It is the most important plant fiber and is used by about 75% of the world's population to make clothing. Cotton is made into more diverse products than any other fiber. The U.S.A. alone uses over 1,814,360,000 kg each year, amounting to about 9 kg for each person.

World production of cotton has increased from about 12 million bales in 1891 to over 51 million bales in 1970.* The U.S.A. is by far the largest producer, and in 1970 produced about 10 million bales. The next largest production area is U.S.S.R. with $8\frac{3}{4}$ million and China

---

* Includes estimates for minor-producing countries also and allowances for countries where data are not yet available.

about $6\frac{1}{2}$ million. Twenty-three other countries produce cotton, each in excess of 150,000 bales, throughout the world in tropical and sub-tropical regions.* Before the Civil War in the U.S.A., cotton ranked as the most important crop of the South, and continues today as a major agricultural commodity.

A. General Uses of Cotton

The most important product from cotton is the lint or fiber used for many kinds of cotton textiles including clothing and other wearing apparel. It has strength and toughness adequate for work clothes, and also may be spun into fine cloth for dressy gowns.

Cottonseed oil is used as the base of many food products, such as margarine, salad oil, shortening and substitute ice-cream. Cotton meal, which remains after the oil is removed from the seed, contains a high percentage of protein and is a valued food for some farm animals. The hull, or outer covering of the cottonseed, is used for feed, fertilizers and chemicals.

The linters or short fuzz on the seeds are used in making lacquers, varnishes and adhesives. Many industries use chemically treated linters as raw material for such products as plastics, photographic film, paper, and even rayon, a competing fiber. Explosives manufacturers use linters in guncotton. The stalks and leaves of the cotton plant after harvest are incorporated into the soil as organic matter.

B. Kinds of Cotton

There are four general types of cotton: 1. American Upland, 2. Egyptian, 3. sea-island, and 4. Asiatic. All have similar plant growth habits but differ in such characteristics as shape and size of bolls, details of flower structure and fiber properties. Each type has several varieties characterized by fiber length, fiber strength or adaptability to irrigated land, mechanical harvesting or to certain climates.

In cool temperate regions cotton is planted annually but in the hot, moist tropical regions cotton plants may live and bloom for several years and grow to as high as 3 m or more.

American Upland cotton is cultivated in almost every producing country. This hardy plant produces a large yield under various growing conditions. Upland cotton makes up almost two-thirds of the world's cotton crop. The Upland cotton plant may grow from 30 cm to 2 m tall. It has creamy-white flowers, and strong, white fibers 2·3–3·2 cm long.

* The Cotton Situation, January 1971, ERS, USDA.

The origin of Upland cotton is unknown; its nearest cultivated relatives today are found in Mexico, Guatemala, and the Bahama Islands.

Egyptian cotton is probably native to South America, but the Egyptians first cultivated it commercially. Its unusually long fibers, 3·8 cm, are suitable for fine fabrics. Growers in the U.S.A. and Peru have developed several varieties called American-Egyptian cottons. An example is Pima variety, which thrives in Arizona, home of the Pima Indians. Its fibers have a uniform length of about 3·8 cm. Characteristics of Egyptian cotton are lemon-colored flowers and long, silky, light-tan fibers. It is especially suitable for thread, dresses, shirts, balloon cloth, and typewriter ribbons.

Sea-Island cotton first grew on the Sea Islands off the coast of South Carolina, Georgia, and northern Florida. This type was once grown in the U.S.A. but its culture was abandoned because of the boll weevil. Today, most sea-island cotton is produced in the West Indies in the Lesser Antilles, where the boll weevil does not occur. Sea-island cotton is one of the most valuable and costly cottons. Its long, silky fibers (4·5 cm) are used to make high-quality textiles. However, its poor yield, small bolls and slow growth make it expensive to produce. The plant has brilliant yellow flowers and white lint. Sea-island is closely related to Egyptian cotton, but growers consider it a separate type, because of its commercial importance.

Asiatic cottons grow in China, India, and Pakistan; they are characterized by short, coarse, harsh fibers and produce low yields. Principal uses are for blankets, padding, filters, and coarse fabrics. Varieties of American Upland cotton are rapidly replacing the Asiatic cottons, because of their superior fibers.

C. Prospects for Cotton in the Future

Cotton yields fluctuated during the years 1866–1934, with little tendency to increase. Since 1934, yields have been increased greatly by improved varieties and chemicals. These changes in production practices have tended to concentrate cotton production on large tracts of relatively flat land. Government programs, especially in the U.S.A., have sought to control or support prices by restricting the acreage. This attempt to control production has rather resulted in a steady increase in overall production (Lewis and Richmond, 1966).

Although outstanding improvements have been made in product quality and efficiency of production, the American cotton industry is faced with competition from other countries and with synthetic fibers. Despite this, cotton still has an important role to play in the economical and sociological future of the U.S.A. and other producing countries.

Current research programs of public agencies and private seed companies are making significant advances in improving cotton quality for competition in modern textile markets. Also, manufacturers may combine cotton fibers with wool, linen, or synthetic fibers to produce cloth with special qualities. New weaves and chemical finishes have made possible the production of textured cotton, cotton tweeds and suiting materials, silky cottons, waterproof cottons, tufted cotton for rugs, mildew resistant cotton, and many other kinds of textiles. Thus the fiber of antiquity, the crop that people in the southern U.S.A. once called "King Cotton", will continue to be used by modern man.

II. Important Plant-Parasitic Nematodes Affecting Cotton Production

Several species of plant-parasitic nematodes cause serious damage to cotton. Certain populations of the root-knot nematode, *Meloidogyne incognita*, the sting nematode, *Belonolaimus longicaudatus*, the reniform nematode, *Rotylenchulus reniformis*, and the lance nematode, *Hoplolaimus galeatus* are now recognized as the most important. The first three of these are also important in disease complexes, increasing the incidence and severity of wilt diseases and of post-emergence damping-off caused by various pathogenic fungi.

Several other nematode species parasitize cotton to a much lesser degree. Some found in association with cotton decline have not been shown to be parasitic.

A. The Root-knot Nematode, *Meloidogyne incognita*

Root-knot, caused by *Meloidogyne incognita*, is the most important nematode disease of cotton (Fig. 1). Most of the populations of this species are pathogenic on cotton and are widespread throughout the production areas of the world. Furthermore, this species is involved in disease complexes on cotton.

Root-knot of cotton was first described by Atkinson (1889). In this classical work, and during the period when this group of nematodes was referred to as *Heterodera radicicola*, Atkinson listed cotton as one of 36 different host plants. As root-knot nematodes were studied more extensively, evidence was found that certain populations had specific host preferences. For example, Sherbakoff (1939) reported considerable root-knot injury to cotton, *G. hirsutum*, grown on land previously planted to cotton, but observed no injury to cotton grown on land previously planted to tomatoes, *Lycopersicon esculentum*, even though the tomatoes had been severely injured by root-knot. This was probably the first evidence that root-knot nematodes differ in their host preference. At this time all root-knot nematodes were taxonomically classified as *H. marioni*.

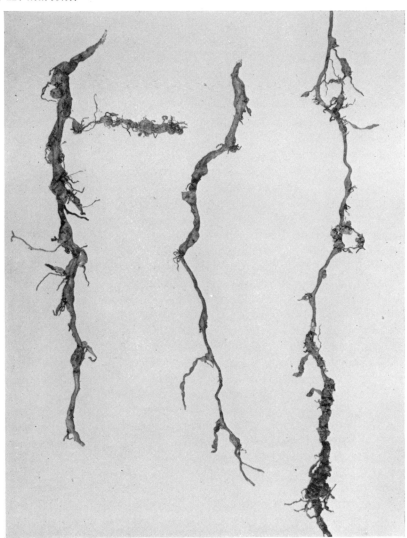

FIG. 1. Severe root-knot on cotton caused by *Meloidogyne incognita*. (By courtesy of H. R. Garriss.)

Christie and Albin (1944) and Christie (1946) established experimentally that populations of *H. marioni* differed in their host–parasite relationships. Among fourteen populations studied, these investigators demonstrated that host specialization may be shown by (1) distinct

G*

host preference, or (2) the character of root galling on susceptible hosts. One of the populations used by these investigators was collected from cotton near Shafter, California, and in subsequent greenhouse tests cotton was highly susceptible to this population.

Chitwood (1949), after a morphological study, removed the root-knot nematodes from the genus *Heterodera* and reassigned them to *Meloidogyne*. The subspecies *M. incognita acrita* was designated the cotton root-knot nematode (Buhrer, 1954).

1. Infectivity and Ability of *Meloidogyne* spp. to Reproduce on Cotton

Sasser (1954), using single species cultures identified according to Chitwood's morphological characters, found cotton to be susceptible to a population designated as *M. incognita acrita*, but not infected by a population designated as *M. incognita incognita* or by *M. hapla, M. javanica* or *M. arenaria*. Neither of the two subspecies of *M. incognita* used in these studies infected peanut, *Arachis hypogaea*.

Martin (1954) found differences in parasitism on cotton among isolates identified as *M. incognita*, and *M. incognita acrita*, ranging from non-parasitism to severe parasitism.

Triantaphyllou and Sasser (1960) could not differentiate morphologically the two subspecies of *M. incognita* and suggested that all populations exhibiting perineal patterns ranging from *incognita* to *acrita* type be considered as one, *M. incognita*. In a recent monograph, Whitehead (1968) recognized *M. incognita* but not *M. incognita acrita*.

Minton *et al.* (1960) studied response of resistant and susceptible cotton varieties to *Meloidogyne* populations in the field. A wide range of root-knot resistance was observed. Monthly (May–October) nematode counts from soil samples from the plant rhizosphere showed populations of *M. incognita* increasing slowly under the more resistant cottons including a selection of *G. barbadense*, a Mexican selection and Auburn 56 of *G. hirsutum*. Increases were more rapid under Empire and Rowden, susceptible varieties of *G. hirsutum*. In September, however, when the populations were highest under all varieties, no appreciable differences were observed between any variety.

Minton (1962) found that *M. incognita acrita* reproduced on all varieties of cotton tested, whereas *M. incognita incognita* reproduced only on some varieties.

Brodie and Cooper (1964a) found that one isolate of *M. incognita* reached egg-laying maturity in seedlings of Coker 100 WR within 30 days whereas another isolate of *M. incognita* failed to develop beyond the second larval stage. All species they used penetrated the roots of cotton seedlings. These investigators concluded that damage to cotton is not related to the ability of *Meloidogyne* to reproduce.

More recently, Sasser (1966), working with populations from various parts of the world, found considerable pathogenic variability on cotton within the *M. incognita* group. Most of the injury to cotton throughout the world, however, appears to be caused by populations of this single species.

2. Host–Parasite Relations

Larvae of *M. incognita* entered the roots of the resistant variety, Auburn 56, as readily as those of the susceptible variety, Stoneville 62, but development of the larvae in roots of the resistant seedlings was sharply retarded (Brodie *et al.*, 1960). Galls on resistant seedlings had fewer egg-laying females.

Minton (1962) studied the host–parasite relations of two varieties (Rowden and Auburn 56) and one wild selection (*G. barbadense*) at various intervals following inoculation with *M. incognita*. Non-infected roots were also compared morphologically. Movement of larvae through the roots was inter- and intra-cellular in all varieties. Migration was away from the root tips, and because of continued root growth and tissue differentiation, the larvae became established in differentiated tissues. During the early stages most of the larvae in all varieties were within and parallel to the stele. After 21 days many nematodes in Rowden and Auburn 56 were oriented outside and at an angle to the stele in contrast to their general orientation within and parallel to the stele in *G. barbadense*. Giant cells started to develop in the vascular cambium and pericycle of the roots of all varieties 2 days after inoculation and were mature after 7 days. Some nematodes, especially in *G. barbadense* roots, failed to stimulate giant cells. A direct correlation existed between giant cell development and nematode maturation. Hyperplasia in the vicinity of the nematodes was extensive in roots of Rowden and Auburn 56, but was rare in roots of *G. barbadense*. The absence or small size of galls on *G. barbadense* roots was apparently related to poor development of hyperplasia and hypertrophy within the root tissues.

3. Factors Influencing Resistance in Cotton to Root-knot Nematodes

Brodie *et al.* (1960) found that resistance in the seedling stage of Auburn 56 and in five breeding lines of cotton to *M. incognita* was associated with three kinds of host response: root necrosis, retarded gall development, and failure of the majority of nematodes to reach maturity.

Minton (1962) concluded that root weight, root diameter, and number of lateral roots were not related to root-knot resistance in the varieties Rowden, Empire and Auburn 56. Furthermore, resistance was not associated with plant morphological differences, or to root barriers that prevented nematode penetration. Minton further concluded that resistance appeared to be a physiological response that restrained larval development. Histopathological studies suggested two explanations for resistance. First, hypersensitivity of the root tips to the penetrating larvae produced tissue necrosis which inhibits development, and second the failure of root cells to respond to the nematode in producing "giant cells" necessary for root-knot nematode survival. Jones and Birchfield (1967) concluded that resistance to root-knot in Bayou and Auburn 56 was related to reduced production of nematode eggs in these varieties.

4. Interrelationship of *Meloidogyne* spp. and the *Fusarium* Wilt Fungus

The association of *M. incognita* infection with the severity of *Fusarium* wilt of cotton, caused by *F. oxysporum* f. *vasinfectum*, has been recognized since Atkinson (1892) described the disease. Observing the consistent association of root-knot with severe wilt outbreaks in wilt-resistant varieties, Orton (1910) concluded that severe root-knot nematode infection apparently "broke" resistance to *Fusarium* wilt. Later, Miles (1939) showed a general varietal correlation between root-knot infection and wilt resistance. This interaction has been pointed out by Smith (1954) who warned that cotton breeding programs must give strict attention to this complex for useful resistance. He had noted that yield, wilt and root-knot resistance, root-knot index and wilt percentage were positively correlated. Young (1938) recorded that wilt incidence could usually be reduced in cotton by the application of potassium. When root-knot nematodes were present, however, neither potash applications nor varietal resistance were effective in control. Smith (1941) found that wilt-resistant varieties developed in the eastern U.S.A., an area of heavy root-knot infestation, were also relatively resistant to root-knot. This relationship was not evident in wilt-resistant varieties developed elsewhere.

Martin *et al.* (1956) studied development of *Fusarium* wilt in cotton growing in steam-sterilized soil artificially infested with different populations of *M. incognita* alone, with *F. oxysporum* f. *vasinfectum* alone, and with combinations of each nematode population and the *Fusarium*. The nematodes reproduced abundantly and caused severe injury on cotton varieties Deltapine 15 (wilt-susceptible) and Coker 100 (wilt-resistant), and significantly increased the incidence of wilt in the two varieties. There appeared to be differences in the isolates of *Meloidogyne* in their ability to increase the incidence of wilt.

The effects of *M. incognita* alone and in combination with *Fusarium* on Rowden, a wilt-susceptible variety, in three soil types of the southeastern U.S.A., including Norfolk sandy loam, Hartsell sandy loam, and Lloyd sandy loam were studied in the greenhouse (Minton and Minton, 1966). Seedling emergence was not affected by high populations of the nematode or the fungus, or by combinations of the two. Severe damage developed later, however, in young seedlings inoculated with the fungus and the nematode. Incidence of wilt was not greatly affected by soil type. Reduction in shoot weight occurred in all soil types, but was greatest in plants receiving nematodes and the fungus.

Cooper and Brodie (1963) in comparing the *Fusarium* wilt indices of cotton varieties and selections with root-knot indices (*M. incognita*) during 1959-1960 found positive correlations, 0·708 to 0·427. Their data was thus in agreement with that of Miles (1939) and Smith (1941) in linking root-knot resistance and *Fusarium* wilt resistance. The extent of vascular browning in *Fusarium*-infected plants has been shown to be correlated with *Meloidogyne* infection (Hollis, 1958). In Peru, it has been shown that populations of *M. incognita* increased damage by the fungus causing *Verticillium* wilt of cotton (Bazan and Aguilar, 1955). Workers in Arizona, U.S.A., have shown that 1000 and 2000 larvae of *M. incognita* combined with 30–120 ml of a suspension of *Verticillium albo-atrum* per 15 cm pot significantly increased the number of plants infected by the fungus (Khoury, 1970).

5. *Relation of Parasitic Nematodes to Post-emergence Damping-off of Cotton*

An increase in post-emergence damping-off of cotton, caused by *Rhizoctonia solani*, was associated with increasing incidence of *M. incognita* and with decreasing plant vigor and survival (Reynolds and Hanson, 1957). Where nematodes were controlled by fumigation the loss in stand from *Rhizoctonia* also decreased. These investigators concluded that root-knot nematodes were debilitating parasites which weakened the plant and made it more susceptible to attack by *Rhizoctonia*.

Brodie and Cooper (1964b) studied the relation of three *Meloidogyne* species—*M. arenaria*, *M. hapla* and *M. incognita*—to post-emergence damping-off of cotton caused by *R. solani* and/or *Pythium debaryanum*. Two isolates of *M. incognita* were used: one that reproduced on cotton and one that did not. It was found that all *Meloidogyne* spp. tested increased the incidence of damping-off caused by *R. solani*. Prolonged susceptibility to *P. debaryanum* was obtained only with seedlings grown in soil infested with either the isolate of *M. incognita* which reproduced on cotton, or *M. hapla*. It was concluded by these investigators that predisposition of cotton seedlings to damping-off by *P. debaryanum* is apparently a function of the physiological response of the host to nematode infection rather than the association with reduction in plant growth.

Norton (1960) showed that reduction in stands of cotton was greater when a fungus—*F. oxysporum* f. *vasinfectum*, *R. solani* or *P. debaryanum*

—and root-knot nematodes were combined than when either acted alone. Damage was more than an additive effect of the pathogens and was greatest at the lower temperatures.

Thus in summarizing, serious damage to cotton is caused primarily by certain populations of *M. incognita* alone or in concert with another pathogen in which case the overall damage is greater than the additive effect of the two pathogens. Other *Meloidogyne* species may enter the roots of cotton and even prolong susceptibility to damping-off fungi, but overall damage resulting from species of *Meloidogyne* other than *M. incognita* is relatively small.

B. The Sting Nematode, *Belonolaimus longicaudatus*

The sting nematode, *Belonolaimus longicaudatus*, is perhaps the most devastating nematode parasite of cotton and is widespread throughout the Coastal Plains of the southeastern U.S.A. This nematode was first observed as a parasite of cotton in the vicinity of Holland, Virginia, U.S.A., during the growing season of 1950 (Owens, 1951). The first extensive studies of this parasite of cotton were done in South Carolina (Graham and Holdeman, 1953). Field symptoms of sting nematode damage on cotton include severe stunting of plants frequently accompanied by early death. Infested areas in the field are variable in size and shape but their margins are usually well marked. Growth of cotton on fumigated and non-fumigated plots in a field heavily infested with *B. longicaudatus* is shown in Fig. 2.

1. Pathogenicity of Sting Nematode to Cotton

Using populations of approximately 600 nematodes per 15 cm pot of sterilized soil, data was taken on growth of cotton, root injury and population increase of the sting nematode (Graham and Holdeman, 1953). Soil temperatures were maintained at 85–90°F (29–32°C) in the greenhouse. After three weeks the inoculated plants were significantly smaller than the controls. Careful observation of plants grown in infested soil in early stages of attack showed minute, dark, shrunken lesions along the root axis or at the root tip. Lesions often advanced laterally to girdle the roots, which soon broke off, and in other cases advanced lengthwise and persisted for some time without severing the roots. The latter case appeared to be the result of repeated attacks along the same root. These shrunken discolored roots gave the entire root system of affected plants a generally unhealthy appearance (Fig. 3). Final data was taken after 76 days and a direct correlation was found between number of sting nematodes and the amount of plant damage.

2. Population Trends of the Sting Nematode on Cotton

Greenhouse studies conducted by Holdeman and Graham (1953b) gave preliminary information which may be of value in planning field work and rotation trials. Seeds or seedlings were planted in pots containing naturally infested field soil. After 60–75 days a soil sample from each pot was examined for nematode populations. The sting nematode consistently increased to high populations on cotton including Coker 100 WR.

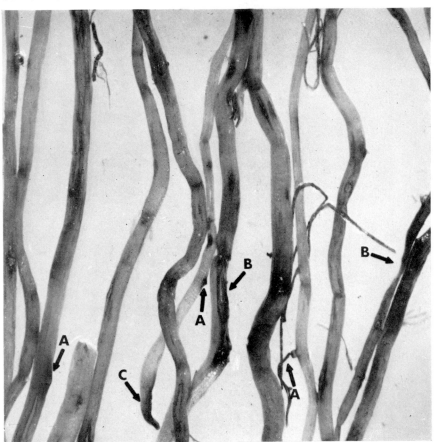

FIG. 2. Root lesions of sting nematodes, *Belonolaimus longicaudatus*, on cotton. The arrows point to: A, points where secondary roots were broken off; B, discolored and shrunken lesions; C, a shriveled root tip. Reproduced with permission from T. W. Graham.

3. Effect of the Sting Nematode on Expression of *Fusarium* Wilt on Cotton

Preliminary experiments demonstrated that the wilt-resistant cotton, Coker 100, was susceptible to wilt when the sting nematode was present (Holdeman and Graham, 1952, 1953a). In later studies (1954) these

FIG. 3. Resulting growth of cotton on fumigated and non-fumigated plots in a field heavily infested with sting nematode, *Belonolaimus longicaudatus*. Plots A, B, and C fumigated with Dowfume W-40 at rates of 112–224 litres/ha. D, untreated plot with weeds. Reproduced with permission from T. W. Graham.

authors used Coker 100 wilt resistant to compare with wilt susceptible (Hurley's Rowden) in greenhouse tests with *Fusarium* and/or sting nematodes in various combinations. The percentage of plants showing external symptoms of wilt 70 days after seeding showed that *Fusarium* inoculum alone caused no wilt in the resistant Coker 100 whereas in the susceptible Rowden, 5% wilt was found (average of 3 tests). The combination of sting nematode–*Fusarium* inoculum caused 60% in the wilt-resistant variety and 78% in the susceptible variety. This demonstrated that *B. longicaudatus* greatly facilitates the development of *Fusarium* wilt in both susceptible and resistant cotton plants. Other workers have shown that *B. longicaudatus* predisposes both resistant and susceptible varieties to *Fusarium* wilt (Cooper and Brodie, 1963).

Minton and Minton (1966) found that soil types were not factors in the nematode–wilt complex as it affected seedling emergence of Rowden (wilt susceptible) in the three soils—Norfolk sandy loam, Hartsell sandy loam, and Lloyd sandy clay. Neither *B. longicaudatus* nor *Fusarium* affected seedling emergence. Wilt occurred, however, in all three soil types receiving the *Fusarium* and the nematode. Furthermore, the disease complex progressed rapidly. For example, on the 14th day, 28, 10 and 14% of the plants exposed to both organisms were affected in the Norfolk, Hartsell and Lloyd soils, respectively.

B. longicaudatus, although much more restricted in its distribution in relation to cotton culture, is a devastating parasite of cotton, often killing all or most of the plants in large areas of infested fields. When it occurs in the same field with the *Fusarium* wilt fungus, almost total crop failure can result.

C. The Reniform Nematode, *Rotylenchulus reniformis*

The reniform nematode, *Rotylenchulus reniformis*, was first identified as a parasite of cotton by Steiner (Smith, 1940). It has since been found infecting cotton in many locations (Martin and Birchfield, 1955; Peacock, 1956; Jones *et al.*, 1959; Sasser *et al.*, 1962). In several parishes in Louisiana *R. reniformis* was found associated with estimated reductions in cotton yield as high as 40–60% in several thousand hectares (Birchfield and Jones, 1961). In addition to reducing cotton yield, this nematode also caused a delay in maturity, a reduction in size of boll, and in some years a reduction in lint percentage (Jones *et al.*, 1959).

1. Host–Parasite Relations and Pathogenicity

Comprehensive studies of the host–parasite relations of *Rotylenchulus reniformis* on cotton were made by Birchfield (1962). Histological preparations of young roots of Coker 100 wilt and Delta Pine Smoothleaf cotton varieties showed that young females initiated infection by destroying epidermal cells and causing a slight browning and necrosis of the surrounding cortical cells as they collapsed. Phloem cells near the nematode head stained darker than normal tissues and were greatly enlarged in the young roots. According to Oteifa (1970) this pest is capable of causing giant cells in the pericycle zone which inhibit the formation of lateral roots. This apparent damage to the phloem can extend several cells along the root axis. Damage to epidermal and cortical parenchyma cells occurs with heavy infection although nematodes do not feed in these areas. Necrosis in the phloem and parenchyma causes severe pruning of seedling roots and consequent stunted growth. Seedlings grown in naturally infested soil harbored an average of 900 egg masses per plant after 30 days and the life-cycle required 17–23 days.

2. *Rotylenchulus reniformis* and the incidence of *Fusarium* Wilt of Cotton

The first indication of the association of *Rotylenchulus reniformis* with *Fusarium* wilt of cotton was given by Smith (1940). Smith and Taylor (1941) reported that a pronounced infestation of this nematode occurred in the regional wilt plots on the roots of cotton collected from Baton Rouge, Louisiana. Neal (1954) did the first experiments to show the relationship of *R. reniformis* to *Fusarium* wilt. A wilt-resistant variety, Delfos 425-920, was compared with the susceptible, Half and Half, in soil naturally infested with both organisms. Additional wilt inoculum was added to the field test plot. The same varieties were also planted in sterilized soil reinoculated with the wilt organism. In soil naturally infested with wilt and nematodes (4370/pt of soil) the susceptible variety, Half and Half, developed 81·4% wilt after 99 days, while the resistant variety developed only 3·1%. In soil containing only the wilt fungus, the susceptible variety developed 10% wilt and none of the resistant varieties were infected. Neal thus showed that in the presence of the wilt organism and parasitism by *R. reniformis*, overall damage to cotton is greatly increased, especially in susceptible varieties. Resistant varieties, developed in Louisiana and elsewhere, have been shown to withstand the combined attacks of *Fusarium* and the reniform nematode. Among these are Auburn 56 and Plains from Alabama; Delfos 425-920, Delfos 9169, Deltapine 6, Deltapine 6 × Delfos 6102, Cook 307-6 × Delfos 9169, Louisiana Hybrid 33 × 14-312, Louisiana 33, Roxie, and Stoneville 2B strains from Louisiana; and Coker 100 wilt and derivatives from South Carolina.

In general, *R. reniformis*, like *B. longicaudatus*, has a more restricted distribution than *M. incognita*. Nevertheless, it is a serious pathogen of cotton and must be controlled for profitable cotton production. Its interrelationship with the *Fusarium* wilt fungus makes it even more important to apply control measures.

D. The Lance Nematode, *Hoplolaimus galeatus*

Examination of soil from the rhizosphere of severely stunted, yellowed and nearly defoliated cotton plants revealed a high population of *Hoplolaimus galeatus* (2631/litre of soil), compared with 125 from plants free of the above symptoms (Krusberg and Sasser, 1956). This apparent association of the lance nematode with diseased cotton plants in the field led to further studies to determine the host–parasite relationship of this species on cotton.

1. Pathogenicity and Population Increase in Greenhouse Tests

Lance nematodes were observed feeding on roots of cotton seedlings grown in autoclaved sand in test tubes. Nematodes, added to the soil at time of planting, did not affect germination. Microscopic observations through the test tube walls after two weeks showed that nematodes migrated freely from root to root, feeding as ecto- and endoparasites. Root lesions were associated with the feeding positions. A brownish-yellow discoloration was observed in the epidermal cells at the point of nematode entry and to a depth of 2–5 cell layers along the path of nematode movement.

The rate of reproduction of lance nematodes on cotton and their effect on plant growth were determined in greenhouse tests at 75–85°F. Cotton was planted in eighteen 20 cm pots with soil that had been fumigated with methyl bromide. Nine pots were infested with 500 lance nematodes each at time of planting, and nine pots without nematodes served as controls. When the first planting of cotton matured, the shoots were excised at the soil level and cotton again planted. This procedure was repeated three times during a total growth period of 344 days. The nematode population was determined after 112 days and after 344 days. The first planting of cotton showed no stunting, but slight stunting was observed in subsequent plantings. The population levels of the lance nematode increased to 741 per pot after 112 days and to 12,883 (average of three pots) after 344 days.

2. Host–Parasite Relations

Lance nematodes penetrate the root cortex causing considerable damage and become situated 2–3 cell layers inside the epidermis and parallel to the longitudinal axis of the root (Fig. 4C). Sometimes the anterior portion of the body and lip region curve into the conducting tissue with the remainder of the body in the cortex (Fig. 4D).

Fig. 4. (A, B) Comparison of areas of a cotton field heavily (A) and lightly (B) infested with lance nematodes, *Hoplolaimus galeatus*. (C–E) Photomicrographs of lance nematode injury in cotton roots: (C) lance nematode in cortical tissue (see arrow); (D) nematode (arrow) feeding in phloem tissue; (E) membrane across penetrated cortical cell (see arrow). Reproduced with permission from Krusberg and Sasser (1956).

The phloem parenchyma and phloem elements were found punctured by the nematode stylet; as a result the contents appear disorganized, and stained greenish or reddish, whereas undamaged phloem stained green. Abnormal division of phloem parenchyma and of the xylem elements was also observed. Tyloses were found in xylem cells and appeared to be associated with nematode injury.

H. galeatus does not appear to be widely distributed in the cotton producing regions of the world. Furthermore, there are no reports that this species predisposes cotton to infection by *Fusarium* or other pathogenic fungi. It has been demonstrated, however, that it can cause serious damage to cotton under field conditions and precautions should be taken to avoid the build-up of high populations in cotton fields.

E. Root Lesion Nematodes, *Pratylenchus* spp.

Cotton is a host for certain root lesion nematode species (Graham, 1951; Jensen, 1953; Endo, 1959; Oteifa, 1962). *P. brachyurus* overwinters in dead roots of cotton and also in the surrounding soil (Graham, 1951). Development of *Pratylenchus* populations on roots of cotton was slow during the summer but populations were consistently high (60/2g of roots) in the late summer and fall. This persistent build-up during the fall carried large populations into the winter, and the tobacco following cotton usually developed severe root rot.

Cotton, variety Coker 100 WR, is a suitable host for *P. brachyurus* but not for *P. zeae*, which may occur in the field on corn and crabgrass (Endo, 1959).

Although cotton is a host for certain other *Pratylenchus* species, there is not sufficient evidence that this group of endoparasites cause appreciable damage to cotton.

1. Interaction with *Fusarium*

Smith (1940) suggested that the distribution of root-lesion nematodes might be correlated with severe wilt development in certain areas of cotton fields in Georgia, U.S.A. Soil treated with carbon bisulfide controlled nematode populations (presumably root-knot and root lesion nematodes) sufficiently to establish wilt-resistant and tolerant varieties (Taylor *et al.*, 1940). The role of *Pratylenchus* species in increasing the incidence of *Fusarium* wilt on cotton, however, has not been clearly demonstrated.

F. Other Nematode Species Associated with Cotton

1. *Trichodorus christiei*

Alhassan and Hollis (1966) conducted tests to define, measure and evaluate some of the quantitative relationships between *Trichodorus christiei* and cotton seedlings. Auburn 56 plants grown in steam-sterilized soil were inoculated with 0, 100, 400, or 1600 nematodes. Three weeks after germination the plants were harvested to record root growth responses and the behaviour of nematode populations. Young root tips of cotton seedlings were attacked by *T. christiei*, causing an overall reduction in size of the root system. There was marked stunting of top growth but an absence of root necrosis. Seedling weights were related inversely to both initial and final populations of the nematode. In the 1600-nematode treatment, root systems were significantly reduced in weight, volume, area and number of branch roots when compared with control plants. High initial populations of *T. christiei* resulted in a slow increase of nematode population; whereas low initial populations increased rapidly. The nematode feeds in the region of the root tip, including the root cap, the meristematic region, and the region of elongation. No feeding was observed in the root hair region.

These results suggest that low numbers of *T. christiei* (approx. 80/500 cc soil) might cause measurable damage to young cotton seedlings in the field. Percentage increases in nematode populations were related inversely to initial numbers and to plant damage, and with the absence of necrosis, this indicates that the relationship of *T. christiei* to the host is one of a balanced parasitism.

Brodie and Cooper (1964a) in greenhouse pathogenicity studies, found that *T. christiei* reproduced around cotton seedling roots, but did not cause appreciable root injury or measurable reduction in seedling growth.

High populations of an undescribed species of *Trichodorus* were found to occur in certain Red River deposited soils that had been planted primarily with cotton (Martin, 1956). In controlled infestations on cotton in the greenhouse, this nematode propagated abundantly on Coker 100 WR without causing measurable differences in stand, height, or green weight. In one of the two tests with cotton variety Deltapine 15, green weight appeared to be reduced as a result of nematode infestation.

2. *Tylenchorhynchus* spp.

The cotton variety Coker 100 WR is a favorable host for *Tylenchorhynchus claytoni* (Krusberg, 1959). This was based on an increase in nematodes from 200 per 15 cm pot to 942 in 90 days. Brodie and Cooper (1964a) reported that *T. claytoni* increased in Coker 100 WR but that no appreciable root injury or measurable reduction in seedling growth occurred after 24 days, even though the population had attained a level of 3650 nematodes per 500 g of soil. These investigators also found that *T. claytoni* populations of 3650 per 500 g of soil did not lengthen the period of susceptibility to post-emergence damping-off of cotton caused by *R. solani* and *P. debaryanum*. Tests conducted in the greenhouse and in the open on growth of Hopi M cotton have shown that *T. cylindricus* reduced shoot and root growth resulting in moderate stunting (Reynolds and Evans, 1953). The stunt nematode, *T. latus*, is capable of causing serious yield reductions on cotton, *G. barbadense* var. *ashmouni*. Yield increases, resulting from nematicide treatments, were correlated with nematode control (Elgindi and Oteifa, 1967).

III. Cultural and Environmental Influences on Parasitism of Nematodes on Cotton

Under the dry climatic conditions which exist in southern Arizona, U.S.A., a system of clean fallowing without irrigation offers a very effective means to control root-knot nematodes (King and Hope, 1934). Yields per hectare of Pima seed cotton before and after 2 years of clean fallow were as follows: before fallow 780, 1532 and 1945 kg, and after fallow in the same respective plots, yields were 1786, 2599 and 2505 kg. During this period the soil was frequently stirred, being plowed three times and disk cultivated eight or ten times without application of water.

It has been reported that light sandy soils favor the spread of and damage caused by plant-parasitic nematodes (Bessey, 1911; Endo, 1959). Soil fumigation experiments for the control of *M. incognita* on Pima S1 and Acala 44 cotton consistently resulted in increased yields on sandy soils but not on soils of finer texture (O'Bannon and Reynolds, 1961). These observations led to investigations to obtain more exact information on the influence of soil texture as related to incidence of root-knot and the effect on cotton yields. They found heavy root-knot populations developed on cotton in coarse-textured soils and relatively light populations in fine-textured soils. Soils classified as loamy sand and sandy loams with 50% or higher sand content were found most favorable for root-knot nematode activity. Less activity was found in silt loam and loam having sand contents of 38·3 and 29·0% respectively. Thus, damage to cotton by the root-knot nematode was directly correlated with soil texture. Highly significant yield increases were obtained from fumigation of fields having loamy sands and sandy loam; yield increases were not obtained with fumigation in a medium-textured soil (silt loam).

Water consumption is not inhibited in Pima S2 cotton infected with *M. incognita* when supply is not limited but is inhibited when the soil water content is allowed to fall below field capacity at intervals (O'Bannon and Reynolds, 1964).

The interrelationship between fertilizer levels, population trends of the stunt nematode *Tylenchorhynchus latus* and cotton yields reveal that fertilizers increase the reproduction of the nematode (Oteifa *et al.* 1965). Unfertilized plants showed the least degree of tolerance despite the low reproductive rate of the nematode. Fertilized plants exhibited more tolerance despite the increase in nematode reproductive rate. An increase in potassium fertilizers, alone or in combination with other elements, provided host plants with the highest level of tolerance.

IV. Control Methods and Practices

A. Resistant Varieties

Resistance in cotton varieties to *Meloidogyne* spp. has been well documented (Miles, 1939; Smith, 1941; Turcotte *et al.*, 1963). Auburn 56, Clevewilt-6, La. Mexico wild, and *G. barbadense* var. *darwinii* are some of the more highly resistant types being utilized in cotton-breeding programs in the U.S.A. Auburn 56, once a widely planted commercial variety, was developed in Alabama (Tisdale, 1953) from crosses involving Cook 307, Coker 100, and Coker 100 Wilt. Its resistance to *Meloidogyne* and to *Fusarium* wilt has been repeatedly confirmed (Jones *et al.*, 1958; Minton, 1962; Brodie *et al.*, 1970). Clevewilt-6 is an old wilt-resistant strain of Upland cotton which Jones *et al.* (1958) and Minton (1962) reported as being slightly more resistant to root-knot nematodes than is Auburn 56. A primitive strain of *G. hirsutum*, La. Mexico wild, collected in Mexico, is highly resistant to root-knot nematodes (Jones *et al.*, 1958; Minton, 1962) and a primitive cotton, *G. barbadense* var. *darwinii*, also is highly resistant to root-knot nematodes (Smith, 1954; Wiles, 1957; Minton, 1962; Turcotte *et al.*, 1963).

The inheritance of resistance to root-knot nematodes in the F_3 generation of a Clevewilt-6 Deltapine 15 cross was studied by Jones *et al.* (1958). Resistance was found to be inherited as a quantitative character and controlled by probably two or three gene pairs. Resistance to root-knot in *G. barbadense* var. *darwinii* is controlled by a pair of recessive genes in crosses with susceptible cultivated types of *G. barbadense* (Turcotte *et al.*, 1963).

A new experimental cotton variety, Bayou, has resistance to root-knot nematode comparable to or exceeding that of Auburn 56 when tested in the field and in the greenhouse (Jones and Birchfield, 1967). Root-knot larvae counts from replicated field plots of five varieties of cotton at three sampling dates revealed that the average root-knot population associated with Bayou was significantly lower than that for Auburn 56, Deltapine 45, Carolina Queen, and Deltapine Smoothleaf.

A moderate amount of root-knot resistance is associated with

Fusarium resistance in varieties developed in the southeast U.S.A. and selected on wilt- and nematode-infested soils. In the process of selecting for *Fusarium* resistance and high yields in segregating progenies, moderate resistance also to root-knot was obtained. This resistance contributes to the *Fusarium* resistance and to the yield of such lines. For example, Tisdale and Smith (mimeographed report) conducted a cotton variety and strain test in 1951 on soil severely infested with both the *Fusarium* wilt fungus and nematodes. The yield of lint of the varieties and strains tested ranged from 541 kg to only 43 kg per hectare. Only 3·4% of the plants were infected with wilt in the highest-yielding varieties. Those with severe wilt were most severely infected with root-knot nematodes. The five varieties or strains that showed approximately 10% or less of wilt were developed in areas where nematode infestation was generally high and in varieties where a tolerance to the wilt-nematode complex had developed.

Cotton varieties combining both *Fusarium* resistance and moderate root-knot resistance are Stonewilt, Coker 100 wilt, and Plains. Several varieties considered susceptible to root-knot are Rowden, Miller, Deltapine, Bobshaw, Stoneville, Empire and Pandora.

There are no known breeding programs or sources for resistance to other nematodes.

B. Crop Rotation

Crop rotation of cotton with a crop that is a less suitable host for the nematode is an effective and widely used land management practice for reducing nematode populations in the soil. Growth of a resistant crop for only one year decreases nematode populations but is generally inadequate; 2 years of resistant crops between susceptible crops may give fair to good control. Since the most important nematode affecting cotton is *M. incognita*, varying degrees of control can be obtained by planting highly resistant crops for 1, 2 or 3 years between each planting of cotton. Crops which are resistant to *M. incognita* include Rescue grass, *Bromus catharticus*; Coastal Bermuda grass, *Cynodon dactylon*; Crab grass, *Digitaria sanguinalis*; Weeping love grass, *Eragrostis curvula*; Switch grass, *Panicum virgatum*; dallis grass, *Paspalum dilatatum*; Wilmington bahia grass, *Paspalum natatum*; and Starr millet, *Pennisetum spicatum* (McGlohon *et al.*, 1961). Other resistant crops include small grains, corn, peanut, *Crotalaria*, velvetbeans, certain clones of alfalfa (such as African), nematode-resistant cowpea, Laredo soybeans and NC95 tobacco.

Satisfactory yields of cotton may be maintained on root-knot nematode infested areas by rotating every 2 or 3 years with alfalfa (King and

Hope, 1934). Although some varieties of alfalfa are a host for the root-knot nematode, rotation with a highly resistant alfalfa increases cotton yields as readily as does clean fallow.

Belonolaimus longicaudatus populations are reduced by tobacco, watermelon and *Crotalaria* (Holdeman and Graham, 1953c). There are no reports, however, concerning the effectiveness of these crops in rotation with cotton in reducing the sting nematode population. The sequence in which certain crops should be grown for greatest effectiveness in minimizing population increases of certain nematodes has been studied (Brodie *et al.*, 1970). In continuous row-crop rotations (cotton-corn–peanut), cotton and corn favored rapid increase of *B. longicaudatus*, while peanut greatly reduced the population. The inability of the population of *B. longicaudatus*, which occurs in the state of Georgia, U.S.A., to increase on peanut had been previously reported (Good, 1969). This is in contrast to the behavior of this nematode in the Carolinas and Virginia where it is highly pathogenic on cotton, corn and peanuts.

Several non-hosts have been reported for *Rotylenchulus reniformis* (Birchfield and Brister, 1962). These include barnyard grass, *Echinochloa crusgalli*; dallis grass, *Paspalum dilatum*; mustard, Florida, *Brassica nigra*; oats, fulghum, *Avena sativa*; onion, Evergreen, *Allium cepa*; pepper, Sweet Bell, California Wonder, *Capsicum annuum*; pepper, red, hot, *Capsicum annuum* var. *fasciculatum*; rice, Blue Bonnet, *Oryza sativa*; sorghum, sweet, *Sorghum vulgare*; sugar-cane, C.P. 44-101, *Saccharum officinarum*; turnip, Purple Top, White Globe, *Brassica rapa*; and wild barley, *Hordeum pusillum*.

Hoplolaimus galeatus populations decreased in greenhouse tests with alfalfa, dallis grass and Pensacola bahia grass (McGlohon *et al.*, 1961).

Theoretically, the planting of resistant and non-host crops for the various nematode pathogens discussed above should reduce the populations of these nematodes. Few studies, however, have been conducted to determine the effectiveness of rotation in controlling nematodes in cotton.

C. Chemical Control

Chemical control of plant pathogenic nematodes in cotton is by far the most expedient and widely used method. Ethylene dibromide (Dowfume W-40) applied to a soil heavily infested with nematodes and the *Fusarium* wilt fungus gave increased yields of Upland cotton and reduced wilt (Smith, 1948; Presley, 1950). The high cost of fumigant nematicides led to trials with row applications where fewer litres per hectare are required. In row application the root system is protected in early season growth until the roots are well established. Infections in late season appear to have little effect on ultimate yields. In tests conducted in the Mesilla Valley of New Mexico, U.S.A., best control of root-knot in Upland cotton was obtained with 112 litres/ha of Dowfume W-40 or of Dowfume N/A applied in the row (Leding, 1950). More recent tests with fumigants in the Cotton Belt have shown that as little as 67–90 litres of DD or 23 litres of Dowfume W-85 per hectare have been satisfactory to control root-knot (Brown and Ware, 1958).

Yields of wilt-resistant Plains Cotton were significantly greater in sting nematode infested plots fumigated with Dowfume W-40, Nemagon, Dorlone and DD (dichloropropane-dichloropropene), than in non-fumigated plots (Good and Parham, 1957). Average seed cotton yields were 1800, 1754, 1150, 991 and 228 kg per hectare, respectively, for the four nematicides and the control. DBCP and EDB increased the yield of cotton the first year seven times over that of the non-fumigated controls (Good and Steele, 1958). Plots treated with DBCP (1,2-dibromo-3-chloropropane) and EDB (ethylene dibromide) had larger and more vigorous plants than other treatments and produced twice as many mature bolls per plant as the controls. The incidence of *Fusarium* wilt was minimal, whereas in control plots and those treated with DBCP-EDB mixture, and DD mixture, the disease became more severe as the season progressed. DBCP and EDB treated plots gave significant increases in corn yields the following year, but only DBCP significantly controlled sting nematodes in the second year.

In large-scale fumigation tests, fumigants applied as pre-planting treatments gave highly significant increases in yield of lint above those in post-plant treatments and above the controls (Reynolds, 1958). DBCP at 11 litres/ha significantly increased yield compared with that of other fumigant treatments regardless of rate of application.

Combination treatments of pentachloro-nitrobenzene (PCNB) and DBCP reduced reniform nematode infection of cotton seedlings more than did DBCP alone (Birchfield and Pinckard, 1964). DBCP at 7·23 kg of active ingredient per hectare applied in the row, and a seed treatment (NDT) consisting of a mixture of DBCP (25%), PCNB (15%), and dieldrin (7·5%), significantly increased yield in variety DDL Smoothleaf Cotton. The yield of seed cotton was 1670, 1588 and 1317 kg/ha for the DBCP, NDT and control, respectively.

The non-volatile nematicide, Shell SD7727, gave excellent control of root-knot nematodes on "Deltapine Smoothleaf" cotton for an entire growing season. Bioassays of soil samples collected at the end of the growing season indicated that SD7727 controlled root-knot nematodes for at least 10 months, whereas Nellite and DBCP did not (Reynolds and O'Bannon, 1966). Satisfactory control of the reniform nematode can be obtained by soil fumigation using DD at 101 and 146

litres per hectare, ethylene dibromide (83%) at 45 litres/ha, and DBCP
at 5·6 litres/ha (Jones *et al.*, 1959). In California, nematicides currently
recommended for root-knot control on cotton include DD mixture,
Telone and Vidden D at 101 litres/ha of technical material in the row
or 225 litres broadcast; Nemagon and Fumazone at 8·4 litres/ha in the
row or 11–14 litres/ha broadcast; and Dowfume W-85 at 28-34 litres/ha
in the row or 56 litres/ha broadcast. These pre-plant fumigants are
applied at a depth of 30–35 cm below the top of the listed bed (Radewald
et al., 1967).

V. Disease Losses and Economics of Nematode Control in Cotton

Estimated reductions in cotton yields in the U.S.A. for the 1970
crop due to nematodes ranged from 0·23 to 10% in the fourteen producing states.* Estimates from other cotton-producing countries are not
available. The 2·56% average reported for the U.S.A. is probably a
conservative estimate and does not include additional losses resulting
from wilt diseases which are frequently more severe when nematodes
are not controlled. The economics of nematode control on cotton are
related to many variables, such as species and population density of
nematodes present, presence of other disease organisms, especially wilt
pathogens, soil texture, production potential of the land, cost of treatment, price of cotton, and other factors. While the reported overall
percentage reduction in cotton yields due to nematodes appears low,
losses from individual fields can be very high. For example, in several
parishes in Louisiana, U.S.A., *Rotylenchulus reniformis* was found
associated with estimated reductions in cotton yield as high as 40–60%
in several thousand hectares (Birchfield and Jones, 1961). Furthermore,
expenditures for control are sizable. For example, it is estimated that
141,600 hectares of cotton were treated with a nematicide in 1970 in
the U.S.A. alone, at a cost of $29·65/ha or a total expenditure of
$4,198,440.† Growers, as a rule, do not engage in farm practices which
are not profitable. One striking example of results which can be
obtained from chemical treatment of cotton has been reported from
California (Raski and Allen, 1953). Yield values were increased by
$549/ha by the application of 225 litres/ha of DD broadcast 4 weeks
prior to planting for the control of root-knot. An increase of $423/ha
resulted from the application of 135 litres/ha applied in the row 2 weeks
prior to planting. At the time of these tests the costs of the fumigant

* Cotton Disease Council; 1970 Cotton Disease Loss Estimate Committee
Report.
† Personal communication, J. M. Good.

was approximately $83.00 for the broadcast treatment with 225 litres/ha of DD and $50.00 for the 135 litres/ha in the row. Thus, the return to the grower was substantial in either case.

In fields where losses are as much as 10% or more, control practices usually result in yield increases sufficient to justify the cost. For example, in a field where nematodes are causing a 10% reduction in yield, a grower who can average 1120 kg of lint cotton per hectare by fumigating to control nematodes would make a profit of $37.00/ha from the extra 45 kg, assuming cotton is worth $0·66/kg and that the cost of treatment is $37.00/ha. If the decrease in yield resulting from nematode attack is 30%, this same grower would make a profit of $185.00/ha. Similarly, if such losses from nematodes amounted to 60%, as has been reported for some fields, this same grower would make a profit of $408.00/ha. Good growers could profit even more under the conditions described above, while growers who average less lint per hectare may not find it profitable to treat for nematodes.

The obvious and measurable effects of nematode control on cotton are increased yields and quality. In addition, there are other important benefits derived from nematode control not so easily measured and perhaps not fully appreciated or associated with nematode control, but which nevertheless contribute to the economical production of the crop. The following are some of the most important:

1. Nematode control often results in control of several other diseases of cotton caused by soil-inhabiting organisms present in the soil, namely pathogenic fungi, which express themselves only in the presence of certain plant-parasitic nematodes. Also, various saprophytic fungi and bacteria which accelerate the root decay process through invasion of lesions, galls, or other malformations caused by parasitic nematodes, do not become established if nematodes are controlled.

2. Healthy root systems resulting from nematode control make possible maximum utilization of moisture and mineral elements in the soil. Cotton roots, for example, damaged by the feeding activities of nematodes, cannot utilize the moisture in the soil efficiently and when water is limited, plants infected with nematodes will often show symptoms of wilting while in the same field other plants growing in soils relatively free of nematodes do not wilt. Also, mineral elements instead of being taken up by the plant will tend to accumulate in the soil or leach out and are therefore wasted.

3. Nematode control usually results in increased stands and more uniform growth of the crop. This is a benefit of special importance in cotton since exact cultural and harvesting schedules are followed in the production of this crop.

4. Nematode control assures greater protection of the overall investment in the cotton crop. Growers customarily fertilize, cultivate, spray for insect and disease control, and often irrigate. Yet, if nematodes are not controlled money spent on these practices may to a large extent be wasted since the yields and quality may be greatly reduced. On the other hand, if nematodes are controlled, a healthy and fibrous root system develops and the crop has an excellent chance to grow to maturity.

References

Alhassan, S. A. and Hollis, J. P. (1966). *Phytopathology* **56**, 573–574.

Atkinson, G. F. (1889). Alabama Polytech. *Inst. Agric. Exp. Stn Bull.* **9** (N.S.).

Atkinson, G. F. (1892). *Alabama agric. Exp. Stn Bull.* **41**, 1–65.

Bazan, De Segura, C. and Aguilar, F. P. (1955). *Pl. Dis. Reptr* **39**, 12.

Bessey, E. A. (1911). *U.S. Dept Agric. Bull.* **217**.

Birchfield, W. (1962). *Phytopathology* **52**, 862–865.

Birchfield, W. and Brister, L. R. (1962). *Pl. Dis. Reptr* **46**, 683–685.

Birchfield, W. and Jones, J. E. (1961). *Pl. Dis. Reptr* **45**, 671–673.

Birchfield, W. and Pinckard, J. A. (1964). *Phytopathology* **54**, 393–394.

Brodie, B. B. and Cooper, W. E. (1964a). *Phytopathology* **54**, 1019–1022.

Brodie, B. B. and Cooper, W. E. (1964b). *Phytopathology* **54**, 1023–1027.

Brodie, B. B., Brinkerhoff, L. A. and Struble, F. B. (1960). *Phytopathology* **50**, 673–677.

Brodie, B. B., Good, J. M. and Marchant, W. H. (1970). *J. Nematol.* **2**, 135–138.

Brown, H. B. and Ware, J. O. (1958). "Cotton." 3rd edition. McGraw-Hill Book Co., Inc. New York.

Buhrer, Edna M. (1954). *Pl. Dis. Reptr* **38**, 535–541.

Chitwood, B. G. (1949). *Proc. helminth. Soc. Wash.* **16**, 90–104.

Christie, J. R. (1946). *Phytopathology* **36**, 340–352.

Christie, J. R. and Albin, F. E. (1944). *Proc. helminth. Soc. Wash.* **11**, 31–37.

Cooper, W. E. and Brodie, B. B. (1963). *Phytopathology* **53**, 1077–1080.

Elgindi, D. M. and Oteifa, B. A. (1967). Preliminary studies on the control of the cotton nematode, *Tylenchorhynchus latus* by DD and DBCP. *Bull. Fac. Agric. Cairo Univ.* **XVIII**, No. 2.

Endo, B. Y. (1959). *Phytopathology* **59**, 417–421.

Good, J. M. (1969). *In* "Tropical Nematology" (G. C. Smart, Jr., and V. G. Perry, eds), pp. 113–138. University of Florida Press, Gainesville.

Good, J. M. and Parham, S. A. (1957). *Phytopathology* **47**, 312.

Good, J. M. and Steele, A. E. (1958). *Pl. Dis. Reptr* **42**, 1364–1367.

Graham, T. W. (1951). Nematode root rot of tobacco. *S.C. agric. Exp. Stn Bull.*, 390.

Graham, T. W. and Holdeman, Q. L. (1953). *Phytopathology* **43**, 434–439.

Holdeman, Q. L. and Graham, T. W. (1952). *Phytopathology* **42**, 283–284.

Holdeman, Q. L. and Graham, T. W. (1953a). *Phytopathology* **43**, 475.
Holdeman, Q. L. and Graham, T. W. (1953b). *Phytopathology* **43**, 291.
Holdeman, Q. L. and Graham, T. W. (1953c). *Pl. Dis. Reptr* **37**, 497–500.
Holdeman, Q. L. and Graham, T. W. (1954). *Phytopathology* **44**, 683–685.
Hollis, J. P. (1958). *Phytopathology* **48**, 661–664.
Jensen, H. J. (1953). *Pl. Dis. Reptr* **37**, 284–387.
Jones, J. E. and Birchfield, W. (1967). *Phytopathology* **57**, 1327–1331.
Jones, J. E., Newsom, L. D. and Finley, Etta L. (1959). *Agric. J.* **51**, 353–356.
Jones, J. E., Wright, S. L. and Newsom, L. D. (1958). *Cotton Impr. Conf. Proc.* **11**, 34–39.
Khoury, F. Y. (1970). The influence of *Rhizoctonia solani* (Kuhn) and of *Meloidogyne incognita acrita* Chitwood on the infection of cotton plants by *Verticillium alboatrum* Reinke and Berth. Ph.D. dissertation, Univ. of Arizona.
King, C. J. and Hope, C. (1934). U.S. Dept Agriculture Circular No. 337.
Krusberg, L. R. (1959). *Nematologica* **4**, 187–197.
Krusberg, L. R. and Sasser, J. N. (1956). *Phytopathology* **46**, 505–510.
Leding, A. R. (1950). Agric. Exp. Stn, New Mexico, Press Bull. 1036.
Lewis, C. F. and Richmond, T. R. (1966). *In* "Advances in Production and Utilization of Quality Cotton: Principles and Practices". The Iowa State University Press, Ames, Iowa, U.S.A.
Martin, W. J. (1954). *Pl. Dis. Reptr Suppl.* **227**, 86–88.
Martin, W. J. (1956). *Phytopathology* **46**, 20.
Martin, W. J. and Birchfield, W. (1955). *Pl. Dis. Reptr* **39**, 3–4.
Martin, W. J., Newsom, L. D. and Jones, J. E. (1956). *Phytopathology* **46**, 283–289.
McGlohon, N. E., Sasser, J. N. and Sherwood, R. T. (1961). *N.C. agric. Exp. Stn Tech. Bull.* 148.
Miles, L. E. (1939). *Phytopathology* **29**, 974–978.
Minton, N. A. (1962). *Phytopathology* **52**, 272–279.
Minton, N. A. and Minton, E. B. (1966). *Phytopathology* **56**, 319–322.
Minton, N. A., Cairns, E. J. and Smith, A. L. (1960). *Phytopathology* **50**, 784–787.
Neal, D. C. (1954). *Phytopathology* **44**, 447–450.
Norton, D. C. (1960). *Texas agric. Exp. Stn Misc. Publ.* 412.
O'Bannon, J. H. and Reynolds, H. W. (1961). *Soil Sci.* **92**, 384–386.
O'Bannon, J. H. and Reynolds, H. W. (1964). *Soil Sci.* **99**, 251–255.
Orton, W. A. (1910). *U.S. Dept Agric. Farmer's Bull.* **333**, 1–24.
Oteifa, B. A. (1962). *Pl. Dis. Reptr* **46**, 572–575.
Oteifa, B. A. (1970). *Parasitology* **56**, 255.
Oteifa, B. A., Elgindi, D. M. and Diah, K. A. (1965). *Potash Review*, July, 1965.
Owens, J. V. (1951). *Phytopathology* **41**, 29.
Peacock, F. C. (1956). *Nematologica* **1**, 307–310.
Presley, J. T. (1950). *Down to Earth* **6**(1). The Dow Chemical Co., Midland, Michigan.

Radewald, J. D., Thomason, I. J., Allen, M. W. and Hart, W. H. (1967). "California Plant Diseases"—28, University of California Agric. Ext. Service.

Raski, D. J. and Allen, M. W. (1953). *Pl. Dis. Reptr* **37**, 193–196.

Reynolds, H. W. (1958). *Pl. Dis. Reptr* **42**, 944–947.

Reynolds, H. W. and Evans, H. M. (1953). *Pl. Dis. Reptr* **37**, 540–544.

Reynolds, H. W. and Hanson, R. G. (1957). *Phytopathology* **47**, 256–261.

Reynolds, H. W. and O'Bannon, J. H. (1966). *Pl. Dis. Reptr* **50**, 512–515.

Sasser, J. N. (1954). *Md. agric. Exp. Stn Bull.* A-77.

Sasser, J. N. (1966). *Nematologica* **12**, 97–98.

Sasser, J. N., Vargas Gonzales, O. F. and Martin, A. (1962). *Pl. Dis. Reptr* **46**, 171.

Sherbakoff, C. D. (1939). *Phytopathology* **29**, 751–752.

Smith, A. L. (1940). *Phytopathology* **30**, 710.

Smith, A. L. (1941). *Phytopathology* **31**, 1099–1107.

Smith, A. L. (1948). *Phytopathology* **38**, 943–947.

Smith, A. L. (1954). *Pl. Dis. Reptr Suppl.* **227**, 90–91.

Smith, A. L. and Taylor, A. L. (1941). (Abstr.) *Phytopathology* **31**, 771.

Taylor, A. L., Barker, H. D. and Kime, P. H. (1940). *Phytopathology* **30**, 710.

Tisdale, H. B. (1953). *S. Seedman* **16**, 17, 57.

Triantaphyllou, A. C. and Sasser, J. N. (1960). *Phytopathology* **50**, 724–735.

Turcotte, E. L., Reynolds, H. W., O'Bannon, J. H. and Feaster, C. V. (1963) *Cotton Impr. Conf. Proc.* **15**, 36–44.

Whitehead, A. G. (1968). *Trans. zool. Soc. Lond.* **21**, 263–401.

Wiles, A. B. (1957). *Phytopathology* **47**, 37.

Young, V. H. (1938). *Arkansas Univ. agric. Expt Stn Bull.* 358.

10
Nematode Diseases of Citrus

Eli Cohn

The Volcani Institute of Agricultural Research
Bet Dagan, Israel

I. The Citrus Industry

Citrus is a multifarious evergreen fruit tree crop comprising several thousands of kinds—species, varieties and hybrids. The commonly cultivated citrus fruits belong to three genera, *Citrus*, *Fortunella* and *Poncirus* of the family Rutaceae (Swingle and Reece, 1967). In general, the citrus fruits of principal commercial importance fall into four reasonably well-defined horticultural groups: the oranges, the mandarins, the pummelos (grape-fruits) and the common acid group (citrus, lemons and limes).

The citrus-producing regions have tropical and subtropical climates occupying a belt extending around the world at both sides of the equator to a latitude of 35°N and 35°S. Conditions limiting its distribution in these areas are soil type, sufficient moisture to sustain tree growth, and lack of severe frost. In regions such as the Mediterranean basin and California, citrus orchards are irrigated during the summer and local frosts are common in the winter; there is a distinct harvest season, fruit yields are high and of good quality. In tropical regions, there is a tendency to produce more than one crop a year, fruit quality is poorer and most of the production is directed to local consumption. In regions with intermediate climatic conditions, such as Florida and Brazil, effective cultural practices result in high yields and excellent quality fruits.

An estimate of the world citrus planted area for 1968, based on several sources (Burke, 1967; Burke, 1969; Gonzalez-Sicilia, 1969; Mendel, 1969; Oberholzer, 1969; Singh, 1969; Spurling, 1969) is presented in Table I. About 61% of the total citrus area in the world is concentrated in the Mediterranean and North and Central America; the Far East accounts for about 21% of the world total, South America about 12%, and other countries, principally South Africa and Australia, about 6%. Approximately 60% of the total world plantings are commercial orchards, and almost 80% of these are concentrated in the Mediterranean region and in North and Central America.

Citrus production is one of the world's largest agricultural industries, world trade in citrus being second only to bananas and more than double the volume of apples. In the U.S.A. alone, more oranges are produced than any other fruit crop, comprising one-third of the total fruit tonnage in that country (Hedlund, 1969). Total world citrus production in 1967, based on FAO data (Anon., 1969a) was approximately 29 million metric tons (Table I). About 81% of this volume were oranges and mandarins, about 8% grapefruits, and about 11% lemons and other citrus fruits. Of the total citrus production, about 32% was produced in North and Central America and 30% in the Mediterranean. The increase in citrus production in recent years has been spectacular. In the early 1950s world citrus production totalled 16 million metric tons; by 1965 it had reached 26·5 million metric tons, attaining 29 million metric tons in 1967. Increased production has, in turn, led to increased trade, and the export of fresh citrus from the producing countries has risen from an average of 2·5 million metric tons between 1950–1954 to 4·5 million metric tons in the 1965–1966 season; this represents a rise in exports of 80% as against a rise in production of 65% for the same period (Levin, 1969). FAO studies indicate that the demand for citrus fruit—both for consumption as fresh fruit and for processing—is expected to reach 33–36 million metric tons by 1975 (Anon., 1968).

II. Nematode Diseases of Citrus

Although many pests and diseases of citrus are of economic import- ance to the industry and millions of dollars have been spent in combating Tristeza disease and fruit flies, recognition of nematodes as a cause of

TABLE I. World citrus area and citrus production (1968)

Region and major producing countries	Total citrus area (1000 hectares)	Total citrus production (1000 metric tons)
NORTH & CENTRAL AMERICA		
U.S.A.	405	7555
Mexico	101	1062
Others	28	613
Total	534	9230
MEDITERRANEAN REGION		
Spain	155	2197
Italy	127	2160
Israel	41	1082
U.A.R. (Egypt)	45	705
Morocco	55	629
Turkey	39	545
Algeria	45	400
Greece	25	305
Lebanon	13	238
Others	35	582
Total	580	8843
SOUTH AMERICA		
Brazil	101	2747
Argentina	57	912
Peru	10	270
Paraguay	8	243
Ecuador	8	230
Others	34	466
Total	218	4868
FAR EAST		
Japan	109	1945
India	105	1370
China (Mainland)	101	650
Pakistan	24	399
Thailand	31	228
Others	11	276
Total	381	4868
OTHER COUNTRIES		
South Africa	32	673
Australia	30	223
Others	43	424
Total	105	1320
World Total	1818	29,129

losses in citrus production has been slow to emerge. Nematodes are root pathogens of citrus so that the resultant above-ground symptoms on host plants are usually non-specific, and diagnosis and proof of pathogenicity are difficult to establish. Although some nematodes were discovered on citrus roots at the turn of the century, it was not until the mid 1950s that they were recognized as causing economic damage to the citrus industry, and research was intensified on developing means for their control.

The first record of an association between a nematode and citrus appears to be that of Neal (1889), who found *Heterodera radicicola* (= *Meloidogyne* sp.) parasitizing citrus roots in Florida. The number of species of plant-parasitic nematodes known to be associated with citrus by 1949 was 8, by 1959, 28, and by 1968, 189 belonging to 39 genera (DuCharme, 1968). However, most of these nematodes are not known pathogens of citrus and their true relationship with their host plant still remains to be established. This section will be devoted to the comparatively few cases of nematode–citrus relationships that have been adequately described. These include two citrus diseases of recognized economic significance—slow decline caused by *Tylenchulus semipenetrans*, and spreading decline caused by *Radopholus similis*.

A. Slow Decline

"Slow decline" of citrus is a diseased condition of trees with symptoms similar to those caused by drought and malnutrition. Affected trees exhibit reduced vigour, chlorosis and falling of leaves, twig dieback and, consequently, reduced fruit production. This decline of the tree is gradual and persists until the crop is so small that tree maintenance may become uneconomical.

T. semipenetrans was discovered in 1912 in California on the roots of citrus trees that exhibited a "mottled" appearance (Thomas, 1913). It was described by Cobb (1913) a year later and by 1914, it had already been reported parasitizing citrus roots in Florida, Malta, Spain, Israel, Australia and South America (Cobb, 1914). Since that time its occurrence on citrus roots has been reported from all the major citrus-growing regions in the world and its ubiquitous association with the crop has earned it the common name of the "citrus nematode".

1. Life History and Habits of *T. semipenetrans*

The larvae hatch as the second stage and the male larva undergoes three additional moults within 7–10 days without the need to feed. The non-feeding adult male has an insignificant stylet, a degenerate, non-functional oesophagus and apparently plays no role in the disease syndrome. The female larva is capable of persisting in the second stage for several years, and cannot develop without feeding. In the presence of a host plant, it penetrates the outermost root cell layers, where it undergoes the three additional moults. The nematode usually enters the 4–5 week old "feeder" roots (Cohn, 1964), and becomes permanently established with its anterior end embedded within the plant tissue and its posterior end protruding from the root. The mature female lays eggs into a gelatinous matrix which covers almost the entire protruding part of the female's body. Reproduction is parthenogenetic, and unfertilized females lay eggs that hatch into larvae of both sexes.

The life-cycle of *T. semipenetrans* from egg to egg is completed at temperatures of 24–26°C within 6–8 weeks (Van Gundy, 1958; Cohn, 1964).

2. Effect on Host

Feeding of *T. semipenetrans* is limited to the cortex of host roots where a permanent feeding site consisting of three to four parenchyma cell layers around the nematode head is formed. The head itself is located in a cavity formed from one cell, and is free to move in different directions (Van Gundy and Kirkpatrick, 1964). The "nurse cells" around the nematode head are not unlike the normal adjacent parenchyma cells in shape or size, but differ in their reaction to stains. Starch has been shown to be depleted in these cells as a result of nematode feeding. As the parasite continues to feed, the cells in the feeding site break down, and appear as a mass of disorganized tissue. Subsequently, secondary micro-organisms invade the tissue along the path of nematode penetration and develop in the feeding site causing dark necrotic lesions within the cortex (Cohn, 1965a).

Heavily infested feeder roots of citrus may harbour over a hundred nematodes per centimetre of root. Such roots bear numerous lesions, which give them a darkened appearance. Furthermore, soil particles usually cling tightly, even after washing, to the gelatinous egg masses which cover the protruding part of the nematode body. In extremely heavily infested roots, the entire cortex may separate from the vascular stele (Figs 1 and 2).

The role of secondary organisms in the disease syndrome caused by *T. semipenetrans* is significant and histological studies show that the major part of tissue destruction in roots can be attributed to such organisms invading the nematode feeding site (Cohn, 1965a). Various bacteria and weak pathogenic fungi have been isolated from the feeding sites. Van Gundy and Tsao (1963) demonstrated a greater reduction in citrus seedling growth due to *T. semipenetrans* and *Fusarium solani* combined, than to either alone. However, the exact nature of the relationship between root-rot fungi and the nematode is still unclear.

There is no evidence of a systemic factor being induced by the nematode in the roots and transported through the plant. As the nematode feeds and reproduces, a large proportion of the feeder roots of citrus trees, particularly in the upper soil layers, is inactivated or destroyed, the uptake of water and minerals from the soil is reduced, and the symptoms appear in the above-ground tree parts.

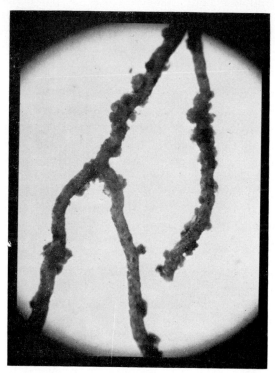

Fig. 1. *Tylenchulus semipenetrans* infection in citrus.
Nematode-infected feeder roots covered with females, egg masses and adhering soil particles (× 12).

3. Host Range

T. semipenetrans is one of the more host-specific plant parasitic nematodes. An attempt to compile a complete list of hosts was made by Vilardebo and Luc (1961) and they recorded 29 species of *Citrus*, 21 citrus hybrids, and 11 other rutaceous species, as hosts. The number of non-rutaceous hosts was six, while seven rutaceous species, all of them other than *Citrus*, were considered proven non-hosts. One additional *Citrus* species, three other rutaceous species, and two non-rutaceous species, have since been reported as hosts. No species of *Citrus* is known to be immune to the nematode. Hence, the evidence is presumptive that all species and hybrids of *Citrus* may act as hosts to *T. semipenetrans*, while the host range among non-*Citrus* forms is limited.

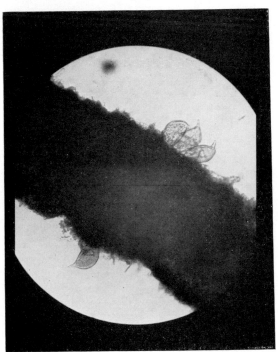

FIG. 2. *Tylenchulus semipenetrans* infection in citrus.
Adult females attached to the roots (× 125).

Hosts of *T. semipenetrans*, including species of *Citrus*, vary considerably in their host status, with some affording a more rapid nematode build-up on their roots than others (Cohn, 1965b). Different populations of *T. semipenetrans* have been shown to exhibit different host preferences, suggesting the existence of biotypes (Baines *et al.*, 1969). Furthermore, different hosts may react differently to parasitism by *T. semipenetrans*. Van Gundy and Kirkpatrick (1964), identified three host reactions in citrus varieties resistant to the nematode—a hypersensitive cell reaction to the feeding of the nematode, a formation of wound periderm in the root cortex and a toxic factor in the root juice.

4. Ecology

The pathogenicity of *T. semipenetrans*, and therefore the manifestation of decline symptoms, is closely related to nematode density. Tree performance deteriorates when the nematode infestation attains a critical level; in Israel this level is approximately 40,000 larvae per 10 g of feeder roots (Cohn *et al.*, 1965). Over the years, populations build up to maximum "ceiling" levels. Although ceiling population levels may vary from one climatic region to another, they are usually attained in orchards in Israel 12–17 years after infested seedlings are planted in virgin soil (Cohn *et al.*, 1965). This is a relatively long time in comparison with other endoparasitic nematodes. However, the duration of the life-cycle of *T. semipenetrans*, even under optimal temperature conditions, is almost twice as long as that of most other endoparasites. Also, the invasion process of *T. semipenetrans* is relatively slow (Cohn, 1964). Hence, citrus trees in their first 2 years in the orchard still usually harbour few nematodes in their roots even if the population of the free-living stages in the soil is high.

Some fairly specific environmental factors are important in determining the rate and extent of nematode build-up. Optimal nematode reproduction occurs at soil temperatures of 28–31°C (Kirkpatrick *et al.*, 1965) and the nematode tolerates wide extremes of soil types. In California, nematode reproduction occurred in soils with a clay content of 5–50%, although optimum reproduction occurred with 10–15% clay (Van Gundy *et al.*, 1964). In fine-textured soils, reproduction was favoured by dry conditions, probably because of an oxygen deficiency when soil moisture was high. No significant differences in population levels were observed in soils of varying texture in Morocco (Vilardebo, 1963). A pH range of 5·6–7·6 has been found favourable for nematode reproduction (Van Gundy and Martin, 1961), and irrigation with water-containing salts has even been reported to favour population build-up (Machmer, 1958). There is some evidence that build-up of *T. semipenetrans* is suppressed in calcareous soils and in orchards irrigated with sewage water (Cohn *et al.*, 1965).

Environmental and cultural conditions, however, may influence the expression of nematode pathogenicity directly, and not only by determining the rate of nematode reproduction. In general, the effect of the nematode on tree performance has been found to be more marked under conditions marginal to citrus cultivation. Martin and Van Gundy (1963) showed that plant growth was inhibited to a greater degree by *T. semipenetrans* when the soil phosphorus level was below the optimum for healthy plant development. Oxygen deficiency in the soil has a more adverse effect on the nematode-infested roots than on nematode reproduction (Stolzy *et al.*, 1962). Also, damage to citrus has been observed to be more severe in wet soils, although the nematode reproduction was better in drier soils (Van Gundy *et al.*, 1964). Finally, the temperature level appears to have a direct influence on disease expression, and workers in California reported that there was greater decrease in weight of *T. semipenetrans*-infested plants at 30°C than at 25°C (Stolzy *et al.*, 1962).

5. *Spread and Survival*

Spread of *T. semipenetrans* is effected primarily by movement of infested plant material and soil. Of these two carriers, there is no doubt that the former is the more effective. The widespread distribution of *T. semipenetrans* throughout the citrus-growing regions of the world was undoubtedly achieved mainly by the transfer of infested citrus planting material. Movement of soil accounts for more local and short distance spread of the free-living stages of the nematode, and its efficiency is dependent on the capacity of the nematodes to survive adverse environmental conditions in the soil. Agricultural implements, animals and man, winds and water are common agents in spreading the nematode in soil. The re-use of sub-soil drainage water in irrigation supplies has led to a widespread contamination of orchards with *T. semipenetrans* in Australia (Meagher, 1969).

Some 70% of a population of free-living stages of the nematode survived storage in water at 10°C for 24 months, 85% of these being second stage female larvae (Cohn, 1966). In soil, the nematode may remain viable in the absence of a host for as long as 9 years (Baines *et al.*, 1962), and can withstand temperatures as high as 45°C for several hours (Feldmesser and Rebois, 1963).

B. Spreading Decline

This disease differs symptomatically from "slow decline" primarily in the increased rate and severity of tree deterioration, in the quick local spread, and in its limited international distribution. An area of spreading decline in a citrus grove has been described as "a group of trees that show the same degree of decline and the area increases in size each year" (Poucher *et al.*, 1967). Affected trees have fewer, smaller leaves, and dead twigs and branches are abundant. The tree appears to be under-nourished although no specific nutritional deficiency symptoms are evident, and it wilts readily during periods of mild moisture stress. Seasonal flushes of growth are weak and although bloom is often profuse, fruit set is sparse and yields are low. The trees do not generally die and they often show temporary recovery after rainy periods. However, normal productivity is never regained (Fig. 3).

Spreading decline was first observed in Polk County, Florida about 1928, increasing in severity in the state during the next decades. By 1957, 2800 ha and by 1966, 6000 ha of citrus were estimated to be

H*

affected (Suit and DuCharme, 1957; DuCharme, 1968). In 1953, the nematode *Radopholus similis* was implicated as the causal agent of spreading decline (Suit and DuCharme, 1957).

FIG. 3. *Radopholus similis* infection in citrus.
Aerial view of citrus orchards showing "spreading decline". (By courtesy of the Florida Department of Agriculture, Bureau of Plant Industry.)

R. similis was first discovered parasitizing banana roots in Fiji in 1893 and was described by Cobb in the same year. It is widely distributed throughout virtually all the tropical and some subtropical regions of the world, but is not known as a pathogen of citrus outside Florida. On account of the cavities and tunnels it produces in the root tissues of its host, *R. similis* is often referred to by its common name— "the burrowing nematode".

1. Life History and Habits of *R. similis*

All larval stages and adults of *R. similis* are vermiform and possess powers of locomotion. It is an endoparasite and may spend its entire life within the host root. Larvae and females penetrate the young and succulent citrus root tips and once inside the root, the nematodes reproduce rapidly. Eggs are laid singly inside the root and larvae hatch within 3–7 days. Males are not capable of penetrating roots and probably do not feed. Females are capable of producing viable eggs in the absence of males, from which colonies of males and females develop. The life-cycle, from egg to egg, requires 18–20 days at 24–26°C (DuCharme and Price, 1966).

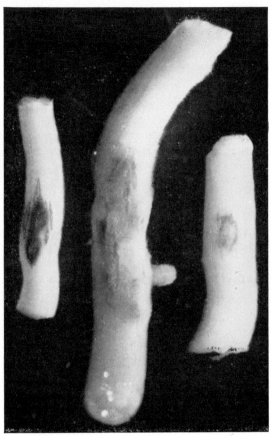

Fig. 4. *Radopholus similis* infection in citrus.
Citrus feeder roots with lesions caused by *R. similis*. (By courtesy of E. P. DuCharme.)

Migration from roots into the soil occurs as a result of population density, food shortage, putrefaction by secondary root invaders, fouling of the habitat from accumulation of nematode waste products, and deposition of wound gum by the plant (DuCharme, 1968).

2. Effect on Host

The penetration and histopathology of *R. similis* on citrus was studied in detail by DuCharme (1959). Females and larvae enter growing feeder roots near their tips in the region of cell elongation and root hair production. Upon penetration, the nematode feeds on the cortical parenchyma cells and gradually burrows towards the stele, creating tunnels and cavities in the tissue. Large numbers of nematodes accumulate in the phloem cambium ring region. This part of the root is often completely destroyed, leaving a cavity filled with nematodes. Cell reaction involves hypertrophy and, when the nematode penetrates the pericycle, hyperplasia and tumour formation occur, and wound gum accumulates in the parasitized tissues.

Infested roots bear numerous lesions (Fig. 4) from which nematodes have been extracted from citrus feeder roots in extremely large numbers, e.g. 1–739 in individual lesions (Poucher *et al.*, 1967). Under controlled conditions, a single female created a colony of 47,000 nematodes in 85 days. Such high populations are not attained in the field because of the activity of secondary organisms which invade the roots after nematode penetration. Hence, populations are usually higher in roots of trees marginal to the diseased area, than on trees in a state of advanced decline.

Spreading decline in a tree is due to the combined activity of the nematode and other organisms that are always found in the parasitized roots. Although it is recognized that the nematode is the primary cause of disease, pathogenic fungi such as *Fusarium, Sclerotium* and *Thielaviopsis* are commonly found in lesions soon after nematode invasion. Subsequently, bacteria and cellulose-destroying fungi such as *Penicillium* and *Aspergillus* invade the root, causing additional destruction. Finally, oligochaete worms, mites and other soil-inhabiting organisms enter the decaying roots and feed on the fungi and the decomposing root tissues. As the nature of the root tissue changes, the nematodes leave their habitat and migrate to new healthy roots (DuCharme, 1968).

The decline symptoms in the above ground parts of the tree are a result of the destruction of feeder roots. Symptoms of decline appear about one year after initial infection of roots. Thus, under orchard conditions, *R. similis*-infected trees that appear healthy occur one to three rows in advance of trees with visible decline symptoms.

3. Host Range

R. similis has a large host range. In 1967, 244 species, both herbaceous and woody plants, were recorded as hosts. In Florida, lists of non-hosts are also periodically compiled to aid growers in selecting cover crops in *R. similis*-infected areas; the non-host list in 1967 included only forty plant species (Poucher *et al.*, 1967). 1275 different kinds of citrus were found to act as hosts of *R. similis* to varying degrees in Florida (Ford *et al.*, 1960).

Two physiological races of *R. similis* have been identified: the "citrus race" which parasitizes both citrus and banana, and has been recorded only in Florida, and the "banana race" which parasitizes banana but not citrus (DuCharme and Birchfield, 1956).

4. Ecology

Populations of *R. similis* are high in the autumn and low in late spring in spreading decline-affected orchards. These fluctuations appear to be primarily a temperature effect. The optimum temperature for nematode reproduction and root invasion is 24°C, the minimum 12°C and the maximum 29·5–32·5°C (DuCharme, 1969).

Population levels of *R. similis* at individual sites vary considerably. More nematodes are found on roots of trees newly infected than on trees infected for 2 or more years (DuCharme and Price, 1966). Unlike *T. semipenetrans*, which is concentrated in the upper soil layers of the grove, *R. similis* is rarely found in the top 15 cm of soil. Highest populations are found between 30 and 180 cm and nematodes have been found feeding on roots at a depth of 4 m (Suit *et al.*, 1953). In trees affected by spreading decline, 25–30% of the feeder roots at depths of 25–75 cm, and 90% of the feeder roots below 75 cm, are destroyed (Ford, 1953). Spreading decline occurs predominantly in areas with sandy soils, where spread of disease and nematode movement are rapid. One nematode was shown to travel through sand to a distance of 100 mm in 96 h (Tarjan and Hannon, 1957), and new areas have been found to become infested with nematodes at a rate of 15·24–21·59 cm per month (Feldmesser *et al.*, 1960).

5. Spread and Survival

The spread of *R. similis* to new regions is effected, as in the case of *T. semipenetrans*, by the movement of soil and infected plants. The limited international distribution of spreading decline, as compared with slow decline, appears to be partly due to a lesser ability of *R. similis* as compared with *T. semipenetrans* to withstand adverse conditions. Populations of *R. similis* do not survive more than about 6 months in soil free of citrus and other hosts, and even in such cases the nematode probably survives in pieces of living root scattered in the soil (Hannon, 1963).

Unassisted spread of *R. similis* is somewhat dramatic in comparison with that of other plant-parasitic nematodes, and the name of the disease it causes is indicative of this characteristic. The nematode has been reported to infect new groves from existing centres of infestation by crossing under clay and asphalt roads up to over 30 m wide, and even under a railway line (DuCharme, 1968). Spread occurs in all directions. The average yearly rate of spread in Florida groves has been calculated as 1·6 trees, or 15·2 m, per year. Sub-soil drainage and topography influence the rate of spread (DuCharme, 1955), and in one instance the nematode was found to spread downhill at a rate of 66 m in one year, while the uphill rate was less than 8 m. The rapid unassisted spread of the nematode has led to efforts to create barriers in an attempt to halt its movement.

C. Other Nematode Pathogens

Soil samples from around citrus trees throughout the world have revealed the presence of numerous species of nematode but only a few of these have been confirmed as being pathogens. Information on other nematodes found attacking citrus is summarized in Table II. Most of the genera listed are fairly widespread, and several species are known as pathogens of other crops of economic importance. It is characteristic, however, that most of the data contained in Table II represent knowledge obtained during the last decade; and the damage these organisms do to citrus still needs to be evaluated.

III. Control

There is no single method of controlling all nematode pests of citrus. Nematodes differ in their biology and parasitism and so require different methods of control. Furthermore, the diversity of climatic, edaphic and cultural conditions under which such a broad spectrum of citrus varieties and rootstocks are cultivated throughout the world necessitates different and sometimes highly specific solutions to nematode control problems.

It should be emphasized from the outset that prevention is generally cheaper and more effective than cure. Once the nematode becomes established, its complete extermination is virtually impossible. Most of the ways and means of controlling nematodes known to us today are directed towards reducing populations to a minimum, and thereby creating optimum conditions in which the citrus trees can thrive. These means are sub-divided in this section under two fairly broad titles—chemical and cultural control measures.

A. *Tylenchulus semipenetrans*

1. Preventive Measures

The most common measures taken to prevent contamination of citrus trees by *T. semipenetrans* are directed at precluding the spread

TABLE II. Nematodes known to attack citrus

Nematode	Feeding habit	Root symptoms	Occurrence	Reference
Aphelenchoides citri	endoparasite		Hungary (greenhouse)	Andrassy (1957)
*Belonolaimus longicaudatus**	ectoparasite	blinded tips, terminal lesions	Florida, U.S.A.	Standifer and Perry (1960)
Criconema australis	ectoparasite		Queensland, Australia	Colbran (1963)
C. civellae	ectoparasite		Maryland, U.S.A. (greenhouse)	Steiner (1949)
Criconemoides citri	ectoparasite		Florida, U.S.A.	Steiner (1949)
Helicotylenchus multicinctus	endoparasite		Canary Islands	Guiran (1962)
*Hemicycliophora arenaria**	ectoparasite	swellings or galls on tips	California, U.S.A.	Van Gundy and McElroy (1969)
H. nudata	ectoparasite	terminal galls	Queensland, Australia	Colbran (1963)
Meloidogyne sp.*	endoparasite	galls and browning	Taiwan; India	Chitwood and Toung (1960)
M. exigua	endoparasite	galls	Surinam	Ouden (1965)
M. incognita	endoparasite	galls	Queensland, Australia	Colbran (1958)
M. indica	endoparasite	galls	India	Whitehead (1968)
M. javanica	endoparasite	galls	Israel	Minz (1956)
*Pratylenchus brachyurus**	endoparasite	lesions	Florida, U.S.A.	Brooks and Perry (1967)
*P. coffeae**	endoparasite	lesions	Florida, U.S.A.	Feldmesser and Hannon (1969)
Rotylenchus reniformis	semi-endoparasite		Ghana	Peacock (1956)
Sphaeronema minutissimum	semi-endoparasite		Indonesia	Goodey (1958)
*Trichodorus christiei**	ectoparasite	stubby-root	Florida, U.S.A.	Standifer and Perry (1960)
*Xiphinema brevicolle**	ectoparasite	browning and lesions	Israel	Cohn and Orion (1970)
X. coxi	ectoparasite		Florida, U.S.A.	Tarjan (1964)
*X. index**	ectoparasite	browning and lesions	Israel	Cohn and Orion (1970)
X. vulgare	ectoparasite		Florida, U.S.A.	Tarjan (1964)

* Proven pathogens.

of the nematode—on a local or international level—through infested nursery stock. In most countries quarantine precautions are taken to ensure that imported, rooted citrus seedlings are free of the nematode. In several countries, regulations exist prohibiting the sale of nematode-infested plants, although enforcement of such laws has proved difficult. Nematode-free seedlings are produced in nurseries in virgin and/or fumigated soil remote from existing citrus orchards. Prior to their transfer to orchards, roots of seedlings can be dipped in hot water at 45°C for 25 min; this treatment kills the nematode without injuring the roots (Baines, 1950). Aqueous emulsions of various systemic organophosphates are effective control agents (O'Bannon and Taylor, 1967).

Other preventive measures are directed against the spread of nematodes in soil. Farm implements and machinery used in infested areas should be cleaned preferably with disinfectants, before being transferred to other regions. Re-use of drainage water from infested regions for irrigation of citrus orchards should be avoided (Meagher, 1969).

2. Chemical Contro

Essentially, chemical control of *T. semipenetrans* is being practised by two different approaches, pre-plant and post-plant fumigation which, accordingly, have two different aims.

(a) *Pre-plant Fumigation.* The aim of pre-plant treatments is to kill the free-living stages of the nematode (larvae and males) and, if possible, the eggs present in the soil. These treatments, done in the absence of the host, can be drastic. They are applied primarily on land where young trees are to be introduced in place of old infested trees which have been removed. The chemicals selected for such procedures are often products not only highly nematicidal, but also generally biocidal; methyl bromide and chloropicrin are the most common chemicals of this type, but they are relatively expensive and, being highly volatile and toxic, usually require application under gas-tight covers for good results. Recently some applicators have been devised for applying them without need for covers (Amstutz, 1968). Less costly nematicides such as DD (dichloropropane-dichloropropene), ethylene dibromide or DBCP (1,2-dibromo-chloropropane) applied by soil injection or drench techniques can be used for pre-plant fumigation, but their effects in improving tree growth are not as impressive or as persistent as those of the overall biocides. It is sometimes convenient and economical to confine treatments of replants to small areas around the planting sites, rather than fumigating the entire area, unless the replanted area is large.

A temporary inhibition of growth of citrus has been observed on seedlings following either soil fumigation (Cohn *et al.*, 1968) or heat treatment of soil (Martin *et al.*, 1963). Consequently, it is advisable to extend the interval between fumigation and planting beyond that normally maintained for other crops. Under subtropical climatic conditions, the most suitable procedure is to fumigate in the late summer or in autumn, prior to the introduction of replants in the following spring or early summer.

Pre-plant soil fumigation usually results in a marked increase in growth and development of replants. It must be emphasized, however, that the effect of such treatment is to reduce the nematode population temporarily and thereby give the young citrus trees a good start in the orchard. Nematode populations in re-set orchards build up relatively rapidly and critical population levels are usually attained within 4–7 years after planting (Cohn *et al.*, 1965).

(b) *Post-plant Fumigation.* The aim of post-plant fumigation is to kill both the free-living stages of *T. semipenetrans* in the soil and, especially, the numerous females attached to the feeder roots of the citrus tree, without adversely affecting plant growth. Only DBCP, the least phytotoxic of the fumigants at present on the market, has been used successfully for this purpose. Excellent nematode control, following DBCP treatments in established orchards, usually results in yield increases. Rates of DBCP applied (ranging between 11 and 67 litres active ingredient per hectare) and modes of application differ in accordance with varying local conditions. In general, three principal modes of application, modified and improvised to suit local needs, have been used: application as a soil drench in irrigation basins or flood systems, application by chisel injection, and application through sprinkler irrigation systems. Drench treatments have usually resulted in the best nematode control (Baines *et al.*, 1960; Cohn and Minz, 1965), but not the most favourable increase in yields, probably because of a phytotoxic effect. This application method is, of necessity, limited to flood-irrigated groves, or groves where preparation of basins is feasible from the standpoint of existing cultivation methods, topography and low labour costs. Application methods based on chisel injection are popular in the U.S.A. and have resulted in satisfactory nematode kill and increased fruit yields (Baines *et al.*, 1963; O'Bannon and Tarjan, 1969), but these methods are liable to cause some root damage where citrus trees have shallow root systems. Sprinkler application methods have so far been the least effective in controlling *T. semipenetrans* (Baines *et al.*, 1960; Baines and Small, 1969), apparently because of poor penetration of the chemical to the site of action and its adsorption on to soil particles. Improved nematode control by sprinkler application has been achieved by using a special drill-perforated irrigation pipe which produces water trajectories closer to the ground, thereby enabling larger quantities of water to reach under the tree canopy (O'Bannon and Tarjan, 1969).

Despite the many convincing results of increased yield and size of citrus fruit following nematode control, there have also been reports on a lack of tree response to DBCP treatments, although nematode population levels were adequately reduced (Cohn et al., 1968; Mendel et al., 1969). Although our knowledge of these factors limiting the success of DBCP fumigation is still fragmentary, some factors have been isolated. Citrus appears to be more susceptible to the phytotoxicity of DBCP than many other plant species (Cohn et al., 1968), although different rootstocks may differ in their degree of susceptibility (Mendel et al., 1969). Consequently, it is always necessary to experiment with different doses and application techniques of DBCP under various local conditions, before embarking on a particular control programme. Also, in fine textured soils, DBCP fumigation should be avoided, as it results in marked phytotoxicity as well as poor nematode kill. Finally, old trees, especially if they are in an advanced state of decline, often do not respond to DBCP treatments, since DBCP is specifically nematicidal and has little effect on the micro-organisms involved in the disease complex caused by T. semipenetrans. It is feasible that the fumigant will be less effective in curing old trees in long-standing orchards, where high populations of these organisms have become established.

As a result of these limitations of DBCP fumigation, chemicals less phytotoxic than DBCP have been tested for use in established orchards, and the systemic compounds Furadan, Mocap and Temik have given promising results in nematode control (Baines and Small, 1969).

3. Cultural Measures

Work was recently begun in California on the breeding of citrus rootstocks resistant to T. semipenetrans. These are intergeneric hybrids of Poncirus trifoliata and several species of Citrus. Preliminary results are promising, with most of the selected hybrids showing initial resistance in greenhouse pot tests and about half of them showing high resistance in the field (Cameron et al., 1969). Different selections of P. trifoliata vary in their tolerance to T. semipenetrans (Feder, 1968) and a further problem in the practical use of nematode-resistant rootstocks is the possibility of resistant breaking biotypes among field populations of T. semipenetrans (Baines et al., 1969).

Tichinova (1957) reported that two applications of manure ($\frac{2}{3}$ cow manure $+\frac{1}{3}$ water, diluted by 1 : 10 before application) in 2 months controlled T. semipenetrans and improved tree growth in Uzbekistan. This finding is corroborated by the report of Cohn et al. (1965) who observed that in an orchard in Israel irrigated with sewage water, nematode populations were very low, while in neighbouring fresh water-irrigated orchards, nematode infestation was heavy. The nature of this effect is not clear, but organic amendments have been shown to reduce soil populations of *T. semipenetrans* apparently by increasing the microbial activity which is unfavourable for survival of the nematode (Mankau and Minteer, 1962).

B. *Radopholus similis*

The control of *R. similis* has proved infinitely more difficult than that of *T. semipenetrans*. This is undoubtedly due to the true endoparasitic habit of the nematode, its concentration in the lower soil layers, the complex and more acute nature of spreading decline disease, and its rapid rate of natural spread through the soil. On the other hand, the limited occurrence of the disease to Florida in comparison with the world-wide distribution of slow decline, has made it possible to devise more uniform control measures applicable to the specific local conditions. A state-wide control programme is based on (*a*) preventing the nematode from becoming established in new areas, (*b*) eliminating the nematode in commercial orchards, and (*c*) slowing down the rate of natural spread through the soil to areas adjacent to infested orchards. This control programme (Poucher *et al.*, 1967), administered by the Florida Department of Agriculture, is included in the following review.

1. Preventive Measures

All citrus-growing countries have special regulations, especially for plant material entering from Florida, to preclude the entry of *R. similis*, and particularly the "citrus race" of the nematode. Stringent intra-state regulatory measures exist also in Florida itself to prevent spread of the nematode. Citrus nursery sites and the movement of material from clay, soil or sand pits to citrus producing areas in the state are subject to approval by the authorities. *R. similis*-infested seedlings are treated before planting in approved sites within an infested region by immersing bare roots of seedlings in hot water at 50°C for 10 min and then immediately cooling them in cold water for 10 min. Nursery trees thus treated require more care and more frequent watering than non-treated trees. More recently, O'Bannon and Taylor (1967) have shown that bare-root dips of citrus seedlings in 250–600 ppm of B-68138 (ethyl 4-(Methylthio)-*m*-tolyl isopropyl phosphoramidate) and 1000 ppm of Cynem, Dasanit and Mocap, gave 100% control of the nematode, without phytotoxic effects.

Movement of infested soil is minimized by cleaning cultivation equipment after use in infected groves. Disinfectants, such as 1% caprylic acid and 2·6% sodium hypochlorite, can be used in the wash water (Tarjan, 1957).

The migration of the nematode from an infested site to adjacent groves is slowed by treating strips of land (buffer zones or barriers) around an infested area. These buffers are at least 4·8 m wide and are placed six rows in advance of the visible decline symptoms in a grove. They are fumigated with ethylene dibromide at 560 litres/ha during initial application and 280 litres/ha every 6 months thereafter, while the surface is maintained clean of weeds and cover crops by cultivation and herbicide treatment.

2. Chemical Control

Nematicidal control of spreading decline has not been very successful. The only effective treatment found to date is the procedure known as "Push and Treat". This involves pushing out all infested trees and the two uninfested outside rows of trees, stacking and burning them, raking the infested field to remove as many roots as possible, levelling the land and treating it with DD at the rate of 1120 litres/ha with injections spaced no farther than 45 cm apart to a depth of 25–30 cm, maintaining the land free of all vegetation for a minimum of 6 months, and releasing the land for replanting of citrus not earlier than 2 years after the original soil treatment. The entire "Push and Treat" programme in Florida is under direct state supervision.

Post-plant treatment to *R. similis* under field conditions has failed, apparently because of poor soil penetration and diffusion of the fumigant throughout the subsoil. Experiments are still under way to achieve successful field control by improving methods of DBCP application (Suit, 1969).

3. Cultural Measures

A large variety of methods and materials including soil conditioners and amendments, exposure of trees to electricity and radiation, "buck horning" or cutting back of trees, deep ploughing and frequent cultivation have been tested throughout the years as a possible cure for spreading decline. Generally, those treatments with some fertilizer value tend to promote a slight temporary improvement in growth, but no treatment successfully controls the disease or even reduces nematode numbers.

Attempts at biological control—either by the use of nematode-trapping fungi (Tarjan, 1961) or by growing *Tagetes* as a cover crop (Tarjan, 1960)—have proved ineffective.

The only effective cultural measure available for checking spreading decline is the use of tolerant or resistant rootstocks. Since 1951 more than 1400 different kinds of citrus have been screened, and in 1964 three rootstocks were formally released to the industry in Florida for planting on land in spreading decline areas which had been pushed and treated (Ford, 1964). "Estes", a rough lemon, is a rootstock which supports a relatively high level of *R. similis* but is classified as "tolerant" because its growth is not reduced to an appreciable degree. "Ridge pineapple", a sweet orange variety, is resistant to *R. similis* and, where planted, existing nematode populations gradually diminish. "Milam" lemon, a citrus hybrid of unknown parentage, also is resistant to *R. similis* and gradually eliminates nematode populations from the soil by preventing egg development within the root cortex. It is, however, susceptible to nematode penetration and feeding. All three rootstocks are tolerant to several virus diseases, but contain no resistance to *T. semipenetrans*. Recently Carrizo citrange (navel orange × *Poncirus trifoliata*) rootstock has been shown to eliminate *R. similis* populations within 2 years, although suffering some growth reduction during this period, but is also tolerant to *T. semipenetrans* and produces good fruit quality on scion varieties (Ford and Feder, 1969). It is recommended to fumigate the old grove soil before replanting with any of the nematode-tolerant or resistant rootstocks.

The use of *R. similis*-resistant rootstocks as biological buffers or barriers has been advocated. This involves the planting of four rows of "Milam" or "Ridge Pineapple" as a biological barrier zone and maintaining a narrow chemical root-killing strip between these trees and the nematode-infested grove. Trials have shown that over a 5-year period so far tested, such barriers have effectively contained *R. similis* migration (Ford and Feder, 1969).

C. Other Nematodes

Although little or no direct work has so far been carried out on the control of other nematodes on citrus, many of the methods discussed above are probably effective also against other nematode species. Several reports indicate sharp reductions of various species of ectoparasitic nematodes following soil fumigation for *T. semipenetrans* or *R. similis* control in citrus orchards.

Hemicycliophora arenaria, an ectoparasite native to the desert valleys of California, was eradicated, with no damage to citrus seedlings, in hot water (46°C for 10 min) and VC-13 dip treatments. Soil fumigation with DD, DBCP and methyl bromide at standard application rates gave complete kill of the nematode to a depth of 1·5 m. Of twelve common citrus rootstocks tested as hosts, six were resistant, namely: sweet orange, sour orange, Marsh grapefruit, Trifoliate orange, Troyer citrange and Carrizo citrange (Van Gundy and McElroy, 1969).

Post-plant treatments of DBCP and the organo-phosphate chemicals Mocap, Dasanit and B-68138 reduced the populations of *Pratylenchus brachyurus* and *P. coffeae* to varying degrees, and increased the growth of rough lemon seedlings in comparison with untreated controls (Tarjan and O'Bannon, 1969).

The only other nematode on citrus for which specific control measures have been considered is *Meloidogyne* sp. ("the Asiatic Pyroid citrus nema") in Taiwan. Chitwood and Toung (1960) suggested as an interim measure the use of several non-hosts as cover and trap crops in citrus orchards, until long-term economic control measures are developed.

IV. Economics

An accurate appraisal of the economic effects of nematodes is dependent on the results of surveys and experiments. Since the severity of nematode damage to plants varies in relation to environmental factors, this information must be based on studies done under a wide range of ecological conditions. Since prices and production costs are prone to regional and seasonal fluctuations, estimates of average annual losses must be made over a range of conditions and seasons. It is not surprising, therefore, that present data, both on measurements of nematode damage and on their conversion to estimates of regional losses, are extremely meagre. Only the economics of slow decline and spreading decline will be considered here. Owing to the scarcity of data, some generalizations in place of more valid estimates are inevitable.

The economic effects of nematodes fall into three broad categories which will be discussed separately:

A. Reduction in quantity and quality of fruit
B. Increase in production costs
C. Indirect losses to the community

A. Reduction in Quantity and Quality of Fruit

LeClerg (1964) recognizes two phases in the study of losses in crops due to disease: determination of disease intensity and the establishment of the relationship between intensity and loss of production.

1. Disease Intensity

The prevalence of the causal organism usually serves as an indicator of disease intensity. Surveys of the occurrence of *T. semipenetrans* indicates that the nematode is present in all citrus-growing countries in the world. Furthermore, distribution within those countries is widespread, and estimates range between a low of 53% of all groves in Florida (Tarjan, 1967) to a high of over 90% in Spain (Scaramuzzi and Perrotta, 1969). On the strength of published surveys, a figure of

70–80% of all citrus trees throughout the world could be realistically considered to be infested to some degree with *T. semipenetrans*.

On the other hand, spreading decline is much more limited in its occurrence. Surveys in Florida have established that in 1967, the disease occurred in about 1·9% of the total citrus acreage of the state (6000 out of 325,000 hectares). Since Florida plantings constitute about 18% of the world citrus acreage, this represents only 0·35% of the total world citrus plantings infected with spreading decline disease.

2. Relationship between Intensity and Loss

The effect of the disease on the crop is dependent not only on the infestation level, but also on the susceptibility and age of the host and the influence of environmental conditions. Consequently, losses are variable according to different conditions and must be evaluated on a local basis. Estimates of loss can be attained by measuring yield increases as a result of disease elimination through successful nematode control, and by comparing the performance of infested with that of uninfested trees. Data are available on yield increases following *T. semipenetrans* control with DBCP under diverse conditions: 5–40% in Mediterranean countries (Scaramuzzi and Perotta, 1969), 6–34% in Florida (O'Bannon and Tarjan, 1969) and 10–30% in Australia (Meagher, 1969). In most cases the increased yields persist for approximately 3 years after a single treatment. In Arizona, U.S.A., where more than 970 hectares, representing about 5% of the total citrus acreage of the state, have been treated, yield increases of 12–38% have been reported (Reynolds, 1969). In California, where research on *T. semipenetrans* control has been more intensive than anywhere else, citrus yields following DBCP treatments have increased by 10–50%, with an average of 27% (Baines and Small, 1969). On the basis of these studies, a world average range of 20–30% increase in citrus yield resulting from the control of *T. semipenetrans* appears likely. To this estimate must be added the effects of secondary organisms which are not controlled by nematicide treatments and allowance must also be made for the mild phytotoxicity of DBCP which prevents treatments from attaining their full potential effect. Accordingly, a final estimate of 25–35% as an average world yield reduction due to *T. semipenetrans* on citrus would be more accurate.

Comparison of yields of infested with those of uninfested trees has proved far more difficult and less practical. However, it has been established that a critical infestation rate of *T. semipenetrans* exists below which tree performance is hardly affected. Surveys in Israel have revealed that 35% of all citrus trees in that country are infested with critical and higher levels of *T. semipenetrans*; this figure is an average of five regions, differing markedly in prevailing climatic and edaphic conditions (Cohn, 1969). Similar information from other countries is not available. On the assumption that this figure is close to the world average, the actual reduction in world citrus yields due to *T. semipenetrans* could be estimated at 8·7–12·2% (25–35% of the world total).

Local crop losses due to spreading decline are appreciably higher than those caused by slow decline, and have been estimated at 50–80% for grapefruits and 40–70% for oranges (DuCharme, 1968). However, in terms of the world citrus industry, these losses have only a relatively minor impact. Since Florida produces about 18% of the world's citrus, and spreading decline affects only 1·9% of the state's orchards, the total reduction in world citrus yields due to spreading decline would be 0·14–0·27%.

Reduction in fruit quality is also a consequence of nematode pathogenicity. Thus, for instance, DBCP-treated grapefruit trees produced 510% and 242% more fruit size 40s and 48s or larger, respectively, than untreated trees during a 7-year period in Arizona (Reynolds, 1969). However, since the industries for citrus-juice and other citrus by-products provide markets for fruit of reduced quality, such losses, although sometimes fairly substantial, particularly in citrus-exporting countries, do not significantly affect the estimates given above in terms of total world production.

The total annual reduction in the world citrus crop due to both slow decline and spreading decline is therefore estimated at 8·8–12·5%.

B. Increase in Production Costs

Nematodes cause losses also by increasing the costs of production primarily through increased maintenance costs of infested orchards, and extra expense incurred by nematode control measures.

1. Maintenance Costs

Nematode-infested groves demand additional care and investment which increases with time. In inoculation trials with *T. semipenetrans*, 10–50% weight reductions in citrus seedlings were obtained (Baines and Clarke, 1952). Additional amounts of fertilizers and water must be supplied to nematode-infested trees in order to maintain a profitable production level. Ironically, the increase in maintenance costs is probably a more important factor in orchards with slow decline than in spreading decline areas, not only because of its wider occurrence, but because disease expression is not so dramatic and diseased orchards are maintained for longer periods, whereas spreading decline is recognized comparatively early and drastic control measures are taken sooner.

Reliable figures from citrus-producing countries on the losses resulting from increased maintenance costs are lacking and an estimate on a world-wide basis would be impractical because of the extreme variability of both production costs and citrus production values in different parts of the world. Some national data on maintenance expenses in citrus orchards, however, are available, and might be useful in attempting to evaluate the part played by nematodes in increasing costs. Data in Israel, for instance, indicate that all cultivation costs (predominantly labour) constituted an average of about 25% of the annual national citrus production value over a 5 year period. Materials (mainly water and fertilizers) constituted only 6·8% of the production value (Anon., 1969b). The extent of losses due to *T. semipenetrans* is not known, but growers have often claimed that 20–50% additional amounts of these materials (the equivalent of 1·4–3·4% of the production value) are needed to maintain the thriftiness of infested orchards. Since 35% of the citrus trees are infested with critical and higher population levels, as indicated earlier, the net loss due to *T. semipenetrans* in Israel could be roughly estimated at 0·5–1·2% of the annual production value. Similar estimates could also be made for other countries where profitability surveys on citrus have been done.

Similarly an estimate of the increasing maintenance costs in Florida due to *R. similis* could be calculated. However, because its occurrence is restricted to Florida, such losses in terms of the world citrus industry would assume much smaller proportions than those caused by *T. semipenetrans*.

2. Nematode Control Costs

Pre-plant fumigation of *T. semipenetrans*-infested orchard soil is usually a once-only operation and so is relatively unimportant. On the other hand, DBCP treatment for control of *T. semipenetrans* on existing trees is a recurrent expense item, the cost of which depends primarily on the rate of the chemical used and the application method employed. These costs can be estimated on a yearly basis by considering a single treatment effective for an average of 3 years, before retreatment is necessary. Reynolds (1969) has calculated that the average cost of chisel application of DBCP in Arizona is about U.S. $0.10–0.12 per tree per year, which is equivalent to approximately U.S. $50–62 per hectare. In nematode control trials in California, costs ranged between U.S. $47–124 per hectare per year, depending on application doses (Baines *et al.*, 1960). The average cost per year of DBCP treatment by chisel application in Israel is estimated at U.S. $35–47 per hectare.

Basin application usually requires more labour expenses, and has been reported to cost an average of U.S. $0.80 per tree in Italy (Scaramuzzi and Perrotta, 1969), or approximately U.S. $131 per hectare per year. The average yearly cost of basin treatment in Israel is estimated at U.S. $57–74 per hectare.

The profitability of nematicide treatments on a regional basis can best be appraised by considering the costs in relation to the production value. In Israel, for example, the cost per hectare per year of chisel treatment represents about 1·2–1·7% of the average citrus production value per hectare for the 1967/68 season (Anon., 1969a), when taking into consideration that only about 35% of all trees in the country require treatment. The actual loss to the world citrus industry as a result of *T. semipenetrans* control costs can be estimated only if the total citrus acreage treated annually is known. Such data are not readily available, but reports indicate that treatments on a commercial scale are still very limited. Thus, in California, only about 1% of the total citrus acreage of the state was treated with DBCP during 1970 (Anon., 1971).

Control of *R. similis*, in terms of cost per unit area, has proved infinitely more expensive than control of *T. semipenetrans*. Indeed, the programme in Florida for eradicating spreading decline stands out as a stern warning of the potential danger of nematodes to world agriculture. Between 1954 and 1955, $70,000 was released by the Florida state authorities for the purpose of identifying diseased areas. In 1955, $1,756,300 was appropriated for controlling or containing the disease (Suit and DuCharme, 1957). Between 1955 and 1966 nearly 3240 hectares were pushed and treated, 360 of them having been double-treated with DD; more than one million citrus nursery trees were treated in the central state-owned hot-water treating tank alone; and 356 km of buffers were installed, within which there were 3250 hectares of infested groves (Poucher *et al.*, 1967). Owing to the preventive nature of these drastic measures, the losses accountable to nematode control within the affected areas of Florida represent more than a total loss of production, but it has been estimated that more than 21,500 hectares would have become infested by 1967 if these measures had not been taken (Poucher *et al.*, 1967).

C. Indirect Losses to the Community

The economic effects of nematodes extend also to fields of activity beyond the citrus industry. This is particularly manifest with *R. similis*. Nematodes reduce soil use, and, therefore, more land is required for citrus production, usually at the expense of other crops. The wide host range of *R. similis* and the resultant need to control the growth of such potential hosts in Florida, has affected other aspects of agriculture. The stringent quarantine regulations existing in most countries of the world, particularly those aimed at preventing importation of *R. similis*, have curtailed international trade of different agricultural products. Thus, the losses caused by nematodes on citrus are by no means confined to the producer, consumer and trader of citrus produce. In fact, although not always immediately identifiable, and largely still immeasurable, these losses are also suffered by the community at large, which, in turn, is a partner in bearing the burden of financing the investigation and the control of the microscopic organisms which cause them.

References

Amstutz, F. C. (1968). *Down to Earth* **24**, 15–16.
Andrassy, I. (1957). *Nematologica* **2**, 238–241.
Anon. (1968). Agricultural Commodities—Projections for 1975 and 1985. Vol. 1, FAO, Rome.
Anon. (1969a). Production Yearbook, 1968. Vol. 22, FAO, Rome.
Anon. (1969b). Citrus Sector Accounts, 1962/63–1967/68. Output, Input, Income and Profitability. Israel Central Bureau of Statistics, Special Series No. 284, Jerusalem.
Anon. (1971). Pesticides used in California in 1970. California Department of Agriculture.
Baines, R. C. (1950). *Calif. Agric.* **4**, 7.
Baines, R. C. and Clarke, O. F. (1952). *Calif. Agric.* **6**, 9, 13.
Baines, R. C. and Small, R. H. (1969). *Proc. First Intern. Citrus Symp.* Riverside, Calif. **2**, 973–977.
Baines, R. C., Stolzy, L. H. and Small, R. H. (1963). *Calif. Citrogr.* **48**, 186, 207–211.
Baines, R. C., Miyakawa, T., Cameron, J. W. and Small, R. H. (1969). *Proc. First Intern. Citrus Symp.* Riverside, Calif. **2**, 955–956.
Baines, R. C., Martin, J. P., De Wolfe, T. A., Boswell, S. B. and Garber, M. J. (1962). *Phytopathology* **52**, 723.
Baines, R. C., Stolzy, L. H., Small, R. H., Boswell, S. B. and Goodall, G. E. (1960). *Calif. Citrogr.* **45**, 389, 400, 402, 404–405.
Brooks, T. L. and Perry, V. G. (1967). *Pl. Dis. Reptr* **51**, 569–573.
Burke, J. H. (1967). *In* "The Citrus Industry" (W. Reuther, H. J. Webber and L. D. Batchelor, eds), University of California, Vol. **1**, pp. 40–189.
Burke, J. H. (1969). *Proc. First Intern. Citrus Symp.* Riverside, Calif. **1**, 135–140.
Cameron, J. W., Baines, R. C. and Soost, R. K. (1969). *Proc. First Intern. Citrus Symp.* Riverside, Calif. **2**, 949–954.
Chitwood, B. G. and Toung, M. C. (1960). *Pl. Dis. Reptr* **44**, 848–854.
Cobb, N. A. (1893). *N. S. Wales Dept. Agr. Misc. Publ.* **13**, 1–59.
Cobb, N. A. (1913). *J. Wash. Acad. Sci.* **3**, 287–288.

Cobb, N. A. (1914). *J. agric. Res.* **2**, 217–230.

Cohn, E. (1964). *Nematologica* **10**, 594–600.

Cohn, E. (1965a). *Nematologica* **11**, 47–54.

Cohn, E. (1965b). *Nematologica* **11**, 596–600.

Cohn, E. (1966). *Nematologica* **12**, 321–327.

Cohn, E. (1969). *Proc. First Intern. Citrus Symp.* Riverside, Calif. **2**, 1013–1017.

Cohn, E. and Minz, G. (1965). *Phytopath. medit.* **4**, 17–20.

Cohn, E. and Orion, D. (1970). *Nematologica* **16**, 423-428.

Cohn, E., Feder, W. A. and Mordechai, M. (1968). *Israel J. agric. Res.* **18**, 19–24.

Cohn, E., Minz, G. and Monselise, S. P. (1965). *Israel J. agric. Res.* **15**, 187–200.

Colbran, R. C. (1958). *Qd J. agric. Sci.* **15**, 101–135.

Colbran, R. C. (1963). *Qd J. agric. Sci.* **20**, 269–274.

DuCharme, E. P. (1955). *Proc. Fla St. hort. Soc.* **68**, 29–31.

DuCharme, E. P. (1959). *Phytopathology* **49**, 388–395.

DuCharme, E. P. (1968). *In* "Tropical Nematology" (G. C. Smart and V. G. Perry, eds), pp. 20–37. University of Florida Press, Gainesville.

DuCharme, E. P. (1969). *In* "Nematodes of Tropical Crops" (J. E. Peachey, ed.), *Tech. Commun. Commonw. Bur. Helminth.*, No. 40, pp. 225–237.

DuCharme, E. P. and Birchfield, W. (1956). *Phytopathology* **46**, 615–616.

DuCharme, E. P. and Price, W. C. (1966). *Nematologica* **12**, 113–121.

Feder, W. A. (1968). *Israel J. agric. Res.* **18**, 175–179.

Feldmesser, J. and Hannon, C. I. (1969). *Pl. Dis. Reptr* **53**, 603–607.

Feldmesser, J. and Rebois, R. V. (1963). *Phytopathology* **53**, 875.

Feldmesser, J., Cetas, R. C., Grimm, G. R., Rebois, R. V. and Whidden, R. (1960). *Phytopathology* **50**, 635.

Ford, H. W. (1953). *Proc. Am. Soc. Hort. Sci.* **61**, 73–76.

Ford, H. W. (1964). *Univ. Florida Agric. Stn Circ.* S-151.

Ford, H. W. and Feder, W. A. (1969) *Proc. First Intern. Citrus Symp.* Riverside, Calif. **2**, 941–948.

Ford, H. W., Feder, W. A. and Hutchins, P. C. (1960). *Univ. Florida Citrus Expt. Stn.* Mimeo Series 60–13.

Gonzalez-Sicilia, E. (1969). *Proc. First Intern. Citrus Symp.* Riverside, Calif. **1**, 121–134.

Goodey, J. B. (1958). *Nematologica* **3**, 169–172.

Guiran, G. (1962). *C.r. Acad. Agric. France* **48**, 388–390.

Hannon, C. I. (1963). *Pl. Dis. Reptr* **47**, 812–816.

Hedlund, F. F. (1969). *Proc. First Intern. Citrus Symp.* Riverside, Calif. **1**, 169–173.

Jensen, H. J. (1953). *Pl. Dis. Reptr.* **37**, 384–387

Kirkpatrick, J. D., Van Gundy, S. D. and Tsao, P. H. (1965). *Phytopathology* **55**, 1064.

LeClerg, E. L. (1964). *Phytopathology* **54**, 1309–1313.

Levin, M. M. (1969). *Proc. First Intern. Citrus Symp.* Riverside, Calif. **1**, 145–164.

Machmer, J. H. (1958). *J. Rio Grande Vall. hort. Soc.* **12**, 57–60.

Mankau, R. and Minteer, R. J. (1962). *Pl. Dis. Reptr* **46**, 375–378.

Martin J. P. and Van Gundy, S. D. (1963). *Soil Sci.* **96**, 128–135.

Martin, J. P., Baines, R. C. and Page, A. L. (1963). *Soil Sci.* **95**, 175–185.

Meagher, J. W. (1969). *Proc. First Intern. Citrus Symp.* Riverside, Calif. **2**, 999–1006.

Mendel, K. (1969). *Wld Crops* **21**, 110–114.

Mendel, K., Meyrovitch, A. and Cohn, E. (1969). *Israel J. agric. Res.* **19**, 171–178.

Minz, G. (1956). *Pl. Dis. Reptr* **40**, 971–973.

Neal, J. C. (1889). *Bull. U.S. Bureau of Ent.* No. 20.

O'Bannon, J. H. and Tarjan, A. C. (1969). *Proc. First Intern. Citrus Symp.* Riverside, Calif. **2**, 991–998.

O'Bannon, J. H. and Taylor, A. L. (1967). *Pl. Dis. Reptr* **51**, 995–998.

Oberholzer, P. C. J. (1969). *Proc. First Intern. Citrus Symp.* Riverside, Calif. **1**, 111–120.

Ouden, H. (1965). *Surinaamse Landbouw.* **13**, 34.

Peacock, F. C. (1956). *Nematologica* **1**, 307–310.

Poucher, C., Ford, H. W., Suit, R. F. and DuCharme, E. P. (1967). *Fla Dept Agric. Bull.* No. 7.

Reynolds, H. W. (1969). *Proc. First Intern. Citrus Symp.* Riverside, Calif. **2**, 969–971.

Scaramuzzi, G. and Perrotta, G. (1969) *Proc. First Intern. Citrus Symp.* Riverside, Calif. **2**, 956–960.

Singh, D. (1969). *Proc. First Intern. Citrus Symp.* Riverside, Calif. **1**, 103–109.

Spurling, M. B. (1969). *Proc. First Intern. Citrus Symp.* Riverside, Calif. **1**, 93–102.

Standifer, M. S. and Perry, V. G. (1960). *Phytopathology* **50**, 152–156.

Steiner, G. (1949). *Proc. Soil Sci. Soc. Fla* 4-B, 72–117.

Stolzy, L. H., Van Gundy, S. D., Labanauskas, C. K. and Szuszkiewicz, J. E. (1962). *Soil Sci.* **96**, 292–298.

Suit, R. F. (1969). *Proc. First Intern. Citrus Symp.* Riverside, Calif. **2**, 961–968.

Suit, R. F. and DuCharme, E. P. (1957). *State Plant Board of Florida*, Vol. 2, Bull. 11.

Suit, R. F., Brooks, T. L. and Ford, H. W. (1953). *Proc. Fla St. hort. Soc.* **64**, 46–49.

Swingle, W. T. and Reece, P. C. (1967). *In* "The Citrus Industry" (W. Reuther, H. J. Webber and L. D. Batchelor, eds), Vol. I, pp. 190–430. Univ. Calif. Press.

Tarjan, A. C. (1957). *Proc. Fla St. hort. Soc.* **70**, 85–90.

Tarjan, A. C. (1960). *Phytopathology* **50**, 577.

Tarjan, A. C. (1961). *Soil Crop Sci. Soc. Florida, Proc.* **21**, 17–26.

Tarjan, A. C. (1964). *Proc. helminth. Soc. Wash.* **31**, 65–76.

Tarjan, A. C. (1967). *Pl. Dis. Reptr* **51**, 317.

Tarjan, A. C. and Hannon, C. I. (1957). *Univ. Florida Agric. Expt. Stn Ann. Rep.* 212–213.

244 ELI COHN

Tarjan, A. C. and O'Bannon, J. H. (1969). *Pl. Dis. Reptr* **53**, 683–686.

Thomas, E. E. (1913). *Circ. Calif. agric. Exp. Stn* No. 85.

Tichinova, L. V. (1957). *In* "Eelworms of Agricultural Crops in the Uzbek SSR." *Tashkent: Akad. Nauk Uzbekskoi SSR*, pp. 101–131.

Van Gundy, S. D. (1958). *Nematologica* **3**, 283–294.

Van Gundy, S. D. and Kirkpatrick, J. D. (1964). *Phytopathology* **54**, 419–427.

Van Gundy, S. D. and Martin, J. P. (1961). *Phytopathology* **51**, 228–229.

Van Gundy, S. D. and McElroy, F. D. (1969). *Proc. First Intern. Citrus Symp*. Riverside, Calif. **2**, 985–989.

Van Gundy, S. D. and Tsao, P. H. (1963). *Phytopathology* **53**, 488–489.

Van Gundy, S. D., Martin, J. P. and Tsao, P. H. (1964). *Phytopathology* **54**, 294–299.

Vilardebo, A. (1963). *Al Awamia* **7**, 57–69.

Vilardebo, A. and Luc, M. (1961). *Fruits* **16**, 445–454.

Whitehead, A. G. (1968). *Trans. zool. Soc. Lond.* **31**, 263–401.

11

Nematode Diseases of Banana Plantations

C. D. Blake*

Department of Plant Pathology
University of Wisconsin
Madison, Wisconsin, U.S.A.

I. Crop Production

Bananas, plantains, and Manila hemp (abaca) are all tall growing, herbaceous monocotyledons belonging to the genus *Musa* (family Musaceae). Four species are commonly grown throughout the tropics and subtropics where the banana is often either a staple food or an important cash crop. These species include the Gros Michel or Jamaica banana, the shorter growing Cavendish or Chinese banana, the more starchy cooking banana or plantain, and the Manila hemp which produces seeds and scanty, non-edible fruit.

The 50 inch isohyte and the 60°F (15·6°C) winter isotherm, which approximate the 30° lines of latitude north and south, define the geographic limits to the world's banana production. Exceptions are the producing areas in Israel, Taiwan, and New South Wales, Australia. Within these broad limits there are many different climates some of

* Present Address: Riverina College of Advanced Education, Wagga Wagga, New South Wales, Australia.

which are more favorable than others for banana production. For example, bananas are rarely grown at altitudes above 3000 ft. The Cavendish varieties are considerably more tolerant than other species to low temperatures and are grown commercially in areas where winter temperatures fall to 50°F (10°C). In Australia, Cavendish bananas are also grown at their southern geographic limit and to avoid chilling and frost damage, plantations are established on steep, sunny north-eastern hillsides. It is usual, however, for bananas to be grown on alluvial or volcanic flats although in some of the Caribbean islands, hillside cultivation is common. In most areas, bananas benefit from irrigation and supplementary watering is critical, for example, in the Jordan Valley, Israel, in parts of Australia, and in the arid regions of Jamaica and Colombia.

The true stem, which bears a terminal inflorescence (or bunch) grows from a rhizome (sometimes called a corm or the mat), through the pseudostem, and emerges after the last leaf has unfurled. In the tropics, bunches are produced annually, but in more temperate regions up to 18 months may elapse between bunches. While a particular plant (or stem) is maturing a bunch, axillary buds at the base give rise to new plants (usually termed "suckers" at first), one or two of which are allowed to persist and mature successive bunches. Thus, although individual stems have a life span of only 12–18 months, the rhizome is potentially perennial.

The banana plant has a shallow but extensive adventitious root system which arises from within the cortex or from buried leaf bases. In either case, the roots are white, fleshy and unbranched and normally radiate from the rhizome for 3–4 m and penetrate the soil to about 1 m (Summerville, 1939). As the bunch approaches maturity, the plant becomes increasingly top-heavy and without a sound root system it is readily uprooted, especially during periods of high winds and/or heavy rain.

Bananas are propagated vegetatively by planting either detached suckers (spearpoints) that arise from axillary buds or pieces of a rhizome taken after destroying an established plant. In either case, the unit of propagation (usually termed a "set" or "bit") is trimmed free of roots before planting.

The commonest cause of root disfunction in bananas is probably parasitic nematodes, but root rots can also be incited by fungi (*Fusarium oxysporum cubense, Marasmius semiustus, Armillaria mellea, Clitocybe* sp., *Corticium* (*Rhizoctonia*), *Ceratocystis* (*Thielaviopsis*) *paradoxa, Botryodiplodia theobromae, Pythium* sp., and *Sclerotium rolfsii*) and rarely by bacteria.

Bananas are perennial plants that are grown in warm, more or less tropical areas that are characterized by torrential rains and/or strong winds. They are vegetatively propagated, have a shallow root system, and much of a plant's weight is borne terminally. All these conditions are conducive to rapid multiplication and dissemination of nematodes, unthriftiness and toppling of bunched plants, and the development of a nematode problem of major economic proportions. Luc and Vilardebo (1961a) recorded some fifteen different nematodes from the roots or soil around *Musa* species or varieties (Table 1).

Wehunt and Edwards (1968) listed some thirty-eight species of nematodes belonging to twenty genera as possible parasites of bananas. The burrowing nematode, *Radopholus similis*, is the most important pest and the root-knot nematodes (*Meloidogyne* spp.), which are widely distributed in banana plantations, are considered usually of minor importance. Other nematode parasites of bananas that are of economic importance are *Helicotylenchus multicinctus* and possibly the root lesion nematode, *Pratylenchus coffeae*.

II. Nematode Diseases

A. The Burrowing Nematode, *Radopholus similis*

Cobb (1893) first described this nematode as *Tylenchus similis* from necrotic root lesions of *Musa sapientum* growing in Fiji. The nematode was renamed *R. similis* by Thorne (1949) and recently confirmed by Sher (1968) as the type species of the emended genus. Sher considers *Radopholus* to be indigenous to Australia and New Zealand because nine of the eleven nominal species in this genus are found mainly in that area.

The disease of bananas caused by *R. similis* is known throughout the world by different names, the most common of which are *Radopholus* root rot, blackhead, blackhead toppling disease, and decline (Blake, 1969). There is clear evidence that infested planting material was introduced to Australia from Fiji between 1890 and 1920 and from these introductions the Australian banana industry was established. Subsequently, the nematode has been widely distributed and is now a major cause of loss of yield (Colbran, 1955; Blake, 1961a). In other parts of the world, the distribution of *R. similis* is more or less coincident with that of the banana, with the exception of the Jordan Valley (Minz *et al.*, 1960). The burrowing nematode is also a major economic problem in the Pacific Islands (Cobb, 1893), Ivory Coast (Luc and Vilardebo, 1961a), and throughout Central and South America (Stover and Fielding, 1958; Wehunt and Holdeman, 1959) and the Caribbean islands (Leach, 1958; Phillips, 1965). In many areas, the nematode has assumed economic importance following the replacement of the Gros Michel by Lacatan bananas to control Panama wilt. There are few

I

TABLE I. Nematodes found in roots of, and in soil around, different banana species and varieties examined in Ivory Coast (from Luc and Vilardebo, 1961a)

HOST SPECIES AND VARIETIES	Radopholus similis	Helicotylenchus multicinctus	Helicotylenchus n.sp.	Hoplolaimus proporicus	Tylenchus n.sp.	Aphelenchoides sp.	Meloidogyne incognita	Tylenchorhynchus spp.	Trophurus imperialis	Hemicycliophora oostenbrinki	Criconemoides onoensis f. kindia	Criconemoides peruensis	Criconema octangulare	Xiphinema ensiculiferum	Xiphinema spp.
Musa acuminata (AA)	+	+								+					
(AA) var. Pahang	+	+	+		+					+					
(AA) var. Siam	+		+				+								
(AA) var. Figue-sucree (Mignonette)	+	+	+		+		+			+					
Musa acuminata (AAA) group *sinensis*											+				
(AAA) var. Poyo	+	+	+	+	+	+	+		+	+				+	+
(AAA) var. Lacatan	+	+		+			+			+					
(AAA) var. Grande-Naine	+	+		+			+			+					

Species / variety	1	2	3	4	5	6	7	8	9	10
Musa acuminata (AAA) group *sinensis*										
(AAA) var. Petite-Naine	++	++	++	++	++		++		+	+
(AAA) var. Seredou	++	++	++		++				+	
(AAA) var. Monte-Christo	++		++				++	++	+	
Musa acuminata (AAA) group Gros Michel										
(AAA) var. Gros-Michel	++	++	++	++	++	++	++		+	+
(AAA) var. Guineo	++	++	++		++		++		+	
(AAA) group red var. Figue-rose	++	++	++				++		+	
Musa balbisiana BB	++		++		++		++		+	
Musa acuminata × *balbisiana* (AAB)										
(AAB) var. plantain-corne	++	++	++				++		+	
(AAB) var. Jock-corne	++	++	++		++					
(AAB) var. Madre del Platano			++		++		++			
(AAB) var. Figue-pomme	++	++	++		++				+	
(AAB) var. Cachaco	++		++		++		++		+	
(AAB) var. argentee	++	++	++		++				+	
Musa basjoo	++				++			+	+	+
Musa ornata	++		++		++				+	
Musa textilis			++				++		+	
Ravenala madagascariensis			++						+	+

++ Nematodes observed in roots.   + Nematodes in soil around roots but pathogenicity not proven.

reports of the nematode causing loss in bananas in southeast Asia (Larter and Allen, 1953) and on the Indian subcontinent (Nair *et al.*, 1966).

1. Biology of *R. similis*

The nematode has a migratory, semi-endoparasitic habit and although the stages remain vermiform throughout, sexual dimorphism is apparent with adult males being somewhat degenerate and probably non-parasitic. Eggs are normally laid in infested tissue over 7–8 days at the rate of about 4 eggs per day. The life-cycle from egg to egg extends over 20–25 days with eggs taking 8–10 days to hatch and the larvae 10–13 days to mature (Loos, 1962).

2. Histopathology

Pre-adult and adult female nematodes puncture epidermal cells along the entire length of the root by repeated stylet thrusts. The nematode enters the root through the wounds and occupies an intercellular position within the cortex. The nematodes feed on the cytoplasm of nearby cells causing the cell walls to rupture, and the nucleus either is ingested or degenerates. Such cell destruction causes cavities in the cortex. If necrosis accompanies cell breakdown, the nematodes either migrate through the tissue in advance of the necrosis or move out of the root into the soil to re-infest the root at another site (Blake, 1961a). Eggs are often found in necrotic tissue.

Any form of wounding to banana roots usually induces a characteristic discoloration of the affected tissue and where nematodes cause this wounding, reddish-brown cortical lesions develop and are diagnostic of the disease. These lesions are clearly seen when an affected root is split longitudinally and examined immediately. Raised lenticular cracks above an area of cortex are usually evident when the surface of an infected root is examined. The strongly suberized endodermis seems to serve as an effective barrier to the invasion of the stele by *R. similis* (Blake, 1966). Extension of necrosis, for any reason, from the cortex to the stele is a significant feature of the field syndrome because it leads to disfunction of the vascular tissue and eventually to atrophy of the root distal to the site of stelar invasion. This may decrease the root system to a few short stubs.

3. Interactions with Other Micro-organisms

By wounding the host and inducing histological change nematodes may provide infection sites for other micro-organisms and thereby affect the etiology, particularly of root diseases (Pitcher, 1965). Newhall (1958) for example, showed that the incidence of Panama wilt (*Fusarium oxysporum* f. *cubense*) was doubled in the presence of *R. similis* during an experimental period of 3 months. When Gros Michel bananas, infected with *R. similis*, were inoculated with *F. oxysporum* f. *cubense* the period between inoculation and the onset of wilt was considerably shortened, though neither the final expression of the symptoms nor their severity was affected at high inoculum levels (Loos, 1959).

Rishbeth (1960) suggested that nematodes caused the breakdown of resistance to Panama wilt in Lacatan bananas. In pot experiments it was noted that *F. oxysporum* f. *cubense* readily established in portions of rootlets of resistant Cavendish bananas invaded by unspecified nematodes but that the fungus seldom invaded intact roots. Blake (1966) found that *F. oxysporum* alone failed to cause root lesions, but colonized parenchyma cells of the cortex that had been wounded either mechanically or by *R. similis*. Sequeira *et al.* (1958) found that a layer of cortical parenchyma two or three cells thick prevented invasion of the root by *F. oxysporum* f. *cubense* and also suggested that *F. oxysporum* is unable to invade living, unwounded cells in lateral roots of bananas. Many isolates of *F. oxysporum*, which are weakly parasitic, are known to be pioneer colonists of newly exposed substrates. Such substrates are probably provided when *R. similis* or mechanical wounds rupture cell walls to expose cytoplasm which may exude from the lesions into the rhizosphere or the intercellular spaces of the cortex.

Lesions formed after inoculation with both *R. similis* and *F. oxysporum* were more extensively necrotic and increased in size more rapidly than when *R. similis* alone was used (Blake, 1966). In the presence of *F. oxysporum* and necrosis, *R. similis* tended to congregate at the periphery of the lesion and then to migrate either into healthy tissue and so enlarge the lesion, or into the soil. The migration of *R. similis* through necrotic tissue caused by the fungus colonizing the affected cells explains the more rapid increase in the size of the lesions noted when roots were co-inoculated. In such plants, it was significant that *F. oxysporum* was able to grow through the endodermis, colonize stelar cells, and cause necrosis of vascular cells. This extension of necrosis to the stele, as noted above, is a significant feature of the field syndrome.

Pitcher (1965) suggested that the presence of endoparasitic nematodes in the cortex of roots creates a food base for the weaker, unspecialized fungi and thereby increases their invasive potential. The available evidence clearly suggests that *R. similis* is a "pathogen", as defined by Mountain (1960), and the cause of banana root rot. However, in its association with *F. oxysporum* and probably some other fungi which are also pioneer colonists, *R. similis* may fulfil a role similar to that suggested by Pitcher. The extensive wounds caused to rhizomes by *R. similis* did not increase the incidence of Panama wilt (Loos, 1959).

There are two likely consequences of large numbers of roots being killed. First, if the root system is severely decreased, the plants are less securely anchored in the soil and when supporting a maturing bunch, are prone to topple, especially during wet and windy weather (Figs 1 and 2). Second, because the absorptive efficiency of the root

FIG. 1. A banana plant affected by *Radopholus similis*. Note the insecure anchorage, premature defoliation and the small bunch (from Blake, 1961a).

system is impaired, bunch weights, the size and possibly the number of leaves, and the rate of growth of suckers is decreased, and the sensitivity of the plant to water stress is increased. The ultimate effect is to decrease the economic life of the plantation from being indefinite to as little as 1 year, a situation that has come to be regarded as normal in some countries where nematode control is not practised. This results in annual replanting and a consequent drastic increase in costs or loss in production, or both.

FIG. 2. Part of a plantation of Cavendish bananas infected with *Radopholus similis* showing the uprooting of bunched plants (from Blake, 1961a).

4. Biotypes and Host Range

Soon after Cobb (1915) named *R. similis* as the cause of necrosis on *M. sapientum* roots in Fiji, *R. similis* was found infecting sugar-cane (*Saccharum officinarum*) in Kauai, Hawaii and subsequently to have a wide host range amongst *Musa* spp. *M. cavendishii* (Mallarmarie, 1939), *M. nana* (Birchfield, 1956), *M. ornata* (Whitlock, 1957), *M. paradisica* (Cardenosa-Barriga, 1948), *M. rosa* (Mumford, 1960), and *M. textilis* (van der Vecht, 1950) are hosts to *R. similis* in addition to those listed in Table I (Luc and Vilardebo, 1961a).

DuCharme and Birchfield (1956), van Weerdt (1957), and Blake (1961a), using differential indicator hosts, distinguished at least two biotypes of *R. similis*—one that attacks citrus and the other that attacks

Fɪɢ. 3. A young corm cut longitudinally showing necrosis of the cortex of the corm resulting from infection by *Radopholus similis*.

bananas. The status of a third biotype that attacks both citrus and bananas is in doubt. Adults of both biotypes are morphologically similar with no significant dimensional differences (Sher, 1968). The biotype that attacks pepper (van der Vecht, 1950) has not been determined. Unlike the citrus biotype that attacks many plant species belonging to diverse families (Brooks, 1954, 1955), the one that attacks banana seems to be restricted to *Musa* spp., *Ipomoea batatas*, and *Pueraria phaseoloides javanica*, and possibly to a few other plants.

The nematode does not seem to have any special adaptations for survival in the absence of a susceptible host. Free-living stages and eggs do not survive fallow or storage in plastic bags for more than about 12 weeks. After a plantation is destroyed, the corms may remain succulent for several months in the more temperate regions and nematodes in these corms probably survive for as long as the corms remain succulent.

5. Dissemination

Some lateral roots of bananas arise from deep within the corm. The endogenous parts of these roots become infected by nematodes that move laterally along the root. Such nematodes may move from the root and infect the surrounding-corm tissue. The surface of a corm in

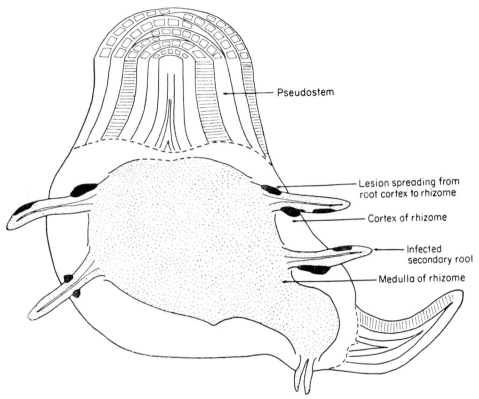

Pseudostem

Lesion spreading from root cortex to rhizome

Cortex of rhizome

Infected secondary root

Medulla of rhizome

Fig. 4. Diagram of a longitudinal section through a corm infected with *Radopholus similis*. Note that the secondary roots arise from the medulla and that infected tissue may occur within the cortex of the rhizome (from Blake, 1969).

contact with soil may be invaded by nematodes directly from the soil and the infected tissue, like that of the root, becomes extensively necrotic (Fig. 3). These corm-borne infections are of particular significance in the introduction of *R. similis* into new plantations (Loos and Loos, 1960; Blake, 1963) (Fig. 4). The roots that arise from sets after planting usually do so in the vicinity of pre-existing roots and so are likely to be invaded by nematodes migrating from existing corm infections. Trimming sets, which is often done before they are planted,

I*

is usually not sufficiently rigorous to remove infections that extend deeply into a set.

Nematodes may be carried by water draining down a slope from an infested plantation above. *R. similis* may be disseminated by water that drains from infested areas and re-enters either irrigation channels or streams from which water is being taken to irrigate bananas, as it is known that other nematodes are disseminated (Faulkner and Bolander, 1966). Soil that adheres to implements, tyres of motor vehicles, and shoes of plantation workers, may also spread *R. similis*. This nematode is not adapted for wind dispersal, nor is there any evidence of its being spread by insects.

B. The Root-knot Nematodes, *Meloidogyne* spp.

Root-knot nematodes appear to be ubiquitous in the warm, moist, and friable soils of banana plantations. At least four species have been reported and all form conspicuous root galls within which female nematodes, in various stages of maturation, are found. At maturity, the female is saccate and has associated with it a few large multinucleate "giant" or "nurse" cells on which the females feed. Eggs are laid singly within a gelatinous matrix which forms an egg sac from which the eggs are eventually liberated into the soil following the breakdown of the root. The life-cycle of the nematode and the histopathology and etiology of the disease do not differ significantly on bananas from that reported in other hosts. Full descriptions of the disease on other hosts have been given by Krusberg and Nielsen (1958), Crittenden (1958), and others.

The loss of yield caused by *Meloidogyne* spp. has not been assessed experimentally for bananas, but the consensus of opinion is that root-knot nematodes usually do not depress yields significantly. There are at least four reasons why this might be so. First, necrosis is rarely associated with root-knot infections and so it is unlikely that necrosis would spread to the stele and so cause a root to atrophy. Second, when the water or nutrients available to a plant limit growth the decreased efficiency of a root to conduct decreases yield or vegetative growth. Good cultural practices, however, aim to remove such limitations. Third, Luc and Vilardebo (1961a) have shown that the rate at which *Meloidogyne* spp. reproduce is limited in the presence of other nematodes. It is probable that the necrosis associated with the parasitism of either *R. similis* or *Helicotylenchus* spp. would render root-knot nematodes relatively innocuous pathogens because banana roots are rarely attacked by only one nematode species. Fourth, the individual roots of the prolific adventitious root system are short-lived in bananas relative to the length of the life-cycle of the nematode.

Root-knot nematodes have a wide host range, especially on dicotyledonous plants, and are usually present in most soils into which banana sets are planted. Thus disinfecting sets and even crop rotation are unlikely to decrease significantly the incidence of root-knot in established plants. Where control of this nematode alone is justified, then nematicides offer the only practical possibilities. These nematicides are not normally selective and they can therefore be equally well applied for the control of *R. similis*, spiral nematodes and for the root-knot nematode.

Neither Newhall (1958) nor Loos (1959) could demonstrate an increased incidence of Panama disease in banana plants co-inoculated with root-knot nematodes and *F. oxysporum cubense*.

C. The Lance and Spiral Nematodes, Family Hoplolaimidae

Like the root-knot nematodes, members of the family Hoplolaimidae are also ubiquitous in banana plantations. *Helicotylenchus multicinctus* is probably the most important and causes serious decline of bananas in the Jordan Valley, Israel (Minz et al., 1960). There, as well as in Australia (Blake, 1961a), Ivory Coast (Luc and Vilardebo, 1961a), Cuba (Venning, 1958), and Honduras (Wehunt and Holdeman, 1959), these nematodes incite discrete, relatively shallow, necrotic lesions on banana roots. Other members of this family that have been recorded either from root lesions or from soil around banana plants or both, include *H. africanus* in the Canary Islands (de Guiran and Vilardebo, 1963), *H. dihystera* in Australia (Colbran, 1955) and Honduras (Stover and Fielding, 1958), *H. erythrinae* in Ecuador, Costa Rica, and Guatemala (Wehunt and Holdeman, 1959) and in Nicaragua (Holdeman, 1960), and *Hoplolaimus pararobustus* in Tanzania (Whitehead, 1959) and Ivory Coast (Luc and Vilardebo, 1961a).

The host–parasite relationships of *H. multicinctus* on banana have been studied in more detail than that of other species and are probably typical. Adults and larvae penetrate the epidermis of the root within 36 h of inoculation. The females, and probably the males, feed directly on parenchyma cells, and after 4 days, the nematodes are usually wholly within the cortex to a depth of 4–6 cells. The head of each nematode is usually oriented parallel to the long axis of the root, but the posterior portion is curved and usually occupies several contiguous cells. The nematode withdraws cytoplasm from surrounding cells, the cell walls are distorted or ruptured, and the nucleus, if intact, is enlarged.

Evacuated cells together with those near "nests" of eggs, soon become discoloured and necrotic. The characteristic macroscopic brown flecks or discrete necrotic lesions then appear (Luc and Vilardebo, 1961a; Zuckerman and Strich-Harari, 1963; Blake, 1966).

Development of root lesions caused by *H. multicinctus* is slow relative to those produced by *R. similis*, and the lesions usually do not spread. However, in Israel where *R. similis* is not known to occur on bananas, *H. multicinctus* causes extensive root necrosis, dieback, and disfunction leading eventually to debility of an entire plant. The lesions caused by spiral nematodes are colonized by fungi, especially by species of *Fusarium*, *Rhizoctonia*, and *Cylindrocarpon*, but a positive interaction has not been established.

H. multicinctus has a wide host range (Goodey *et al.*, 1965) but whether it reproduces equally well in all hosts and whether biotypes exist is not known with certainty. There is scant information on the survival of *H. multicinctus* in the absence of susceptible hosts. *H. buxophilis* declined from 250 to 2 nematodes in potted soil in 8 months and none could be found after 12 months (Golden, 1956). *H. nannus* survived for at least 6 months when stored in soil contained in plastic bags (Ferris, 1960).

D. The Root Lesion Nematodes, *Pratylenchus* spp.

Root lesion nematodes cause lesions similar to, but less extensive than, those caused by *R. similis*. The initial entry of the nematode produces in the cortex a reddish elongated fleck which enlarges as the nematode and its offspring feed. The older parts of the lesion turn black and shrink while the advancing margin remains red. Neither the nematode nor the eggs are found beyond this red margin and both are rare in the older portions of the lesion. The life-cycle is completed within the root; where the nematodes feed on cortical parenchyma the larvae and mature females remain motile but seldom enter the vascular tissue.

Five species of *Pratylenchus* have been reported attacking bananas. *P. coffeae* seems to be the most widespread and there are many reports of this species, particularly from plantations in Central America (Wehunt and Edwards, 1968). *P. goodeyi* and *P. thornei* have been reported from the Canary Islands (de Guiran and Vilardebo, 1963) whereas *P. penetrans* and *P. scribneri* have both been reported from Israel (Minz, 1957).

The genus is not usually regarded as being an important debilitating one on bananas and there have been few systematic studies to develop specific controls. Studies with other crops, however, suggest that each species has a number of common hosts so that the nematode is likely to be already present in many soils in which bananas are planted. It is likely, too, that root lesion nematodes are clone-borne and that sets can be disinfected by the same methods used to exclude *R. similis*. Nematicides used to control *R. similis* and *Meloidogyne* in soil also should be useful in the control of these nematodes.

III. Control Measures

Three methods are used to control *Radopholus* root rot. First, so as to avoid introducing inoculum into a new plantation, sets may be disinfected either by paring, heat therapy, or chemical treatment, or by a combination of these methods; second, the number of nematodes in soil may be decreased either by fallowing or by crop rotation; and third, the number of nematodes in infested soil may be decreased before planting or during the life of a plantation by the application of nematicides. Pre-planting nematicides may be applied generally or to specific planting sites in a plantation.

Genetic resistance to *R. similis* is known amongst the *Musa* spp. and attempts are being made to incorporate such genes into new hybrids along with genes for resistance to Panama wilt and Sigatoka disease (*Mycosphaerella musicola*).

A. Disinfecting Banana Sets

1. By Paring

The black or reddish-brown discoloration of infected root and corm tissue is a clear and reliable indicator of the extent of nematode infection (Fig. 3). The removal of such tissue along with some of the surrounding healthy tissue under conditions that avoid recontamination will normally disinfect sets (Loos and Loos, 1960). The paring is best done with a cane knife or machete, or some other stout cutting instrument with the set resting on a wooden block. Leaf bases and the outer regions of the corm are removed first, and then deeper tissue if necessary, to remove all traces of infected tissue no matter how deeply this tissue may be located. When paring is complete, all exposed tissue should be white and undiscolored. Paring destroys all axillary buds, but not the enclosed apical growing point. Pared sets should be handled so as to avoid recontamination by *R. similis* and, because pared sets are prone to fungal attack, Loos and Loos (1960) suggested dipping the pared sets in a Bordeaux Mixture—DBCP paste made by mixing 20 kg hydrated lime, 20 kg copper sulphate, 1288 ml 70% DBCP and 455 litres water. The sets are planted as soon as the paste has dried.

Vilardebo and Robin (1969) examined the useful technique of soaking sets for a few seconds in 550 ml DBCP plus 40 litres clay plus 50 litres water. This treatment, termed "pralinage", completely coated the set in a persistent nematicidal preparation. After 2 years, plants from sets so treated showed improved growth, greater vegetative development, and produced heavier bunches. Some phytotoxicity was evident, and in the first year there was no difference in the yield from treated and untreated plots. However, in the second year the yield from the treated plots was much improved and the multiplication and spread of R. similis retarded.

2. Heat Therapy

Hand paring, though effective, is tedious and time-consuming and therefore only likely to be practised on a small scale or in areas where labour is cheap and plentiful. Heat treatment of sets has the potential to provide the large numbers of disinfected sets needed for commercial plantings, especially when a mechanized and central treatment plant can be constructed. In vitro, adults and larvae of R. similis are killed almost instantaneously when immersed in water at 52°C (Birchfield, 1954; Blake, 1961a). Large banana sets, with their dense tissue and high moisture content, are well insulated and heat penetration is slow. However, by using relatively small sets and long immersion periods (for example, 55°C for 25 min) temperatures high enough to kill embedded nematodes without causing significant set losses can be attained.

The great bulk of material that needs to be treated for commercial plantings makes the maintenance of an accurate temperature control during heat treatment of banana sets a major problem. In the large, well insulated treatment tank, the water must be circulated rapidly so that heat lost through immersing a mass of relatively cold sets can be quickly replaced (Blake, 1961a). A desirable organization is for cut and trimmed sets to be delivered to a centrally located treatment plant, washed with high pressure hoses to remove all adhering soil and other debris, and placed in wire baskets which can then be lifted by a mechanical hoist into the treatment tank at 55°C. After 25 min, they are removed and the sets placed on a nematode-free area to drain and cool. Post-treatment losses are decreased when sets are planted into soil free of R. similis as soon as possible and preferably the same day. This method of disinfection is highly effective provided that only small sets are treated and that embedded nematodes receive a full exposure to the heat treatment. The more thoroughly the sets are trimmed before treatment the greater is the likelihood of their being disinfected. Fungicides applied after treatment would probably decrease any post-treatment losses caused by the fungi. Blake (1961b) attempted to disinfect sets by immersing them in dips of nematicides, insecticides, or fungicides. All failed to disinfect the sets and many were phytotoxic. It is probable that all failed to penetrate the tissues of the dormant corm sufficiently to kill nematodes deeply embedded in the tissue.

3. Certification Scheme

Theoretically, such a scheme consists of the establishment of a supply of planting material certified by an appropriate authority to be free from *R. similis*, planting nematode-free sets, and inspecting them regularly thereafter. The nematode-free sets could be obtained by either paring, heat treatment, or selection. In the quarantine area, abnormal cultural techniques can be followed that encourage multiple suckering and close planting. Sets produced in this way can be planted directly provided that care is taken to avoid their becoming infected after leaving the quarantine area.

B. Crop Rotation and Cultivation

The biotype of *R. similis* that attacks bananas apparently has a narrow host range and limited survival in soil in the absence of a host. If these observations are confirmed, then the destruction of an infested area either mechanically with a bulldozer or by injection of a herbicide into the pseudostem (Jeater *et al.*, 1951) and replanting the area with nematode-free sets probably within 12 months would be a practical approach to the rehabilitation of infested plantations. This approach might be particularly useful in those areas where the application of nematicides is impracticable because of terrain or of cost. The population of *R. similis* in soil appears to decline at different rates depending on how the infected stools are destroyed. Where herbicides are injected into the pseudostem, aerial parts may collapse quickly, but the rhizome and roots may remain succulent for several months and continue to support an active population of nematodes. By contrast, the population decreases rapidly when the whole plant, including the rhizome, is pushed from the ground with a bulldozer (Fig. 5).

Colbran (1964) examined the effects of cover crops on the nematode population in Queensland after an infested plantation was destroyed. Over a 32-week sampling period he found that green panic (*Panicum maximum* var. *trichoglume*) was not a host to the nematodes that infect bananas and that Siatro (*Phaseolus atropurpureus*) supported neither *R. similis* nor *Meloidogyne javanica*.

Loos (1961) planted sugar-cane in soil following the destruction of infected bananas infected with *R. similis* and, after 10 weeks, found that *R. similis* was eliminated. In another experiment, plants grown

from nematode-free sets in an area under sugar-cane for the previous
4 months, were free of *R. similis* 9 months later. Panamanian soils that
had been flooded for 5 months for the control of *Fusarium* wilt were
freed of *R. similis* (Loos, 1961) and these results were later confirmed in
Honduras. The population of the nematode-trapping fungi *Dactylaria
thaumasia* and *Arthrobotrys musiformis* were increased by adding
organic matter to soil but they failed to reduce the number of *R.
similis* (Tarjan, 1961).

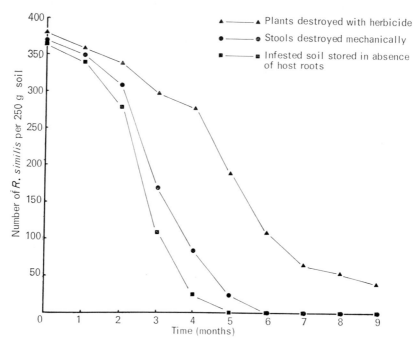

Fig. 5. A comparison of the rates of decline of populations of *Radopholus similis* in soil
when stored in the absence of a host, and following the destruction of infected plants
with a herbicide or with a bulldozer. Each point is the mean of five replicates (from
Blake, 1969).

C. Chemical Soil Treatment

Nematicides offer the best prospect for success where it is necessary
to plant bananas in infested soil or to decrease the nematode population
during the life of a plantation. In detailed experiments made on the
Ivory Coast, Luc and Vilardebo (1961b) evaluated three nematicides.
Two of these, dichloropropane-dichloropropene (DD) and ethylene
dibromide (EDB) were applied at 300 litres/ha and 150 kg/ha, respectively. Growth was at first retarded, but by the end of the first year,
production increased 30–50%. After 1 year, further treatments with

nematicides were necessary, but the response to treatment declines. The third and most satisfactory nematicide tested was 1,2 dibromo-chloropropane (DBCP). Luc and Vilardebo (1961b) recommended that this be applied at 40 litres/ha at planting during May and June and 25 litres/ha in October, and at 15 litres/ha in March each year thereafter. This program is profitable, gave excellent control of nematodes, and has been widely adopted by commercial producers of bananas in the Ivory Coast. Working with Cavendish bananas, Price (1960) obtained vegetative responses 9 months after the application of 75% emulsifiable concentrate of DBCP, and yield increases of about 56%.

At present, there is only limited use of nematicides on bananas outside the Ivory Coast and Jordan Valley, but experiments are proceeding in most countries. Factors that limit the use of nematicides include the costs of application and likely financial returns, the pressures and indiscipline of small scale subsistence farming, and the difficulties of using mechanized soil applicators other than hand injectors in many areas because of the steepness of the terrain, surface obstructions and litter, and the failure of plants to remain aligned after the first or second year. The incorporation of DBCP into irrigation water holds some promise (Ziv, 1961; de Guiran and Vilardebo, 1963). Evaluation of other nematicides, especially those belonging to the carbamate and organo-phosphate groups, and granular formulations suitable for broadcast application, are proceeding. Unfortunately, some of the more promising materials have high mammalian toxicity and none has been recommended for field use.

D. Resistant Varieties

The susceptible Gros Michel variety has been replaced with Lacatan varieties (e.g. Valery) throughout most of Central America and the Caribbean area because of the ravages of Panama wilt. The Lacatan varieties are susceptible to *R. similis* and consequently the incidence of root rot and toppling has increased dramatically. Attempts are being made, particularly through the Jamaican banana breeding scheme (Shepherd, 1968), to incorporate the resistance of the Gros Michel banana into the genome of new hybrids. Wild, diploid bananas are being used as pollen parents and Gros Michel or Highgate as triploid female parents. Selections from the resulting tetraploids show considerable resistance to *R. similis*, as well as to Panama wilt and Sigatoka (*Mycosphaerella musicola*). A new variety Bodles Altaport, a cross between Gros Michel and Pisang Lilin (Osborne, 1963), has been released for growing on the hillsides of Jamaica. This variety has an extensive root system, withstands wind and is more resistant to *R. similis* than Lacatan.

Little is known of the genetic basis of resistance to *R. similis* in bananas but there is some evidence to suggest that it is controlled by a single dominant gene. If this is so, the development of resistant varieties seems likely although little is known of the genetics of the nematode, especially of its ability to produce resistance-breaking biotypes.

IV. Economics of Control

Whether or not to institute specific measures to decrease or avoid nematode infestations in banana plantations is essentially an economic decision, although continuity of production might be a paramount consideration where the banana is a staple food or essential markets must remain supplied. To assess the economic feasibility of a control measure, the additional costs incurred by the treatment must be balanced against the value of the additional fruit produced or the amount of fruit saved. In most situations regardless of whether bananas are grown in large commercial plantations or in small peasant plots, *R. similis* will cause unthriftiness, reduction in bunch weights, toppling, and the need for frequent replanting. Yield increases of 30–60% have been obtained following the control of the nematode over a range of growing conditions. Such yield increases have been measured when infested and non-infested plots have been compared over 1 or 2 years, but the results of such experiments may not express the full effects of the loss. In practice, because of the incidence of infestation, replanting is done routinely every 1 or 2 years or when the effects of parasitism are evident, but long before the yield is minimal. In addition to the value of the yield lost, one should also consider the costs of labour and materials, of replanting and the normal trends towards increasing yields as a plantation matures.

It is difficult to generalize about the relative economic advantages of the different approaches to control because of the widely different economic conditions under which bananas are grown. Those methods that are labour intensive and require relatively unskilled labour, such as hand paring of sets, would seem most suited to some Central American, Caribbean, and African situations. Other methods for disinfecting sets, such as bulk heat treatment and certification schemes, can be adapted for use in those areas where labour is more costly. Where land is expensive or the value of bananas high, fallow or crop rotation would not be recommended.

Table II gives the yield of fruit from plants grown from infected and pared sets in replicated trials in Honduras and Panama. In non-infected plants, individual bunches were on average up to 7·25 kg heavier than those from the infected plants and the total yield of fruit per acre was greater by as much as 17,000 lb per acre (19,050 kg/ha) per year. The differences in yield were greater in Panama than in Honduras (Wehunt and Edwards, 1968).

TABLE II. Comparison of the yields of bananas established in Honduras and Panama from sets infested with *Radopholus similis* and sets disinfested by paring; 1 lb = 454 g; lb/acre = 1·12 kg/ha (Based on Wehunt and Edwards, 1968.)

Experiment	Age in months	Variety	Treatment	Average weight of bunch (lb)	Total weight (lb) bunch fruit per acre per year	Pounds difference per plant
Honduras 1	36	Cocos	Pared sets	100·8	41,362	66·7
			Infested	97·8	32,996	
Honduras 2	45	Valery	Pared sets	109·0	64,644	39·7
			Infested	105·0	60,700	
Panama	33	Lacatan	Pared sets	92·1	41,869	116·7
			Infested	75·7	24,039	

Nematicides probably will prove to be the most useful of the methods available to control nematodes in banana plantations. Their present use is limited by the need to treat plants singly by hand-operated injectors, to apply the nematicides annually or more frequently, and to a lesser extent by their phytotoxicity and costs. Formulations suitable for broadcast application and/or those that possess long residual action together with techniques for applying nematicides to large areas mechanically, would expand the use of nematicides and substantially increase the production of bananas.

Minz (1960) and Minz *et al.* (1960) reported that banana plantations in Israel usually declined after the second crop because of infestation by *H. multicinctus*. From experiments in which 2 year old plants were treated in spring with 10 litres per dunam (approximately 0·6 hectare) of DBCP, bunches from the treated plots averaged 29·9 kg compared to 24·7 kg from untreated plots. In a 3 year old planting, 10 litres per dunam of DBCP in irrigation water raised the average bunch weight from 15 to 18·6 kg, but applications of nematicides before irrigation did not increase bunch weight.

Because bananas are either a staple food or a valuable earner of export income for a significant proportion of the world's population, disease, and especially that caused by plant-parasitic nematodes, is of great social and economic importance. The epidemiology of nematode diseases in bananas is now well understood and a choice in effective control programme is available. Many factors will have to be considered when making the choice but the level of economic development of a producing country will probably be the most important factor in this choice.

References

Birchfield, W. (1954). *Proc. Fla Hort. Soc.* **67**, 94–96.
Birchfield, W. (1956). *Pl. Dis. Reptr* **40**, 866–868.
Blake, C. D. (1961a). *Nematologica* **6**, 295–310.
Blake, C. D. (1961b). *Report Sixth Commonw. Mycol. Conf.*, Lond. 53–54.
Blake, C. D. (1963). *Agric. Gaz. N.S.W.* **74**, 526–533.
Blake, C. D. (1966). *Nematologica* **12**, 129–137.
Blake, C. D. (1969). *In* "Nematodes of Tropical Crops" (J. E. Peachey, ed.) Tech. Commun. Commonw. Bur. Helminth., No. 40, St Albans.
Brooks, T. L. (1954). *Proc. Fla Hort. Soc.* **57**, 81–83.
Brooks, T. L. (1955). *Pl. Dis. Reptr* **39**, 309.
Cardenosa-Barriga, R. (1948). *Notas agron. Estac. agric. exp. Palmira* **1**, 15–29.
Cobb, N. A. (1893). Macleay Memorial Vol., Linn. Soc. N.S.W. 252–308.
Cobb, N. A. (1915). *J. agric. Res.* **4**, 561–568.
Colbran, R. C. (1955). *J. Aust. Inst. agric. Sci.* **21**, 167–169.
Colbran, R. C. (1964). *Qd J. agric. Sci.* **21**, 233–236.
Colbran, R. C. (1967). *Qd J. agric. Sci.* **24**, 353–354.
Crittenden, H. W. (1958). *Phytopathology* **48**, 461.
de Guiran, G. and Vilardebo, A. (1963). *Fruits* **17**, 263–277.
DuCharme, E. P. and Birchfield, W. (1956). *Phytopathology* **46**, 615–616.
Faulkner, L. R. and Bolander, W. J. (1966). *Nematologica* **12**, 591–600.
Ferris, J. M. (1960). *Phytopathology* **50**, 635.
Golden, M. (1956). *Bull. Md agric. Exp. Stn* A-85, 28 pp.
Goodey, J. B., Franklin, M. T. and Hooper, D. J. (1965). The nematode parasites of plants catalogued under their hosts. Tech. Commun. Commonw. Bur. Helminth., No. 30, St Albans.
Holdeman, Q. L. (1960). *United Fruit Co. Newsletter* **7**, 12–19.
Jeater, J. G., Cann, H. J. and Eastwood, H. W. (1951). *Agric. Gaz. N.S.W.* **62**, 140–144.
Krusberg, L. R. and Nielsen, L. W. (1958). *Phytopathology* **48**, 30–39.
Larter, L. N. H. and Allen, E. F. (1953). *Malay. agric. J.* **36**, 36–43.
Leach, R. (1958). *Nature, Lond.* **181**, 204–205.
Loos, C. A. (1959). *Proc. helminth. Soc. Wash.* **26**, 103–111.
Loos, C. A. (1961). *Pl. Dis. Reptr* **45**, 457–461.
Loos, C. A. (1962). *Proc. helminth. Soc. Wash.* **29**, 43–52.
Loos, C. A. and Loos, S. B. (1960). *Phytopathology* **50**, 383–386.
Luc, M. and Vilardebo, A. (1961a). *Fruits* **16**, 205–219.
Luc, M. and Vilardebo, A. (1961b). *Fruits* **16**, 261–279.

Mallarmarie, M. (1939). *Agron. Colon.* **254**, 33–75.

Minz, G. (1957). *Pl. Dis. Reptr* **41**, 92–94.

Minz, G. (1960). *Proc. First FAO/OCTA International meeting on banana production*, Adidjan, Ivory Coast.

Minz, G., Ziv, D. and Strich-Harari, D. (1960). *Ktavim* **10**, 147–157.

Mountain, W. B. (1960). *In* "Nematology: Fundamentals and Recent Advances with Emphasis on the Plant Parasitic and Soil Forms" (J. N. Sasser and W. R. Jenkins, eds), pp. 419–421. University of North Carolina Press, Chapel Hill.

Mumford, M. (1960). *U.S. Dept Agric., Agric. Res. Service, Pl. Quarant. Div.* 86 pp.

Nair, M. R. G. K., Das, N. M. and Menon, M. R. (1966). *Indian J. Ent.* **28**, 553–554.

Newhall, A. G. (1958). *Pl. Dis. Reptr* **42**, 853, 856.

Osborne, R. E. (1963). *J. agric. Soc. Trin. Tobago* **63**, 35–45.

Phillips, C. A. (1965). *Winban News* **1**, 17–19.

Pitcher, R. S. (1965). *Helminth. Abstr.* **34**, 1–17.

Price, D. (1960). *Trop. Agric., Trin.* **37**, 107–109.

Rishbeth, J. (1960). *Emp. J. exp. Agric.* **28**, 109–113.

Sequeira, L., Steeves, T. A., Steeves, M. W. and Riedhart, J. M. (1958). *Nature, Lond.* **182**, 309–311.

Shepherd, K. (1968). *Pes.. Art. News. Summ.* Sec. **B14**, 370–379.

Sher, S. A. (1968). *Proc. helminth. Soc. Wash.* **35**, 219–237.

Stover, R. H. and Fielding, M. J. (1958). *Pl. Dis. Reptr* **42**, 938–940.

Summerville, W. A. T. (1939). *Qd agric. J.* **48**, 376–392.

Tarjan, A. C. (1961). *Nematologica* **6**, 170–175.

Thorne, G. (1949). *Proc. helminth. Soc. Wash.* **16**, 37–73.

van der Vecht, J. (1950). *In* "De plagen van de cultuur gewassen in Indonesie", pp. 16–41. W. van Hoeve. 's-Gravenhage.

van Weerdt, W. (1957). *Pl. Dis. Reptr* **41**, 382–385.

Venning, F. D. (1958). *Proc. Am. Soc. hort. Sci.* Caribbean Region, Sixth Annual Meeting, 17–20.

Vilardebo, A. and Robin, J. (1969). *In* "Nematodes of Tropical Crops" (J. E. Peachey, ed.) Tech. Commun. Commonw. Bur. Helminth., No. 40. St Albans.

Wehunt, E. J. and Edwards, D. I. (1968). *In* "Tropical Nematology" (G. C. Smart and V. G. Perry, eds), pp. 1–19. University of Florida Press, Gainesville.

Wehunt, E. J. and Holdeman, Q. (1959). *Proc. Soil Crop Sci. Fla* **19**, 436–442.

Whitehead, A. G. (1959). *Nematologica* **4**, 99–105.

Whitlock, L. S. (1957). *Pl. Dis. Reptr* **41**, 814.

Ziv, D. (1961). *Hassadeh* **41**, 416–419.

Zuckerman, B. M. and Strich-Harari, D. (1963). *Nematologica* **9**, 347–353.

12
Nematode Pests of Coffee

Luiz Gonzaga E. Lordello

Department of Zoology
Escola Superior de Agricultura "Luiz de Queiroz"
University of São Paulo
Piracicaba, Brazil

I. Introduction

Half a dozen species of *Coffea* are grown commercially in the world, and the adaptability of the coffee plant has facilitated coffee culture in a wide geographical area. Coffee is grown throughout the tropical belt, but it extends beyond the tropics in such areas as Taiwan in the Northern and Brazil in the Southern Hemisphere. *Coffea* species are found at elevations from slightly above sea-level to 2400 m (Ecuador). This great adaptability is probably unequalled by any other commercially important perennial tropical crop (Krug and Poerck, 1968).

The most important species is *C. arabica*, which is grown either exclusively or in association with one or more of the other species. Several varieties of *C. arabica* are available. The countries in which the most important species of *Coffea* are produced are shown in Table I.

Coffee is produced on some 3–4 million farm units in the world and, contrary to common belief, it is mainly a smallholder's enterprise and of considerable economic importance (Krug and Poerck, 1968). These authors state that "Only rarely it is a cash crop of relatively secondary value. In the majority of cases it is the main and often the only cash crop needed for subsistence of farmers. Coffee has contributed substantially, much more than cocoa or any other crop in the tropics, to raising the standard of living of millions of people, particularly in Africa. It is, therefore, a crop of considerable social importance."

Brazil is the principal exporting country and in 1959 exported over one million metric tons (Coste, 1961). Other important exporting countries are Colombia, Mexico, and the Ivory Coast.

TABLE I. World distribution of the economically important species of *Coffea* (Krug and Poerck, 1968)

Species of *Coffea*	Countries in which it is produced
C. arabica	Western Hemisphere and higher regions of Africa: Ethiopia, Sudan, Angola, Rwanda, Burundi, Democratic Republic of Congo, Uganda, Kenya, Tanzania, and Malawi.
C. liberica	Guyana, Surinam, Malaysia, Fernando Pó, Philippines, San Tomé, and Liberia.
C. canephora	Dahomey, Togo, Gabon, Angola, Cameroon, Democratic Republic of Congo, Guinea, Ivory Coast, Tanzania, Uganda, Madagascar, Indonesia, the Republic of Vietnam, and Western Hemisphere (Antilles, Ecuador and Panama).
C. dewevrei var. *excelsa*	Central African Republic, and Republic of Vietnam.
C. racemosa	Mozambique
C. stenophylla	Sierra Leone
C. abeokutae	Ivory Coast

II. Nematode Pests

Different nematode species have been reported to attack coffee trees and the species recorded as harmful to coffee in one region are not necessarily harmful to the same species in another region, where they are known to occur.

In recent years, nematode pests of coffee have been described by Sylvain (1959), Lordello *et al.* (1961), Lordello (1965), and Schieber (1968). Lordello's contributions concern the nematode problems in Brazil, while those of Schieber concern the problems in Guatemala and other Central American countries. The main nematode species that attack coffee are listed in Table II.

The nematodes listed as *Tylenchus* spp., obtained from coffee in Colombia, India and Hawaii, may have been one of the lesion nematodes mentioned in Table II (Cassidy, 1931; Obregon, 1936).

TABLE II. Nematode species known to attack coffee roots

Nematode	Host	Countries	Selected references
Meloidogyne exigua	*Coffea arabica*	Brazil, Colombia, Peru, Costa Rica, Guatemala, El Salvador, Dominican Republic	Salas and Echandi (1961) Lordello (1965) Schieber (1966) Schieber and Grullon (1969)
M. incognita	*C. arabica* *C. canephora*	Guatemala, Cuba, Brazil, Ivory Coast	Chitwood and Berger (1960) Luc and Guiran (1960) Decker and Garcia (1966) Decker (1968) Lordello and Mello (1970)
M. javanica	*C. canephora* var. *robusta*	Congo	De Coninck (1962)
M. hapla	*C. canephora* var. *robusta*	Congo	De Coninck (1962)
M. inornata	*C. arabica*	Guatemala	Schieber and Sosa (1960)
M. africana	*C. arabica* *C. canephora* var. *robusta*	Kenya, Congo	Whitehead (1959) De Coninck (1962)
M. coffeicola	*C. arabica*	Brazil	Lordello and Zamith (1960)
M. decalineata	*C. arabica*	Tanganyika	Whitehead (1968)
M. megadora	*C. canephora* *C. arabica* *C. congensis* *C. eugenioides*	Angola	Whitehead (1968)
M. oteifai	*C. canephora* var. *robusta*	Congo	Elmiligy (1968)
Pratylenchus coffeae	*C. arabica*	Java, Guatemala, El Salvador, Costa Rica, West Indies, Congo	Zimmermann (1898) Bredo (1939) Vayssière (1955)
P. brachyurus	*C. arabica* *C. canephora* *C. excelsa*	Peru, Brazil, Ivory Coast	Krusberg and Hirschmann (1958) Luc and Guiran (1960) Lordello *et al.* (1968)

TABLE II (*continued*)

Nematode	Host	Countries	Selected references
Radopholus similis	*C. arabica* *C. canephora* *C. excelsa* *C. quillon* *C. canephora* var. *robusta*	Java, El Salvador	Bally and Reydon (1931) Sledge *et al.* (1963)
Helicotylenchus spp. *Rotylenchus* spp.	*C. arabica* *Coffea* sp.	Java, India, Angola, Costa Rica, Brazil	Whitehead (1960)
Rotylenchulus reniformis	*C. arabica*	India, Puerto Rico	Ayala (1962) Sekhar (1963) Ayala and Ramires (1964) D'Souza (1965) D'Souza and Sreenivasan (1965)
Hemicriconemoides sp.	*C. arabica* *C. canephora* var. *robusta*	India	Kumar and D'Souza (1969)
Xiphinema radicicola	*C. arabica*	Peru, Guatemala	Krusberg and Hirschmann (1958) Chitwood and Berger (1960a, 1960b)
X. americanum	*C. arabica*	Guatemala	Thorne and Schieber (1962)
X. brevicolle	*C. arabica*	Brazil	Coste (1961) Monteiro (1970)
X. krugi	*C. arabica*	Brazil	Monteiro (1970)
X. insigne	*C. arabica*	India	Somasekhar (1959)
X. basilgoodeyi	*C. arabica*	Congo	Coomans (1964)

A. Biology

1. Root-knot Nematodes, *Meloidogyne* spp.

Meloidogyne exigua was the first nematode species found in coffee roots, according to a report by Jobert (1878). A few years later, Goeldi (1887) published a description of this species as part of his classical report on a disease of coffee trees in the so-called Province of Rio de Janeiro, Brazil. *M. exigua* is quite widespread in the coffee growing regions in Brazil, and Lordello and Zamith (1958a) have given a detailed description of its morphology (Curi *et al.*, 1969). Apparently it has not been recorded on coffee outside the Americas (Salas and Echandi, 1961). D'Souza's record (1965) of *M. exigua* in S. India in certain weeds grown in coffee plantations, but not on the coffee roots themselves, is of special interest. A form found attacking coffee in El Salvador and referred to by Abrego and Holdeman (1961) as *M. javanica* may, in fact, have been *M. exigua*.

Typically the *Meloidogyne* are gall-forming nematodes, but *M. exigua* in many instances does not form galls on coffee roots. In these instances, the mature females usually break through the root surface, so that they protrude from the root and may then be seen as globular whitish bodies with yellowish or brownish egg masses attached (Lordello, 1964, 1968). Under these conditions, the parasite is more injurious than in those cases where a smooth, uncracked gall is formed. "This is due to the fact that roots cracked or opened by the action of this nematode are at once invaded by a whole group of secondary agents, including other nematodes, fungi, and bacteria and this usually leads to quick decay. Plants thus attacked naturally suffer much more than those which form 'smooth', uncracked or, if such an expression is permitted, 'healthy galls' " (Steiner, 1949) (Fig. 1).

The egg stage is unquestionably the most resistant to adverse environmental conditions (Steiner, 1949; Christie, 1959). The eggs hatch into the preparasitic or migrant larvae, which occur free in the soil. These larvae usually are in the second larval stage, and after entering the host plant induce the giant cells. The larvae feed, begin to swell rapidly, first becoming sausage-shaped and then growing to a pear-shaped whitish or brownish body, which is the adult female.

There is strong evidence of the existence of three physiological races of this nematode (Curi *et al.*, 1970a) and all three races are believed to occur in the state of S. Paulo, Brazil. Former authors reported the possible existence of only two races in S. Paulo (Lordello, 1968). There are no records of physiological races occurring in other countries or other states in Brazil (Chebabi and Lordello, 1968). The races differ from one another in their effects on the plant, one of them being much more severe than the other two. Although the differences in behaviour may be due to different environmental conditions, investigations suggest the existence of distinct races.

M. exigua is known to attack coffee, tea and sweet pepper. Attempts to infest several other plant species, including tomato var. Rutgers, have failed (Lordello, 1964). *M. exigua* induced the formation of galls on the roots of cucumber, but no adult females were found inside the tissue.

M. coffeicola females usually lay their eggs outside the root, through

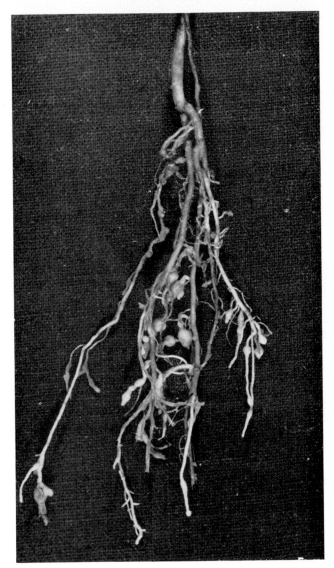

FIG. 1. Roots of an arabica coffee seedling showing galls caused by *Meloidogyne exigua.*

cracks that they have induced in the root tissue. Infected roots show numerous dark spots, each of which is an ootheca of the nematode. Very few females lay eggs inside the root tissues. In some instances, the position in the roots of the long-necked mature females of *M. coffeicola* resembles parasitism by a *Nacobbus* species, such as *N. dorsallis* (Thorne and Allen, 1944).

Cross-sections of roots attacked by *M. coffeicola* do not show giant or nectarial cells so clearly as do those sections of roots attacked by other species (e.g. *M. exigua*). Actually the secretion of *M. coffeicola* oesophageal glands seems to be much less active in promoting the formation of the giant cells. There are no data referring to the biology of the other species associated with coffee, such as *M. africana, M. incognita, M. inornata, M. javanica, M. hapla, M. oteifai, M. decalineata* and *M. megadora.*

2. Root Lesion Nematodes, *Pratylenchus* spp.

The root lesion nematodes are migratory parasites and there is no stage of development which can be called an infective stage, because adults and larvae of various ages constantly migrate into and out of the roots (Christie, 1959). The favoured place of entrance is not at the root tip, but slightly behind the elongating zone in the piliferous region (Linford, 1939). Godfrey (1929) confirmed this for *P. brachyurus* attacking roots of pineapple. He also observed that root tips were sometimes invaded and destroyed.

As previously stated, *P. brachyurus* is the only *Pratylenchus* species known to infect coffee roots in South America (Peru and Brazil).

P. brachyurus reproduces more rapidly at soil temperatures of 26–32°C than in lower temperatures (Graham, 1951) and optimum conditions for development occur in the summer.

In South America, *P. brachyurus* is usually found in association with a number of grasses, such as the "gordura" and "jaraguá" grasses (*Melinis minutiflora* and *Hyparrhenia rufa*, respectively), in the roots of which reproduction takes place when other hosts are not available (Lordello and Mello, 1969b). Feldmesser *et al.* (1960) found that the longevity of *P. brachyurus* in fallow soil is about 21 months.

3. Burrowing Nematode, *Radopholus similis*

The burrowing nematode enters the root and feeds within the cortex, causing extensive lesions and cavities which may result in root disintegration (Christie, 1959).

The burrowing nematode is essentially a lesion nematode, its injuries to roots being much the same as, though often more extensive than, those inflicted by *Pratylenchus* species.

This is one of the most destructive nematodes in the tropical regions. However, most data available refer to this nematode parasitizing other host plant species such as sugar-cane, citrus, etc. There is no information on *Radopholus* sp. as a parasite of coffee beyond that of Zimmerman (1898) and Bally and Reydon (1931).

Muir and Henderson (1926) wrote that in sugar-cane roots grown under laboratory conditions, one generation of *Radopholus similis* requires about 4–5 weeks. Suit and DuCharme (1953) found that in citrus roots, the cycle requires around 21 days at a temperature of 24°C.

R. similis ability to survive in the absence of plants has been investigated by several authors, the data available being contradictory. Its longevity in fallow soil is between 3 months (DuCharme, 1955; Feldmesser *et al.*, 1960) and 6 months (Tarjan, 1960).

4. Other Nematodes

Only the females of *Rotylenchulus reniformis*, the Reniform nematode, are parasites of the coffee roots. The females commence egg laying about 9 days after entering the root tissues, and each female lays about 120 eggs, which are deposited in a gelatinous matrix. The eggs hatch within 8 days of being deposited and larval development requires 8 days (Linford and Oliveira, 1940; Peacock, 1956).

After hatching, larvae of both sexes develop rapidly in the soil, passing through three moults. Apparently, during larval development the individuals do not feed. Finally, the female larva becomes established either completely within the root or having the posterior portion of body protruding. The body enlarges progressively and soon acquires a characteristic kidney shape.

The spiral nematodes (*Helicotylenchus dihystera*, *H. erythrinae* and *H. pseudorobustus*) also are found associated with coffee but little is known of their biology.

B. Symptoms

Seedlings infested by *M. africana* are generally stunted and chlorotic, while feeder roots and often main roots are "blinded", resulting in the formation of numerous branch rootlets behind the affected root tip (Whitehead, 1959). Heavy attacks of mature coffee trees by *Meloidogyne* spp. are generally associated with unthriftiness. Little is known of the symptoms caused by *M. decalineata* and *M. megadora* (Whitehead, 1968).

On the root of arabica coffee trees, *M. exigua* produces rather small elongated galls, which may be easily overlooked, particularly if the root material is permitted to dry (Fig. 1). Necrotic areas also occur on the roots (Lordello and Zamith, 1958b, 1958c). Galls are located mainly at root tips although in many instances roots attacked by this species do not gall; instead the roots exhibit crackings and detachment of cortical tissue. The root system is conspicuously reduced and the rootlets and root hairs practically non-existent (Reyes, 1957).

The above-ground parts of the plant show chlorosis, premature leaf drop and general decline. Many trees die, particularly during droughts and freezing because affected plants lack vigour and have reduced ability to withstand adverse conditions. Affected trees show an increased tendency to develop symptoms of nitrogen and zinc deficiencies in their leaves and branches.

Arruda (1960b) infested with *M. exigua* coffee seedlings raised in sheet wood containers. At 6 months, the average height of the infected seedlings was $19 \cdot 1 \pm 0 \cdot 4$ cm, whereas that of the same number of non-infected control seedlings was $24 \cdot 9 \pm 0 \cdot 3$ cm. Stunting of the coffee seedlings was attributed to a primary depressive effect induced by the nematodes found in the seedling roots.

Further investigations by Thomaziello (unpublished), using a more destructive race of *M. exigua* than did Arruda, showed the infected seedlings to be much more stunted and disfigured before they were finally killed.

Coffee trees attacked by *M. coffeicola* are chlorotic, showing strong defoliation and lack of vigour (Fig. 2). Death occurs within a variable

FIG. 2. An arabica coffee tree showing decline and severe defoliation as a result of an attack by *Meloidogyne coffeicola*. (By courtesy of S. M. Curi.)

period of time. Roots are not properly galled, but are slightly thickened and heavily cracked (Fig. 3). Cortical tissues are detached, resulting in a typical "rough" root. Symptoms in the aerial parts are similar to those produced by trees infected with *M. exigua*, but are much more severe, the trees becoming completely defoliated before dying.

FIG. 3. Roots from a coffee tree infected with *Meloidogyne coffeicola* which show the typical symptoms of an irregularly thickened root, and rough cortex owing to the cracking and detachment of cortical tissues. Reproduced with permission from Lordello (1968).

During the rainy season, there is a general improvement in the coffee plantation affected. But during the dry periods or immediately after harvesting the plantation declines.

M. coffeicola is much more detrimental to coffee cultivation in South America than is *M. exigua*. Fortunately, *M. coffeicola* is limited to the states of Paraná and S. Paulo in Brazil (Lordello, 1967; Curi, 1968; Curi *et al.*, 1970a).

Reporting on "pyroid nemas" (*Meloidogyne* spp.) found in Guatemala, Chitwood and Berger (1960b) stated that "symptoms included lower stem and root galling with surface breakage, severe stunting, chlorosis, and even death of trees. Roots exhibited cracking and galling. Seedlings in nurseries are extremely stunted, the chlorotic leaves often drop and the plants die. Older trees, even mature ones, may be killed. External, lower stem galling and root cracking provides an entry for all types of other organisms including normally saprophytic fungi."

Abrego (1959) and Abrego and Holdeman (1961) investigated a nematode complex (root-knot and root lesion species) attacking coffee in El Salvador where the diseased young trees were stunted and showed thin stems and complete destruction of secondary roots and rootlets. The root systems of seedlings were poor and the plants were easily pulled out of the ground. Abrego and Holdeman (1961) confirmed the pathogenicity of *Pratylenchus coffeae* previously reported from Java as a very destructive species (Zimmermann, 1898; Vayssière, 1955).

In Brazil, young coffee plants attacked by *P. brachyurus* are stunted and have a poor root system. Affected trees in plantations are severely chlorotic and stunted, death occurring within a variable period of time. In some coffee trees a corky region at the base of the trunk results from severe infestations by *P. coffeae* (Schieber, 1968). Large populations of this nematode are present in this affected bark tissue.

Schieber (1968) investigated the nematode pests of the coffee areas of Central America. In severe infestations, the following characteristics developed: " . . . absorption rootlets are destroyed, principally by the lesion and dagger nematodes. Galls up to 10 mm in diameter are produced by the root-knot nematodes. The galls may develop at the tip of rootlets or at other sites along the roots. Fine rootlets infected with lesion nematodes turn brown and in severe infestations the rootlets dry completely. *Xiphinema americanum* punctures and destroys root cells during the feeding process, leaving lesions through which fungi and bacteria enter. In severe infestations in the nursery, the main root is also completely destroyed by the different nematode species. Leaves of infested adult plants show chlorosis and necrosis. During dry seasons, the typical symptom is complete wilting of the foliage. In severe attacks, total defoliation occurs."

D'Souza and Sreenivasan (1965) stated that *Rotylenchulus reniformis* caused the following symptoms in coffee: (*a*) almost complete absence of feeder roots; (*b*) a poorly developed tap-root; (*c*) yellowing and wilting of above-ground parts. In heavily infested fields (10 individuals per 50 cc of soil) arabica coffee fails to establish, in spite of good cultivation and manurial treatment.

III. Cultural and Environmental Influences

The active migration of coffee root-knot nematodes through soil is due solely to the movement of the pre-parasitic larvae before they enter the roots. This process is slow and in some species has been shown to be 1 cm per day or about 30 cm a month (Taylor and McBeth, 1941).

Root-knot and possibly other nematodes are spread from infested fields in the soil adhering to tillage implements, feet of animals, etc. This method of spreading the infestation is greatest during the rainy days when the soil sticks to implements, feet of farm animals, etc., and does not dry quickly. Water running from one field to another (either as run-off or as irrigation water) transports considerable numbers of nematodes. Man is the principal agent of this process of active transportation of nematodes from one farm or field or country to another.

In the main coffee growing regions in Brazil, coffee seedlings usually are raised in sheet wood containers and if infested soil is used in the containers the seedlings carry nematodes in their roots and in adhering soil into the plantations. *M. exigua* and *Pratylenchus brachyurus* are spread extensively in this way and *Pratylenchus* spp. especially are carried over long distances in soil on nursery stock (Oostenbrink, 1957; Wallace, 1963).

Lordello and Mello (1969a) recorded pangola grass (*Digitaria decumbens*) to be a host of two root lesion nematode species, namely *P. zeae* and *P. brachyurus*. According to these authors, pangola grass is responsible for active dissemination of both nematode species as rooted material is currently used for propagation. In South America only a few cases of severe infestation of coffee roots by *P. brachyurus* are known.

The raising of coffee seedlings in infested soil in nurseries and their subsequent transplanting into plantations is a common way for long range spread of the many nematode pests of coffee.

IV. Control

Coffee farmers should be encouraged to produce seedlings for their own use as this would avoid introducing nematodes from infested nurseries. Nursery owners are urged to use soil properly treated with nematicides such as methyl bromide, DD (dichloropropane-dichloropropene), EDB (ethylene dibromide), etc. In certain countries (e.g. Brazil), regulatory measures have been proposed, according to which owners of coffee nurseries have to make proper use of nematicides.

Nursery soil should be fumigated under a plastic cover for 3–4 days, and then aerated, particularly if less volatile nematicides are used (DD, EDB), for about 10 days before seeding.

K

Nurseries should not be located close to plantations that are suspected of being infested, particularly if the plantation is higher than the nursery and so enabling water to drain into the nursery (Gonçalves et al., 1968).

Infected seedlings should be burned and under no circumstances should they be planted into an area free of parasitic nematodes.

The Indian Coffee Board (1955) adopted the following practices in order to prevent the dissemination of nematodes to free areas: (a) dig trenches up to a metre deep, around the affected area; (b) divert drains away from healthy coffee growing areas; (c) prevent movement of soil from infested to healthy areas; and, (d) sterilize implements used in affected areas before using them in others.

Studies on the disinfestation of equipment capable of disseminating certain nematodes have been done in the U.S.A. The commonly recommended sterilizing practices may be used. In coffee plantations, a minimum cleaning has been adopted, consisting of the removal of all plant parts, especially roots, followed by the removal of soil from equipment before it leaves the diseased plot.

Fluiter (1949) and Lordello (1968) stated that the replanting of old nematode infested plantations be preceded by a fallow period of at least 2 years after removing the old coffee roots. During the 2 years no coffee or other potential host plants should be allowed to grow in the area. This requires complete knowledge of the host range of the nematode species present. A period of 3 or more years free from host plants seems to be best for controlling such nematodes as *Meloidogyne* spp.

Roots of infected coffee trees should be collected and burned within the area and not dragged to another location.

The chemical treatment of infested trees is not feasible because of the phytotoxicity of most of the commonly used nematicides, and the cost of treatment of coffee plantations would be very high owing to the large size of the plantations.

Several authors (e.g. Schieber, 1968) emphasize that fertilizers (N, P, etc.) help the diseased coffee trees to recover. These authors state that fertility and cultural practices are essential in a control programme. Nevertheless, such treatment is short-lived in its effect and does not prevent spread of the pests. Such increased fertilizer application is not successful in decreasing the destructive effect of *M. coffeicola* and certain races of *M. exigua*.

Fluiter (1949) stated that mulching or manuring the trees with organic material will reduce damage inflicted by nematodes. In certain coffee growing areas it is difficult to obtain large quantities of organic matter for mulching or manuring.

The best method of preventing nematode damage is by using varieties or rootstocks which are resistant to nematode attack (Taylor, 1953). Schieber and Sosa (1960) reported that robusta coffee is highly resistant to *P. coffeae* and *M. exigua*.

A project testing a great number of coffee types for resistance to *M. exigua* is in progress in Peru (Sylvain, 1959). New sources of resistance to *M. exigua* and *P. coffeae* are being investigated also in Guatemala. A similar project, including species of *Coffea* and varieties of arabica coffee, is in progress in Brazil. In this country, research into varieties resistant to nematodes is concerned mainly with two species, namely *M. exigua* and *M. coffeicola* (Curi, 1969; Curi *et al.*, 1970b).

It is hoped that in the near future some varieties will be found to be resistant to the main nematode pests. These resistant varieties will be used as rootstocks or as parents in breeding programmes to replace susceptible rootstocks now growing in infested areas.

V. Economic Importance

By the end of the last century *M. exigua* had caused considerable damage in the state of Rio de Janeiro, Brazil (Goeldi, 1887) as extensive areas of coffee plantations were wiped out by attacks of this nematode. This crop was then replaced by other crops, mainly sugar-cane.

In Indonesia, Cramer (1957) considered nematodes as the most destructive coffee pest and in Java, *P. coffeae* destroyed over 95% of arabica coffee plantations (Vayssière, 1955).

Lordello and Zamith (1960) found that in the state of Paraná, Brazil, *M. coffeicola* caused the death of over 15,000 trees in a single plantation.

In several coffee growing countries, nematodes are fully recognized as agents of heavy losses. But data on exact losses resulting from their infestation are very few. Possibly, the most useful information on this aspect is contained in papers by Arruda (1960a, 1960b) and by Arruda and Reis (1962).

Arruda (1960a) and Arruda and Reis (1962) studied the effect of a seedling infection by *M. exigua* on the growth and yield of the coffee plant after transplanting. One year after transplanting the seedlings into the field, height measurements showed that the non-infected controls grew by an average of 54·7 cm whereas the plants that had been infected with the nematode grew only 37·5 cm. This difference in growth was highly significant. Also it was found later (Arruda and Reis, 1962) that the yield of the non-infected plants was more than twice that of the plants artificially inoculated with *M. exigua* (104 and 50 kg, respectively).

References

Abrego, L. (1959). *Bol. Inf. Inst. Salvadoreño Inv. Café*, suppl. **1**, 13–17.

Abrego, L. and Holdeman, Q. L. (1961). *Bol. Inst. Salvadoreño Inv. Café*, suppl. **8**, 1–16.

Arruda, H. V. de (1960a). *Bragantia* **19**, 15–17.

Arruda, H. V. de (1960b). *Bragantia* **19**, 179–182.

Arruda, H. V. de and Reis, A. J. (1962). *Biológico* **28**, 349.

Ayala, A. (1962). *J. agric. Univ. P. Rico* **46**, 73–82.

Ayala, A. and Ramirez, C. T. (1964). *J. agric. Univ. P. Rico* **48**, 140–161.

Bally, W. and Reydon, G. A. (1931). *Arch. Koffiecultuur* **5**, 23–216.

Bredo, H. J. (1939). *Bull. Agric. Congo belge* **30**, 266–307.

Cassidy, G. (1931). *Hawaii Plant Rec.* **33**, 305–339.

Chebabi, A. and Lordello, L. G. E. (1968). *Rev. Agric.*, *Piracicaba* **43**, 140.

Chitwood, B. G. and Berger, C. A. (1960a). *Pl. Dis. Reptr* **44**, 841–847.

Chitwood, B. G. and Berger, C. A. (1960b). *Phytopathology* **50**, 631.

Christie, J. R. (1959). "Plant Nematodes, their Bionomics and Control." University of Florida, Gainesville, Fla.

Coomans, A. (1964). *Nematologica* **10**, 581–593.

Coste, R. (1961). "Les caféiers et les cafés dans le monde." Éditions Larose, Paris.

Cramer, P. J. S. (1957). "A Review of Literature of Coffee Research in Indonesia." Turrialba, Costa Rica.

Curi, S. M. (1968). *Biológico* **34**, 20–21.

Curi, S. M. (1969). *Biológico* **35**, 21–22.

Curi, S. M., Lordello, L. G. E., Bona, A. de and Cintra, A. F. (1969). *Biológico* **35**, 41–44.

Curi, S. M., Lordello, L. G. E., Bona, A. de and Cintra, A. F. (1970a). *Biológico* **36**, 26–28.

Curi, S. M., Carvalho, A., Moraes, F. P., Monaco, L. C. and Arruda, H. V. de (1970b). *Biológico* **36**, 293–295.

Decker, H. (1968). *Wiss. Z. Univ. Rostock* **17**, 421–438.

Decker, H. and Garcia, R. C. (1966). *Bol. Ci. Tecnol. Univ. Central Marta Abreu, Las Villas* **1**, 19–29.

De Coninck, L. (1962). *In* "Bijdragen tot de kennis der plant emparasitaire en der vrijlevende nematoden van Kongo", Ghent, Rijksuniversiteit, pp. 1–9.

D'Souza, G. I. (1965). *Indian Coff.* **29**, 22–23.

D'Souza, G. I. and Sreenivasan, C. S. (1965). *Indian Coff.* **29**, 11–13.

DuCharme, E. P. (1955). *Proc. Fla St. hort. Soc.* **68**, 29–31.

Elmiligy, I. A. (1968). *Nematologica* **14**, 577–590.

Feldmesser, J., Feder, W. A., Rebois, R. V. and Hutchins, P. C. (1960). *Anat. Rec.* **137**, 355.

Fluiter, H. J. de (1949). *Bergcultures* **18**, 138–139, 141, 143, 145.

Godfrey, G. H. (1929). *Phytopathology* **19**, 611–629.

Goeldi, E. A. (1887). *Arch. Mus. Nac. Rio de Janeiro* **8**, 7–123.

Gonçalves, J. C., Thomaziello, R. A., Pessenda, C. E. and Matuo, T. (1968). "Nematóides nocivos ao cafeei ro." Secretaria da Agricultura, Campinas, Brasil.

Graham, T. W. (1951). *S.C. agric. Exp. Sta. Bull.* **390**, 25 pp.

Indian Coffee Board, Res. Dept. (1955). *Indian Coff.* **19**, 122–126.

Jobert, C. (1878). *C.r. Acad. Sci. Paris* **87**, 941–943.

Krug, C. A. and Poerck, R. A. De (1968). "World Coffee Survey." *F.A.O.*, Rome.

Krusberg, L. R. and Hirschmann, H. (1958). *Pl. Dis. Reptr* **42**, 599–608.

Kumar, A. C. and D'Souza, G. I. (1969). *Pl. Dis. Reptr* **53**, 15–16.

Linford, M. B. (1939). *Proc. helminth. Soc. Wash.* **6**, 11–18.

Linford, M. B. and Oliveira, J. M. (1940). *Proc. helminth. Soc. Wash.* **7**, 35–42.

Lordello, L. G. E. (1964). *Anais. Esc. sup. Agric. "Luiz de Queiroz"* **21**, 181–218.

Lordello, L. G. E. (1965). *In* "1ª Reunión Técnica Internacional sobre plagas y enfermedades de los cafetos", pp. 100–108, *Inst. Inter-americano de Ciencias Agricolas, San José, Costa Rica, public.* misc. No. 23.

Lordello, L. G. E. (1967). *Rev. Agric. Piracicaba* **42**, 162.

Lordello, L. G. E. (1968). "Nematóides das plantas cultivadas." Livraria Nobel S.A., São Paulo, Brazil.

Lordello, L. G. E. and Costa, C. P. da (1961). *Rev. Brasil Biol.* **21**, 363–366.

Lordello, L. G. E. and Mello, Filho, A. de T. (1969a). *Rev. Agric. Piracicaba* **44**, 122.

Lordello, L. G. E. and Mello, Filho, A. de T. (1969b). *Solo* **61**, 27.

Lordello, L. G. E. and Mello, Filho, A. de T. (1970). *Rev. Agric. Piracicaba* **45**, 102.

Lordello, L. G. E. and Zamith, A. P. L. (1958a). *Proc. helminth. Soc. Wash.* **25**, 133–137.

Lordello, L. G. E. and Zamith, A. P. L. (1958b). *Pl. Dis. Reptr* **42**, 199.

Lordello, L. G. E. and Zamith, A. P. L. (1958c). *Rev. Agric. Piracicaba* **33**, 59–62.

Lordello, L. G. E. and Zamith, A. P. L. (1960). *Rev. Brasil. Biol.* **20**, 375–379.

Lordello, L. G. E., Monteiro, A. R. and D'Arce, R. D. (1968). *Rev. Agric. Piracicaba* **43**, 79–82.

Lordello, L. G. E., Monteiro, A. R., Oliveira, A. J. de and da Costa, C. P. (1961). *Divulg. agron. Shell* **4**, 2–11.

Luc, M. and Guiran, G. de (1960). *L'Agron. trop.* **15**, 434–449.

Monteiro, A. R. (1970). "*Dorylaimoidea* de cafèzais paulistas (Nemata, Dorylaimida)." Escola Superior de Agricultura "Luiz de Queiroz", Piracicaba, Brazil, doctor's thesis.

Muir, F. and Henderson, G. (1926). *Hawaii. Plant. Rec.* **30**, 242–245.

Obregon, R. (1936). *Rev. cafet. Colomb.* **6**, 2040–2042.

Oostenbrink, M. (1957). *Z. PflKrankh. PflPath. PflSchutz* **64**, 484–490.

Peacock, F. C. (1956). *Nematologica* **1**, 307–310.

Reyes, M. (1957). *Bol. trim. Exp. agropec.* **6**, 14–16.

Salas, L. A. and Echandi, E. (1961). *Café* **3**, 21–24.

Schieber, E. (1966). *Turrialba* **16**, 130–135.

Schieber, E. (1968). Nematode Problems of Coffee. *In* "Tropical Nematology" (G. C. Smart and V. G. Perry, eds), pp. 81–92, *Univ. Fla Press*, Gainesville, U.S.A.

Schieber, E. and Grullon, L. (1969). *Turrialba* **19**, 513–517.

Schieber, E. and Sosa, O. N. (1960). *Pl. Dis. Reptr* **44**, 722–723.

Sekhar, P. S. (1963). *Café* **5**, 1–4.

Sledge, E. B., Denmark, J. and Esser, R. P. (1963). A tentative list of the host plants of the burrowing nematode, *Radopholus similis* (Cobb, 1893) Thorne, 1949, including a list of plants considered as non hosts. State of Florida Dept Agric., U.S.A.

Somasekhar, P. (1959). *Pl. Prot. Bull. F.A.O.* **7**, 78–79.

Steiner, G. (1949). "Plant nematodes the grower should know." State of Florida Dept of Agriculture, Tallahasse.

Suit, R. F. and DuCharme, E. P. (1953). *Pl. Dis. Reptr* **37**, 379–383.

Sylvain, P. G. (1959). *Coff.* **1**, 2–13.

Tarjan, A. C. (1960). *Phytopathology* **50**, 656.

Taylor, A. L. (1953). *U.S. Dep. Agric.* Year Book, 129–134.

Taylor, A. L. and McBeth, C. W. (1941). *Proc. helminth. Soc. Wash.* **8**, 53–55.

Thorne, G. and Allen, M. W. (1944). *Proc. helminth. Soc. Wash.* **11**, 27–31.

Thorne, G. and Schieber, E. (1962). *Pl. Dis. Reptr* **46**, 857.

Vayssière, P. (1955). *In* "Les caféiers et les cafés dans le monde" (R. Coste, ed.), **1**, 233–318, Paris.

Wallace, H. R. (1963). "The Biology of Plant Parasitic Nematodes." Edward Arnold, London.

Whitehead, A. G. (1959). *Nematologica* **4**, 272–278.

Whitehead, A. G. (1960). *Rev. Café Port.* **7**, 5–16.

Whitehead, A. G. (1968). *Trans. zool. Soc. Lond.* **31**, 263–401.

Zimmermann, A. (1898). *Lands Plantentuin* **27**, 1–67.

13

Nematode Pests of Tea

P. Sivapalan

Tea Research Institute of Ceylon
Talawakele, Ceylon

I. Introduction

Tea (*Camellia sinensis* L.) is cultivated in about twenty-five countries in the world of which about ten could be considered as major tea producers. More than half of the one million hectares of tea cultivation is in India and Ceylon, on areas ranging from 225,000 to 350,000 hectares. Between them, these two major producers account for more than three-quarters of the world's export trade of 600,000 metric tons of this commodity. This apparently disproportionate share in the export trade is because in these two countries, domestic consumption is far outstripped by total production, leaving a large exportable surplus. The sizable tea production in some other countries is often insufficient to meet their domestic needs.

Agricultural and manufacturing practices vary greatly in the different tea growing areas of the world but the following is a brief outline of the practices commonly adopted for tea growing in the tropics.

Tea, in its natural state, is a tree that grows to a height of 9 m or more; but when grown for commercial purposes, it is pruned regularly so that it does not grow higher than about 1 m and takes the shape of a bush. The young shoots, each consisting of two leaves and a bud, commonly referred to as the flush, are plucked regularly by hand. The frequency of harvest varies from 5 to 14 days and is determined primarily by climatic conditions. A young healthy tea field yields on average 10,000–12,500 kg of flush/ha/year (corresponding to about 2000–2500 kg of made tea). After continuous harvesting the bush shows signs of losing vigour by producing flower and very little flush. When this stage is reached the bushes are pruned to remove old wood and the bush is given new vigour. The frequency of pruning (pruning cycle) varies from 2 to 5 years and is determined by the climatic conditions that influence growth rate. Tea grows best in acid soils (pH 4·5–5·0). The harvested flush is transformed to the consumable tea by a lengthy and skilled manufacturing process.

The information available on nematode pests of tea is not extensive and is restricted almost entirely to those countries where tea has been produced for a long period. Such information as is available on the species of nematodes affecting tea, diagnostic symptoms, cultural and environmental influences, methods of control and economics of the pests and their control will be dealt with in the succeeding sections.

II. Nematode Diseases

Root-knot nematodes are the ones most frequently encountered in the tea growing regions and most of the species of *Meloidogyne* except *M. brevicauda* are associated only with nursery plants. The most destructive and most intensively investigated nematode pest of tea is the root lesion nematode *Pratylenchus loosi*, which is equally destructive to both young and old tea plants and is a problem, therefore, in nurseries and in mature tea fields. It is widely distributed in tea fields in Ceylon and Japan and causes extensive crop loss. *Hemicriconemoides kanayaensis* is a common pest of tea in Japan and Taiwan.

There has been no positive evidence of pathogenicity for the majority of plant-parasitic nematodes that have been associated with tea. Some of the species have been recorded in the course of routine surveys and in tea displaying no presumptive symptoms. The world distribution of nematodes associated with tea is shown in Table I.

TABLE I. The world distribution of nematodes associated with tea (*Camellia sinensis*)

NEMATODE SPECIES	N.E. India (Anon. 1957; Das, 1958; Banerjee, 1966; Basu, 1967; Anon., 1968)	S. India (Rau, 1962; Venkata Ram, 1962, 1963; Rao, 1968; Rao (Personal communication))	E. Pakistan (Akbar, Personal communication)	Ceylon (Loos, 1953a; Hutchinson, 1960b; Hutchinson and Vythilingam, 1963b; Sivapalan, 1968a)	Japan (Kaneko and Ichinohe, 1963)	Indonesia (Senmangun, personal communication)	E. Africa (Hainsworth, 1970; Hainsworth, Personal communication)	Central Africa (Martin, 1962)	Taiwan (Wu, Personal communication)	Queensland, Australia (Colbran, Personal communication)
Tylenchus agricola	●									
Tylenchus sp.							●			
Ditylenchus sp.							●			
Tylenchorhynchus					●					
Tylenchorhynchus sp.	●			●			●			
*Pratylenchus loosi				●	●					
*P. brachyurus	●									
Pratylenchus sp.	●	●				●			●	
*Radopholus similis				●		●				
Pratylenchoides sp.							●			
Rotylenchulus sp.				●						
Rotylenchus sp.	●						●			
*Helicotylenchus dihystera				●		●				
*H. erythrinae					●				●	
Helicotylenchus sp.		●	●	●			●			
Scutellonema brachyurum	●			●						
Scutellonema sp.							●			
Hoplolaimus columbus	●									
Hoplolaimus sp.			●	●			●			
Boleodorus sp.				●						
**Meloidogyne javanica*	●	●		●		●	●	●		●
**M. incognita*	●				●			●		
**M. arenaria*				●				●		
**M. hapla*	●			●						
*M. brevicauda		●		●						
Meloidodera floridensis	●									
*Hemicriconemoides kanayaensis						●			●	
*Paratylenchus curvitatus	●			●	●					
Paratylenchus sp.		●					●			
Macroposthonia ornata (?)	●									
Trophotylenchulus sp.							●			
Aphelenchus agricola	●									
Aphelenchoides composticola	●						●			
Xiphinema insigne	●									
Xiphinema sp.			●				●			
Longidorus sp.			●							

* The association of these nematodes with tea is covered in some detail in the text.
** Nematodes occurring in tea nurseries only.

K*

A. Root-knot Nematodes, *Meloidogyne* spp.

1. Root-knot Nematodes of Young Tea

The first report of root-knot nematode in young tea was from South India, where large numbers of tea seedlings were found infected (Barber, 1901). Large-scale failures in the nurseries of Ceylon were attributed to severe root-knot nematode infection, which sometimes destroyed entire nurseries (Stuart Light, 1928). Several species of *Meloidogyne* have been found associated with tea (Table I).

Young nursery plants, both seedling and vegetatively propagated clonal plants, are severely damaged by root-knot nematodes. Seedling plants in which both the tap-root and lateral roots are severely attacked suffer greater damage than do clonal tea plants of equivalent age, probably because seedling tea plants possess less than half the root bulk of clonal plants of similar age (Kerr, 1963b). The susceptibility of young plants to root-knot nematodes decreases with age. A marked increase in resistance is observed between 8 and 15 months (Gadd and Loos, 1946; Loos, 1951; Loos, 1953a; Kerr, 1963b).

Some root-knot nematode larvae enter young feeder roots of mature tea bushes but fail to cause giant cells and are apparently unable to complete the moult between the second and third instars (Gadd and Loos, 1946). The subsequent interplanting of mature stands of seedling tea with root-knot susceptible cover crops and shade trees has resulted, in many areas, in a marked increase in the nematode population. Nevertheless, mature tea bushes growing in such areas show no signs of attack (Loos, 1953b).

2. Root-knot Nematodes of Mature Tea

In certain areas, root-knot nematodes cause severe galling of the roots of mature tea bushes with resultant decline. This was first suspected to be a specialized race of *M. javanica* that had adapted to mature tea (Gadd and Loos, 1946) but later was confirmed as a separate species, *M. brevicauda* (Loos, 1953b). Tea plants of all ages are susceptible to infection by this nematode which has been recorded in only Ceylon and South India. In Ceylon, it has been recorded in only three estates, all bordering the same jungle, at altitudes of 1500–2000 m (Hutchinson and Vythilingam, 1963b).

The above-ground symptoms of attack by *M. brevicauda* are similar to those of the root lesion nematode *Pratylenchus loosi*. The infected bushes grow more slowly and the leaves tend to become smaller, dull and yellow in appearance. The roots show the characteristic swelling and pitting that is typical of *M. brevicauda* (Fig. 1).

The effects of infection on the growth of tea bushes is most evident during the period of recovery from pruning. Healthy bushes are usually harvested about 16 weeks after pruning; but the harvesting of severely infected bushes is necessarily delayed up to a further 6 months. The production of new branches by infected bushes is very restricted, and as a consequence the cropping capacity of the bush is reduced. The most severely affected bushes fail to recover from pruning (Loos, 1953b).

Fig. 1. The root system of a mature tea bush infected with *Meloidogyne brevicauda*.

The extent of damage and the consequent crop-loss caused by this nematode is comparable to that caused by *P. loosi*. However, living nematodes are rare in the severely galled roots and only a few mature females and males are detected. The larval population in the soil is also low. Tea is probably a poor host to this nematode, yet it appears to be hypersensitive to infection (Hutchinson and Vythilingam, 1963b).

Young clonal tea planted in areas heavily infested with *M. brevicauda* has not been adversely affected (Sivapalan, 1969c) as this nematode species seems to require relatively long periods to establish in the roots of young tea plants (Venkata Ram, 1963).

B. Root lesion Nematodes, *Pratylenchus* spp.

1. *Pratylenchus loosi*

This is the most serious nematode pest of tea plantations in Ceylon (Gadd, 1939; Gadd and Loos, 1946; Loos, 1953a; Hutchinson, 1961a) and a severe pest of tea in Japan (Kaneko and Ichinohe, 1963; Takagi, 1969). In Ceylon, it is very widely distributed in tea plantations at altitudes of 900–1800 m, causing heavy crop loss in mature tea (Fig. 2) and extensive failures in newly planted young tea fields (Hutchinson and Vythilingam, 1963a). In Japan, where the tea is cultivated at altitudes of 0–300 m, damage caused by *P. loosi* is common in a large number of plantations (Kaneko, personal communication).

FIG. 2. Debilitation of mature tea caused by *Pratylenchus loosi*. Affected bush on the right.

The largest soil populations of *P. loosi* are at depths of 10–15 cm (Hutchinson and Foster-Barham, 1963; Kaneko and Ichinohe, 1963). The percentage of males in both soil and roots is about 37–40%. However, crowded conditions tend to shift the population towards more males, as may be observed in the periphery of root lesions (Gadd and Loos, 1941). A marked periodic fluctuation in the soil populations of these nematodes is often observed and these changes appear to be closely linked to rainfall patterns (Sivapalan, unpublished data).

The visual symptoms of injury caused by *P. loosi* consist of patches of unthrifty tea, with the affected bushes appearing "thin" due to a deficiency of maintenance foliage (Fig. 2). Such bushes produce very little new growth, the leaves tend to become smaller and yellow in appearance and there is a pronounced tendency to flower and seed. The affected bushes become progressively less thrifty and the more susceptible ones die, particularly during adverse conditions, such as a period of drought, or following pruning.

P. loosi attacks mainly the cortex of the small feeder roots of tea. As long as there are sufficient roots from which to obtain nutrient, the major proportion of the nematode population remains within the roots and only a small number is found free in the soil. When the parasitized roots decay, the nematodes migrate through the soil in search of fresh feeder roots (Gadd and Loos, 1946). The efficiency of the root system is thus diminished and the attacked bush begins to show above-ground symptoms, fails to grow normally and becomes unthrifty. The stage ultimately arrives when the feeder roots are destroyed as fast as they are formed and the bush becomes moribund.

P. loosi also attacks the large storage roots; as a result of their feeding the nematodes cause the tissues to die, and form discrete necrotic areas or lesions which are exposed when the bark is peeled. The nematodes within such large roots move inwards towards the cambium, away from the dying tissue and only rarely move outwards into the soil. Large numbers of these nematodes are found crowded at the periphery of such lesions (Gadd and Loos, 1941). The lesions eventually expand and girdle the root, and as a consequence the portion of the root distal to the lesion dies.

Young tea, newly planted in areas that were previously infested with *P. loosi*, suffers extensive damage. The young plants are stunted and remain dormant (Fig. 3) so that large patches of declining tea appear.

2. Pratylenchus brachyurus

This species has been reported only from northeast India (Anon., 1957), where it is restricted mainly to the plains of Assam. Investigations so far indicate that it is pathogenic primarily to 3 year old seedlings (Anon., 1968). The nematodes penetrate deep into the roots of seedlings causing dark red lesions at the point of invasion. Because of incomplete surveys, it is not possible to conclude whether mature tea is resistant to this nematode.

C. Burrowing Nematode, *Radopholus similis*

This nematode was first reported as a pest of tea in Indonesia (Java) (Zimmerman, 1899). The suitability of tea as a host to this nematode has been established (Steiner and Buhrer, 1933) and in Ceylon, young tea plantings at altitudes of 500–1000 m are severely attacked by it. Premature flowering and fruiting, together with a general stunted

appearance of young plants is associated with large numbers of
Radopholus similis in the roots and in the surrounding soil (Sivapalan,
1968a).

I have found this nematode associated with decline of young nursery
plants in a few nurseries at the above altitudes. In one instance they
were found associated with *Pratylenchus loosi*. Experiments with a
mixed inoculum of *R. similis* and *P. loosi* have indicated that, in clonal
tea, the latter multiplied very much faster, and in time, the root
population shifted towards *P. loosi*.

Fig. 3. A field of young clonal tea badly affected by *Pratylenchus loosi*.

D. Sheath Nematode, *Hemicriconemoides kanayaensis*

This nematode is found widely distributed in the tea fields of Japan
and Taiwan, where it is one of the more serious pests of tea (Nakasono
and Ichinohe, 1961; Kaneko and Ichinohe, 1963; Wu, personal communication). It is an ectoparasite and attacks mainly the succulent
feeder roots of the tea plant. Sloughing of the root cortex from the
stele and the brownish discoloration of the latter, are characteristic
symptoms of damage (Takagi, 1969). The greatest numbers of *H. kanayaensis* are recovered from the deeper layers of soil, approximately 30
cm below the surface, in the accumulation of decaying infected roots
(Kaneko, 1963).

E. Spiral Nematodes, *Helicotylenchus* spp.

Helicotylenchus dihystera is by far the commonest plant-parasitic nematode in each of the three major climatic regions in which tea is grown in Ceylon (Hutchinson and Vythilingam, 1963b). Populations of these nematodes are greatest in soil samples collected in the vicinity of shade trees such as *Erythrina lithosperma* (dadaps) and *Gliricidia sepium*, which are probably host plants (Gadd, 1943; Hutchinson and Vythilingam, 1963b). Although inoculation experiments with pot plants have indicated that tea is an unlikely host, this nematode has frequently been recovered in small numbers from tea roots collected from the field (Hutchinson and Vythilingam, 1963b).

Helicotylenchus dihystera and *H. erythrinae* have frequently been encountered in the tea plantations of Japan (Kaneko and Ichinohe, 1963). Recent surveys have shown that *Helicotylenchus* spp. are important parasites of tea in East Africa (Hainsworth, 1970).

Host range studies indicate that coffee is a good host of *H. dihystera* (Hutchinson and Vythilingam, 1963b). Another unidentified species of *Helicotylenchus*, commonly found in old coffee lands planted with young tea, is associated with plant failures in south India. The affected patches look unthrifty and the plants, which are generally stunted, have sparse and yellowing foliage (Rao, 1968). This species is also encountered in nurseries (Rao, personal communication).

F. Pin Nematode, *Paratylenchus curvitatus*

This is one of the most prevalent plant-parasitic nematodes in the tea plantations of Ceylon, Japan and northeast India, where it is sometimes recovered in large numbers from the feeder roots of both young and mature tea bushes (Hutchinson and Vythilingam, 1963b; Kaneko and Ichinohe, 1963; Anon., 1968). Despite the large numbers of these nematodes found distributed in the soil occupied by the feeder roots (Anon., 1968), there is no positive evidence of pathogenicity to tea.

III. Cultural and Environmental Influences

Little is known of the influence of cultural and environmental factors on nematodes associated with tea. Such information as exists appears to be almost wholly confined to observations made in Ceylon on *Pratylenchus loosi*.

A. Cultural

Tea is a crop of very long duration but, nevertheless, certain cultural practices in tea growing are modified from time to time or omitted.

Many of these cultural practices have either a direct or indirect bearing on the occurrence, distribution and prevalence of nematode pests.

In this section those factors which tend to increase the harmful effects of nematodes have been arbitrarily selected for discussion, while those which are suitably modifiable as a means of mitigating these effects are discussed in the control section.

1. Natural Susceptibility and Tolerance

Most tea fields are planted from seed and hence the individual bushes differ from one another genetically; the degree of natural susceptibility to nematode attack is variable. As a consequence, the spread of nematode infection is not uniform and severe nematode attacks are usually found limited to groups of bushes or individual bushes, scattered at random. When a nematode-susceptible, vegetatively-propagated clone is planted in nematode infested areas, the spread of the pest is rapid and uniform. Consequently, the effects of nematode infection on clonal tea fields is more striking than on seedling fields.

The use of non-susceptible clones for replanting in areas where there is a risk of nematode infestation is currently a favoured practice (Kerr and Vythilingam, 1967b).

2. Soil Movements

In earlier plantings, tea bushes were planted at right angles to the contour and this method favours a free movement downwards of soil and water. Regular weeding of tea lands with weed-scrapers and periodical forking loosens the soil which tends to move downhill, transporting nematodes with it, especially during heavy rains when the soil is trapped in silt pits or reverse-slope drains. When these silt pits or drains are cleaned out, the silt is thrown back onto the field and the contained parasitic nematodes infect the neighbouring tea bushes (Gadd, 1939). It is quite common to see patches of nematode infected declining tea at the bottom of such slopes. Infection of new plantings situated at the bottom of slopes have often been traced to movement of soil and water from infected tea situated above such clearings (Kerr and Vythilingam, 1966). Contour planting and commencement or replanting along the contour at the top sections are current practices that mitigate against this type of spread (Hutchinson, 1962a).

The frequent movement of tea pluckers from field to field during regular harvesting results in infested soil being transferred on the feet to hitherto uninfested areas. This is particularly so during wet weather.

3. Cultivation of Shade Trees and Cover Crops

In most tea growing regions, it is customary to interplant tea with different varieties of shade trees and leguminous cover crops. These cover crops provide shade, aid in nitrogen fixation and provide a regular addition of organic mulch. This mulch increases the saprophytic organisms and natural predators of nematodes that help to keep the parasitic nematode population under check and also facilitates a better root growth which enables the tea bush to "tolerate" nematode parasitism. On the other hand, some cover crops are hosts to nematodes that are tea pests and increase the incidence of these pests (Loos, 1951; Visser, 1959a; Hutchinson and Vythilingam, 1963b; Banerjee, 1967).

Tephrosia vogelli, a common leguminous cover crop grown in the tea plantations of Ceylon, is a good host of *Pratylenchus loosi* and its cultivation is known to sustain and increase the incidence of this pest. The cultivation of this cover crop is now discouraged in areas where these nematodes occur (Visser, 1959b). The cultivation of *Tephrosia* and dadaps, both of which are excellent hosts of *Meloidogyne* spp., increases the occurrence of these nematode pests. Although the mature tea is not adversely affected by these nematodes, movement of such infested soil into silt pits, soil from which is occasionally used as a source of nursery soil, poses a threat to young nursery plants. Likewise, the interplanting of shade trees (dadaps and *Gliricidia sepium*) results in a substantial build-up of spiral nematodes, *Helicotylenchus* spp. (Hutchinson and Vythilingam, 1963b). Although the pathogenicity of these to tea is not proven, circumstantial evidence suggests that tea could be affected (Rao, 1968).

4. Fertilizer Applications

In experiments using potted plants with different nitrogen levels it has been observed that while the plants in the low nitrogen treatment show more evident symptoms of infection, relatively small numbers of *P. loosi* are recovered from the roots. At higher levels of nitrogen, although symptom manifestation is diminished, nematode counts are greater (Sivapalan, 1968b). This effect arises through increased root production which benefits both the host plant and nematode parasite. In contrast, populations of the ectoparasitic nematode *Hemicriconemoides kanayaensis* decline with increased levels of nitrogen (Kaneko and Ichinohe, 1963).

An increase in the potash and phosphate in the fertilizer mixture did not improve growth of young tea plants infected with *P. loosi* (Kerr, 1963b). However, in a subsequent experiment under similar conditions, analyses of leaves and roots showed that the potash content is inversely related to nematode pathogenicity (Sivapalan, 1968b).

5. Pruning

The pruning cycle for tea bushes varies from 2 to 5 years depending on climatic conditions and growth rates. When the tea bush is pruned, a proportion of the feeder roots decay and the endoparasitic nematodes, like *P. loosi*, migrate into the soil (Gadd and Loos, 1946). During recovery from pruning when the fresh feeder roots are developing, the nematodes re-invade the new roots. As long as the tea bush is sufficiently vigorous to produce significantly more feeder roots than are affected by nematodes, the bush recovers normally and is ready for harvesting within the usual 14–16 weeks. On the other hand, the relatively few feeder roots put out by weaker bushes are readily attacked by nematodes and these bushes take as long as 10–12 months to recover. The very badly affected bushes fail to recover. Lengthening of the pruning cycle and delaying post-prune harvesting in the heavily infected low yielding tea fields has helped to reduce casualties from pruning.

The effect on nematode populations and tea growth of soil exposure at pruning, and the increased organic matter through feeder root decay and pruned leaf litter, are factors that remain to be examined.

6. Uprooting of Old Infested Tea Fields

Old, infested, uneconomic tea fields are replanted, following suitable pre-planting treatments. Root fragments left behind in the soil maintain living nematodes as focal points for the infection of subsequently planted tea after uprooting operations occur (Hutchinson, 1962a) even through a 2 year period under a non-host "rehabilitation" grass (cover crop grown for improving soil structure and adding organic mulch). Populations of over 7500 bushes per hectare are normal. The bushes are deep-rooted and the normal uprooting procedures are insufficient to ensure complete extraction of the roots. Winching out of old tea, although preferred to other forms of manual uprooting (Sivapalan, 1967b) still results in substantial numbers of residue roots in the soil (Sivapalan, 1968b). Large-scale failures of newly replanted tea areas have been traced to inefficient uprooting (Sivapalan, 1967a).

7. Transport of Infected Planting Material

The transport of young tea plants from infected nurseries to planting sites is one of the prime causes of spread of *Pratylenchus loosi* infection in the tea areas of Ceylon (Hutchinson and Vythilingam, 1963a; Kerr and Vythilingam, 1966). The spread of the sheath nematode *Hemicriconemoides kanayaensis* in the tea plantations of Japan is believed to be mainly via infected seedlings (Kaneko, 1963; Takagi, 1969). Hence, an infected tea nursery serves also as a distribution centre for infection (Hutchinson, 1960b).

Even in nurseries that routinely fumigate their soils, subsequent infections are caused by contaminated irrigation water (Kerr, 1964). In general, for purposes of irrigating nursery plants, water courses that pass through various sections of the plantation are channelled into reservoirs that hold water temporarily and so permit the nematodes to settle.

To reduce the risk of spreading infection from plantation to plantation, the transport and sale of nursery plants is now forbidden by law and every plantation is expected to have its own nursery.

B. Environmental

1. Soil Temperature

Pratylenchus loosi is found in the tea growing regions of Ceylon in all planting districts and at all elevations but it is most abundant at altitudes of 900–1800 m where it causes the most obvious damage (Hutchinson and Vythilingam, 1963a).

In an experiment where plants infected with *P. loosi* were exposed to environmental conditions of three elevations, the largest increase in number of nematodes was at 1200 m, with a lower rate at 2100 m while at 60 m numbers rapidly declined (Hutchinson and Vythilingam, 1963a).

When plants uniformly inoculated with *P. loosi* were maintained at four different temperatures in controlled environment chambers, the number of nematodes per gram of root increased at both 15·6°C and 21·1°C, but declined at 11·5°C and 28·0°C (Sivapalan, 1969c) (Fig. 4).

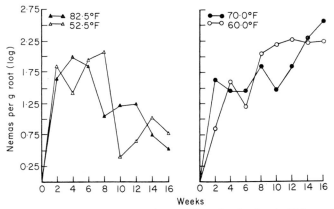

Fig. 4. The effect of temperature on numbers of *Pratylenchus loosi* within roots of young infected tea plants in pots.

Soil temperature varies with elevation, and the favourable temperature range for *P. loosi* reproduction (15·6–21·1°C) occurs in soils at elevations between 900 and 1800 m. At higher and lower elevations the soil temperatures (10–12°C and 27–30°C respectively) cause a much slower rate of multiplication of *P. loosi*.

Meloidogyne brevicauda is restricted in Ceylon to a few estates at elevations of 1500–2000 m, while in south India (where climatic conditions are almost identical), the only report is from one estate at an elevation of 900–1000 m (Venkata Ram, 1963).

2. Soil Type

Soils suitable for tea culture fall into two distinct classes in Ceylon. At the highest and lowest elevations they are of podzolic origin, while in the median regions they are mainly lateritic (Panabokke, 1967). *P. loosi* is not confined to either group but although both soils are of a similar loamy texture, occasional localized variations do influence the multiplication and build-up of *P. loosi*.

In a pot experiment, the greatest build-up of *P. loosi* and the greatest damage to plants occurred in clay soils, while the least build-up was in the gravel soils and the least damage to tea plants was in the sandy soils (Sivapalan, 1971).

3. Soil Moisture

Within the appropriate elevation range, *P. loosi* is found in greater concentrations and more widely distributed in areas retaining a uniform and adequate soil moisture. Surveys have shown that the greatest abundance of this pest is in areas receiving a high and well-distributed rainfall (Hutchinson and Vythilingam, 1963a). Soil populations of *P. loosi* examined at regular intervals at different sites revealed that the population levels fluctuated widely at different times of the year. This was closely associated with prevailing soil moisture and rainfall patterns. The population increased with increased precipitation and declined markedly during the dry months (Sivapalan, unpublished data).

4. Predators and Competitors

Little is known of the predators and parasites affecting tea nematodes. A sporozoan endoparasite has occasionally been recorded from *P. loosi* (Gadd and Loos, 1946) but it is doubtful whether this has any economic significance.

Hemicriconemoides kanayaensis is frequently attacked by a Phycomycete. Kaneko and Ichinohe (1963) recorded an instance where as many as 30% of the adult females were killed by this fungus.

The addition of compost and other forms of organic matter was reported to suppress soil populations of *P. loosi* (Visser, 1959a), but subsequent findings did not support this view (Hutchinson, 1961b).

IV. Control

The lack of proper surveys has encouraged the belief that the nematode problem in many of the tea growing regions is mainly a nursery problem. Consequently, the majority of the control programmes are confined to controlling and eradicating these pests in the nursery. Even where nematodes are known to be severe pests of mature tea, chemical control measures are seldom adopted. Tea is a perennial crop and so a single nematicidal application is ineffective for long-term control and, on the other hand, routine applications are uneconomical. Improved cultural practices that maintain the tea bushes in a healthy and vigorous condition seem to be the only practical and economic means of minimizing nematode damage in mature tea plantations.

A. Chemical

1. Nurseries

Very good, if not complete eradication of nematode pests is possible at the nursery stage. While every effort is made to procure nematode-free soil for nurseries, common nursery practice is the routine chemical disinfestation. Treatment of bulked soil in storage sheds, nursery beds *in situ* or of soil-filled bags *in situ* on nursery beds, are alternative procedures. To eliminate possible phytotoxic effects, an appropriate period of time is allowed to lapse before planting seeds or cuttings irrespective of the nature of the treatment.

The different nematicides and their respective dosage rates used in the nurseries of different tea growing regions are presented in Table II.

2. Pre-planting Control in the Field

Pre-planting chemical control of intended planting sites is becoming increasingly popular. Prior to large-scale replanting of tea, the planting site is either completely fumigated or, planted to a non-host cover crop for a period of 1–2 years, followed by fumigation confined to planting rows. As an alternative to soil fumigation, granular nematicides are incorporated into planting holes at planting time.

The different nematicides and their respective dosage rates used in planting sites are presented in Table III.

3. Control in Mature Tea

It is only in a few tea growing regions, where reasonably comprehensive surveys of nematode distribution have been done, that chemical treatments of infected mature tea have been attempted. Mature tea fields infected with *P. loosi* and fumigated with EDB (335 litres/ha), ethylene chlorobromide (335 litres/ha), DBCP (22, 56, 112 or 225 litres/ha) or DD (450 litres/ha) did not show any improvement in growth. In fact, EDB, ethylene chlorobromide and DD produced toxic symptoms on tea (Loos, 1955). Fumigation with DBCP at 90–112 litres/ha

TABLE II. Nematicides used in nurseries of the different tea growing regions

Nematicide	Dose	References
DBCP (1,2 dibromo, 3 chloropropane)	22–112 litres/ha	Loos (1955), Visser (1957), Venkata Ram (1962), Akbar and Ali (1965b), Mukerjee (1966), Kaneko (1967), Anon. (1969)
DD (1,2 dichloropropane and 1,3 dichloropropene mixture)	280–560 litres/ha	Loos (1953a), Hutchinson (1960b), Kerr (1963a), Akbar and Ali (1965a), Rao (1966), Kaneko (1967), Anon. (1969)
EDB (83% E.C.) (Ethylene dibromide)	78 litres/ha	Mukerjea (1966)
EDB (30% E.C.)	300 litres/ha	Masuda and Kaneko (1963)
Methyl bromide	1 kg/6 m³	Kerr and Vythilingam (1967a), Sivapalan (1969a)
Chloropicrin	—	Takagi (1969)

TABLE III. Nematicides used in planting sites

Nematicide	Dose	Type of treatment	References
EDB or DD	225–335 litres/ha	Complete fumigation	Masuda and Kaneko (1963) Takagi (1969)
Methyl bromide	236 kg/ha	Complete fumigation	Sivapalan (1969b)
Di-Trapex (DD+methyl isothiocyanate 80:20 v/v)	225 litres/ha	Complete fumigation	Sivapalan (1969b), Anon. (1971)
Di-Trapex	56 litres/ha	Row-fumigation	Anon. (1971)
Terracur-P (Fensulfothion 5% granules)	14g/plant	Planting holes	Anon. (1971)

reduced the original population levels of *P. loosi* by 90%, but this decrease in population lasted for only about 7 months (Visser, 1957).

In an experiment where pre-treatment yields were maintained in a field infested with *P. loosi*, a 12% increase in yield followed a post-prune treatment with DBCP (75% E.C.) at 112 litres/ha (Hutchinson, 1962b). Injection of DBCP (80% E.C.) around the base of unpruned mature tea bushes gave temporary control and consequent increase in yield in tea (Venkata Ram, 1964; Ichinohe, 1967).

Masuda and Kaneko (1963) showed that the application of DBCP (80% E.C.) by the furrow irrigation method at 50–100 litres/ha effectively controlled nematodes. DBCP and DCIP (dichloro di-isopropyl ether) are used extensively in nematode infested tea plantations in Japan (Takagi, 1969). Both nematicides are used either as the emulsifiable concentrate (75% E.C.) or as the granular formulation (20% and 30% respectively). The emulsion is poured into shallow trenches (15–20 cm deep) cut around the base of the tea bush; the granular formulation is dibbled into shallow trenches or broadcast on the surface.

B. Cultural

A vigorous tea bush tolerates nematode parasitism better than a weak bush. Cultural methods that generally promote a better growth of tea are usually carried out with particular care in nematode infested fields. Marginal tea fields, where yields have declined, but not to uneconomic levels, benefit by such practices.

1. Incorporation of Organic Matter

Incorporation of organic matter into nematode infested soil is known to suppress the incidence of parasitic nematodes on tea (Eden, 1949; Loos, 1953a; Visser, 1958; Anon., 1969; Takagi, 1969). Shade trees such as dadaps that were severely affected by *Meloidogyne javanica* benefited by the addition of compost (Eden, 1949). The incorporation of large quantities of green manure such as loppings from dadaps, *Tephrosias* and marigolds, resulted in population decline of *P. loosi* in mature tea, and the number of non-parasitic nematodes increased more than sixfold (Visser, 1958). On a field scale, however, the incorporation of compost at the rate of 50 metric tons/ha raked into deep-forked soil did not significantly depress the population of *P. loosi* but there was a sharp temporary rise in the numbers of saprophagous nematodes (Hutchinson, 1960c, 1961a).

The incorporation of compost and stable manure does not necessarily decrease the number of parasitic nematodes but it benefits the tea mainly by promoting vigorous growth and thereby enabling the tea bush to better tolerate nematode parasitism (Visser, 1959b; Hutchinson, 1961a). In this respect this practice is beneficial.

2. Regular Fertilizer Application and Forking

As long as the tea bushes produce new roots in excess of those damaged by nematodes, they will continue to appear healthy, even though the crop yield is diminished. If, however, nematode damage exceeds the rate of production of new roots, the bushes show symptoms of decline and yields decrease.

Infected fields receiving low dosages of fertilizer generally show an accelerated rate of decline, which can be arrested by an increased application of fertilizer.

Nematode damage is often severe in the heavy clay loams and poorly-drained soils (Kerr and Vythilingam, 1966; Sivapalan, 1971). With time, such soils often become panned, a condition which leads to poor soil aeration and consequent poor root growth (De Silva, 1967). Nematode infected tea bushes growing under such conditions decline rapidly. Under these circumstances, the response to fertilizer application is also poor.

Regular forking helps to break the hard pan and promote better soil aeration, and consequently improves root growth (Portsmouth, 1956). Tea bushes infected with *P. loosi* show a significant improvement following forking. Symptoms such as a pronounced yellowing on leaves and profuse flowering and fruiting disappear gradually and response to fertilizer application improves steadily.

3. Cultivation of Marigolds (*Tagetes* spp.)

The cultivation of marigold (*Tagetes erecta* and *T. patula*) as a pre-plant crop considerably reduces the population of *P. loosi* and *Meloidogyne* spp. The reduction in the nematode population is significantly greater than when the land is left fallow (Visser and Vythilingam, 1959; Kerr, 1963b).

The interplanting of marigold in tea has resulted in a 7% increase in yield compared to 12% increase that followed soil fumigation with DBCP (Hutchinson, 1962b). The marigolds do not compete unduly with the mature tea for soil moisture or nutrients. However, they do compete with the shallow-rooted young tea plants (Hutchinson, 1960a). A similar observation has been made for young seedling tea in East Africa (Hainsworth, 1970).

The nematicidal compounds in marigolds which have been identified as a-terthienyl and its analogues (Uhlenbroek and Bijloo, 1958) kill nematodes that enter the roots. *P. loosi* is attracted to marigold roots, but upon entry the nematodes die within a week. Roots of marigold in the vegetative stage are effective in this way, but roots of plants in full bloom or post-bloom are less effective. Consequently, it is preferable to maintain the plants in a non-flowering state by periodic lopping (Hutchinson, 1961b, 1962b).

4. Replanting

When a tea field is heavily infected with nematodes and the yield is below marginal, it is seldom possible to improve the field through better bush and soil management. When the percentage of dead bushes and consequent vacancies are high, resupplying becomes impractical and replanting of the entire field is preferred.

(a) *Uprooting.* As mentioned in the previous section, root removal operations are critical during uprooting of old infested tea fields. Exhaustive root removal operations follow uprooting in order to minimize the carryover of infection to the replanted tea. Roots are removed by deep-forking to depths up to 75 cm before planting with a rehabilitation crop (Sivapalan, 1967a).

(b) *Rehabilitation.* Following the uprooting of old tea, it is customary to plant the area with a soil reconditioning crop such as Guatemala grass (*Tripsacum laxum*) or Mana grass (*Cymbopogon confertiflorus*), for one or more years to rest the soil and to help consolidate and improve the soil structure (Tolhurst, 1956). These rehabilitation crops are also resistant to *P. loosi* and *Meloidogyne* spp. (Visser, 1959a; Hutchinson, 1962a; Banerjee, personal communication). Consequently, the soil populations of these parasitic nematodes decline rapidly (Fig. 5). The decline in the soil population of *P. loosi* in soils under Guatemala grass is very much greater than when the land is left fallow (Visser, 1958; Kerr, 1963b; Kerr and Vythilingam, 1966).

(c) *Replanting with Nematode Tolerant Tea Clones.* It is practically impossible to eradicate nematodes from an infested tea land and a reduction in numbers to an economically acceptable level is all that can be achieved. The young tea that is planted in previously infested areas becomes parasitized (Hutchinson, 1962a; Hainsworth, 1970), although the more vigorous the plants the more tolerant they are to parasitic nematodes.

Experiments in pots have shown that some tea plants are more tolerant than others to attack by *P. loosi*. The susceptible plants are severely debilitated within relatively short periods, while others grow normally with hardly any symptoms of damage (Gadd, 1944). With the introduction of the technique of vegetative propagation in tea, a large number of selections of clones have been made from seedling bushes apparently tolerant to this nematode (Loos, 1953a; Visser, 1958;

Hutchinson, 1960a). Selections from the TRI series (Tea Research Institute) 62/9, 2025 and 2142, and other estate selections such as Norwood 2, Drayton 1 and 95, Kirkoswald 145 are recommended for use in areas infested with *P. loosi* (Kerr and Vythilingam, 1967b). Selection of tea clones for tolerance to nematodes is also in progress in East Africa (Hainsworth, 1970). The use of such tolerant tea clones has greatly increased yield in nematode infested areas.

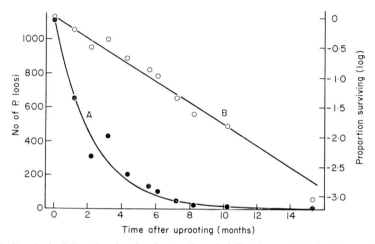

FIG. 5. Survival of *Pratylenchus loosi* over the rehabilitation period: A—the number of *P. loosi* extracted from 32 soil samples, each of 100 gm; B—logarithm of the proportion of survivors. Reproduced with permission from Kerr and Vythilingam (1966).

Crop rotation is one of the most potent defensive measures available to the grower of annual or bi-annual crops, and this is denied to the cultivator of a perennial crop like tea.

V. Economics

Nematode pests of tea have not attracted much attention until recent years. The available data is very meagre and grossly inadequate as a basis for the consideration of economic aspects. In most countries, even the preliminary surveys of nematode distribution in tea are as yet incomplete. It is only in Ceylon, where the major nematode pest of tea, *Pratylenchus loosi* has been investigated in some detail, that attempts have been made at evaluating crop loss and the economics of control measures. Much importance has been attached to such investigations, because the tea crop is of vital economic importance, contributing more than 60% of the total foreign exchange earnings of Ceylon.

A. Crop Loss

Injury caused to tea in Ceylon by *Pratylenchus loosi* is most severe on tea at elevations above 900 m. This zone constitutes a third of the total tea acreage and the produce here attracts the best prices. Of about 87,500 ha of high-grown tea in Ceylon, approximately 5% is known to suffer obvious injury from *P. loosi*. The reduction in yield in such areas has been estimated to average between 225 and 350 kg made tea/ha/year (Visser, 1958). At an average market price of Rs. 6·50 per kg, this would amount to a total annual loss of at least Rs 6 million* (approximately U.S. $1,000,000). Since this nematode is present to a lesser extent in at least another 4,000 ha of high-grown tea, a certain amount of loss is being sustained in these areas too. However, the most important fact is that the entire high elevation tea area is liable to infection and if the pest spreads unchecked, losses could amount to at least 20 times their present figure. Such damage would approximate 10% of the present total value of the Ceylon tea crop. For a country that is so heavily dependent on tea for its foreign exchange earnings, this would be a catastrophic loss.

B. Control Measures

1. Nurseries

An infected tea nursery could serve as the distribution centre to contaminate hitherto uninfested tea lands. Adequate control measures at this point are, therefore, vitally important.

A moderate control of the root-knot nematode sufficient to produce healthy plants is adequate, as they develop complete resistance by about the 12th–15th month. Common soil fumigants such as DBCP, DD and EDB, at 112, 335 and 335 litres/ha respectively give very satisfactory results. At these rates of application, the cost of treatment amounts to only 0·1–0·2 cents/plant. This is only a negligible part of the cost of producing a nursery plant (Visser, 1959b).

More drastic control measures are required for soils infested with *P. loosi*, as both young and mature tea plants are vulnerable to attack. The few nematodes carried with the apparently healthy-looking plants could contaminate a hitherto uninfested field and would serve as a locus for spread. In order to achieve almost complete eradication of this pest, high rates of application of the more common potent soil fumigants, such as methyl bromide, are commonly used. In Ceylon where *P. loosi* is quite prevalent, the use of methyl bromide at the rate of 1 kg/6 m³ of soil is the current recommendation for nurseries. The

* One U.S. dollar = 5·95 rupees (Rs).

cost of treatment is a little over 1 cent/plant. This is still quite economical for the average tea estate. The increased rate of growth of the young nursery plants in treated soil is an advantage as they are ready for the field much earlier, and this curtails the cost of nursery maintenance.

2. Pre-planting Control in the Field

It is customary to plant a grass cover crop such as Guatemala on uprooted tea land. During the 2-year period considered necessary for a proper soil improvement by the grass, populations of parasitic nematodes decline rapidly. As the primary aim is soil rehabilitation, the incidental nematode control can be considered to have been achieved at no special cost. If proper root removal has been achieved initially and if the area is to be planted with nematode tolerant varieties, this measure of control is adequate. This is the most widely accepted method of pre-planting nematode control.

In certain instances, when the soil condition is exceptionally good and, therefore, does not warrant any reconditioning, the period of rehabilitation under grass is sometimes omitted and the land is planted with tea soon after uprooting the old tea field. If such land has a history of nematode infestation such immediate replanting is discouraged unless adequate control measures are adopted. In such instances, three alternatives are available: (a) direct replanting following complete soil fumigation (area fumigation); (b) 2 years rehabilitation under grass, with no fumigation; and (c) one year rehabilitation under grass followed by nematicidal treatment of planting rows only.

In the first instance, the cost of fumigating the entire area is around Rs 4200/ha. The 2-year unproductive period under grass is thereby avoided. While the cost of maintaining grass is relatively low at Rs 1500/ha for 2 years, the added cost of area fumigation is adequately compensated for by the 2 years' extra production, the latter becoming economical. With alternative (c), costs of grass maintenance and of row nematicidal treatment amount to about Rs 2200. The tea then comes into production one year earlier than when the land is kept under grass for the full term of 2 years.

While there is no obvious preference for any one of these alternatives, particular conditions on individual plantations determine the choice. As all nematicidal chemicals are presently imported, their use involves foreign exchange costs. The use of grass, by obviating this, is therefore currently attractive.

3. Post-planting Control

In order to improve the trend of the decline in yield in the tea fields infested with *Pratylenchus loosi*, the field is either treated with suitable nematicides, or the relatively weak patches of tea are uprooted, the sites cleaned, chemically disinfested and resupplied with high yielding, nematode tolerant clonal tea.

Marked drop in yield in nematode infested tea fields is due to three main causes: (a) vacancies caused by complete death of bushes; (b) existence of obviously unthrifty "passenger" bushes, yielding hardly any flush; and (c) presence of apparently healthy tea bushes yielding sub-optimally. Conditions (a) and (b) are the ones that are primarily responsible for such drops in yield. Improvements in yield from fields treated with nematicides is most likely to occur from tea bushes in category (c). "Passenger" bushes are almost on the verge of complete collapse; as such, one could hardly expect these to yield better following chemical treatment.

In a tea field infested with *Pratylenchus loosi*, a 12% increase in yield was obtained in the first year following treatment with DBCP (75% E.C.) at 112 litres/ha (Hutchinson, 1962b). This beneficial effect was clearly transient. In this field, yielding a cycle average of 1000 kg/ha, even if it is presumed that this improvement could be maintained for as long as 3 years, the monetary gain would have been, at the most, only Rs 2200/ha, while the initial cost of treatment was Rs 2350/ha. This method of control, therefore, is uneconomical.

An approximate increase in yield of 7% was obtained by sowing marigolds in an adjoining block of the above infested field (Hutchinson, 1962b). Even assuming that this improvement in yield lasted only during the first year, the monetary gain corresponded to Rs 310/ha, while the cost of sowing and maintaining marigolds amounted to only about Rs 135/ha and this was economical.

In general, uprooting and resupplying weak patches of tea, which include the dead and obviously unthrifty bushes, is the best and most long-lasting remedial measure. The obvious "passenger" bushes yield hardly any flush and their removal usually does not affect the existing yield trends of the field.

Assuming that the dead and weak bushes account for 20% of the total stand of tea bushes/ha, the entire cost of operations, inclusive of treatment and resupplying such an area, would amount to about Rs 2750/ha. The yield return during the first year of harvest by itself could adequately compensate for this cost of resupplying.

Fields with over 40% vacancies or unproductive bushes are unprofit-able to maintain. Complete replanting of such fields is more practical than any attempts at large-scale resupplying. It will be apparent that even where the vacancies are lower, they may be so evenly distributed over the entire area as to make spot treatment less feasible than re-supplying the whole extent. A "Tea Replanting Scheme" provides subsidies for the replanting of contiguous extents exceeding 0·1 ha. Many estates, therefore, make use of this facility.

Apart from the obvious financial losses occasioned by the debilitation of infected tea and the cost of pest control measures, there is a third and hidden loss. This is the sub-optimal return obtained from cultural and other cost inputs. This latter cannot easily be estimated accurately.

Considering the nature of the pest and crop, it is improbable that complete nematode control will ever be achieved. Rather, the future will lie in attaining a favourable balance between host and parasite by the use of tolerant or resistant varieties and the adoption of those measures which are economically feasible for the attainment of some degree of control over depredations by the nematode pest.

References

Akbar, K. and Ali, M. A. (1965a). *Tea J. Pakistan* **3**, 14–18.

Akbar, K. and Ali, M. A. (1965b). *Tea J. Pakistan* **3**, 19–25.

Anon. (1957). *Rep. Tocklai Exp. Stn Indian Tea Ass.* 1957, 129–130.

Anon. (1968). *Rep. Tocklai Exp. Stn Indian Tea Ass.* 1967–68, 63–65.

Anon. (1969). *Tea (J. Tea Boards E. Africa)* **10**, 11–13.

Anon. (1971). "Replanting Tea in Eelworm Infested Areas." *Tea Research Institute of Ceylon—Advisory Circular* No. 12/71. N3.

Banerjee, B. (1966). *Rep. Tocklai Exp. Stn Indian Tea Ass.* 1966, 75–76.

Banerjee, B. (1967). *Two and a Bud.* **14**, 82–83.

Barber, C. A. (1901). *Bull. Dep. Ld Rec. Agric. Madras*, **Branch 2**, (45), 227–234.

Basu, S. D. (1967). *Two and a Bud.* **14**, 84–85.

Das, G. M. (1958). *Rep. Tocklai Exp. Stn Indian Tea Ass.* 1958, 229.

De Silva, R. L. (1967). *Tea Q.* **38**, 340–343.

Eden, T. (1949). The work of the Agricultural Chemistry Department of the Institute. *Monograph on Tea Production in Ceylon* **No. 1**, 1949, pp. 78.

Gadd, C. H. (1939). *Tea Q.* **12**, 131–139.

Gadd, C. H. (1943). *Bull. Tea Res. Inst. Ceylon* **No. 25**, 26.

Gadd, C. H. (1944). *Bull. Tea Res. Inst. Ceylon* **No. 26**, 23–25.

Gadd, C. H. and Loos, C. A. (1941). *Ann. appl. Biol.* **28**, 39–51.

Gadd, C. H. and Loos, C. A. (1946). *Tea Q.* **18**, 3–11.

Hainsworth, E. (1970). *Tea (J. Tea Boards E. Africa)* **11**, 15–17.

Hutchinson, M. T. (1960a). *Tea Q.* **31**, 13–18.

Hutchinson, M. T. (1960b). *Tea Q.* **31**, 119–120.

Hutchinson, M. T. (1960c). *Rep. Tea Res. Inst. Ceylon.* 1960, 70–74.

Hutchinson, M. T. (1961a). *Tea Q.* **32**, 129–132.

Hutchinson, M. T. (1961b). *Rep. Tea Res. Inst. Ceylon.* 1961, 84–92.

Hutchinson, M. T. (1962a). *Tea Q.* **33**, 138–140.

Hutchinson, M. T. (1962b). *Rep. Tea Res. Inst. Ceylon.* 1962, 70–86.

Hutchinson, M. T. and Foster-Barham, C. B. (1963). *Tea Q.* **34**, 34–37.

Hutchinson, M. T. and Vythilingam, M. K. (1963a). *Tea Q.* **34**, 68–84.

Hutchinson, M. T. and Vythilingam, M. K. (1963b). *Tea Q.* **34**, 119–126.

Ichinohe, M. (1967). *Agric. Asia.* Special Issue No. **5**, 93–98.

Kaneko, T. (1963). *Study of Tea* **28**, 31–41.

Kaneko, T. (1967). *Agric. Asia.* Special Issue No. **5**, 56–61.

Kaneko, T. and Ichinohe, M. (1963). *Jap. J. appl. Ent. Zool.* **7**, 165–174.

Kerr, A. (1963a). *Tea Q.* **34**, 150–151.

Kerr, A. (1963b). *Rep. Tea Res. Inst. Ceylon.* 1963, 95–102.

Kerr, A. (1964). *Rep. Tea Res. Inst. Ceylon.* 1964, 67–72.

Kerr, A. and Vythilingam, M. K. (1966). *Tea Q.* **37**, 67–72.

Kerr, A. and Vythilingam, M. K. (1967a). *Tea Q.* **38**, 22–28.

Kerr, A. and Vythilingam, M. K. (1967b). *Tea Q.* **38**, 42–51.

Loos, C. A. (1951). *Tea Q.* **22**, 27–30.

Loos, C. A. (1953a). *Tea Q.* **24**, 34–38.

Loos, C. A. (1953b). *Proc. helminth. Soc. Wash.* **20**, 83–91.

Loos, C. A. (1955). *Bull. Tea Res. Inst. Ceylon.* No. **37**, 51–52.

Martin, G. C. (1962). *Rhodesia agric. J.* **59**, 28–35.

Masuda, K. and Kaneko, T. (1963). *Study of Tea* **28**, 26–30.

Mukerjea, T. D. (1966). *Trop. Agric., Trin.* **43**, 335–340.

Nakasono, K. and Ichinohe, M. (1961). *Jap. J. appl. Ent. Zool.* **5**, 273–276.

Panabokke, C. R. (1967). "Soils of Ceylon and Fertilizer Use." Metro Printers Ltd., Colombo.

Portsmouth, G. B. (1956). *Tea Q.* **27**, 67–69.

Rao, G. N. (1966). *Rep. Tea Scient. Dep. Un. Plrs' Ass. Stn India.* 1965–66, 43–59.

Rao, G. N. (1968). *Rep. Tea Scient. Dep. Un. Plrs' Ass. Stn India.* 1968, 37–45.

Rau, S. A. (1962). *Bull. Un. Plrs' Ass. Stn India* No. **21**, 2–6.

Sivapalan, P. (1967a). *Tea Q.* **38**, (Conference proceedings), 260–268.

Sivapalan, P. (1967b). *Rep. Tea Res. Inst. Ceylon.* 1967, 79–87.

Sivapalan, P. (1968a). *Pl. Dis. Reptr* **52**, 528.

Sivapalan, P. (1968b). *Rep. Tea Res. Inst. Ceylon.* 1968, 70–78.

Sivapalan, P. (1969a). *Tea Q.* **40**, 111–114.

Sivapalan, P. (1969b). *Tea Q.* **40**, 115–118.

Sivapalan, P. (1969c). *Rep. Tea Res. Inst. Ceylon.* 1969, 97–107.

Sivapalan, P. (1971). *Tea Q.* **42**, 131–137.

Steiner, G. and Buhrer, E. M. (1933). *Z. ParasitKde* **5**, 412–420.

Stuart Light, S. (1928). *Tea Q.* **1**, 19–22.

Takagi, K. (1969). *JARQ.* **4**, 27–29.

Tolhurst, J. A. H. (1956). *Tea Q.* **27**, 60–66.

Uhlenbroek, J. H. and Bijloo, J. D. (1958). *Rec. Trav. Chim. Pays-Bas Belg.* **77**, 1004–1009.

Venkata Ram, C. S. (1962). *Bull. Un. Plrs' Ass. Stn India* No. **21**, 7–20.

Venkata Ram, C. S. (1963). *Rep. Tea Scient. Dep. Un. Plrs' Ass. Stn India.* 1962–63. Appendix, 31–36.

Venkata Ram, C. S. (1964). *Rep. Tea Scient. Dep. Un. Plrs' Ass. Stn India.* 1963–64, 21–35.

Visser, T. (1957). *Bull. Tea Res. Inst. Ceylon* No. **39**, 47–50.

Visser, T. (1958). *Rep. Tea Res. Inst. Ceylon.* 1958, 67–73.
Visser, T. (1959a). *Tea Q.* **30**, 96–107.
Visser, T. (1959b). *Tea Q.* **30**, 143–149.
Visser, T. and Vythilingam, M. K. (1959). *Tea Q.* **30**, 30–38.
Zimmerman, A. (1899). *Teysmannia.* **10**, (3/4), 523–531.

L

14

Nematodes of Forest Trees

John L. Ruehle

Forestry Sciences Laboratory, U.S.D.A.
Athens, Georgia, U.S.A.

I. Forest Resources

History tells us that man evolved from the Stone Age to the Bronze Age and then to the Iron Age. It certainly seems that we are now in the cellulose age, as almost everything in our daily use contains cellulose. The increased use of such products is causing the pulpwood and construction industries to expand dramatically to keep pace with the enlarging population, and this is motivating the plywood and lumber mills to intensify their demands. This ever-increasing requirement for raw products puts more and more pressure on foresters to produce the maximum yield of timber per acre in the shortest possible time.

Our forests must be developed to keep pace with increasing demands. The quantities of wood products needed are far greater than can be supplied without greatly intensifying forest management on many millions of acres. It is estimated that in the year 2000, in the southern U.S.A. alone, the timber cut will be 2·3 times the current cut, and the annual growth to meet this demand must provide nearly 70% more softwood and 40% more hardwood than in 1968 (Anon., 1969).

The foresters' problems are compounded by the shrinking area on which merchantable timber can grow. As the population grows, demands will increase for more recreational use of forest lands and additional forest acreage will have to be preserved and managed as watershed areas for our growing urban areas. Also such continually expanding encroachments of civilization as dams, highways and urban housing developments decrease the acreage of productive timberland.

In order to meet these demands, foresters are changing their concepts of forest management. Foresters are now beginning to orient their thinking in much the same way as that of agricultural farmers—improve planting stock, change cultural techniques to include fertilization, and effectively control insect pests and diseases. The change from naturally regenerated, mixed forest stands to forests planted with a single species will give soil-borne pathogens (previously held in check naturally) an opportunity to build up to damaging levels. That is, management of the forest "crop" is likely to change the balanced natural soil environment to a specialized environment favorable for proliferation of soil-borne pathogens. *Fomes annosus* root rot, mimosa wilt, and littleleaf disease are well known soil-borne diseases of forest trees caused by fungi.

Diseases caused by nematodes, another group of soil-borne organisms, are a relatively new forestry problem. It is misleading to suggest that nematodes are new to forestry. We have just recently begun to recognize them as being important.

The importance of a nematode disease largely depends on whether population limits exceed the level at which economic damage occurs to a crop plant. Forest monoculture may permit nematode population densities to exceed the tolerance level.

Trees in forest nurseries are particularly vulnerable to soil-borne diseases caused by nematodes. The continuous cultivation of the same or closely related plant species within the same area, irrigation to maintain soil moisture levels for optimum plant growth, and the maintenance of high soil fertility levels all create an environment that is beneficial to nematode development. Nursery seedlings are lifted and shipped to distant areas for planting and so root infections caused by nematodes may cause sale or distribution of the crop to be restricted by quarantine regulations.

Nematodes may damage forest plantations which are usually established on poorly stocked natural forest areas, on planted areas which have been recently clear-cut, or on abandoned farm lands. Changing to closely-spaced planting of tree species susceptible to nematodes may promote an increase in nematode populations that result in an unacceptably large decrease in rate of growth. Nematode diseases were probably one of the causes of poor crop yield on the old farm lands and contributed to their abandonment. Many of the species of nematodes which damage corn, cotton, peanuts, and other agricultural crops also damage forest trees.

Our knowledge of nematode diseases of natural forest stands is sadly lacking. Large trees growing throughout vast areas on diverse sites and mixed with various other plant species are difficult to study. The undisturbed soil in natural woodlands is a biotic complex that tends to persist indefinitely in equilibrium. The growth rate of mature trees is slowed by age, and individual trees are commonly attacked by wood-destroying fungi and insects. The mixture of tree species and age classes within species creates conditions that tend to discourage the build-up of nematodes specific to a given host. Parasitic nematodes in natural woodland soils rarely find conditions suitable for their development beyond maintenance levels, and they only occasionally become sufficiently abundant to damage trees seriously. Adverse changes in temperature and moisture over extended periods, however, could decrease the growth of trees and accelerate the activities of certain nematodes to a point where serious problems could evolve. Also, soil-borne pathogens could act in concert with nematodes to cause a serious disease complex problem.

II. Nematode Associations in Forests

Nematodes important to forestry are diverse in type and form. One group of nematodes parasitizes forest insect pests; some nematodes feed directly on the aerial parts of the plant, others feed only on other organisms inhabiting the rhizosphere of the tree. One particular association is the group of "fungus feeders" that parasitizes mycorrhizal symbionts. Finally, there are nematodes that feed directly on or in tree roots.

A. Nematodes Parasitic on Tree Insects

Massey (1966) reviews many of the nematode parasites of forest insects. These nematode parasites are important to the forest industry because they kill or weaken the timber boring beetles and the defoliating larvae of Hymenoptera and Lepidoptera. These nematodes may be useful in biological control of these important insect pests (see Chapter 19).

B. Nematodes Parasitic on Aerial Parts of Trees

Nematodes parasitic on the aerial parts of forest trees are usually limited to tropical regions, but in Belgium an infection by *Aphelenchoides fragariae* in cuttings of *Fiscus* spp. in greenhouse propagation beds watered daily was reported by De Maeseneer (1964b). Infections by *Aphelenchoides besseyi* on *Fiscus elastica* in open fields in southern Florida caused discolored lesions in the interveinal leaf tissue and rendered unsalable those plants sought for their attractive, distinctive foliage (Marlatt, 1966). In Mysore, India an unidentified species of *Aphelenchoides* caused a bud rot of *Areca* palms in which infected trees became thin-crowned, gradually shed their leaves, and finally died (Thirumalacher, 1946). Van Hoof and Seinhorst (1962) reported *Rhadinaphelenchus cocophilus* consistently associated with "littleleaf" symptoms of oil and coconut palm in Surinam and British Guiana where nematodes were commonly retarding the development of the young leaves in which they occurred. This nematode also causes a severe disease of oil palm, *Elaeis guineensis*, in Venezuela (Oostenbrink, 1963).

In parts of the West Indies *R. cocophilus* causes a disease of coconut palms called "red ring disease" (Fenwick and Maharaj, 1963). Red ring disease is typically a disease of trees less than 20 years of age causing non-specific external symptoms of yellowing and browning of leaves followed by withering and death of the growing point and finally tree mortality. Fenwick (1962) concluded that this disease was mainly soil-borne with the nematodes entering the roots, producing a spongy and discolored cortex, moving into the bole of the tree, and causing a red discoloration that spreads outward and upward to form the diagnostic red-ring cylinder. Blair and Darling (1968) later proved that although root infections do occur, the crown was the principal infection court and concluded that the palm weevil, *Rhynchophorus palmarum*, which oviposits in the internodes between the petioles and the stem, acts as the principal vector of the nematode. Control measures include eradication of diseased trees, control of the palm weevil, and discontinuation of cultural practices which damage roots and spread nematode-infested tissue from diseased to healthy trees.

C. Nematodes Inhabiting the Rhizosphere of Trees

1. Myceliophagous Nematodes Feeding on Mycorrhizal Symbionts

Riffle (1967) found that a species of *Aphelenchoides* associated with mycorrhizae of southwestern tree species greatly decreased the growth of aerial and substrate mycelium of the symbiotic fungi *Suillus granulatus* and *Mycelium radicis atrovirens* in laboratory culture. *Aphelenchus avenae* also feeds on various mycorrhizal fungi and inhibits formation of mycorrhizae on red pine in pure culture (Sutherland and Fortin, 1968). This nematode apparently does not enter roots and destroy established mycorrhizae but probably indirectly affects plant development as it inhibits mycorrhizal formation by reducing the amount of fungus inoculum in the rhizosphere.

Marx and Davey (1969) proved that mycorrhizal symbionts protect roots from attack by soil pathogens. In suppressing mycorrhizal formation, mycophagous nematodes could indirectly contribute to root disease by removing the biologically protective symbiont and leaving the feeder roots vulnerable to pathogenic fungi.

Myceliophagous nematodes that damage mycorrhizae may reduce the beneficial effect of fungal symbionts and become important in health and growth of trees in areas of marginal rainfall and low fertility.

Saprophytic nematodes inhabiting the rhizosphere of trees also occur in the necrotic tissue of moribund roots. These saprophytes release metabolic products which may either kill nearby plant cells, thus predisposing the plant cells to invasion by damaging micro-organisms, or support the growth of harmful micro-organisms.

2. Parasitic Nematodes Feeding Directly on Roots

My review of literature for information concerning diseases caused by nematodes feeding on roots of forest trees was limited by the lack of publications on this subject. Reports of nematodes damaging to forest trees are restricted generally to nursery investigations. Nematode diseases in plantations and natural woodlands were virtually unexplored until the 1950s, but there is an increasing awareness of such problems in these areas.

III. Nematode Diseases in Forest Nurseries

Forest nurseries with the continuous and intensive cultivation of the same or similar plant species within the same area are similar to certain agronomic monoculture systems which have been plagued by nematode diseases. The crop of tree seedlings is as valuable as tomatoes or tobacco grown in seedbeds, and considerable attention and capital can be economically invested to control nematode diseases in such areas.

A. Association

It is not clear when the first nematode investigation was made of nursery seedlings. However, in 1932 Yamaguchi recorded his discovery of a *Pratylenchus* sp. parasitizing the roots of *Picea* spp. in a Japanese nursery, and in the early 1940s Steiner (1949) found nematodes invading roots of pine seedlings in several nurseries in Florida, U.S.A. Since then, both endoparasitic and ectoparasitic nematodes have been found commonly associated with unthrifty and stunted nursery seedlings in various parts of the world. Table I lists many of the more common nematode genera associated with diseased forest tree seedlings in nurseries.

B. Effect on Growth

Henry (1953) found a few parasitic nematodes in the roots of *Pinus elliottii, P. palustris, P. echinata*, and *P. taeda* seedlings in Mississippi, U.S.A., and he stated that there were too few nematodes to account for the amount of root damage. However, when the nematicidal fumigant ethylene dibromide (EDB) was applied to the nursery bed soil prior to planting, the disease was successfully controlled. He made no mention of assaying the soil for nematodes and it is possible that a disease complex involving ectoparasites and pathogenic fungi may have been overlooked.

During the next few years nematodes were investigated in European nurseries. Weischer (1956) found brown root lesions on conifers growing in nematode-infested areas. Stunted and yellow seedlings of *Picea, Pinus* and *Larix* species in nurseries in certain areas of Austria and Germany (Immel, 1957; Donaubauer, 1959) were probably due to nematode damage. *Rotylenchus robustus, Pratylenchus penetrans, Tylenchorhynchus* sp., and *Ditylenchus* sp. were commonly associated with diseased seedlings by Nolte and Dieter (1957) and these same species of nematodes caused root damage on *Pinus montana mughus, Picea pungens, P. abies*, and *P. sitchensis* (Rühm, 1959). *Rotylenchus robustus* also caused damage to *Pinus* sp. seedlings in nurseries in the Netherlands (Oostenbrink, 1958).

Hopper (1958) made a survey of thirty-five nurseries in the southern U.S.A. to obtain information on the relationship between parasitic nematodes and the growth of southern pines. *Meloidodera floridensis, Tylenchorhynchus claytoni*, and *Tylenchorhynchus* sp. were associated with stunted and unthrifty pine seedlings. *M. floridensis* was found in the roots of *Pinus clausa, P. nigra, P. palustris*, and *P. taeda*, but caused severe damage only on *P. clausa* in one Florida nursery. Although Chitwood and Esser (1957) failed to prove this nematode pathogenic to *P. elliottii*, it was later proved pathogenic to seedlings of this important pine by Ruehle and Sasser (1962). Seedlings grown for 6 months in potted soil infested with 10,000 larvae per pot of *Meloidodera* had significantly smaller root systems, and mature female nematodes

TABLE I. Plant-parasitic nematodes frequently associated with diseased forest tree seedlings in nurseries

Nematode species	Host genera or species	Location	Reference
Helicotylenchus sp.	*Picea pungens*	U.S.A. (Mich.)	Knierim (1963)
Hoplolaimus sp.	*Picea pungens*	U.S.A. (Mich).	Knierim (1963)
H. galeatus	*Pinus*	U.S.A. (Fla)	Steiner (1949)
Longidorus maximus	*Abies, Alnus, Acer, Pinus, Larix, Carpinus, Robinia, Quercus, Corylus, Picea, Thuja*	Germany	Sturhan (1963)
Meloidodera floridensis	*Pinus*	U.S.A. (Fla)	Hopper (1958)
Meloidogyne sp.	*Acacia*	Japan	Hashimoto (1962)
	Pinus	U.S.A. (Fla)	Donaldson (1967)
		Japan	Hashimoto (1962)
	Cornus, Catalpa	U.S.A. (Ga)	Johnson et al. (1970)
	Anthocephalus	Philippines	Postrado and Glori (1968)
Pratylenchus sp.	*Cryptomeria, Chamae- cyparis, Pinus, Picea*	Japan	Yamaguchi (1932)
	Picea pungens	U.S.A. (Mich.)	Knierim (1963)
P. penetrans	*Picea, Taxus, Juniperus, Cedrus, Thuja, Abies, Pinus, Pseudotsuga*	Belgium	De Maeseneer (1964a)
	Picea	Germany	Rühm (1959)
	Cryptomeria	Japan	Mamiya (1969)
	Picea, Pinus	Canada	Sutherland (1967)
Rotylenchus robustus	*Picea, Acer*	England	Goodey (1965)
	Juniperus, Thuja, Taxus, Cedrus, Picea, Pseudotsuga	Belgium	De Maeseneer (1964a)
	Pinus	Belgium	De Maeseneer and van den Brande (1965)
Trichodorus sp.	*Pinus*	U.S.A. (Ala)	Hopper and Padgett (1960)
	Picea	England	Goodey (1965)
	Acer	Belgium	De Maeseneer and van den Brande (1965)
	Picea pungens	U.S.A. (Mich.)	Knierim (1963)
Tylenchorhynchus sp.	*Picea*	England	Goodey (1965)
	Taxus	U.S.A. (Mich.)	Knierim (1963)
T. claytoni	*Picea, Pinus*	U.S.A. (W. Va)	Sutherland and Adams (1964)
	Pinus	U.S.A. (Ala)	Hopper (1958)
	Cryptomeria, Chamae- cyparis	Japan	Hashimoto (1962)
T. ewingi	*Pinus*	U.S.A. (La)	Hopper (1959)
Xiphinema ameri- canum	*Abies*	U.S.A. (Wis.)	Griffin and Darling (1964)
X. bakeri	*Pseudotsuga*	Canada	Sutherland and Dunn (1968)
X. diversicaudatum	*Abies, Larix, Pinus, Alnus, Carpinus, Acer, Robinia*	Germany	Sturhan (1963)

were visible protruding through the epidermis of both lateral and short roots. As the females developed in the roots they caused compression and collapse of the surrounding cortical cells (Ruehle, 1962a). *Tylenchorhynchus claytoni* and *T. ewingi* were commonly associated with chlorotic and dying *P. elliottii* seedlings in Louisiana and Mississippi nurseries (Hopper, 1959). Affected seedlings had clusters of short, stubby roots on a decreased number of laterals.

Foster (1961) working on black root rot in some of these same southern nurseries concluded that parasitic nematodes in the nursery beds containing diseased seedlings could not cause the typical black swellings of black root rot, and unlike Henry (1953), he failed to control this disease by soil fumigation with EDB (but this may have been due to low initial nematode populations). Hopper and Padgett (1960) found black root rot symptoms appearing 4 weeks before build-up of any suspected pathogenic nematode. The high populations of *Trichodorus christiei* found in autumn near the close of the growing season were probably not related to black root rot, but did cause late root damage. The ability of this nematode to damage the roots of *P. elliottii, P. palustris,* and *P. taeda* seedlings was proved under controlled conditions (Ruehle, 1969a). In soil originally infested with 5000 nematodes per pot the seedlings were severely stunted, lacked fine feeder roots and mycorrhizae were few or lacking.

The increased use of soil fumigants and the accompanying improvement in seedling growth prompted more interest in nematodes as disease-causing agents in forest nurseries. *Longidorus maximus* is a significant parasite in tree nurseries in southern Germany (Sturhan, 1963). This nematode has a long period of development and a low rate of reproduction, and the highest populations occur during the summer or autumn and lowest in the spring. The highest population densities of this nematode occur at depths of 20–40 cm and cause thickenings and contortions of root tips and poor overall development of the root system.

Epstein and Griffin (1962) investigated declining and unthrifty spruce (*Picea pungens* and *P. glauca* var. *densata*) in nurseries in Wisconsin, U.S.A. and found heavy populations of *Xiphinema americanum* and *Criconemoides xenoplax* in soil samples from around trees with reduced annual growth, yellowing of new shoots and damaged and reduced root systems. High populations of *Xiphinema americanum* are also associated with "winterkill" of *Picea* (Griffin and Epstein, 1964), as greenhouse experiments showed that the root growth of *Picea* transplants was decreased after a 4-month exposure to *X. americanum* and, when exposed to winter conditions, significantly increased mortality.

Goodey (1965) found *Rotylenchus robustus, Trichodorus pachydermus,*

L*

and *Tylenchorhynchus* spp. were the common nematode parasites of *Larix leptolepsis, Picea sitchensis, Pinus contorta, P. nigra* var. *poiretiana,* and *Pseudotsuga menziesii* in forest nursery soil in Britain, and in greenhouse studies he proved that *R. robustus* was pathogenic to *Picea sitchensis.* Hijink (1969) also proved *R. robustus* was pathogenic to *Picea abies* in a nursery in the Netherlands.

The major species (*Cryptomeria japonica, Chamaecyparis obtusa, Pinus* sp. and *Acacia*) grown in the nurseries of the Fukuoka Prefecture, Japan were commonly parasitized by *Meloidogyne incognita, Tylenchorhynchus claytoni* and *Pratylenchus* sp. (Hashimoto, 1962). *Cryptomeria* and *Chamaecyparis* were severely stunted and the root tips galled by *M. incognita*; foliage was yellow in infested areas of the nursery. The galls on the roots of *Acacia* sp. seedlings attacked by this nematode were much larger than those on the conifers. *Tylenchorhynchus claytoni* was reported to feed only on root tips and caused browning and withering of feeder roots. Nematode damage was more severe in old nurseries, in nurseries started on vegetable fields, or in plantations of *Acacia* sp. than in nurseries established on newly cultivated lands. *Pratylenchus penetrans* causes considerable damage to coniferous seedlings and has the widest distribution in Japanese nurseries (Mamiya, 1969).

Seed imported to Brazil from Japanese forest trees became heavily infested by *Meloidogyne arenaria* when grown in non-fumigated nursery beds (Lordello and Kanazawa, 1967). Infested seedlings were stunted, had galled and reduced root systems, and approximately 75% of them died. Only *Paulownia kawakani* and *P. tomentosa* showed some degree of resistance.

Seedlings of *Anthocephalus chinensis*, a fast-growing tropical tree grown for paper pulp, are extensively and severely stunted in Philippine nursery beds by *M. incognita* (Postrado and Glori, 1968).

Donaldson (1967) found an isolated case of root-knot nematode damage to *Pinus elliottii* in a Florida nursery. One large circular area in the nursery had severely stunted seedlings. The tap-root of each infected seedling terminated in a large, galled clump and there was an absence of lateral roots. Mature females were dissected from these galls, but larvae were not recovered from the soil.

Sutherland and Dunn (1970) surveyed nurseries in British Columbia, Canada and found *Xiphinema bakeri* consistently associated on *Pseudotsuga menziesii* with a roughened, thickened, and browned root condition termed "corky root" by Bloomberg (1968). *In vitro* studies showed that *X. bakeri* fed on tips of seedling roots and caused the incipient stages of typical corky root disease of *P. menziesii*, but there was still doubt as to whether the nematode was the sole causative agent (Sutherland, 1969). Sutherland and Dunn (1970) isolated *Cylindrocarpon radicicola*, a soil-borne fungus, from diseased roots and it was suspected to be involved in the problem. De Maeseneer (1964a) isolated this fungus as well as *Fusarium oxysporum* from nematode-infested conifer seedlings in Belgium, but concluded that parasitic nematodes were the primary pathogens.

C. Control

Nematode diseases in forest nurseries are soil-borne and the techniques of root-disease control may help alleviate these problems. Prevention of spread of nematodes and the decrease of nematode populations are the two basic control principles used in forest nurseries.

Nematodes may be carried into nurseries by wind, water, animals and birds, but the most common means of nematode spread is the transportation of infested soil and infected plants by nursery operators. Untreated topsoil should not be hauled into the nursery to build new seedbeds. Cultivation tools on farm equipment often carry viable nematodes in the adhering soil. Hence, implements should be cleaned before being moved from one field to another. Transplants from non-fumigated seedbeds should not be lined-out in other nursery areas until they are determined to be free from pathogenic nematodes.

Rivers, ponds, or drainage ditches, common sources of irrigation water, may contain plant-parasitic nematodes and thereby effectively transport the nematodes for relatively long distances. The use of deep wells as a water source effectively precludes water-borne infections.

Shifting cultivation is a simple and inexpensive method of nematode control in seedbeds in agricultural systems. However, in forest nurseries, the large area of the seedbeds, the installation of irrigation and the construction of the necessary buildings entails large investments, and heavy losses would result if the site had to be changed frequently.

Once the nursery is infested with plant-parasitic nematodes it is virtually impossible to eradicate them. So the problem becomes one of decreasing nematode populations to a point where seedlings can be grown profitably.

The rotation of host and non-host crops effectively controls nematode populations when the particular species of parasite involved has a narrow host range. Forest nurseries in the southern U.S.A. are currently rotating soybeans, corn, cowpeas or millet with the crop of southern pine seedlings and in nurseries infested with the pine cystoid nematode, *Meloidodera floridensis*, rotation with any of these crops should effectively decrease the nematode populations as this nematode parasitizes only conifers. However, when two or more parasitic nematodes infest the same nursery, the rotation of crops may not easily solve the problem. In German nurseries *Larix* sp. and *Pseudotsuga* sp. cause an increase and *Pinus* sp. and *Picea* sp. a decrease of populations of *Pratylenchus penetrans*, whereas these four forest tree species have exactly the opposite effect on populations of *Rotylenchus robustus* (Rössner, 1969). Crops of coniferous and hardwood trees should be rotated to avoid root-knot problems on the hardwood crop because *Meloidogyne* spp. rarely attack conifers.

Particular care must be taken in the use of agricultural crops in forest nursery rotations because nematode problems may be increased instead of decreased. Rössner (1969) found that rotation with sunflowers, potatoes, peas and beans frequently increased population densities of *Pratylenchus penetrans* and *Rotylenchus robustus* in German nurseries.

Soil fumigants have a broad spectrum of biological activity against numerous soil pests depending upon the material used and method of application. The use of broad spectrum fumigants such as methyl bromide can usually be economically justified in older nurseries because continuous cultivation for a number of years is generally accompanied by an increase in soil-borne diseases and weed problems as well as nematode problems. In other situations having only nematode problems, applications of selected nematicides (DD or EDB) are more economical.

The most successful and commonly used soil fumigants in forest nurseries for nematode control are the volatile halogenated hydrocarbons (Table II). The best chemical control at present for nematodes is given by the broad spectrum fumigants containing methyl bromide. These materials are available as the following brands of soil fumigants: Brozone®—67% MBr + 33% inert ingredient; Dowfume® MC-33—67% MBr + 33% chloropicrin; Haltox®—98% MBr + 2% trichloronitromethane; and Dowfume® MC-2—98% MBr + 2% chloropicrin. These materials are used most effectively under a plastic cover over the entire nursery area, including pathways and irrigation riserline areas, to prevent recontamination by pathogens during subsequent cultivation. Application methods have now been refined by using equipment designed to apply the fumigants and lay the plastic cover in one operation which results in more uniform pest control.

There are inherent problems in the indiscriminate and continued use of soil fumigants in forest nurseries. A stimulation of tree seedling growth following fumigation has been noted as an added benefit in forest nurseries (Foster, 1961; Weihing *et al.*, 1961), but this stimulation of *Pinus resinosa* seedlings following SMDC (Vapam®) fumigation in a nursery in Wisconsin, U.S.A. produced abnormally large, succulent 3-0 pine seedlings with totally inadequate root systems because of a marked reduction in mycorrhizae and fibrous laterals (Iyer and Wilde, 1965). Danielson and Davey (1969) found a retardation of mycorrhizal development in *Pinus taeda* seedlings following seeding in fumigated nursery beds and determined that applications of non-sterilized pine needle mulch to fumigated beds served as a source of inoculum of mycorrhizal fungi and eliminated the possibility of growing seedlings with poor mycorrhizal development.

We need research to give us dependable procedures for predicting seedling losses based on the kinds and numbers of plant-parasitic nematodes recovered from soil and root samples. On the basis of this information, recommendations could be made as to whether it is economical to fumigate a particular problem area in a nursery and, if so, with what. If nematicidal fumigants are used, we also need information on their side effects—whether soil fertility, mycorrhizal colonization, and soil microflora balance are affected.

Occasionally forest nurseries grow special tree seedlings for retail sales and in certain locations quarantine regulations are placed on these seedlings to control their sale and distribution. Several chemicals are now used as dips to control nematodes on bare-rooted nursery stock effectively enough to satisfy these quarantine regulations. Johnson *et al.* (1970) obtained excellent control of *Meloidogyne incognita* infecting dogwood seedlings in the U.S.A. by dipping bare roots in aqueous emulsions of several phosphatic nematicides.

IV. Nematode Diseases in Forest Plantations

Plantations established on "cut and cleared" forest land have different nematode problems than do plantations established on agricultural land. Nematode populations have reached an edaphic–biological balance in forest soils and parasitic species generally exist in lower numbers than in agricultural soils. Cultivation, fertilization, and intensive cropping tend to encourage high populations of parasitic nematodes. Many abandoned farm lands have deteriorated because of accumulated soilborne diseases, in many instances ones caused by nematodes, and plantations of susceptible tree species established on such sites risk the possibility of unacceptable growth loss. Thus far, complete destruction of a forest plantation by nematodes has not been reported. There are, however, accounts in the literature which implicate nematodes as causal agents of growth retardation, decline, and stunting of trees in forest plantations.

TABLE II. Common chemicals for nematode control by soil fumigation in forest nurseries

Chemical	Tree species	Nematode parasites	Location	Reference
Methyl bromide (Brozone, Haltox, Dowfume MC-2)	*Picea* sp., *Pinus* sp.	*Rotylenchus* sp., *Tylenchorhynchus* sp., *Trichodorus* sp.	Germany	Rühm (1962)
	Pinus taeda	*Xiphinema americanum*, *Helicotylenchus* spp.	U.S.A.	Hansbrough and Hollis (1957)
	Pinus resinosa	*Paratylenchus* sp., *Hoplolaimus galeatus*, *Trichodorus* sp.	U.S.A.	Sutherland and Adams (1966)
	Pinus elliottii	*Hoplolaimus galeatus*, *Criconemoides* sp., *Trichodorus* sp., *Xiphinema americanum*	U.S.A. U.S.A.	Ruehle and Sasser (1962) Ruehle *et al.* (1966)
1,3-dichloropropene-propane mixture (DD, Vidden D)	*Pinus resinosa*	*Paratylenchus* sp., *Hoplolaimus galeatus* *Trichodorus* sp.	U.S.A.	Sutherland and Adams (1966)
	Cryptomeria japonica *Picea sitchensis* *Acer* sp., *Fagus* sp., *Prunus* sp., *Crataegus* sp., *Cotoneaster* sp., *Pinus* sp, *Sorbus* sp.	*Pratylenchus penetrans* *Rotylenchus robustus* *Rotylenchus* sp., *Trichodorus* sp.	Japan England Belgium	Mamiya (1969) Goodey (1965) De Maeseneer and Van den Brande (1965)
	Pseudotsuga menziesii	*Xiphinema bakeri*	Canada	Bloomberg and Orchard (1969)
	Pinus elliottii *Robinia pseudo-acacia*	*Tylenchorhynchus ewingi* *Pardongidorus maximus*	U.S.A. Germany	Shoulders *et al.* (1965) Sturhan (1964)

Treatment	Host	Nematode	Country	Reference
1,3-dichloropropene (Telone)	*Pseudotsuga menziesii, Pinus sylvestris*	(not listed)	U.S.A.	Anderson (1963)
	Anthocephalus chinensis	*Meloidogyne incognita*	The Philippines	Glori and Postrado (1969)
chloropicrin (Picfume)	conifers	*Rotylenchus* sp., *Pratylenchus* sp., *Tylenchorhynchus* sp.	Germany	Rühm (1959)
	Picea sp.	*Rotylenchus* sp., *Ditylenchus* sp., *Tylenchorhynchus* sp.	Germany	Weischer (1956)
	Picea sitchensis	*Rotylenchus robustus*	England	Goodey (1965)
ethylene dibromide (Dowfume W-85)	*Pinus* sp.	(not listed)	U.S.A.	Henry (1953)
	Picea sitchensis	*Rotylenchus robustus*	England	Goodey (1965)
sodium-N-methyldithio-carbamate (Vapam)	conifers	*Rotylenchus* sp., *Pratylenchus* sp., *Tylenchorhynchus* sp.	Germany	Rühm (1959)
1,2 dichloro, 3 bromo-propane (Nemagon)	*Pinus taeda*	*Tylenchorhynchus* sp.	U.S.A.	Rowan and Good (1960)
	Picea pungens	*Helicotylenchus dihystera, Pratylenchus penetrans, Xiphinema americanum*	U.S.A.	Ferris and Leiser (1965)

A. Association

Ruehle and Sasser (1962) surveyed pine plantations in North Carolina, U.S.A. and the nematodes most commonly associated with stunted *Pinus taeda* and *P. elliottii* were *Hoplolaimus galeatus, Meloidodera floridensis, Helicotylenchus dihystera,* and *Xiphinema americanum.* Other parasitic species found in these plantations, but not correlated with stunted trees, were *Belonolaimus* sp., *Criconemoides* sp., *Dolichodorus* sp., *Hemicriconemoides* sp., *Hemicycliophora* sp., *Longidorus* sp., and *Pratylenchus zeae.*

X. americanum was found in 95% of sampled areas of shelter-belt plantings in South Dakota, U.S.A. and was associated with all commonly planted tree species, including *Populus deltoides, Ulmus* spp., *Pinus* spp., *Fraxinus pennsylvanica,* and *Celtis occidentalis* (Malek, 1968). *X. americanum, Tylenchorhynchus acutus* and *Paratylenchus* spp. were commonly associated with unthrifty trees.

Lane and Witcher (1967) compared nematode populations in stands of *Liriodendron tulipifera* and *Pinus taeda* in South Carolina, U.S.A. *Helicotylenchus* sp., *Hemicycliophora* sp., *Criconemoides* sp., *Tylenchus* sp., *Paratylenchus* sp., *Trichodorus* sp., and *Xiphinema* sp. were associated with both tree species. In this survey there were more nematodes in soil around the roots of *L. tulipifera* than around those of *P. taeda.*

In *Platanus occidentalis* plantations in Georgia, U.S.A., Churchill and Ruehle (1971) found *Criconemoides* sp., *Helicotylenchus multicinctus, H. pseudorobustus, Hemicycliophora* sp., *Hoplolaimus galeatus, H. stephanus, Trichodorus* sp., and *X. americanum.*

B. Effect on Growth

Plant-parasitic nematodes usually have uneven distribution patterns in plantations. The common field picture of nematode disease finds stunted trees, characterized by low vigor and having small, distorted root systems, unevenly distributed over the area.

Ruehle and Sasser (1962) found four genera of parasitic nematodes commonly associated with stunted *Pinus taeda* and *P. elliottii* in plantations. Subsequent greenhouse studies proved that *Hoplolaimus galeatus* and *Meloidodera floridensis* were pathogenic, *Helicotylenchus dihystera* caused only slight growth reduction and *X. americanum* was not pathogenic under these test conditions for these hosts.

Lordello (1967) found *Pratylenchus brachyurus* infesting roots of stunted and yellowed young *Eucalyptus* trees in Brazil. Root systems were reduced, roots were necrotic and cracked, and many diseased plants died.

Churchill and Ruehle (1971) found *Hoplolaimus galeatus* in two plantations of *Platanus occidentalis* in Georgia, U.S.A. Although this parasite was not correlated with stunted trees, it proved to be pathogenic in greenhouse tests. This nematode is capable of causing severe damage to seedlings—coarse, darkened roots devoid of laterals with an accompanying marked decreased top growth. *Helicotylenchus pseudorobustus* was recovered from soil samples in seven of ten plantations surveyed and this parasite was not as damaging to sycamore seedlings as was *H. guleatus*, but nevertheless capable of significantly reducing total surface areas of the roots.

C. Control

Ruehle and Sasser (1962) conducted one of the first soil fumigation tests for nematode control in a *Pinus elliottii* plantation in North Carolina, U.S.A. Broadcast application of MBr under plastic covers decreased *Hoplolaimus galeatus*, *Helicotylenchus dihystera*, *X. americanum*, *Meloidodera floridensis*, *Criconemoides* spp., and *Trichodorus* sp. to undetectable levels during the first growing season, and DD injected broadcast decreased the populations of these species to trace amounts during the first year. After 5 years the differences in growth between pine trees growing in fumigated plots and those in non-fumigated plots were still evident (Ruehle and Sasser, 1964). The average height of the trees was significantly greater in plots treated with MBr than in those treated with DD or in non-treated soil. MBr gave excellent nematode control of all but *H. galeatus*. Hansbrough *et al.* (1964) achieved similar results in a *Pinus taeda* plantation in Louisiana, U.S.A.

MBr is far superior to DD but is too expensive for nematode control in plantations. In another plantation fumigation in the row with DD resulted in a significant increase in height growth of *Pinus elliottii*, but with DBCP resulted in phytotoxicity (Ruehle and Sasser, 1962). Also, growth of *P. palustris* in rows fumigated with DD was greater than for trees in check rows, but this difference in growth response may have been greater if certain nematode parasites (particularly *H. galeatus* and *M. floridensis*) had been infesting the soil in this plantation. Ruehle (1969b) later tried in-row fumigation tests with DD and DBCP in a plantation in south Georgia, U.S.A. that had a typical deep sandy soil previously supporting naturally regenerated *P. palustris* and was infested with *Helicotylenchus* sp., *Hoplolaimus galeatus*, *Trichodorus christiei*, *X. americanum* and several other species commonly associated with stunted pines. Only DD significantly decreased the nematode populations during the first growing season; DBCP not only failed to control nematodes, but was phytotoxic. *P. palustris* seedlings in rows fumigated with DD started growing earlier and produced taller trees after 5 years than those in non-fumigated rows.

Row fumigations in plantations infested with parasitic nematodes will satisfactorily control nematodes and decrease root losses during the critical first year. Preliminary findings suggest that continued research on the problem will be fruitful. In order to make nematicidal fumigation in forest plantations economically feasible, we need to find or develop non-phytotoxic nematicides that can be applied at the time of planting.

V. Nematode Diseases in Natural Woodlands

The role of nematodes in disease of trees in natural woodlands is less easily understood. The biotic community of natural forest soil is basically stable but modified from time to time by nature or man. Natural woodlands, usually a mixture of species of uneven age classes, do not afford parasitic nematodes affecting a given tree species opportunity to increase to unusually high and damaging levels, thus a biological balance is maintained between parasite and host. However, changes in climate may affect the biological balance in the soil to such an extent as to influence the tree diseases (Hepting, 1963). Nematodes may have a role in the premature mortality of *Pinus ponderosa* in the lower elevations of the southwestern U.S.A. following periods of prolonged drought (Riffle, 1968). Feeding by nematodes on tree roots over long periods, especially on trees adversely affected by changes in climate, may weaken the trees and make them more vulnerable to environmental changes, other diseases, and insect attack. The potential of nematode damage to trees in natural woodlands is considerable.

A. Association

Only in recent times have nematologists done comparative and quantitative sampling in natural woodlands. Shigo and Yelenosky (1960) surveyed an experimental forest in New Hampshire, U.S.A. and recovered *Ditylenchus, Criconemoides, Helicotylenchus, Pratylenchus, Rotylenchus,* and *Tylenchorhynchus* in both natural woodlands and adjacent clearcut areas. From the results of a survey of cultivated and adjacent woodland areas in New Jersey, Hutchinson and Reed (1959) reported that *Meloidodera floridensis* was parasitizing the roots of *Pinus rigida* and *P. echinata* in natural woodlands. Similar reports of endoparasitic nematodes parasitizing trees in natural woodlands come from other areas in the U.S.A., namely, the Florida everglades with *Tylenchulus semipenetrans* in the roots of sabal palmetto on virgin land (Esser,

1964); Oregon with *Meloidodera* sp. infecting roots of *Pseudotsuga menziesii*; and Arkansas with *Heterodera* sp. (later described as *H. betula* (Hirschmann and Riggs, 1969)) infecting the roots of *Betula niger* growing on a river bank (Riggs and Hamblen, 1967).

The causes of declining or dieback problem areas in natural woodlands are generally considered to be complex in nature, and nematodes are investigated as only one facet of a many-sided problem. Nematode problems in natural woodlands generally have been investigated by surveying diseased areas and comparing nematode populations with those in adjacent healthy areas (Table III). Data from these studies are difficult to interpret because the composition in natural woodlands is usually a mixture of uneven-aged tree species. The result, as reported by Bassus (1962) and Yuen (1966), is that plant-parasitic nematodes are not evenly distributed vertically in the soil profile or laterally within a stand. Consequently, correlations between high nematode populations and diseased areas are often confounded.

B. Effect on Growth

The effects of nematode depredations of forest tree roots are masked by nutritional disorders, fungal diseases, insect pests, flooding, drought, and physical problems relating to soil, such as compaction and erosion. Consequently, measuring the loss of tree growth caused by nematodes is complicated by the general difficulty in recognizing the problem. The nature of root damage can be studied on tree seedlings in the controlled environment of the greenhouse, but relating this to natural forest conditions is difficult. There has been no quantitative study on the specific decrease in tree growth in natural woodlands caused by any one or more of the nematodes associated with these trees, and until such data become available, it will not be possible to evaluate the importance of each nematode species to naturally growing forest trees.

C. Control

The control of nematodes in natural woodlands will be limited to timely silvicultural practices that stimulate biotic factors in the forest environment to decrease populations of nematodes and increase tree vigor. A return to mixed stand composition would help control those parasitic nematodes which have a limited host range. For example, we know that *Meloidodera floridensis* is restricted to conifers in southeastern U.S.A.; therefore, a pure stand of southern pine infected with this nematode would probably benefit by being thinned and managed as a mixed stand of pine and hardwood.

TABLE III. Plant-parasitic nematodes associated with decline problems of trees in natural woodlands

Nematode species	Tree species	Location (U.S.A.)	Reference
Criconema sp.	*Pinus monticola*	Idaho	Nickle (1960)
	Fraxinum americanum	New York	Ross (1966)
Criconemoides sp.	*F. americanum*	New York	Ross (1966)
Helicotylenchus sp.	*Acer saccharum*	New York	Hibben (1964)
	Fraxinus americanum	New York	Ross (1966)
	Pinus ponderosa, P. edulis, Juniperus monosperma, J. deppeana, J. scopulorum	New Mexico	Riffle (1968)
H. platyurus	*Acer saccharum*	Wisconsin	Riffle and Kuntz (1966)
Hemicycliophora sp.	*A. saccharum*	New York	Hibben (1964)
		Wisconsin	Riffle and Kuntz (1966)
H. vidua	*Pinus echinata*	Alabama, Georgia, N. Carolina, S. Carolina, Virginia	Ruehle (1962b)
Heterodera sp.	*P. monticola*	Idaho	Nickle (1960)
Hoplolaimus galeatus	*P. echinata*	Georgia, N. Carolina, Virginia	Ruehle (1962b)
Longidorus sp.	*Fraxinus americanum*	New York	Ross (1966)
Meloidodera floridensis	*Pinus echinata*	New Jersey	Hutchinson and Reed (1959)
Meloidogyne ovalis	*Acer saccharum*	Wisconsin	Riffle and Kuntz (1967)
Pratylenchus sp.	*Fraxinus americanum*	New York	Ross (1966)
Trichodorus sp.	*F. americanum*	New York	Ross (1966)
T. californicus	*Acer saccharum*	Wisconsin	Riffle and Kuntz (1966)
T. elegans	*Pinus monticola*	Idaho	Nickle (1960)
Tylenchorhynchus sp.	*P. monticola*	Idaho	Nickle (1960)
	Fraxinus americanum	New York	Ross (1966)
Xiphinema sp.	*Acer saccharum*	New York	Hibben (1964)
	Fraxinus americanum	New York	Ross (1966)
X. americanum	*Acer saccharum*	Massachusetts	Di Sanzo and Rohde (1969)
		Wisconsin	Riffle and Kuntz (1966)
	Pinus echinata	Georgia, Alabama, N. Carolina, S. Carolina, Virginia	Ruehle (1962b)
	P. ponderosa, P. edulis, Juniperus monosperma, J. deppeana, J. scopulorum	New Mexico	Riffle (1968)

Trees from seedling to physiological maturity are inherently vigorous and apparently tolerate parasitic nematodes in natural woodlands. After maturity the rate of regeneration of feeder roots markedly decreases, and parasitic nematodes may significantly affect the health of such trees. There is a need to recognize the age at which different species of tree reach maturity, and thinnings and commercial cutting should be made before trees become overmature and start to decline.

Controlled burning in natural woodlands may affect nematodes in the soil. The decomposition of duff and organic litter on the surface increases the natural enemies of nematodes. Burning of woodlands decreases this organic matter and probably markedly retards the development of certain biological control agents in the soil.

The chemical control of nematodes in natural woodlands seems impractical at this time because technical problems of applying nematicides have not been overcome and the cost of such measures is prohibitive.

VI. Economic Importance

The measurement and estimation of tree losses caused by nematode disease is dependent on recognition of the problems. Most foresters have little or no knowledge of nematode diseases. Not until the past few decades have nematodes even been considered a problem to forest trees. Specific decrease in seedling growth in nurseries or tree growth in plantations has been recognized and measured in only a few locations. Therefore, any attempt to present figures on losses attributable to nematodes is pure conjecture. To ask anyone today to present an estimate of the impact of parasitic nematode losses on the forest economy is comparable to asking N. A. Cobb to evaluate nematode losses in American agriculture in 1920. Information was too limited to make a good estimate in those days for agriculture; so it is today for forestry.

Practically all reports of nematode damage in forest nurseries indicate that the problems are localized. Typical damage is restricted to a few spots in nursery beds. Methyl bromide fumigation in most forest nurseries in the southern U.S.A. for root rot and weed control has practically eliminated nematode problems. In other areas of the world fumigation is less frequently used, and nematode diseases account for seedling losses.

I know of no report that nematodes kill trees in plantations or natural woodlands. Nematode diseases, like many other soil-borne diseases, cause only limited root mortality each year which may reduce growth rate and cause some stunting. However, if this growth loss is extended over a 20- or 30-year period the decrease in a forest stand may be significant.

It has been estimated that the southern region of the U.S.A. has 80 million hectares of forest land. The net annual growth of sawtimber softwoods in this geographic region is in excess of 17 billion board feet. A conservative estimate of the stumpage value of the current market is approximately U.S. $500 million. If nematodes cause an average of only 1% reduction (a figure 10 times less than that estimated for agriculture in the U.S.A.) in mean annual growth increment, the loss would exceed $5 million per year.

Nevertheless, any effort to calculate the amount of growth loss caused by nematodes in world forestry is impossible. The lack of single infections and pure populations of nematodes in forest areas account for the problem of evaluating losses. Rather, forest trees in both plantations and natural woodlands are attacked by many species of nematodes; one or more species may prevail, but it is not necessarily the most numerous ones that produce the most damage.

There is a general lack of awareness in forestry that nematodes cause growth loss in trees. Not until forest nematology research yields sufficient knowledge, and forest specialists are trained to use that knowledge in diagnosing diseases caused by nematodes, will we have reliable information on which to base estimates of economic losses to nematodes.

References

Anderson, H. W. (1963). *Down to Earth* **19**, 6–8.

Anon. (1969). The South's Third Forest. South. Forest Resource Counc.

Bassus, W. (1962). *Nematologica* **7**, 281–293.

Blair, G. P. and Darling, H. M. (1968). *Nematologica* **14**, 395–403.

Bloomberg, W. J. (1968). *Can. Dep. Forest and Rural Dev., Bi-Mon. Res. Notes* **24**, 8.

Bloomberg, W. J. and Orchard, W. R. (1969). *Ann. appl. Biol.* **64**, 239–244.

Chitwood, B. G. and Esser, R. P. (1957). *Pl. Dis. Reptr* **41**, 603–604.

Churchill, R. C., Jr. and Ruehle, J. L. (1971). *J. Nematol.* **3**, 189–196.

Danielson, R. M. and Davey, C. B. (1969). *Forest Sci.* **15**, 368–380.

De Maeseneer, J. (1964a). *Meded. LandbHogesch. OpzoekStns Gent* **29**, 797–809.

De Maeseneer, J. (1964b). *Nematologica* **10**, 403–408.

De Maeseneer, J. and Van den Brande, J. (1965). *Meded. LandbHogesch. OpzoekStns Gent* **30**, 1454–1460.

Di Sanzo, C. P. and Rohde, R. A. (1969). *Phytopathology* **59**, 279–284.

Donaldson, F. S. Jr. (1967), *Pl. Dis. Reptr* **51**, 455–456.

Donaubauer, E. (1959). *Anz. Schädlingsk.* **32**, 68–69.

Epstein, A. H. and Griffin, G. D. (1962). *Pl. Dis. Reptr* **46**, 17.

Esser, R. P. (1964). *Pl. Dis. Reptr* **48**, 533.

Fenwick, D. W. (1962). *J. Agric. Soc. Trin.* **62**, 27–43.

Fenwick, D. W. and Maharaj, S. (1963). *J. Helminth.* **37**, 21–26.

Ferris, J. M. and Leiser, A. T. (1965). *Pl. Dis. Reptr* **49**, 69–71.

Foster, A. A. (1961). *Georgia Forest Res. Counc. Reptr* **8**, 5 pp.

Glori, A. V. and Postrado, B. T. (1969). *Philipp. Reforestation Admin. Res. Note* 3, 6 pp.

Goodey, J. B. (1965). *Great Brit. Forest Comm. Bull.* **37**, 210–211.

Griffin, G. C. and Darling, H. M. (1964). *Nematologica* **10**, 471–479.

Griffin, G. D. and Epstein, A. H. (1964). *Phytopathology* **54**, 177–180.

Hansbrough, T. and Hollis, J. P. (1957). *Pl. Dis. Reptr* **41**, 1021–1025.

Hansbrough, T., Hollis, J. P., Merrifield, R. G. and Foil, R. R. (1964). *Pl. Dis. Reptr* **48**, 986–989.

Hashimoto, H. (1962). *Jap. Forest. Soc. J.* **44**, 248–252.

Henry, B. W. (1953). *Phytopathology* **43**, 81–88.

Hepting, G. H. (1963). *A. Rev. Phytopath.* **1**, 31–50.

Hibben, C. R. (1964). *Phytopathology* **54**, 1389–1392.

Hijink, M. J. (1969). *Mededel. Rijksfakulteit LandbouwWettensch. Gent* **34**, 539–549.

Hirschmann, H. and Riggs, R. D. (1969). *J. Nematol.* **1**, 169–179.

Hopper, B. E. (1958). *Pl. Dis. Reptr* **42**, 308–314.

Hopper, B. E. (1959). *Nematologica* **4**, 23–30.

Hopper, B. E. and Padgett, W. H. (1960). *Pl. Dis. Reptr* **44**, 258–259.

Hutchinson, M. T. and Reed, J. P. (1959). *Pl. Dis. Reptr* **43**, 801–802.

Immel, R. (1957). *Anz. Schädlingsk.* **30**, 88–90.

Iyer, J. G. and Wilde, S. A. (1965). *J. Forest.* **63**, 703–704.

Johnson, A. W., Ratcliffe, T. J. and Freeman, G. C. (1970). *Pl. Dis. Reptr* **54**, 952–955.

Knierim, J. A. (1963). *Mich. St. Univ. Agric. Stn Quart. Bull.* **46**, 254–262.

Lane, C. L. and Witcher, W. (1967). *J. Forest.* **65**, 898–901.

Lordello, L. G. E. (1967). *Pl. Dis. Reptr* **51**, 791.

Lordello, L. G. E. and Kanazawa, P. S. (1967). *Revta. Agric.* (Piracicaba) **42**, 107–108.

Malek, R. B. (1968). *Pl. Dis. Reptr* **52**, 795–798.

Mamiya, Y. (1969). *Bull. Gov. Forest Exp. Stn* **220**, 121–132.

Marlatt, R. B. (1966). *Pl. Dis. Reptr* **50**, 689–691.

Marx, D. H. and Davey, C. B. (1969). *Phytopathology* **59**, 559–565.

Massey, C. L. (1966). *Bull. ent. Soc. Amer.* **12**, 384–386.

Nickle, W. R. (1960). *Pl. Dis. Reptr* **44**, 470–471.

Nolte, H. W. and Dieter, A. (1957). *Nematologica* **2**, 63–67.

Oostenbrink, M. (1958). *Tijdschr. Plziekten* **64**, 122.

Oostenbrink, M. (1963). *Meded. LandbHogesch. OpzoekStns Gent* **28**, 663–671.

Postrado, B. T. and Glori, A. V. (1968). *Philipp. Reforestation Admin. Res. Note* 2, 5 pp.

Riffle, J. W. (1967). *Phytopathology* **57**, 541–544.

Riffle, J. W. (1968). *Pl. Dis. Reptr* **52**, 52–55.

Riffle, J. W. and Kuntz, J. E. (1966). *Pl. Dis. Reptr* **50**, 677–681.

Riffle, J. W. and Kuntz, J. E. (1967). *Phytopathology* **57**, 104–107.

Riggs, R. D. and Hamblen, M. L. (1967). *Phytopathology*, **57**, 827.

Ross, E. W. (1966). *State Univ. Coll. Forest. Syracuse Univ. Tech. Bull.* **88**, 80 pp.

Rössner, J. (1969). *Z. Angew. Zool.* **56**, 1–64.

Rowan, S. J. and Good, J. M. (1960). *Phytopathology* **51**, 645.

Ruehle, J. L. (1962a). *Phytopathology* **52**, 68–71.

Ruehle, J. L. (1962b). *Pl. Dis. Reptr* **46**, 710–711.

Ruehle, J. L. (1969a). *Forest Sci.* **15**, 130–134.

Ruehle, J. L. (1969b). *J. Nematol.* **1**, 248–253.

Ruehle, J. L. and Sasser, J. N. (1962). *Phytopathology* **52**, 56–68.

Ruehle, J. L. and Sasser, J. N. (1964). *Pl. Dis. Reptr* **48**, 534–536.

Ruehle, J. L., May, J. T. and Rowan, S. J. (1966). *U.S.D.A. Tree Pl. Notes* **76**, 4–7.

Rühm, W. (1959). *Merck. Blätter* **9**, Series 3, 1–16.

Rühm, W. (1962). *Z. PflKrankh. PflPath. PflSchutz* **69**, 278–283.

Shigo, A. L. and Yelenosky, G. (1960). *U.S.D.A. Forest Serv. Northeast. Forest Exp. Stn Res. Note* 101, 4 pp.

Shoulders, E., Hollis, J. P., Merrifield, R. G., Turner, E. E. and Verrall, A. F. (1965). *U.S.D.A. Tree Pl. Notes* **73**, 14–21.

Steiner, G. (1949). *Soil Sci. Soc. Florida Proc.* (1942) **4-B**, 72–117.

Sturhan, D. (1963). *Z. Angew. Zool.* **50**, 129–193.

Sturhan, D. (1964). *Berlin Biol. Bundesanst Land-Forstwirtsch. Mitt.* **111**, 106–112.

Sutherland, J. R. (1967). *Pl. Dis. Reptr* **51**, 91–93.

Sutherland, J. R. (1969). *Phytopathology* **59**, 1963–1965.

Sutherland, J. R. and Adams, R. E. (1964). *Nematologica* **10**, 637–643.

Sutherland, J. R. and Adams, R. E. (1966). *Nematologica* **12**, 122–128.

Sutherland, J. R. and Dunn, T. G. (1968). *Forest Res. Lab. Victoria, B.C. Inform. Rep. BC-X-25*, 13 pp.

Sutherland, J. R. and Dunn, T. G. (1970). *Pl. Dis. Reptr* **54**, 165–168.

Sutherland, J. R. and Fortin, J. A. (1968). *Phytopathology* **58**, 519–523.

Thirumalachar, M. J. (1946). *Nature* **157**, 106–107.

Van Hoof, H. A. and Seinhorst, J. W. (1962). *Tijdschr. PlZiek.* **68**, 251–256.

Weihing, J. L., Inman, R. and Peterson, G. W. (1961). *Pl. Dis. Reptr* **45**, 779–802.

Weischer, B. (1956). *NachrBl. dt. PflSchutzdienst, Stuttg.* **8**, 34–36.

Yamaguchi, S. (1932). *Hokkaido Univ. exp. Forest Bull.* **7**, 209–215.

Yuen, P. H. (1966). *Nematologica* **12**, 195–214.

15

Nematodes of Tree Fruits and Small Fruits

F. D. McElroy

Research Station
Canada Department of Agriculture
Vancouver, Canada

I. Tree Fruits

A. Crop Production

1. Cherry (*Prunus avium, P. cerasus*)

Cherries are grown in the world's temperate regions where local conditions permit. They are cultivated throughout Europe especially in West Germany, Italy and Central European countries. Great Britain is the largest Commonwealth producer followed by Canada, Australia and New Zealand. Large scale commercial production occurs in the U.S.A. and to a lesser extent in Turkey and Japan.

Sweet cherries (*Prunus avium*) and sour cherries (*Prunus cerasus*) comprise the two principal groups of cherries. Several varieties of Hearts and Bigarreaus make up the sweet cherry crop which is sold mainly on the fresh market. The remainder is either canned or brined and later remanufactured into Maraschino and glacé cherries. Morellos and Dukes comprise the sour cherry crop most of which is frozen, canned, salted, or used for brandy.

Cherry trees have exacting climatic requirements in that they fruit early and therefore suffer severe damage from frost at blossom time and from rain during ripening. They produce best on light, well drained loamy soils.

2. Peach (*Prunus persica*)

Two types of peach are grown in the temperate zones. Most of the freestone group is sold fresh and only a small quantity is canned or dried. Most of the clingstone peaches are canned, but a small portion of this and of the freestone group is frozen.

About 4·5 million hectares in the world are devoted to peaches, with the U.S.A. producing 40% of the world total. Australia, Italy, France, Greece, Turkey, Japan and Argentina have increased production in the past decade. Canada, Australia and New Zealand are important Commonwealth producers. Ideal conditions for peach production are warm summers free from fog and high humidity, cool winters, and infrequent spring frosts.

3. Apple (*Malus sylvestris*)

Apples are a widely cultivated deciduous fruit, occupying over 15 million hectares of which Europe accounts for about 22 million acres. The principal apple-growing countries are Italy, U.S.A., Japan, France, Britain, West Germany and Turkey.

Dessert and cooking varieties account for the major part of the crop but processing or dual-purpose varieties are gaining in importance and several European countries grow cider apples.

Apples produce best in areas having a long, cool growing season permitting slow development of the crop. High summer temperatures damage the fruit and winter temperatures must be sufficiently low for winter chilling.

4. Pear (*Pyrus communis*)

World production of dessert pears is about one third that of apples. Europe and the U.S.A. are the major producers with Italy leading in production. In most areas the majority of the pear crop is canned and the rest is sold fresh.

5. Walnut (*Juglans nigra, J. regia*)

The United States leads in walnut production, supplying about 45% of the world total. Other contributing countries are Turkey, Iran and India, and in the last decade France and Italy have expanded production greatly.

B. Nematode Diseases

There are numerous reports of nematodes associated with declining orchard trees and with replant failures, but evidence of direct damage caused by nematode feeding is scant and the precise role the nematode plays in the etiology of replant problems is even less clear. "Decline" describes a situation whereby fruit trees that have previously produced profitable yields no longer grow or produce satisfactorily. Under such conditions a "replant disease" may arise when a second or a following planting of the same or closely related species is planted at the same site and fails to become established (Savory, 1966). Factors known to be involved in replant diseases include parasitic nematodes, weak parasitic fungi and bacteria, toxic chemicals from old roots, unbalanced soil nutrition, and organic matter (Parker *et al.*, 1966).

1. Cherry

Cherries are one of the most susceptible fruit trees to nematode damage and the replant problem is of greatest economic importance. Several nematode genera are present in established orchards but injury has been attributed to *Pratylenchus* spp. in most instances. *P. penetrans*, *P. pratensis* and *P. vulnus* have all been reported as associated with decline or unthrifty growth but *P. penetrans* causes the greatest damage and is the most prevalent. In New York State (U.S.A.), this nematode has been reported to be the most numerous species in cherry and apple orchards and to be the primary cause of poor growth and death of young trees due to damage of small roots (Mai and Parker, 1967).

In greenhouse experiments Mazzard cherry seedlings, inoculated with 6000 *P. penetrans*, showed wilting, paler green leaf color, darker roots, less growth, and a greater number of dead roots when compared to uninoculated trees (Parker and Mai, 1956). Severely damaged roots may either lack feeder rootlets, or may develop witch's broom symptoms, appearing as tufts of short, partially dead roots (Fig. 1).

Cherry orchards in which young trees had made very poor growth or had died, either were fumigated with DD or left untreated and replanted to cherry (Mai and Parker, 1967). After 3 years all trees in the untreated areas were dead. Girth of trees in the treated areas increased rapidly during the first 3 years but was progressively less thereafter. Nematode populations were considerably less in treated than in untreated areas 2 years after fumigation but had begun to increase by the third year. Substantial decrease in growth occurred sooner in trees on Mazzard than on Mahaleb rootstocks which corresponded to their susceptibility to *Pratylenchus*. There was more winter injury on trees in untreated soil than on those in treated soil, and in the laboratory

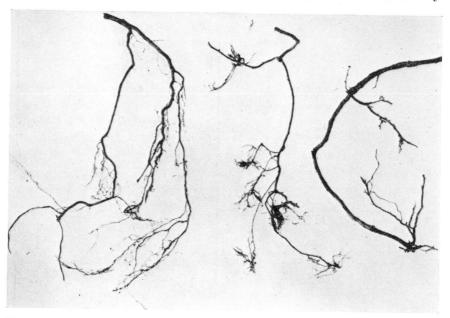

Fig. 1. Mahaleb cherry rootstock. Left—normal root; Centre—root from moderately affected tree, with many witch's broom symptoms developed because of injury to feeder rootlets by nematodes; Right—root from severely affected tree, showing almost complete absence of feeder rootlets. (By courtesy of Department of Plant Pathology, Cornell University, Ithaca, New York.)

twigs from check trees were more susceptible to cold injury than those from treated trees (Edgerton and Parker, 1958).

Knierim (1964) estimated that 20–50% of cherry trees planted on old orchard sites in Michigan (U.S.A.) were unthrifty, exhibiting poor vigor and very little terminal growth. These problem orchards are usually infested with *P. penetrans* and *Xiphinema americanum*. A pre-plant treatment with DBCP gave significantly greater growth than trees growing in other plots.

P. penetrans, *Xiphinema americanum*, and *Paratylenchus* spp. occurred on cherry in problem sites in Ohio State (U.S.A.), but only the first species was judged to be a critical factor in decreasing tree growth (Wilson and Hedden, 1967). Pre-plant nematicide treatments decreased nematode populations and increased tree heights, crown widths and trunk diameters as compared to untreated checks.

P. penetrans is primarily responsible for the cause of the replant problem in the Netherlands. Growth of cherry seedling rootstocks was poor on untreated soil heavily infested with this nematode, but good on similar soil treated with a nematicide (Hoestra and Oostenbrink, 1962).

P. vulnus has been reported associated with dieback of bearing cherry trees in Riverside, California (Serr and Day, 1949). Three different rootstocks were compared for their susceptibility to nematode damage over a 5-year period in heavily infested nursery soil. Roots of Mazzard and Stockton Morello were severely cankered by lesions but Mahaleb roots were symptomless and apparently resistant to both *P. penetrans* and *P. vulnus*.

In California field tests the roots of Mahaleb cherry were lightly galled by a *Meloidogyne* sp. but the roots of Mazzard, Stockton Morello, English Morello and Montmorency were entirely free of galls (Day and Tufts, 1944). This genus does not cause an economic loss in California cherry orchards.

Several *Xiphinema* and *Longidorus* species are reported from cherry orchards but no direct effect of these nematodes has been recorded. However, evidence is accumulating which suggests that these species are important virus vectors (Table II, p. 355).

Several species have proven to be vectors but the exact nature of other nematode–virus relationships remain in question. In Europe, *X. diversicaudatum*, which is a known vector of arabis mosaic virus (AMV) and strawberry latent ringspot virus (SLRV), is associated with a leaf enation symptom in cherry (Cropley, 1961) but transmission by the nematode has not been demonstrated.

In England, surveys of cherry orchards infected with leaf-roll virus (CLRV) revealed an association with species of *Xiphinema* and *Longidorus* although no single species occurred consistently. Transmission from virus-infected to healthy *Chenopodium* was obtained with *X. diversicaudatum* and *X. vuittenezi*, and with a mixed population of *L. profundorum* and a *Xiphinema* sp. Transmission occurred only when infected and healthy plants were grown in the same container in nematode naturally infested soil, with suitable checks to eliminate the possibility of infection by root grafting or contact. These results suggest that CLRV may have more than one vector and that transmission may be due to virus particles carried as short term surface contaminants on the nematode's mouthparts. More critical studies are necessary to establish the exact relationships.

2. Peach

Historically crops often failed for no known reason. The nineteenth century Pennsylvania Dutch blamed these failures on the work of supernatural powers and employed a number of bizarre practices such as pow-wows to exorcise curses on fields, hex signs on barns for fertility, and trees were horse-whipped before breakfast on Good Friday to guarantee a good peach harvest (Anon., 1970). Nematodes were probably the cause in many instances and while our methods of dealing with the problem have improved, our understanding of the role of nematodes is far from complete.

FIG. 2. Young peach trees of the same age growing in soil infested with the root-knot nematode, *Meloidogyne incognita*. Left—on susceptible Lovell rootstock; Right—on Rancho Resistant rootstock. (By courtesy of the Department of Nematology, University of California, Davis.)

Of the many nematodes associated with peach trees the root-knot nematode (*Meloidogyne* sp.) was the first known to cause damage in the form of knotted, swollen roots, reduced root and top growth and leaf chlorosis (Fig. 2). Species of *Meloidogyne* have been reported causing damage in Australia, South Africa, South America, the Middle East, Canada and the U.S.A. (Sharpe *et al.*, 1969).

Until 1949 all root-knot nematodes were considered a single species, *Heterodera marioni*, which caused misunderstanding with regard to differences in rootstock susceptibility. It was found that at least two species had been involved in earlier studies, *M. incognita* and *M. javanica*. Stocks of "Yunnan", "Shalil", and "Bokhara" were found to be susceptible to *M. javanica*. The first two stocks were immune to *M. incognita* and the third showed a 3 : 1 segregation of immune to susceptible. "Okinawa" and "Nemaguard" selections are resistant to both species of *Meloidogyne*. However, the reactions of these stocks to the nematodes differ. The term "immune" designates complete freedom from galls, and "resistant" describes a condition whereby the nematode does not reproduce but does produce small root galls. Thus "Okinawa" and "Nemaguard" seedlings are resistant but not immune to *M. javanica*. "Okinawa" seedlings are immune to *M. incognita* but "Nemaguard" segregates to 3 immune and 1 resistant. These two rootstocks have been used extensively in areas where *Meloidogyne* is a problem.

Temperature tank studies have shown these rootstocks to be infected by *M. arenaria* and *M. hapla*, but there are no reports of widespread field damage by these species. Of more concern, however, is the discovery in Florida in 1966–1967 of a new strain or species of *Meloidogyne* capable of attacking and reproducing on "Okinawa", "Nemaguard" and on most of the advanced breeding lines derived from these and other known resistant stocks (Sharpe et al., 1969). The distribution of this species has not yet been determined.

In addition to attack by *Meloidogyne*, peach is subject to serious decline and replant diseases in which nematodes play a part, although their exact role is not completely understood. Nematodes associated with peach decline or replant problems have been reported from Italy, Australia, Israel, South Africa, Canada and the U.S.A.

Pratylenchus species appear to be the most important nematodes associated with the decline and replant disease. *P. penetrans* has been studied in Ontario, Canada for several years where it is associated consistently with tree damage. It has been found in commercial nurseries and is transported in roots of nursery stock. Pathogenicity has been demonstrated in the absence of bacteria and fungi (Mountain and Patrick, 1959). Under aseptic conditions it invades and kills roots by hydrolysing the cyanophoric β-glucoside, amygdalin. In non-sterile growth room tests, peach seedlings 14 days after inoculation with 800 nematodes showed stunting and leaf chlorosis and more severe damage than under sterile conditions. Hence, it was suggested that the main role of *P. penetrans* in peach replant failure was its ability to incite root degeneration by providing extensive infection courts for soil microorganisms.

P. vulnus has also been associated with decline in California and Georgia (U.S.A.). Peach is a good host for this species and in a Georgia survey was always associated with rapid deterioration of feeder roots and reduced tree vigor (Fliegel, 1969). The nematode was found in large numbers throughout the year with a rapid increase in the soil population during August to December. A high of over 3500 per g of root was attained in April and May. *P. zeae* and *P. brachyurus* were not associated with severe damage to roots even though *P. zeae* reached over 800 per g of root in the fall.

In Australia, however, *P. zeae* was associated with poor growth of peach and in Florida (U.S.A.) *P. brachyurus* was associated with severe damage to peach roots (Stokes, 1967). This latter species had little effect on seedling growth of Lovell, Nemaguard and Okinawa root-stocks. Population increases were about twice as great on Okinawa as on the other stocks but top and root weights after 16 weeks were only slightly less than check plants when inoculated with 1000 nematodes per tree.

A more recent report substantiates the importance of *P. penetrans* and *P. vulnus* and suggests that other species are of limited importance on peach (Barker and Clayton, 1969). The following cultivars were tested as hosts for 6 species of *Pratylenchus*: Nemaguard, Elberta, Lovell, NC73, NC88 and NC240. There were no significant differences among the cultivars to attack by any of the species. Populations of *P. vulnus*, *P. penetrans* and *P. brachyurus* increased by 10·8, 4·6 and 1·1 fold respectively, while *P. coffeae*, *P. scribneri* and *P. zeae* decreased during the 8-month test.

Three species of *Criconemoides* attack peach, namely *C. xenoplax* in New Jersey and California, *C. simile* in Maryland and North Carolina, and *C. curvatum* in New Jersey (U.S.A.). *C. xenoplax* commonly occurs around roots of S-37 and Lovell rootstocks in sandy soils in California in numbers up to 12 nematodes per g of soil (Lownsbery, 1959) and causes stubby fasciculation of the roots. Pre-plant fumigation with DD improved the growth of peach replants.

Disease was produced in Lovell peach seedlings planted in soil containing *C. simile*. Under sterile conditions *C. curvatum* causes extensive lesions and pits, and under non-sterile conditions the lesions and pits are invaded by other organisms which cause general discoloration and loss of vigor of the root system (Hung and Jenkins, 1969).

These studies with *Criconemoides* spp. and *Pratylenchus* spp. suggest that while several species may in themselves act as pathogens they play a big role as a part of a disease complex involving other cultural factors and micro-organisms.

The addition of benzene hexachloride to *Xiphinema*-infested soil resulted in increased tree growth and decreased *Xiphinema* populations (Adams, 1955). There is no further evidence of direct injury due to this nematode but *X. americanum* is a vector of peach yellow bud mosaic (PYBM) (Teliz *et al.*, 1967). In greenhouse tests *X. americanum* transmitted PYBM to roots of "Lovell" peach, "Royal" apricot and "Damson" plum seedlings. Plum and apricot seedlings were tested because of their possible use as virus resistant rootstocks. There was no evidence of the virus moving from the infected roots to the tops.

In another type of interaction study, "Lovell" seedlings were more susceptible to the crown gall causing organism *Agrobacterium tumefaciens* in the presence of *M. javanica* (Nigh, 1966). Not only was there a significantly higher incidence of bacterial infection but also a larger number (27·5% increase) of nematode galls when the two organisms occurred together.

3. Apple

Several genera have been reported in association with apple trees but only *Pratylenchus* species appear to be of major economic importance. The particular species of importance varies with geographical location and, perhaps, soil type. *Pratylenchus penetrans* is the important species associated with diseased apples and has been recorded as a parasite of apple stocks in the U.S.A., Canada, the Netherlands, England and Germany. *P. coffeae* has been found associated with unthrifty growth of apple seedlings in Australia, and *P. vulnus* as a parasite of apple in California and Oregon (U.S.A.).

Several workers in different countries have studied the effect of *P. penetrans* on apple. Malling rootstock VII is susceptible and M. I, II, IV, IX, XI, XVI and seedlings highly susceptible to this species (Hoestra, 1968). More than 1000 nematodes per 10 g of roots was found in all instances. Experiments in Belgium and the Netherlands have shown that severity of damage is proportional to the number of nematodes found in the roots. In Germany the number of *P. penetrans* in the field reached 1500 per 100 ml of soil and in pot experiments nematode numbers increased on apple and seriously affected growth.

In an attempt to determine the exact nature of the damage caused by *P. penetrans*, studies were conducted under sterile conditions and showed that apple feeder roots reacted to nematode invasion by rapid discoloration of the outermost and innermost cortical tissues but by little or no reaction in the intervening cortical parenchyma (Pitcher *et al.*, 1960). The workers suggested that root necrosis results from a reaction of plant phenolic compounds to nematode enzymes, giving rise to products that are toxic to invaded and adjacent cells.

Although the nematode damages apple roots in the absence of other organisms, greater damage occurs under natural field conditions, suggesting an interaction with other organisms or soil factors.

M

P. coffeae is widely distributed in the orchards of the Santhorpe district of Australia where it appears to be the most important factor causing unthrifty growth of replant trees and poor growth and dieback of older trees (Colbran, 1953). Inoculation experiments caused severe stunting of the top growth and of seedlings.

There are two reports of association of *Paratylenchus* spp. with reduced apple vigor. In Australia (Fisher, 1967) *P. neoamblychephalus* has been found in large numbers around roots of apple and cherry trees. Seedlings inoculated with 10,000 nematodes had reduced shoot and root growth.

A survey of the Hudson Valley of New York State showed *P. curvitatus* to be the most numerous parasitic nematode found associated with apple roots (Palmiter *et al.*, 1966). The favorable response of apple trees to DBCP treatments was closely correlated with reduction of this and other nematodes. *P. curvitatus* was considered to be the nematode responsible for most of the decline but *P. pratensis* and other nematodes were believed to be contributing factors.

4. Pear

Plant-parasitic nematodes although prevalent in many pear orchards do not appear to cause economic loss. *P. coffeae*, *P. neglectus*, *P. penetrans*, *P. crenatus*, *P. thornei* and *P. vulnus* parasitize pear but only *P. penetrans* has been associated with pear disease in the Netherlands (Oostenbrink, 1954) and in Germany (Decker, 1960). Pear is attacked by *Meloidogyne incognita* but is resistant to other *Meloidogyne* spp. common in California (U.S.A.) (Day and Tufts, 1944). Pear is resistant also to *Criconemoides xenoplax* (Lownsbery *et al.*, 1964).

An extensive survey of pear orchards in California during 1961–1962 showed *Xiphinema americanum* and *Paratylenchus hamatus* to be present in 90% of the orchards and associated with pear decline disease (French *et al.*, 1964). Pear is a host to both species but neither of them causes plant damage. Reduction of the nematode populations in declining orchards by fumigation with EDB did not improve the growth of Bartlett or Oriental pear seedlings.

X. diversicaudatum has been reported parasitizing pear in Belgium but extent of injury caused was not recorded (Maeseneer, 1962). As more information is gained regarding the extent of involvement of *Xiphinema* species in virus transmission, the importance of the two species parasitizing pear may become more evident.

5. Walnuts

Of the several species of nematodes parasitizing nut trees, *Pratylenchus vulnus* attacking walnut appears to be of greatest economic importance (Lownsbery, 1959). Most of the information concerning this problem comes from work done in California (U.S.A.), the prime producer of walnuts.

The attack of *P. vulnus* on walnut is more severe than on other trees (Lownsbery, 1956). Exposure of the cortex of affected roots reveals lesions which appear as black islands in the healthy light yellow tissue. Dieback of older trees, stunting of replants, and reduced yields are other symptoms associated with this nematode (Fig. 3). Low potassium

Fig. 3. Lesions on walnut root associated with a lesion nematode, *Pratylenchus vulnus.* (By courtesy of the Department of Nematology, University of California, Davis.)

content occurs in the leaves of parasitized trees and may be related to the dieback symptom which is associated with potassium deficiency in walnuts.

Cacopaurus pestis is a severe pathogen of the roots of Persian walnut but is of minor importance in California. This nematode has been associated with a typical decline symptom which eventually results in complete defoliation and death. *Meloidogyne* species also attack Persian walnut rootstocks but these nematodes are not common in California walnut orchards.

C. Cultural and Environmental Influences

In general it appears that only those orchards on soils containing a high percentage of fine sand and little silt or clay are severely affected by nematodes, especially *P. penetrans*. Parasitic species occur in heavy soils but populations are lower and damage is minimal. In the lighter soils root damage is more likely in the top 15–45 cm of soil and better root development occurs below that level.

The reason for greater damage in sandy soils infested with *Pratylenchus* spp. is due, at least in part, to an interaction of soil moisture with soil type (Kable and Mai, 1968b). The higher the silt and clay content of a soil the greater is the moisture tension necessary for satisfactory nematode population growth. In clay soils moisture tensions optimum for population increase exist for a shorter period of time than in sandy soils. In these latter soils in New York (U.S.A.) more *P. penetrans* overwinter at 30 cm than 15 cm but in clay soils the reverse is true. However, the number of nematodes surviving in senescent roots was not significantly affected by soil type or depth. Therefore this species is more active and damaging in sandy than in clay soils and is able to survive adverse conditions best within tree roots.

Temperature is also an important factor in nematode activity and tree damage. Apple seedling inoculation experiments at different temperatures showed greatest reduction in root growth at 29·4°C (Mai, 1960).

As nematodes are concentrated in the top 45 cm of soil the type of cover crop will influence their population levels. Growth of peach trees is significantly increased by planting root-knot resistant cover crops where this nematode is a problem (McBeth and Taylor, 1944). In *P. penetrans*-infested orchards, Sudan grass (*Sorghum vulgare* var. *sudanense*) is a good cover crop because it is a poor host for this nematode and grows sufficiently vigorously to become the dominant cover (Parker *et al.*, 1966). This latter point is important as it precludes weeds which are hosts for this nematode. In Ontario (Canada), 55 weed species were found to be infested by this nematode (Townshend and Davidson, 1960).

Conversely, Egunjobi (1968) observed that nematode populations in apple roots were very high under clean cultivation, but under a cover crop, populations were low in the apple roots and high in the grass roots. This suggests that a cover crop in a *Pratylenchus*-infested orchard may reduce the level of infestation in the tree roots by attracting some of the nematodes to the grass roots. There is no experimental evidence to support this but it is worth further study.

The broad host range of *P. penetrans* is an important factor in its control. In the northeastern U.S.A., it has been reported to be associated with over 60 plant species and is considered one of the most damaging nematode species (Jenkins and Taylor, 1967). A similar report is given by Jensen (1953) for western regions of the Pacific Northwest. Jensen found also that all 33 species and varieties of cover crops tested were hosts to *P. penetrans*. Extensive field tests in the Netherlands showed that 164 cultivated plant species and varieties were hosts, and that certain hosts allow the development of extremely high nematode populations without exhibiting symptoms. The extremely polyphagous nature of this nematode precludes crop rotation as a control measure and necessitates fumigation before planting in infested areas.

It is of interest that periodic top pruning of a plant can greatly increase nematode population build-up on the roots (McDonald and Mai, 1963). Pruning of vetch to 2·5 cm every 21 days for 3 months yielded 1345 nematodes per g of root compared to 5 per g of root in unpruned plants.

The nutritional status of the host also affects the population levels, rate of development and reproduction of nematodes parasitizing that host (Kirkpatrick *et al.*, 1964). A high phosphorus (P)–low potassium (K) nutritional status favors increased populations of *P. penetrans* and *X. americanum*. Since P tends to be residual in the soil compared with nitrogen (N) and K, continued use of NPK complete fertilizers results in a high P-low K situation, resulting in K deficient trees with increased populations of pathogenic nematodes.

Damage by nematodes is more severe on trees planted in soil in which another fruit tree has been previously planted. An insufficient time interval between removal and replanting also increases the problem. Old roots in the soil may harbor nematodes, be a source of toxic chemical substances, and interfere with the action of soil fumigants. After old trees have been removed at least two years is required for the soil condition to be satisfactory for replanting. Planting a poor or non-host cover crop during this period decreases nematode populations.

Nevertheless, all of this is to no avail if nematodes are reintroduced in infected planting stock. Endoparasites such as *Meloidogyne* and *Pratylenchus* are readily transported on tree roots even in the absence of soil. Irrigation water containing drainage water from nematode infested fields probably contains most of the parasitic genera in those fields. They can be carried in a viable state for great distances and contaminate other fields.

D. Control

Control of nematodes attacking fruit and nut trees may be divided into three types of treatments: soil fumigation prior to planting (pre-plant), fumigation of the soil in which trees are established (post-plant), and treatment of planting stock to kill nematodes in and adhering to the roots (disinfection). Any one or all of these may be necessary to obtain maximum production. Generally the control measures in each category are applicable to all trees (Table I) with the only differences being in the economics of treatment.

Pre-plant fumigation is necessary to allow establishment of the trees. Nematicides containing dichloropropene mixtures (DD, Telone, Vorlex, etc.) are most effective for this purpose. Most trees respond significantly to a pre-plant treatment but the long-term effect depends on the kind of tree, the nematode species and the complex cultural and environmental influences.

Frequently the effect of pre-plant fumigation lasts only 1–3 years, after which nematode populations are back to pre-treatment levels and so post-plant treatment is often economically justified for many trees. Presently only DBCP is available for such a treatment and although the number of tests are limited, increased yields have been obtained by treating some types of trees either annually or every 2–3 years. In California peach and walnut yields were increased by up to 64% by such treatments (Lownsbery et al., 1968).

Strict precautions should be taken to ensure that nematode-free trees are planted into a nematode-free field. Dormant infected planting stocks may be treated by a hot-water dip to control *Pratylenchus* and *Meloidogyne* species on apple and peach (Bosher and Orchard, 1963). Tree survival is increased by pre- and post-treatment cold storage of the stocks.

Nematode resistant varieties are not available for most trees. Only peach shows promise along these lines and only against certain *Meloidogyne* species. Since the unspecialized, polyphagous *Pratylenchus* species are the nematodes most frequently involved in decline and replant diseases there is little hope for control by this method.

E. Economics

Many species of nematode attack the roots of tree fruits but *P. penetrans* is of greatest economic concern. Cherries, apples, peaches and pears are damaged by this nematode and show marked improvement in growth when planting sites are fumigated but there is limited information on the resulting increase in yield.

Many reports show results in terms of increases in growth but this is difficult to project in terms of money returned to the grower's pocket. A representative sample of reports dealing with economic information will be discussed.

TABLE I. Control of nematodes parasitizing fruit and nut trees

Trees	Application	Chemical	Rate	Method of application	Reference
Apple	Plant (roots)	Water	46·1°–46·7°C; 30 min	Dip	Bosher and Orchard (1963)
Peach	Plant (roots)	Water	50·0°–51·1°C; 10 min	Dip	Nyland (1955)
Cherry	Plant (roots)	Water	50·0°–51·1°C; 10 min	Dip	Nyland (1955)
All	Soil (Pre-plant)	Telone	393 litres/ha	Chisel	Mai et al. (1970)
All	Soil (Pre-plant)	Vorlex	281 litres/ha	Chisel	Mai et al. (1970)
All	Soil (Post-plant)	DBCP	23–36 litres/ha (annually)	Chisel, flood, sprinkler	Mai and Parker (1967)
Apple	Soil (Post-plant)	DBCP	1·14 litres per tree (every 3 years)	Flood	Palmiter et al. (1966)

A study of the effect of controlling *P. penetrans* on apples and cherries was recently conducted in New York (Mai and Parker, 1970). Trunk circumference increases of apple trees in soil receiving Vorlex and Telone pre-plant treatments were 68–90% greater than those growing in un-treated soil 3 years after treatment. There was a slight decrease in advantage of treatment during the fourth year. The most critical period for protection from nematodes is during the first year after planting, as the roots formed during this period become the frame roots of the trees.

The above results show a growth response to nematicide treatment but it is difficult, with present knowledge, to predict the economic return to the grower. Results from treatment of established apple trees in the Hudson Valley of New York give a somewhat clearer economic picture (Palmiter *et al.*, 1966). A single application of 907 ml of DBCP per tree resulted in decreased populations of *Paratylenchus curvitatus, Pratylenchus pratensis* and *Xiphinema americanum*, and an average increase of 21 boxes of apples per tree over non-treated trees during the 3 years following treatment. At an average price of $1.68 per box (19 kg box), less cost of treatment ($5.00 per tree), the grower realized an approximate $30.00 per tree, per year increase return. In this instance, it appears that treatment would more than pay for itself especially when the alternatives are considered. The cost of pulling out an old orchard and replanting, a pre-plant nematicide treatment to avoid a replant problem, and loss of income while trees become established is considerable.

In Ohio State (U.S.A.), Wilson and Hedden (1967), demonstrated the economic benefits of nematode control of cherry trees. Pre-plant soil treatment with Telone at 450 ml/ha resulted in more than a 150% (6·2 kg per tree) yield increase over trees growing in non-treated soil 8 years after treatment. This treatment reduced *Pratylenchus penetrans* populations by 66% and *Xiphinema americanum* by 83%.

Using a price of 44 cents/kg (average of 4 years) and a tree density of 68 trees/ha this gives an increase of $184.96/ha as a result of fumigation. Cost of fumigation at approximately $320.00/ha spread over the 8 year period amounts to $40.00/ha per year. The economic return due to fumigation more than covers the average cost of treatment for a single year.

Established peach trees also respond to treatment with DBCP. In a 1 year old Georgia orchard containing *Meloidogyne* spp., *Tylenchorhynchus* spp., *Pratylenchus* spp. and *Criconemoides* spp. the effect of fumigation on tree survival and yield was studied. Chisel application of 40·4 litres/ha increased yield by 5·3 dal per tree per year over 2 years, or 354·6 dal/ha. This treatment also increased tree survival by 45%

adding another 157·5 dal/ha to the total yield. Fumigation therefore increased returns by $280.00/ha (at an average price of $0.09/kg). Cost of fumigation at this rate would be between $175.00 and $200.00/ha. This increased return is not high enough to apply the cost of fumigation to a single year and still obtain much profit. The combination of a relatively low value crop with a less effective post-planting treatment gives only a marginal increase in return. However, the 45% increase in tree survival due to treatment increases the number of trees, and hence the production potential of the operation, and should increase the productive life and yield of the trees over a period of time.

In California (U.S.A.), pre- and post-planting soil fumigation in peach, walnut and prune orchards were compared (Lownsbery et al., 1968). Positive growth or yield response resulted in 9 of 11 trails (82%) of the pre-plant applications, and in only 8 of 26 trails (31%) of the post-plant applications. Yield information for pre-plant treatment was presented only for French prune infested with *Pratylenchus vulnus*. Application of DD at 595·2 litres/ha increased yield by 180% 5 years after treatment. However, this amounts to an increase of only $40.50/ha (453·6 kg/ha at 11 cents/kg). Cost of fumigation at $185.00/kg even spread over 5 years amounts to $37.00/ha per year. Hence, the treatment could not be justified on the information available despite the yield increase of 180%.

A significant increase in walnut tree growth, as assessed by trunk circumference and terminal shoot growth, has been obtained from both pre-plant and post-plant fumigation (Lownsbery and Sher, 1958). Increases in trunk circumference range up to 180% while terminal growth may be increased by as much as 300% over non-fumigated trees. Yield increases however, are available only for treatment of established trees. Pre-plant treatments did not appreciably affect the yield results of post-plant treatments. Best results were obtained from an annual application of DBCP at 28 litres/ha. At the end of 5 annual applications yields were increased by 5·3 kg per tree. This amounts to 423·4 kg/ha (at 68 trees/ha) or an increase of $207.90/ha (at 59 cents/kg). Cost of fumigation would be approximately $125.00/ha per year. With the limited information available it is difficult to determine the exact economic advantage of fumigation. Although the increased return of $30.00 per acre is minimal, the many added side effects of growing a healthier tree are difficult to put into dollars and cents.

These studies indicate that information on the economics of fumigating tree fruits is sadly lacking, partly because of the fact that trees require several years to come into full production and remain in production for 1–2 decades. Further, only within the last 10–15 years have nematodes been recognized as a major pest of orchard trees.

M*

Generally pre-plant fumigation gives the highest degree of success. This may be supplemented with post-plant treatments but when used in place of pre-plant treatment much higher rates are necessary to obtain an economic benefit.

Basically the relative value per unit of crop determines the economic soundness of fumigation. Relatively low value crops such as apples, peaches, prunes, pears, apricots for example, require an increased yield in the order of 1960 kg/ha (at an average of 11 cents/kg) to cover the cost of a $175.00 fumigation. Cherries require an increase of only 392 kg/ha (at an average of 59 cents/kg) to cover these same fumigation costs.

No mention has been made of the cost of not applying a nematicide and this presents a far more formidable task. It is not the reciprocal of the increased return due to treatment, because aside from outright destruction of a tree, nematodes transmit viruses, cause a nutrient imbalance, decrease efficiency of nutrient and water uptake and utilization, and render the tree more susceptible to adverse environmental conditions and to attack by other organisms. Included in the cost of non-treatment are the costs of attempting to overcome the disease problem by increased use of fertilizer, water and pesticides. Whether the cost of treatment or of non-treatment is higher must be decided by the grower but it remains with the researcher to give the grower guide lines with which to make that decision. To establish these guide lines more information is needed such as: the effect of pre-plant and post-plant fumigation separately and together on yield; the length of time each treatment remains effective; the effect of treatment on length of life and productivity of the tree; the rate and type of nematicide necessary to give an economic return of the crop in question. This information must be determined for each crop, in each area, for each nematode involved.

II. Strawberry and Raspberry

A. Crop Production

1. Strawberry (*Fragaria* spp.)

Strawberry production is concentrated in Europe (which produces 552,600 metric tons) and North America. The U.S.A. is the leading producer, followed by Poland, Mexico and Japan. Other areas of the world producing strawberries include New Zealand, Australia, Israel, and Argentina.

Strawberry is a highly adaptable plant by virtue of being a hybrid of two highly variable octoploid species. It grows in a wide variety of climatic conditions ranging from extreme cold ($-50°$F.), to the semi-tropics; and from continuous light of the summer Arctic to a 12-hour day at the Equator. The same variety may fruit either continuously, or for only 2 or 3 weeks depending upon the climate.

2. Raspberry (*Rubus* spp.)

Production of raspberries is about one seventh that of strawberries and is more limited in its distribution. The United States leads in production, followed by Germany, Yugoslavia, Great Britain, Hungary and Poland. Other areas of the world producing raspberries include Canada, New Zealand and Australia.

Raspberries grow best where winters are mild and the summers cool and moist. Yields of 12–13 metric tons/ha are common in the main production areas of the world.

B. Nematode Diseases

1. Strawberry

Strawberry roots and above ground parts are attacked by several ecto- and endoparasitic nematodes. Wherever strawberries are grown they are subject to attack by one or more of these nematodes. The nematode species of greatest economic importance depends upon the geographic location.

In areas of mild winters and long growing seasons the root-knot nematode, *Meloidogyne hapla,* is of extreme importance (Plakidas, 1964). *M. javanica* is reported on strawberries in Zambia, Africa. Affected plants are unthrifty, stunted, unproductive and often of pale green color, having numerous small root galls caused by nematode feeding. Under adverse conditions plants may wilt or even die.

The root lesion nematode, *Pratylenchus,* in many regions is of greater importance than *Meloidogyne.* Species of this genus are found throughout the world and one or another of these species is pathogenic in almost every strawberry growing area. *P. penetrans, P. pratensis* and *P. scribneri* have been consistently associated with declining, root-rot affected strawberry plantings. Infested plants are stunted, pale green, have reduced yields and eventually die. Root symptoms appear as distinct brown lesions in early or light infestation. The entire root system eventually becomes black and necrotic as populations increase and secondary organisms invade.

Townshend (1963) demonstrated that *P. penetrans,* in the absence of bacteria and fungi, invaded strawberry roots, causing discoloration and necrosis of certain tissues and hyperplasia in others. Under field conditions this nematode is involved in a disease complex called black root rot. Despite the paucity of experimental evidence it is generally believed that the nematode is the primary causal agent, opening the roots to invasion of fungi and bacteria through its feeding and root penetration.

Several species of dagger nematode, *Xiphinema* spp., have been associated with diseased strawberry plantings. *X. americanum* and *X. chambersi* were the first species shown to be pathogenic to strawberry. Both species cause small indeterminate, reddish-brown, sunken lesions along the roots in early stages of parasitism. The roots eventually become dark brown to black. Stunting of both tops and root is apparent in parasitized plants. Field decline attributed to these nematodes is apparently slow since production is not materially affected until the second or third year.

Of the two species, *X. americanum* is more widespread, occurring throughout the U.S.A. and Canada in strawberry plantations. This nematode transmits tomato ringspot (TomRV) and peach yellow bud mosaic viruses (PYBMV), to various hosts (Frazier and Maggenti, 1962). There are indications that the first virus may be transmitted to strawberry where it causes a severe disease. The second virus, for which wild strawberry may act as a reservoir, has been transmitted to strawberry by *X. americanum*.

X. bakeri has been reported from two areas in the U.S.A., and from British Columbia, Canada with diseased strawberry (McElroy, unpublished data). In the latter area it is native and widespread in strawberry plantings causing disease and in populations which may reach 1–2000 per 500 cc of soil. In preliminary greenhouse tests *X. bakeri* increased four-fold in 12 weeks and completely destroyed the root system of 50% of the test plants. The nematode feeds at root tips causing stunting, swelling and curling of the tips similar to the damage caused by *X. diversicaudatum*.

This latter species of the dagger nematode has wide distribution in strawberries in Europe but is more limited elsewhere. It is a severe pathogen of strawberry causing stunting of roots and tops and decreasing and preventing runner growth. Root symptoms are enlargement and curling of the ends of the roots. This species also transmits two economically important viruses to strawberry, arabis mosaic (AMV) and strawberry latent ringspot virus (SLRV) (Table II).

These viruses and their vector have been reported from most strawberry or raspberry growing areas in Europe. The viruses persist in the nematode about 8 months and also in perennial weeds, which serve as reservoirs. The longevity of adult nematodes in fallow soil is only slightly greater. The combination of a wide host range of this nematode

TABLE II. Nepo-viruses of tree and small fruits

Crop	Vectors	Virus	Distribution	Reference
Cherry	*Longidorus macrosoma*	RRV (European rasp-leaf)	Europe	Harrison (1964)
Cherry	*Xiphinema diversicaudatum* (?)	CLRV	Germany	Fritzsche and Kegler (1964)
Cherry	*X. coxi* (?)	CLRV	Germany	Fritzsche and Kegler (1964)
Cherry	*X. americanum*	TomRV (Yellow bud mosaic)	N. America	Teliz *et al.* (1966)
Cherry	*X. americanum*	TomRV (Eola rasp-leaf)	N. America	Milbrath and Reynolds (1961)
Cherry	*X. americanum* (?)	CRLV	N. America	Nyland *et al.* (1969)
Peach	*X. americanum*	TomRV (Yellow bud mosaic)	N. America	Teliz *et al.* (1967)
Raspberry	*L. elongatus*	RRV	Scotland	Taylor (1962)
Raspberry	*L. elongatus*	SLV	Holland	Meer, van der (1965)
Raspberry	*L. elongatus*	TBRV	Scotland	Harrison *et al.* (1961)
Raspberry	*L. macrosoma*	RRV	England	Harrison (1964)
Raspberry	*X. diversicaudatum*	AMV	England, Scotland	Jha and Posnette (1959)
Raspberry	*X. americanum* (?)	TomRV	N. America	Teliz *et al.* (1967)
Strawberry	*L. elongatus*	RRV	Scotland	Taylor (1962)
Strawberry	*L. elongatus*	TBRV	Scotland	Harrison *et al.* (1961)
Strawberry	*L. macrosoma*	RRV	England	Harrison (1964)
Strawberry	*X. americanum*	TomRV	N. America	Teliz *et al.* (1966)
Strawberry	*X. diversicaudatum*	SLRV	Scotland, England	Lister (1964)
Strawberry	*X. diversicaudatum*	AMV	England	Jha and Posnette (1959)
Blueberry	*X. americanum*	TobRV (necrotic ringspot)	N. America	Griffin *et al.* (1963)
Grape	*X. americanum*	TomRV	N. America	Teliz *et al.* (1967)
Grape	*X. index*	GVFV/YMV (Grape yellow vein)	Europe, N. America, Australia	Hewitt *et al.* (1958)

Key: RRV (Raspberry ringspot virus)
CLRV (Cherry leaf-roll virus)
TomRV (Tomato ringspot virus)
CRLV (Cherry rasp-leaf virus)
SLV (Spoon leaf virus)
TBRV (Tomato black ring virus)

AMV (Arabis mosaic virus)
SLRV (Strawberry latent ringspot virus)
TobRV (Tobacco ringspot virus)
GVFV (Grapevine fanleaf virus)
YMV (Yellow mosaic virus)
(?) indicates suspected vector but not proven

among crops and weeds, and long retention of viruses in the vector and in weeds, provides an effective natural system for the maintenance and propagation of viruses and vector.

Another nematode having a similar distribution in Europe, causing direct damage and transmitting two viruses to strawberry is the needle nematode, *Longidorus elongatus* (Taylor, 1967). This nematode causes stunted and non-vigorous plants with reduced root systems having a stubby root appearance caused by malformed feeder roots with swollen, stubby, blackened tips.

Raspberry ringspot (RRV) and tomato black ring (TBRV) viruses are both transmitted by *L. elongatus* and can cause serious losses in strawberry crops. Both the viruses and the vector have wide host ranges including many weeds common to strawberry plantings. These viruses persist in the nematode for only 9 weeks and therefore infectivity is lost during a winter fallow. The nematode becomes viruliferous in the spring by feeding on virus infected weeds where it is seedborne. Lack of virus persistence in the vector is not related to longevity of the nematode for it can survive at least 2 years without feeding. In strawberry fields most individuals have an annual life-cycle.

While the distribution of RRV and TBRV appears to be limited, *L. elongatus* is not. It has been reported from England, Scotland, the Netherlands, Germany, Belgium, Switzerland, Austria, Hungary, Poland, U.S.S.R., Turkey, Egypt, U.S.A., and Canada.

An ectoparasitic nematode of limited distribution but a serious pest of strawberry is the sting nematode, *Belonolaimus longicaudatus*. It is widespread in the south-central and eastern U.S.A., but appears to be of economic importance on strawberry only in Florida (Plakidas, 1964). Nematode feeding kills the fine feeder roots and the larger root tips and if feeding is prolonged the plant stops growing and eventually dies. Young plants may remain small for several weeks.

The above-ground portions of the strawberry plant are attacked by several nematode species. Strawberry dwarf is caused by at least three species of *Aphelenchoides*. This disease occurs in several countries and is known by many different common names. Dwarf is actually two distinct diseases occurring at different times of the year and caused by different *Aphelenchoides* species but having essentially identical symptoms.

A. fragariae causes spring dwarf, a cool-weather disease of strawberry. This disease has been reported from several areas in Great Britain, continental Europe, New Zealand, and the U.S.A., where it has become established in Massachusetts.

Symptoms appear in the early spring as soon as plant growth starts. Bud and leaf development is retarded and plants fail to develop normal foliage. The crown is a compact mass of crinkled, twisted, distorted leaves with very short twisted petioles. When the apical bud is killed, numerous adventitious buds develop and give rise to multiple-crowned plants. Generally fruit buds are killed and those that escape produce small distorted fruit. New runners are usually infected and in England the petioles of certain cultivars develop a prominent red coloration. Hence, the disease there is called "red plant".

The second dwarf disease, summer dwarf, has very similar symptoms but is caused later in the season by *A. besseyi*. The severely stunted plant has short petioles and elongated, crinkled and asymmetrical leaflets. Leaf margins usually curl upwards in the young leaves and downward in the older ones. As is true with spring dwarf, the main bud is killed, resulting in multiple crowns. This disease is distributed throughout the strawberry growing areas of the southern U.S.A., but is known in other parts of the world.

Aphelenchoides species attacking strawberry live as ectoparasites in buds and leaflets of the growing point or in leaf axils. Only occasionally are specimens found within the leaf tissues. About 14 days is required for mature adults to develop from eggs deposited in the axils or within leaves.

A. fragariae and/or *A. ritzemabosi* occasionally interact with a bacterium, *Corynebacterium fascians*, causing a disease called "cauliflower". Both nematode and bacterium must be present to produce the disease. Diseased plants are stunted with deformed stems, leaves and flowers which in advanced stages resemble tiny cauliflowers. This disease although severe appears to be rare and not of economic importance.

The leaf gall disease of strawberry is caused by the stem nematode, *Ditylenchus dipsaci*. It has been reported from Europe, U.S.S.R. and the Pacific coast of the U.S.A. Infested plants have gall-like swellings on the petioles, leaf bases, fruit stems, fruit clusters and leaves. The foliage is frequently swollen and distorted and covered with red, yellow or brown spots. In severe infestations entire plantings may be lost as a result of reduced foliage and fruit production.

D. dipsaci parasitizes hundreds of species of plants and there are several races within the species which differ in host preference but are morphologically indistinguishable. The strawberry race is reported to be limited to two hosts, strawberry and red clover, but work in England shows that oat-onion-rye, red clover, narcissus, and teasel races also attack strawberry (Goodey, 1951).

2. Raspberry

Pratylenchus is the most frequently reported genus found in association with raspberry. At least six species have been recorded from North

America and Europe. Many of the reports are records of association but not necessarily with unhealthy plants.

The pathogenic effect of *Pratylenchus* species or the numbers necessary to bring about this effect has not been demonstrated because of the difficulty of rearing plants and nematodes free of soil micro-organisms.

P. penetrans causes stunted growth and a poor blackened root system and absence of fine feeder roots (Seinhorst *et al.*, 1956). Treshow and Norton (1958) suggested *P. vulnus* was contributing to raspberry decline in Utah (U.S.A.) by gradually destroying the root system and preventing water and nutrient uptake.

Canby and Newburgh raspberry cultivars are hosts for *Meloidogyne hapla* (Griffin *et al.*, 1968). In studies on the interaction of *M. hapla* and the crown gall bacterium (*Agrobacterium tumefaciens*) they showed that crown galling occurred only when inoculation included the nematode. No nematodes were found in the crown galls but they were recovered from adjoining tissue.

Xiphinema bakeri causes considerable damage to raspberry plantings in British Columbia (Canada). Pathogenicity studies under controlled conditions showed that populations as low as approximately 1 per 5 cc of soil reduced root and top growth of raspberry by 40–50% (Fig. 4). This rather recently discovered indigenous species is widespread in the raspberry growing areas of British Columbia, and poses a considerable threat to the industry. *X. bakeri* feeds mainly at the root tips causing swelling and curving of the root.

Some ectoparasitic nematodes apparently cause little direct damage to raspberry plants but feed enough to transmit economically important viruses (Table II).

Xiphinema diversicaudatum transmits both strawberry latent ringspot virus (SLRV) and arabis mosaic virus (AMV) which commonly occur together and are prevalent in England and Scotland (Taylor and Thomas, 1968). This vector is widely distributed in England, associated mainly with perennial crops such as raspberry and strawberry. It occurs to a limited extent in Scotland, usually on heavier soils. The nematode transmits the viruses to many plant species including cultivated crops and weeds but the virus infection is not always correlated with the suitability of the plant as a nematode host. Both AMV and SLRV persist in the stolons and rhizomes of perennial weeds such as *Mentha arvensis*, *Tussilago farfara*, and *Lamium amplexicaule*. This persistence combined with the wide host range and long retention of viruses in *X. diversicaudatum* provides an effective natural system for the maintenance and propagation of viruses and vector. The feeding of *X. diversicaudatum* causes stunting of top growth as well as galling of the root tips, frequently in a curved fashion (Jenkins and Taylor, 1967).

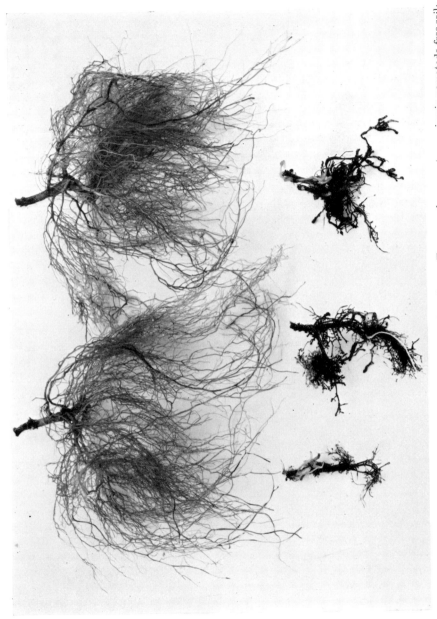

FIG. 4. The effect of *Xiphinema bakeri* feeding on the roots of raspberry. Top—raspberry roots growing in nematode-free soil; Bottom—raspberry roots growing in soil containing X. *bakeri.*

Although raspberry is a poor or non-host for *Longidorus elongatus*, sufficient feeding occurs to permit transmission of raspberry ringspot virus (RRV) and tomato black ring virus (TBRV) by the nematode. Many weed species are not only hosts for RRV and TBRV but also for *L. elongatus* and thus serve as a virus reservoir (Thomas, 1969).

There appears to be a certain specificity between virus and vector whereby strains of viruses other than those occurring naturally with the nematode species are only rarely or never transmitted (Taylor and Murant, 1969). *L. elongatus*, the natural vector of RRV and TBRV in Scotland, is able to transmit English isolates of RRV almost as efficiently but transmits English and German isolates of TBRV only occasionally. Conversely *L. attenuatus* and *L. macrosoma*, natural vectors of TBRV and RRV respectively, only rarely or never transmit the Scottish strains of these viruses. *L. elongatus* ingests several viruses which it does not transmit. Recently electron microscope studies have shown virus-like particles in the lumen of the buccal capsule and in the space between the stylet and the guiding sheath of *L. elongatus* fed on plants infected with RRV and TBRV (Taylor and Robertson, 1969). This association was found only with viruses normally transmitted by the species and it was suggested as an explanation of the specificity of virus transmission. Transmission of viruses not normally transmitted may occur as contamination in the buccal capsule.

X. americanum has been reported on several occasions associated with but not directly affecting raspberry. In a survey of raspberry in the eastern U.S.A., this species occurred most frequently and in the greatest numbers (Golden and Converse, 1965). It transmits tomato ringspot virus, a strain of which causes raspberry ringspot. This disease differs from the ringspot found in Europe and is found in only Canada and the U.S.A. There is evidence that this virus also causes a condition known as crumbly berry which reduces yield and harvestability (Keplinger, 1968). No other symptoms of plant damage due to this nematode feeding have been reported.

C. Cultural and Environmental Influences

The fact that strawberry and raspberry are vegetatively propagated and that many parasites are endoparasitic permits spread of nematodes in planting stock. Many ectoparasites are also transferred with planting stock as it is rarely completely free of soil.

Land planted to strawberry, raspberry, or other suitable nematode hosts for more than 4–5 years permits a significant increase in the nematode populations and makes establishing a new planting difficult, if not impossible. Strawberries planted following removal of an old peach or cherry orchard may be subject to *Verticillium* wilt disease. The severity of this disease increases in the presence of nematodes, especially *P. penetrans*. Apparently certain antifungal treatments applied to infested soil actually increase the wilt incidence because the treatment increases the number of *P. penetrans* (Rich and Miller, 1964).

Weeds serve as reservoirs for viruses and as hosts for many nematode species and so their control is important (Townshend and Davidson, 1960). Many nematode species may be found in the soil to a depth of 30 cm or more which makes control difficult and allows rapid reinfestation of fumigated land. Under these conditions populations are usually back to pre-treatment levels after 1–2 years. This is especially true where the previous crop was deep-rooted, as in the case of fruit trees.

The activity of nematodes infesting the above-ground portions of strawberry plants is dependent upon moisture. *Aphelenchoides* species depend upon a thin film of moisture from rain, dew or high humidity for movement and infection. *Ditylenchus dipsaci* is found in the crowns of plants and under wet conditions enters the young leaves and stems.

D. Control

Pre-plant soil fumigation is essential in many areas in order to obtain a good strawberry crop (Table III) and 1,3-dichloropropene, 1,2-dichloropropane (DD) has proved to be the most effective nematicide. A fall application of DD has given yields almost 60% greater than a similar spring application, and summer treatment gave the lowest increases (Morgan, 1964). Telone controlled both nematodes and *Verticillium* wilt, a common problem in some areas, but dibromochloropropane (DBCP) did not reduce wilt nor did it give as great a yield increase.

There are very few reports on post-plant treatments of strawberries. However, those that do appear look quite promising and suggest a need for further study. *Meloidogyne* spp. were controlled by sidedressing infested plants with DBCP (liquid or granular) up to 12 weeks after planting (Potter and Morgan, 1956; Kantzes and Morgan, 1962). A pre-plant treatment plus a post-plant treatment in the second or third year may boost yields and add an extra year or so to the life of the planting.

Freeing plants of nematodes may be accomplished in several ways. A hot-water treatment is very effective for controlling *Meloidogyne* and *Pratylenchus* species but care must be taken because the margin of safety between killing the nematode and not the plants is very narrow (McGrew and Goheen, 1954). Strawberry plants dipped for 1 min in a 1/4000 solution of iodine were freed of *Xiphinema*, and runnered and fruited well (Staniland, 1963).

TABLE III. Control of nematodes parasitizing small fruit crops

Crop	Application	Chemical	Rate	Application	Reference
Strawberry	Plant (above ground)	Parathion	(15%) 0·45 g/litre water	Spray	Raski and Allen (1948)
Strawberry	Plant (above ground)	Zinophos	(0·8%) 5·0 litres/m² (3 times/year)	Spray	Ibanenko and Metlitski (1965)
Strawberry	Plant (above ground)	Zinophos	300 ppm; 5 min	Dip	Locascio et al. (1966)
Strawberry	Plant (roots)	KI-I solution	1/4000; 1 min	Dip	Staniland (1963)
Strawberry	Plant (roots)	Water	52·8°C; 2 min	Dip	Goheen and McGrew (1954)
Strawberry	Soil (Pre-plant)	DD	169 litres/ha	Chisel	Morgan (1964)
Strawberry	Soil (Pre-plant)	DBCP	326 litres/ha	Chisel	Morgan (1964)
Strawberry	Soil (Pre-plant)	Telone	326 litres/ha	Chisel	Townshend et al. (1966)
Strawberry	Soil (Post-plant)	DBCP	(17·3% gran.) 48·4 kg/ha	Side dress	Smart et al. (1967)
Strawberry	Soil (Post-plant)	DBCP	5–9 litres/ha	Side dress	Potter and Morgan (1956)
Strawberry	Soil (Post-plant)	Zinophos	(10% gran.) 4·5 kg/ha (watered in)	Side dress	Smart et al. (1967)
Cranberry	Soil (Pre-plant)	DBCP	(67%) 34–67 litres/ha.	Drench	Bird and Jenkins (1963)
Cranberry	Soil (Pre-plant)	Zinophos	(10% gran.) 9–18 kg/ha.	Drench	Zuckerman (1964)
Blueberry	Soil (Rooting beds)	VC13	(75% EC) 112–225 litres/ha	Drench	Hutchinson et al. (1960)
Grape	Plant (Rootings)	Water	51·7°C; 5 min	Dip	Lear (1966)
Grape	Soil (Pre-plant)	Telone	360–1540 litres/ha	Chisel	Raski et al. (1965)
Grape	Soil (Pre-plant)	DD	450–1910 litres/ha	Chisel	Raski et al. (1965)
Grape	Soil (Pre-plant)	Vapam	562–843 litres/ha	Chisel	Raski et al. (1965)
Grape	Soil (Post-plant)	DBCP	22·5–28·0 litres/ha	Chisel	Raski et al. (1965)
				Flood	

Nematode-free stock plants may be obtained by: training runners above the soil on slotted stakes; rooting runners in nematode-free soil in containers placed in the row; planting infested mother plants in containers and training runner plants in a fumigated field; or by setting infested plants in very heavy soil and training the runner plants to root in the same soil. The last method depends upon the inability of the nematodes to move easily through heavy soil and the runner plants escape infection by being rooted a short distance from the infested mother plant.

Nematodes infesting the above-ground portions of the plants are most effectively controlled by roguing the planting beds and burning rogued plants. Hot-water treatment of above-ground plant parts is used in England to a limited extent but is not practical and reports from Russia state that such treatments against *Ditylenchus dipsaci* were useless (Ibanenko and Metlitski, 1965).

Parathion is about 80% effective in controlling *Aphelenchoides* species in plant buds. *A. besseyi* is controlled by a Zinophos dip (Locascio *et al.*, 1966), and this chemical sprayed three times per year on *Ditylenchus dipsaci*-infested plants gives good results (Ibanenko and Metlitski, 1965).

Little information is available concerning control of nematodes attacking raspberry, and that which is, is derived from attempts to control spread of viruses by *Xiphinema* and *Longidorus*. Fumigation with DD is effective against both genera and controls virus spread. In areas such as Scotland where *L. elongatus* causes little direct damage, efficient weed control is an inexpensive way of preventing transmission of RRV and TBRV without controlling the nematode (Murant and Taylor, 1965). Aqueous extracts from raspberry canes and roots have a nematicidal activity against *L. elongatus*. Much of the information regarding nematode control of strawberry is applicable to raspberry.

E. Economics

Strawberry is a high value crop and therefore will stand the high cost of soil fumigation. A few such tests are reported here to demonstrate the economic soundness of such treatment in different geographical locations. It is difficult to relate the fumigant effect directly to nematode kill, especially when other soil organisms are affected also by fumigation, but numerous tests suggest that nematode control is essential for increased fruit production.

Horn *et al.* (1956) obtained good control of *Meloidogyne hapla* with ethylene dibromide (EDB) at 33·6 litres/ha. By planting nematode-free strawberries in fumigated soil an increase of 323 crates/ha was obtained.

During that year the average price per crate was $5.00, which gave the grower an increase of about $1612.00/ha. This increase paid for the cost of fumigation (about $175.00/ha) plus a sizable profit. Fumigation also affected plant establishment resulting in no loss and good growth.

Trizone at 224 kg/ha controlled *M. hapla* and gave economical increases in strawberry yields (Riggs and Hamblen, 1962). The root-knot rating was reduced by 86%, plant and fruit production increased by 214% and 56% respectively. While there was a slight decrease in yield per plant the tremendous increase in number of plants more than compensated for this. There was an average increase of 4014 litres/ha. With an average price of 20 cents/litre there was an increase arising from fumigation of $788.00/ha, less $175.00/ha for fumigation costs, or $613.00/ha.

Townshend *et al.* (1966) in Ontario (Canada) using Telone at 326 litres/ha reduced *P. penetrans* in the soil by 92% and the number of straw-berry plants with *Verticillium* wilt by 40%. The stand of plants was increased by almost 84% giving an increase in fruit production of 11027 litres/ha. The cost of applying Telone at $188.00/ha is more than covered by this return.

Fall treatment with DD significantly reduced populations of *P. penetrans*, amount of root-rot, and gave an additional 7164 litres of straw-berries per hectare or $1750.00/ha compared to an untreated planting (Morgan, 1964). This treatment gave 42% and 50% increases in yield over spring and summer treatments respectively.

In areas where NEPO (nematode-transmitted polyhedral) viruses are important, nematode control has given increased returns. Harrison *et al.* (1963) using DD at 844 mg/m^2 obtained a 99% kill of *Xiphinema diversicaudatum*, an almost 100% reduction in AMV infection of straw-berry and an increase in plant height of 40%.

DD and Quintozene have been effective in controlling *Longidorus elongatus* and RRV and TBRV (Taylor and Murant, 1968). Rates of 448 and 67·2 kg/ha respectively killed 95% of the nematodes and increased strawberry yields by 130% in the third year following treat-ment. Studies in the Netherlands have shown that yield reduction due to direct nematode damage may occur from densities of 2 nematodes per 10 g of soil upwards. Therefore, treatment even of low level infesta-tions appears to be of economic benefit. Quintozene is easier and less expensive to use, being almost 20% cheaper than DD.

Very little recent information is available concerning the economic importance of the above-ground nematode diseases. Spring dwarf would be economically important if it were not for its limited distri-bution. Reports in 1938 and 1939 estimated reduction in fruit yields in Massachusetts (U.S.A.) to be between 60 and 70%. While this was an extreme case, due to importation of infested planting stock, it points to the potential of this disease.

Summer dwarf is much more of a problem. Plakidas (1964) reported that dwarf plants yielded 10·7% less per plant than healthy plants. He stated that in 1930 in Louisiana the average percentage of dwarf plants was 10% giving a decrease in yield for the average planting of about 1–2%. This disease is more important in areas of mild winters such as Florida (U.S.A.). In 1929 the loss from this disease was 0–75% in individual fields. Fruit produced by infested plants is worthless, as it comes on late and is of poor quality.

The damage to strawberries by the above-ground nematodes is much less now than when many of the above reports were made; this is due to persistent roguing and selection. However where these practices are not rigorously carried out severe damage still occurs, as is evidenced by reports from strawberry areas in Europe and the U.S.S.R.

Since most of the work with raspberry has been concerned only with surveys there is little information on pathogenicity and nothing on effect of control on yields.

III. Cranberry and Blueberry

A. Crop Production

1. Cranberry (*Vaccinium oxycoccus macrocarpus*)

The cranberry is a highly specialized crop, and its production is restricted mainly to acid soils along the edges of streams, on seashores and in bogs of temperate North America. The cool moist conditions in these areas are ideal for growth. Production is concentrated in the U.S.A. to the northeastern states, Wisconsin and the Pacific Northwest, and in Canada to parts of British Columbia, Quebec and Nova Scotia. Cranberries are used primarily for making into sauce and recently a sizable market has developed for cranberry juice and cocktail.

2. Blueberry (*Vaccinium cyanococcus* spp.)

The true blueberries belong to the genus *Vaccinium* which is native to North America (Varney and Stretch, 1967). The blueberry industry, limited to the United States and Canada, is based on the harvest of three distinct types within the genus *Vaccinium*. Lowbush types, usually 15–45 cm high and represented by *V. angustifolium*, *V. myrtilloides* and *V. brittonii*, are grown in the northeastern U.S.A. and Canada. More than 40,500 ha (1966 figures) of the lowbush species are under cultivation in the U.S.A. with the bulk of the crop going to processing plants.

Highbush types, 1 m high or more, have developed mainly from *V. austrate* and *V. corymbosum*, and are grown primarily in New Jersey, Michigan, North Carolina, Washington (U.S.A.) and southwestern British Columbia (Canada), and to a lesser extent in Indiana, Ohio, Pennsylvania, New York, and Massachusetts (U.S.A). Over 8,100 ha are planted to this type in the U.S.A.

The third blueberry type, *V. ashei*, or the rabbiteye blueberry, attains heights of 9 m and is native to North Carolina, Georgia, northwestern Florida, Alabama, Mississippi, and Louisiana (U.S.A.). More than 1418 ha of this species are in cultivation on upland soils which are unsuitable to the highbush types.

B. Nematode Diseases

1. Cranberry

At least fourteen species of nematodes are known to be parasitic on cranberry, but only five of these are pathogenic (Zuckerman, 1962; Bird and Jenkins, 1964). *Trichodorus christiei*, *Hemicycliophora similis*, *Pratylenchus penetrans*, *Criconemoides curvatum* and *Tetylenchus joctus* all reduce the number of runners formed, total runner length, and fresh top and root weight of cranberry plants. *C. curvatum* does not affect the growth of mature runners. *P. penetrans* is parasitic on cranberry in the laboratory, but is not reported so in the field. Several *Hemicycliophora* and *Helicotylenchus* species were found associated with diseased plants in Wisconsin (U.S.A.) but pathogenicity was not demonstrated (Barker and Boone, 1966).

Nematode distribution in bogs is not uniform but occurs in heavily-infested "pockets" with the concentration gradient perpendicular to the soil surface (Zuckerman *et al.*, 1964). Two seasonal population peaks occur in July and November in Massachusetts (U.S.A.). Total populations decline and nematodes migrate downward during dry periods.

2. Blueberry

As with cranberries, nematodes associated with blueberries have been studied only in the eastern U.S.A. and eastern Canada. Goheen and Braun (1955), in the first survey of highbush blueberries, reported several genera from Maryland, Massachusetts and North Carolina (U.S.A.). Hutchinson *et al.* (1960) reported five genera associated with blueberries in New Jersey and Zuckerman and Coughlin (1960) reported two genera in Massachusetts. Plant-parasitic nemas from twelve genera were found in Michigan blueberry fields (Tjepkema *et al.*, 1967) and one genera (*T. joctus*) in Indiana (U.S.A.). A *Helicotylenchus* species was found in constant association and feeding on non-woody roots of lowbush blueberry in New Brunswick (Canada) (Morgan and Wood, 1962).

Trichodorus christiei, which appears to be the most important species, reduced root growth of cuttings by an average of 67% (range 20–80%) (Zuckerman, 1962). Cuttings attacked by the nematode form a normal callus, but the parasitized new roots fail to grow and the cutting eventually dies. *T. joctus* parasitizes cuttings and seedlings but does not cause easily recognizable symptoms or consistent root growth reduction. *Trichodorus* has less of an effect on larger bushes and would likely be a problem only where growth was limited by other factors. Formation of small terminal root galls and reduced root growth is caused by *Hemicycliophora similis* under laboratory conditions (Zuckerman, 1964a), but little is known of its economic importance in commercial cutting beds and plantings. The blueberry root-knot nematode, *M. carolinensis,* is widespread in North Carolina (U.S.A.) but no report is available regarding pathogenicity (Fox, 1967). *Pratylenchus penetrans* was studied under sterile conditions but found to be non-pathogenic on lowbush blueberry seedlings (McCrum and Hilborn, 1962). These nematodes appear not to cause much direct damage but their importance may lie in providing a means of entry for other soil organisms. Thus, *Xiphinema americanum* does not appear to have any direct economic effect on blueberry but transmits tobacco ringspot virus which is the cause of blueberry necrotic ringspot disease (Griffin *et al.,* 1963).

C. Cultural and Environmental Influences

The unique culture condition of cranberries which allows for prolonged periods of flooding apparently has no adverse effect on nematodes normally found there. Five species of stylet-bearing nematodes were recovered from soil in a bog flooded for 2 years, in numbers similar to those in a bog under normal culture (Bird and Jenkins, 1965). Nematode populations and dissolved-oxygen concentrations remained constant for 145 days in flooded glass crocks containing rooted cranberry cuttings.

D. Control

There is relatively little information concerning nematode control and effect on yields of cranberries. Bird and Jenkins (1963) in a 1 year study obtained a greater number of uprights and fruit buds using DBCP (67% v/v) at 34, 45 and 67 litres/ha and Zinophos (10% granular) at 9, 18 and 36 kg/ha. Zinophos at 36 kg/ha gave 93% nematode kill, 38% uprights and 44% more fruit buds than untreated checks. While no phytotoxicity was observed with either material at these rates, Zuckerman (1964b) in a 3 year study reported a phytotoxic effect with Zinophos at 36 kg/ha which caused reddening of the vines and a significant yield reduction. His results suggested that a rate of 18 kg actual/ha of 10% granular Zinophos applied in the spring every other year would increase the yield of cranberry vines.

Since nematode attack of blueberries is most severe in the cutting beds, sanitary measures must be taken to prevent contamination. Steamed or fumigated soil should be used. If high populations of harmful nematodes are found in a new planting site a pre-plant fumigation with EDB or DD is recommended (Jenkins, 1961). Infected established plants may be side dressed with DBCP.

E. Economics

Sufficient information is not available to make any meaningful statement regarding the economics of controlling nematode-induced diseases of either cranberry or blueberry.

IV. GRAPE *(Vitis vinifera)*

A. Crop Production

A little over one half of the world's vineyards are in Europe where there are 6·6 million hectares. Of these there are 1·8 million hectares in Spain, 1·7 in Italy, and 1·4 in France. Other areas of the world producing grapes include Africa, Australia, U.S.A., Argentina, Chile, U.S.S.R., and Turkey.

The majority of the acreage is devoted to the culture of wine grapes with the remainder being divided between grapes grown for fresh market, and for raisins. Grapes require a wet winter and a hot, mainly dry summer.

B. Nematode Diseases

The importance of parasitic nematodes on grape production has only been appreciated within the last 15 years. Recently surveys have been conducted in most grape producing areas in the world in an attempt to assess the amount of damage caused by nematodes.

Several species of root-knot nematodes, *Meloidogyne* spp., parasitize grape and differ in importance depending upon the geographic location. In California *M. incognita*, *M. javanica*, and *M. thamesi* occur on grape but the first is the most common and important (Raski *et al.*, 1965). Grapevines in Victoria, Australia appear to be infected exclusively by

M. javanica, while *M. incognita* and *M. hapla* occur on crops in other areas (Meagher, 1969).

This nematode induces gall production in young rootlets and heavy infections may completely destroy the root systems of young plants. Certain species stimulate the production of new rootlets above the infection site resulting eventually in an abnormally branched "hairy root" condition.

Pratylenchus vulnus is the most prevalent and destructive of the lesion nematodes in California and Australia, and has been associated with grape in other parts of the world. Damage usually is more severe on heavier soils but it occasionally occurs in lighter ones. Nematode feeding and root penetration causes lesions which often girdle large roots and the infested plants have severely reduced root systems, loss of vigor and reduced production. Young plants in infested soil remain weak and may die because of their inability to produce an adequate root system.

Xiphinema americanum frequently is associated with grape but its effect is not completely understood. *X. index*, however, is highly pathogenic to grapevine, causing gross malformations to rootlets and in high populations killing seedlings outright (Raski *et al.*, 1965). Feeding of the nematode incites swelling, distortion, discoloration and killing of the growing points of feeder roots. The limited distribution of this species in northeastern Victoria, Australia, has been related to the importing of *Phylloxera* resistant rootstock from southern France (Meagher, 1969). A similar situation may exist in California (U.S.A.) where the species is distributed most commonly in coastal areas but also occurs in the San Joaquin Valley.

Of equal importance is the fact that *X. index* transmits the fanleaf, yellow mosaic, veinbanding virus (GVFV-GVYMV) disease complex of grape. The nematode is not only responsible for virus spread within a field, but also for reinfection of new grape plantings on old virus infected grape land (Raski *et al.*, 1965).

The citrus nematode, *Tylenchulus semipenetrans* is common on commercial citrus in most areas of the world where the crop is grown and the proximity of vineyards to citrus orchards influences grapevine infestation. Where grapes and citrus are grown in separate areas, such as in California (U.S.A.), this species is found on grape in only a few localities. In Australia, where citrus groves are found interspersed throughout vineyards, *T. semipenetrans* is the most commonly occurring parasite on grape (Meagher, 1969).

In greenhouse tests grape was a good host for *T. semipenetrans* resulting in very high populations (Raski *et al.*, 1965). Heavily infested roots are discolored, owing to the numerous lesions formed by nematode invasion. These roots generally lack vigor and are smaller than non-infected ones. The effect on grape under field conditions has not been completely evaluated.

Other parasitic nematodes such as *Criconemoides* spp., and *Paratylenchus* spp., occur in large numbers on the roots and in the soil around the roots of grape, but their effect is obscure.

C. Cultural and Environmental Influences

Nematode problems can arise in vineyards in several ways and certain cultural practices can either lessen or enhance the situation. A heavy infestation in an established vineyard progressively weakens the vines and reduces yields each year. This situation may be temporarily alleviated by providing increased moisture, fertilizer and a change in pruning practices to reduce the strain on the plants. In most cases, and especially in light soils, these vines eventually become unproductive and must be replaced.

If a vineyard is to follow vines or fruit trees, an interval of 2 or more years in an annual, non-host crop aids in reducing nematode populations and allows time for breakdown of old infested roots. If the previous grapevines were infected with the fanleaf virus and *X. index*, at least a 5-year rotation is suggested to avoid infestation of subsequent plantings (Raski *et al.*, 1965). Virus-infected roots left in the soil after removal of old plants remain alive for several years, providing food and a source of virus inoculum for the nematodes. This species has been recovered as deep as 360 cm in an old vineyard under fall-rotation for 4·5 years. Fanleaf virus was also recovered from root pieces after the same length of time.

Even if the soil is treated before planting the use of infested stock nullifies that treatment. Endoparasitic nematodes such as *Meloidogyne*, *Pratylenchus*, and *Tylenchulus* are readily spread on infested rootstock.

Another means by which nematodes are spread is by the re-use of drainage water. In many irrigated areas of the world it is common practice to return drainage water to the irrigation supply. Eight genera of parasitic nematodes were found in the irrigation water entering a vineyard in Australia and in 6 years, occurrence of *T. semipenetrans* in vineyard samples increased from 14% to 98% (Meagher, 1969).

D. Control

The most effective chemicals for pre-plant soil treatments contain 1,3-dichloropropene (1,3-D) as the principal active ingredient. Table III lists the chemicals containing this active ingredient and their recommended rates (Raski *et al.*, 1965). The wide range of rates listed is necessary because of variations in previous cropping and in soil type. The higher rates are necessary for nematode control in old vineyards or orchard soils. Medium rates are for heavy soils and lower rates for light soils previously used only for annual crops.

The most effective treatment of an old vineyard is obtained when soil fumigation follows a 2- or 3-year rotation with a non-susceptible crop. If the fanleaf virus is present the rotation should be extended to 5 years.

Post-plant treatment may be warranted if established vineyards show signs of decline and reduced productivity. In many instances it may be necessary and economic to follow a pre-plant fumigation 2–3 years later with a post-plant treatment. The chemical 1,2-dibromo-3-chloropropane (DBCP) can be applied to established plants without phytotoxic effects.

Three methods of application are available: chisel injection, flood irrigation, or sprinkler irrigation. While good results have been reported with chisel injection in Australia (Meagher, 1969), application in flood irrigation water appears to be more reliable (Raski and Schmitt, 1964). A rate of 22·5 litres active ingredient per hectare should be applied in sufficient irrigation water to penetrate 1·2–1·8 m. It is essential that nematode-free planting stock be used to avoid re-contamination of treated soil. To insure this, rootings should come from fumigated nurseries. Rootings infested with either *Meloidogyne* spp. or *Pratylenchus* spp. may be cleaned by use of a hot-water dip (Lear, 1966). Only strong, completely dormant rootings should be used and they should be cooled immediately after treatment.

The use of resistant varieties is the ultimate answer in dealing with nematode problems. Several varieties are available which show resistance to *Meloidogyne, Pratylenchus* and *Xiphinema*. The use of resistant rootstocks however requires careful management as these stocks differ considerably in vigor and in their effects on fruit quality and vine behavior.

Dogridge (*Vitis champini*) and Salt Creek (*Vitis champini*) varieties are both resistant to all three of the above nematodes. Solonis × Othello 1613 is resistant to most *Meloidogyne* species but not to *Pratylenchus* or *Xiphinema*.

E. Economics

Grapevines may remain productive for 30 years or more under favorable conditions. However, declines in growth and yield due to nematodes are a cause of major economic losses to grape growers. Removal of infected vines followed by fumigation and replanting is an expensive operation. Even after such measures there is little guarantee of long-range effectiveness because there is rapid re-invasion by nematodes into fumigated replant soils. Re-invasion comes from nematodes, especially *Meloidogyne* spp., which have been protected from fumigant action in non-decomposed roots deep in the soil.

It is for these reasons that much work has been done with post-plant treatments in an attempt to increase productivity of established vineyards. Warren (1960) treated established Tokay grapes with EDB (67·2 kg/ha) or DBCP (33·6 kg/ha) injected into the irrigation water. The top growth of vines treated with DBCP and EDB was 267% and 283% respectively greater than that of the untreated the season following treatment. There was also a 63% increase in total weight of the fruit produced by vines treated with either chemical.

In a 50-year-old Tokay vineyard, California (U.S.A.) workers (Raski and Schmitt, 1964) obtained an increase of 11424 kg/ha the second year after applying DBCP at 28 litres/ha in irrigation water. That year grapes sold for about $95.00 per 906 kg which gave an increase of almost $1198.00/ha, more than covering the $99.00/ha cost of fumigation. Populations of *X. index* (the predominant species) and *M. incognita* the second year after treating were only about 2% of the untreated populations.

While California workers had little success with chisel application in established vineyards, reports from Australia (Meagher, 1969) give evidence of growth and yield response using this method. Applying DBCP to Sultana vines at 28 litres/ha increased yield of dried fruit by 16%. This amounted to a $233.00/ha increase which more than paid for the cost of treatment at $99.00/ha. *M. javanica*, *T. semipenetrans* and *P. vulnus* were all present in this vineyard.

Experimental and survey results in Australia have shown that plant growth and yield increases do not necessarily indicate that treatment is economic. To be of economic benefit treatment must raise the total yield to a level greater than that required to cover estimated costs of production. In Australia this is equivalent to 2688 kg/ha of dried fruit.

Such observations have led workers to devise a formula as a guide to growers for recommendation of fumigation of mature vines. It is based on average fruit yields before treatment. If yields are 4480 kg/ha of dried fruit the long term economic prospects from fumigation are good; 3360–4480 kg, fair; 2240–3360 kg, poor; and 2240 kg, very poor. Such a formula can only serve as a guide as there are many factors involved in bringing about yield increase in grape production. However, a guide of this type is a step toward putting nematode control on an economic basis.

References

Tree Fruits

Adams, R. E. (1955). *Phytopathology* **45**, 477–479.

Anon. (1970). *Down to Earth* **26**, 22.

Barker, K. R. and Clayton, C. N. (1969). *Phytopathology* **59**, 1017.

Bosher, J. E. and Orchard, W. R. (1963). *Can. J. Pl. Sci.* **43**, 193–199.

Colbran, R. C. (1953). *Aust. J. agric. Res.* **4**, 384–389.

Cropley, R. (1961). *Ann. appl. Biol.* **49**, 530–534.

Day, L. H. and Tufts, W. P. (1944). *Univ. Calif. agric. Exp. Stn Circ.* **359**, 16 pp.

Decker, H. (1960). *Nematologica Suppl.* **2**, pp. 68–75.

Edgerton, L. J. and Parker, K. G. (1958). *Amer. Soc. hort. Sci. Proc.* **72**, 134–138.

Egunjobi, O. A. (1968). *N.Z. Jl agric. Res.* **11**, 386–406.

Fisher, J. M. (1967). *Aust. J. agric. Res.* **18**, 921–929.

Fliegel, P. (1969). *Phytopathology* **59**, 120–124.

French, A. M., Lownsbery, B. F., Ayoub, S. M., Weiner, A. C. and El-Gholl, N. (1964). *Hilgardia* **35**, 603–610.

Fritzsche, R. and Kegler, H. (1964). *Naturwiss.* **51**, 299.

Hoestra, H. (1968). *Meded. LandbHogesch Wageningen* **68**, 13.

Hoestra, H. and Oostenbrink, M. (1962). *Neth. J. agric. Sci.* **10**, 286–296.

Hung, C. L. P. and Jenkins, W. R. (1969). *J. Nematol.* **1**, 12 (Abstr.).

Jenkins, W. R. and Taylor, D. P. (1967). "Plant Nematology." Reinhold, New York.

Jensen, H. J. (1953). *Pl. Dis. Reptr* **37**, 384–387.

Kable, P. F. and Mai, W. F. (1968a). *Nematologica* **14**, 101–122.

Kable, P. F. and Mai, W. F. (1968b). *Nematologica* **14**, 150.

Kirkpatrick, J. D., Mai, W. F., Parker, K. G. and Fisher, E. G. (1964). *Phytopathology* **54**, 706–712.

Knierim, J. A. (1964). *Q. Bull. Mich. St. Univ. agric. Exp. Stn* **46** 527–532.

Lownsbery, B. F. (1956). *Phytopathology* **46**, 376–379.

Lownsbery, B. F. (1959). *Pl. Dis. Reptr* **43**, 913–917.

Lownsbery, B. F. and Sher, S. A. (1958). *Calif. Agric.* **12**, 7–12.

Lownsbery, B. F., Hart, W. H. and Martin, G. C. (1969). *Diamond Walnut News* **51**(1), 5–6.

Lownsbery, B. F., Mitchell, J. T. and Parocer, S. M. (1964). *Hilgardia* **35**, 611–614.

Lownsbery, B. F., Mitchell, J. T., Hart, W. H., Charles, F. M., Gerdts, M. H. and Greathead, A. S. (1968). *Pl. Dis. Reptr* **52**, 890–894.

McBeth, C. W. and Taylor, A. L. (1944). *Proc. Am. Soc. hort. Sci.* **45**, 158–166.

McDonald, D. H. and Mai, W. F. (1963). *Phytopathology* **53**, 730–731.

Maeseneer, J. De. (1962). *Nematologica* **7**, 13.

Mai, W. F. (1960). *Phytopathology* **50**, 237.

Mai, W. F. and Parker, K. G. (1967). *Pl. Dis. Reptr* **51**, 398–401.

Mai, W. F. and Parker, K. G. (1970). *N.Y. State hort. Soc. Proc.* **115**, 207–209.

Mai, W. F., Parker, K. G. and Hickey, K. D. (1970). *Pl. Dis. Reptr* **54**, 792–795.

Milbrath, J. A. and Reynolds, J. E. (1961). *Pl. Dis. Reptr* **45**, 520–521.

Mountain, W. B. and Patrick, Z. A. (1959). *Can. J. Bot.* **37**, 459–470.

Nigh, E. L. (1966). *Phytopathology* **56**, 150.

Nyland, G. (1955). *Pl. Dis. Reptr* **39**, 573–575.

Nyland, G., Lownsbery, B. F., Lowe, S. K. and Mitchell, J. F. (1969). *Phytopathology* **59**(8), 1111–1112.

Oostenbrink, M. (1954). *Verst. Plantenziekt. Dienst. Wageningen* **124**, 196–233.

Palmiter, D. H., Braun, A. J. and Keplinger, J. A. (1966). *Pl. Dis. Reptr* **50**, 877–881.

Parker, K. G. and Mai, W. F. (1956). *Pl. Dis. Reptr* **40**, 694–699.

Parker, K. G., Mai, W. F., Oberly, G. H., Brase, K. D. and Hickey, K. D. (1966). *Bull. N.Y. St. Coll. Agric.* **1169**, 19 pp.

Pitcher, R. S., Patrick, Z. A. and Mountain, W. B. (1960). *Nematologica* **5**, 309–314.

Savory, B. M. (1966). *Commonwealth Bur. Hort. and Plantation Crops, E. Malling Res. Rev.* No. 1, 64 pp.

Serr, E. F. and Day, L. H. (1949). *Proc. Am. Soc. hort. Sci.* **53**, 134–140.

Sharpe, R. H., Hesse, C. O., Lownsbery, B. F., Perry, V. G. and Hansen, C. J. (1969). *J. Am. Soc. hort. Sci.* **94**, 209–212.

Stokes, D. E. (1967). *Nematologica* **13**(1), 153 (Abstr.).

Teliz, D., Lownsbery, B. F., Grogan, R. G. and Kimble, K. A. (1967). *Pl. Dis. Reptr* **51**, 841–843.

Townshend, J. L. and Davidson, T. R. (1960). *Can. J. Bot.* **38**, 267–273.

Wilson, J. D. and Hedden, O. K. (1967). *Ohio Agric. Res. Circ.* **153**, 12 pp.

Strawberry and Raspberry

Frazier, N. W. and Maggenti, A. R. (1962). *Pl. Dis. Reptr* **46**, 303–304.

Goheen, A. C. and McGrew, J. R. (1954). *Pl. Dis. Reptr* **38**, 818–826.

Golden, A. M. and Converse, R. (1965). *Pl. Dis. Reptr* **49**, 987–991.

Goodey, J. B. (1951). *Ann. appl. Biol.* **38**, 618–623.

Griffin, G. D., Anderson, J. L. and Jorganson, E. C. (1968). *Pl. Dis. Reptr* **52**, 492–493.

Harrison, B. D. (1964). *Virology* **22**, 544–550.

Harrison, B. D., Mowat, W. P. and Taylor, C. E. (1961). *Virology* **14**, 480–485.

Harrison, B. D., Peachey, J. E. and Winslow, R. D. (1963). *Ann. appl. Biol.* **52**, 243–255.

Horn, N. H., Martin, W. J., Wilson, W. F. and Giamalua, M. J. (1956). *Pl. Dis. Reptr* **40**, 790–797.

Ibanenko, B. and Metlitski, O. (1965). *Zashch. Rast. at Vredit. Bolez.* **2**, 20–23.

Jenkins, W. R. and Taylor, D. P. (1967). "Plant Nematology." Reinhold, New York.

Jha, A. and Posnette, A. F. (1959). *Nature, Lond.* **184**, 962–963.

Kantzes, J. G. and Morgan, O. D. (1962). *Phytopathology* **54**, 164.

Keplinger, J. A. (1968). *Pl. Dis. Reptr* **52**(5), 386–390.

Locascio, S. J., Smart, Jr. G. C. and Marvel, M. E. (1966). *Proc. Fla St. hort. Soc.* **79**, 170–175.

Lear, B. (1966). *Pl. Dis. Reptr* **50**, 858–859.

Lister, R. M. (1964). *Ann. appl. Biol.* **54**, 167–176.

McGrew, J. R. and Goheen, A. C. (1954). *Down to Earth* **10**, 4–5.

Morgan, G. T. (1964). *Can. J. Pl. Sci.* **44**, 170–174.

Murant, A. F. and Taylor, C. E. (1965). *Ann. appl. Biol.* **55**, 227–237.

Plakidas, A. G. (1964). "Strawberry Diseases." Louisiana State Univ. Press, Baton Rouge. 195 pp.

Potter, H. S. and Morgan, O. D. (1956). *Pl. Dis. Reptr* **40**, 187–189.

Raski, D. J. and Allen, M. W. (1948). *Calif. Agric.* **2**, 23–24.

Rich, S. and Miller, P. M. (1964). *Pl. Dis. Reptr* **48**, 246–248.

Riggs, R. D. and Hamblen, M. L. (1962). *Down to Earth* **18**, 5–7.

Seinhorst, J. W., Klinkenberg, C. H. and van der Meer, F. A. (1956). *Tijdschr. PlZiekt.* **62**, 5–6.

Smart, Jr., G. C., Locascio, S. J. and Rhoades, H. L. (1967). *Nematologica* **13**, 152–153.

Staniland, L. N. (1963). *Pl. Path.* **12**, 91.

Taylor, C. E. (1962). *Virology* **17**, 493–494.

Taylor, C. E. (1967). *Ann. appl. Biol.* **59**, 275–281.

Taylor, C. E. and Murant, A. F. (1968). *Pl. Path.* **17**, 171–178.

Taylor, C. E. and Murant, A. F. (1969). *Ann. appl. Biol.* **64**, 43–48.

Taylor, C. E. and Robertson, W. M. (1969). *Ann. appl. Biol.* **64**, 233–237.

Taylor, C. E. and Thomas, P. R. (1968). *Ann. appl. Biol.* **62**, 147–157.

Thomas, P. R. (1969). *Pl. Path.* **18**, 23–28.

Townshend, J. L. (1963). *Can. J. Pl. Sci.* **43**, 75–78.

Townshend, J. L. and Davidson, T. R. (1960). *Can. J. Bot.* **38**, 267–273.

Townshend, J. L., Ricketson, C. L. and Wiebe, J. (1966). *Can. J. Pl. Sci.* **46**, 111–114.

Treshow, M. and Norton, R. A. (1958). *Utah agric. Exp. Stn Circ.* **140**, 16 pp.

van der Meer, F. A. (1965). *Neth. J. Pl. Path.* **7**, 33–46.

Cranberry and Blueberry

Barker, K. R. and Boone, D. M. (1966). *Pl. Dis. Reptr* **50**, 957–959.

Bird, G. W. and Jenkins, W. R. (1963). *Phytopathology* **53**, 347.

Bird, G. W. and Jenkins, W. R. (1964). *Phytopathology* **54**, 677–680.

Bird, G. W. and Jenkins, W. R. (1965). *Pl. Dis. Reptr* **49**, 517–518.

Fox, J. A. (1967). *Diss. Abstr.* **28**, 1311–1312.

Goheen, A. C. and Braun, A. J. (1955). *Pl. Dis. Reptr* **39**, 908.

Griffin, G. D., Huguelet, J. E. and Nelson, J. W. (1963). *Pl. Dis. Reptr* **47**, 703.

Hutchinson, M. T., Reed, J. P. and Race, S. R. (1960). *New Jers. Agric.* **42**(4), 12–13.

Jenkins, W. R. (1961). *Proc. 29th Ann. Blueberry Open House*, **29**, 6.

McCrum, R. C. and Hilborn, M. T. (1962). *Pl. Dis. Reptr* **46**, 84–85.

Morgan, G. T. and Wood, G. W. (1962). *Pl. Dis. Reptr* **46**, 800.

Tjepkema, J. P., Knierim, J. A. and Knobloch, Natalie (1967). *Mich. St. Univ. agric. Exp. Stn Quart. Bull.* **50**, 37–45.

N

Varney, E. H. and Stretch, A. W. (1967). "Blueberry Culture." Rutgar University Press.

Zuckerman, B. M. (1962). *Phytopathology* **52**, 1017–1019.

Zuckerman, B. M. (1964a). *Pl. Dis. Reptr* **48**, 170–171.

Zuckerman, B. M. (1964b). *Pl. Dis. Reptr* **48**, 172–175.

Zuckerman, B. M. and Coughlin, J. W. (1960). *Mass. agric. Exp. Stn Bull.* **521**, 1–18.

Zuckerman, B. M., Khera, S. and Pierce, A. R. (1964). *Phytopathology* **54**, 654–659.

Grape

Hewitt, W. B., Raski, D. J. and Goheen, A. C. (1958). *Phytopathology* **48**, 586–595.

Lear, B. (1966). *Pl. Dis. Reptr* **50**, 858–859.

Meagher, J. W. (1969). Nematodes and their control in vineyards in Victoria, Australia. *International Pest Control* **11**, 14–18.

Raski, D. J. and Schmitt, R. V. (1964). *Am. J. Enol. Vitic.* **15**, 199–203.

Raski, D. J., Hart, W. H. and Kasimatis, A. N. (1965). *Calif. agric. Exp. Stn Exten. Cir.* 533.

Warren, L. E. (1960). *Down to Earth* **15**, 13–16.

16

Nematode Pests of Vegetable and Related Crops

H. J. Jensen

Department of Botany and Plant Pathology
Oregon State University
Corvallis, Oregon, U.S.A.

Introduction

Vegetables are by far the largest constituent of the diet for most humans although proportions vary in different countries of the world. The population of many underdeveloped countries depends almost entirely upon vegetables, while others more fortunate utilize large quantities of animal protein. Vegetable production in virtually every part of the world is impaired to a greater or lesser extent by nematode pests.

I. Historical Resumé of Nematode Pests of Vegetables

The first record of nematode injury to vegetables was that of root-knot nematode damage to cucumber in an English greenhouse (Berkeley, 1855). Nearly 23 years passed before Kühn (according to Beijerinck, 1883) observed *Ditylenchus dipsaci* as a probable cause of "onion bloat" in Germany. Beijerinck (1883) in Holland and Chatin (1884) in France reported on this disease and by the turn of the century this pest was known in several areas of western Europe. By then its host range included such vegetables as bean, cabbage, carrot, onion, parsnip, radish, tomato, watermelon and yam. Before the turn of the century, cyst nematodes had become serious pests of certain crops and Liebscher (1890) reported a nematode, later called *Heterodera göttingiana*, on peas.

From the turn of the century until the early 1950s additional pests were discovered but most progress was achieved by discoveries which expanded host ranges and distribution records of known pests. Recent advances in nematology associated with vegetable production are: (*a*) recognition that nematodes which rarely invade host tissues and which feed from the plant exterior (ectoparasites) are important plant pests; (*b*) evidence that nematodes are vectors of certain soil-borne viruses; and (*c*) that interaction of nematodes with other disease pathogens (usually bacteria and fungi) often results in a more serious disease than caused by either organism alone.

II. Economic Importance of Nematodes

In more recent years nematode diseases of vegetables have increased greatly in economic importance and have sometimes come near to catastrophic proportions. An almost complete lack of world-wide data of nematode injury to vegetable crops makes an economic appraisal impractical. Aside from the catastrophic example of the "yellows disease" of black pepper in Indonesia (van der Vecht, 1950; Christie, 1959), such information consists of local situations or survey reports.

Recently a committee of the Society of Nematologists under the chairmanship of Dr Julius Feldmesser has made public an estimate of crop damage caused by nematodes in the U.S.A. (Anon, 1971). It is reported that vegetable crops suffer an 11% or $266,989,100 annual loss due to nematodes. This would account for an average loss of $132.57 per hectare for vegetable crops in the U.S.A. The greatest losses on a dollar basis in order of severity are from tomato, bean, cucumber, cantaloup and carrot. On a percentage of crop loss basis, bean, Brussels sprouts, carrot, cucumber and melon are given as 20% and green pepper and tomato as 15%. Other situations are not as well known, but, nevertheless, a considerable decrease in yield of vegetable crops also is caused by interrelationships of nematodes with other plant pathogens such as bacteria and fungi.

III. Ecological Relationships of Nematodes and Cultural Practices

Most vegetable crops are grown in soil, the exception being the hydroponic-type cultures of tomato in Hawaii and in various South Pacific areas. Although the origin of nematode pests, primarily root-knot nematodes, in the hydroponic system is not known, it has been assumed that water which made up the nutrient solution was contaminated. In most instances, however, soil is the common abode for both vegetable hosts and nematode parasites where there is ample opportunity for infection. Usually longer exposure periods synchronized by a continuous food supply from host plants (biennials, perennials, or cyclic growing periods with monoculture or a series of susceptible crops) favor an increase of parasites and result in more severe infections.

Infection sites of most vegetables are normally beneath the soil surface except for *Ditylenchus dipsaci* infections of the garlic and onion family where bulbs, leaves and seed may be infected. Bulbs, rhizomes, roots, tubers or other subterranean portions of plants are invaded by nematodes and infected tissue makes such products undesirable as marketable food items. In other instances the foliage or the root systems are invaded and this has an indirect effect upon the marketability of the crop through a reduction in functional metabolism which ultimately reduces quality and quantity of produce.

Cultural methods used in producing various vegetable crops often contribute to dissemination of nematode pests. Practices which move propagation stocks (bulbs, cuttings, rhizomes, roots, seeds, transplants, and tubers) may spread nematode pests over wide areas either in attached soil or in plant tissues. Other practices of cultivation, irrigation, etc., often distribute crop debris, soil and water containing nematodes. Vegetable processing and storage also may contribute to dissemination of nematode pests by failure to dispose of infected plant tissue or infested soil or water. The close proximity of individual plants in bed or row culture may be an important factor in above- and below-ground spread of nematode pests. Most vegetables grow rapidly in soils with an adequate level of fertility and moisture and this environment of active root growth is ideal for nematode population development and spread. Harvesting vegetable crops may contribute additional opportunities for nematode dissemination by additional distribution or by mixing the diseased produce and crop debris with marketable items.

IV. Distribution of Nematodes in Climatic Zones

Many nematodes are cosmopolitan in distribution, while others appear to be restricted to specific climatic zones. Obviously, it is not the sole factor limiting distribution because of the obligate nature of the parasitic nematodes' association with host plants. Climate, however, affects the environment of the nematode and that of its host plant.

Some nematodes, such as *Radopholus similis*, *Rotylenchulus reniformis* and *Scutellonema bradys*, are associated with warm climate and tropical and semitropical vegetable crops while other nematodes, such as *Ditylenchus dipsaci*, *Pratylenchus penetrans* and *Longidorus elongatus*, favor a cooler climate and temperate zone vegetable crops. Nematode pests often thrive outside of their known climatic range if they and their host plants are established in a controlled or partly controlled environment, e.g. arboretum, conservatory, glasshouse, hot-bed, lath-house, etc.

Many genera of parasitic nematodes are widely distributed throughout the world's many varied climatic and topographic areas, e.g. *Heterodera*, *Longidorus*, *Meloidogyne*, *Paratylenchus*, and *Trichodorus*. Some species, however, tend to be more restricted, e.g. *Heterodera carotae* in carrot from England and Italy, *Longidorus africanus* in lettuce from the Imperial Valley of California, *Meloidogyne inornata* in soybean from Brazil, *Paratylenchus projectus* on parsley in New Jersey (U.S.A.), and *Trichodorus allius* on onions in Oregon (U.S.A.).

Most nematode parasites seek sites for feeding and reproduction near plant growing points where greatest metabolic activity is occurring. Thus, nematodes are probably most active when potential feeding sites (flowers, leaves, stems, and roots) are tender and growing rapidly. Once the nematode establishes a relationship with a susceptible host plant, it can usually tolerate any environmental condition that the host can withstand and an infection can be expected to occur.

V. Nematode Pests

A. Root-knot Nematodes (*Meloidogyne* spp.)

1. Occurrence

Root-knot nematodes are the most important and most cosmopolitan of nematode pests of vegetables. The principal symptoms, root "galls" or "knots", probably were observed many centuries ago but were first reported by Berkeley (1855). Occurrence records had accumulated a world-wide distribution and approximately 2500 kinds of host plants of root-knot nematodes by 1949. The majority of distribution and host records for root-knot nematodes were attributed to one species, and Chitwood's (1949) revision of the genus established several species and invalidated most previous host records. Since then additional species, subspecies and, more recently, bio- or pathotypes have been described or discussed. Whitehead (1968) published a comprehensive account of 23 species. A number of these species have been reported as having either a limited or a wide host-range on vegetables.

2. *Important Species and Principal Host Crops*

Table I summarizes the major species of root-knot nematode affecting vegetables and their principal hosts and localities. Nearly all vegetables have been reported as hosts for root-knot nematodes, and some crops may be invaded by more than one species. Populations of some species show differences in behavior, distribution and host preferences and these groups are called "biotypes" or "pathotypes".

TABLE I. Distribution and hosts of root-knot nematodes (*Meloidogyne* spp.) affecting vegetable crops

Binomial name of pest	Host crops	Distribution
M. acronea	Bean and Tomato	South Africa
M. africana	Cowpea and Clove	East Africa
M. arenaria	Beet, Celery, Ground-nut, Onion, Tomato, Watermelon, and other vegetables	Europe, Israel, South Africa, U.S.A.
M. artiellia	Cabbage and other Brassicas, Broad Bean, and Pea	England
M. ethiopica	Cowpea and Tomato	Tanganyika
M. exigua	Pepper	Brazil
M. graminicola	Bean	U.S.A. (Louisiana)
M. hapla	Most vegetables grown in temperate and semitropical climatic zones	Cosmopolitan
M. incognita	Lima and Hyacinth Bean, Celery, Ginger, Onion, Pea, Cowpea, Pepper, Sweet Potato, Tomato and Yam	Africa, Asia, Australia, Europe, Israel, U.S.A., probably cosmopolitan
M. javanica	Bean, Broccoli, Cabbage, Squash, Tomato, Yam	Cosmopolitan in tropical and semitropical climates
M. kikuyensis	Cowpea (*Vigna sinensis*)	East Africa
M. naasi	Beet and miscellaneous Brassicas	England, U.S.A.
M. thamesi	French Bean and other Legumes, Cabbage	Australia, U.S.A.

3. Symptoms and Nematode Behavior

Ordinarily, the root and underground portions of the plant are the favorite infection sites, but nematodes may occur in leaves and shoots that are in contact with the soil. Host tissues of high metabolic activity, such as apical meristems, appear to be preferred infection sites. Following hatching, second-stage larvae seek host plants and invasion may occur on an individual or mass basis. Some larvae may penetrate host tissue, fail to develop a feeding site and actually leave the host. Larvae which do not find a host and an acceptable feeding site continue to search until their energy supply is exhausted when they die. Once active feeding commences, the nematodes begin a sedentary mode of existence. Host tissue reaction to penetration and feeding and subsequent giant cell formation (syncytium) are well documented and although nematode invasion may take place, development of the nematode to maturity does not occur when the host does not produce giant cells.

4. Interrelationships

Root-knot nematode association with the host plant is often accompanied by other pathogens, usually bacteria and fungi, and under these circumstances disease complexes may develop. The nematode may cause: (a) a loss of fungal resistance to the host plant (e.g. *Fusarium* wilt resistance in cotton, tomato and tobacco); (b) an additive effect resulting from combined pathogenicity to pests; (c) a synergistic effect resulting in an accentuated degree of pathogenicity; (d) a suppression of symptoms; or (e) an earlier appearance of symptoms. According to Pitcher (1965), *Meloidogyne* is the most important incitant genus and *Fusarium* the most frequent partner, although each pathogen can cause disease without the other being present. Bacterial associations are not as common, although Lucas *et al.* (1955) reported that a combination of *M. incognita acrita* and *Pseudomonas solanacearum* developed symptoms sooner and more extensively than with the bacterial pathogen alone. Later, Libman *et al.* (1964) reported an increased incidence and severity of wilt in tomato when *M. hapla* was combined with *P. solanacearum*. Thus far, virus associations with *Meloidogyne* spp. are not known although both pathogens frequently infect the same host.

Mayol and Bergeson (1970) measured the pathogenic effect of normally non- or low grade-pathogenic micro-organisms and *M. incognita*. Combinations of secondary microbial invasion of tomato plants previously inoculated with this nematode caused 75% and 48% weight reduction of foliage and roots, respectively. Nematodes alone, in aseptic conditions, caused 37% reduction of foliage and a 50% increase in root weight.

Sometimes such interactions result in an unfavourable environment for the nematode partner. Pitcher (1965) noted that the obligate plant parasite, as initiator of the complex, seems especially vulnerable to competition. Effect of competition is inevitable when two types of pathogens seek nourishment from the same individual host plant and often the same organ of that host, according to Pitcher (1965). Christie (1959) refers to the incompatibility of plant-parasitic nematodes and decay-promoting organisms. Powell and Nusbaum (1960) noted the destruction of giant cells by fungal hyphae. Such effects prevent the maturation of female *Meloidogyne*.

5. Control

Control of root-knot nematodes is difficult to accomplish by such cultural methods as crop rotation, trap cropping, summer fallow, flooding, etc., because of the vast host ranges, nematodes' ability to survive in adverse conditions, and uneconomic loss of production with idle land. The most commonly used control methods are soil fumigation with one of the several nematicides or by the use of resistant varieties of vegetables. Resistant varieties are known for the following crops: bush type snap bean, lima bean, pole bean, cantaloup, carrot, gherkin, okra, pepper (bell or hot), sweet potato, tomato and watermelon. Performance of resistant varieties may vary according to: species, subspecies or biotypes of the nematode; growing conditions of the crop; and the bacterial or fungal disease complexes involved.

B. False Root-knot Nematodes (*Nacobbus* spp.)

1. Occurrence

At least one species of this genus is a potential vegetable parasite and it has been reported from scattered localities in the Great Plains Area of the U.S.A. (Colorado, Kansas, Montana, Nebraska, and Wyoming).

2. Important Species and Principal Hosts

Although sugar-beet is the major host of *Nacobbus batatiformis*, Thorne and Schuster (1956) conducted host range studies to demonstrate its potential as a plant parasite. In their study the following vegetables were attacked: beet, broccoli, cabbage, carrot, cucumber, lettuce, pea, pumpkin, radish, rutabaga and turnip. Tomato also has been reported as a host of *Nacobbus* sp. in England.

3. Symptoms and Nematode Behavior

Symptoms are similar to those described for root-knot nematodes with evidence of unthrifty above-ground growth and galls or knots on
N*

the roots. Galls differ, however, in that those caused by *Nacobbus* often bear numerous secondary roots, they appear spindle-shaped and usually are larger. According to Schuster and Thorne (1956) pathological history was similar in some respects to that described for *Meloidogyne* galls. Thorne (1961) noted that the nematode typically inhabited light sandy soils and behavior and development were similar to that of *Meloido-gyne*. A notable exception is that immature males and females (third stage) migrate to larger roots, initiate additional gall formation and complete the life-cycle.

4. Control

There are many economic non-hosts available for crop rotation; also, standard nematicides give satisfactory results. Thorne (1961) also mentions that lamb's quarters, mustards, Russian thistle and other weeds are hosts and therefore must be considered as obstacles in a control program.

C. Cyst Nematodes (*Heterodera* spp.)

1. Occurrence

Nearly all *Heterodera* favor temperate climatic zones with some species cosmopolitan and many others known only from limited localities. It is not unusual for several species to occur in a limited area but most often only one or two species inhabit a specific locality.

2. Important Species and Principal Hosts

Approximately 50 species of *Heterodera* have been described; however, Esser (personal communication) regards only 26 as valid species. Many of these are important vegetable pests and are listed in Table II along with their principal hosts and distribution.

3. Symptoms and Nematode Behavior

Above-ground evidence of the nematode varies according to the species, its population density, and development and tolerance of the host plant. Typical symptoms include off-color foliage, stunting and severe wilting during times of abnormal water stress such as during hot weather. Wilt symptoms are observed best during the warmest part of the day because hosts may temporarily recover during the night or during cooler weather. Severe stunting may occur in the seedling stage and such plants are smothered by uninfected plants or weeds. Below-ground symptoms vary in degree of severity depending upon population density of the pest and the type and vigor of the host plant. Root development may be severely stunted, characterized by a reduction in size of tap or large roots. Frequently, excessive secondary root development accompanies infection. Host tissue usually responds to nematode feeding by producing hypertrophied multinucleate syncytia. Although this results in a type of giant cell formation, according to Pitcher (1965), it lacks the hyperplastic reaction commonly associated with *Meloidogyne*. Initially, slight swellings on roots may result from histological host reactions, but most evidence of infection is the appearance of white pearl-like cysts upon the roots.

TABLE II. Distribution and hosts of cyst nematodes (*Heterodera* spp.) affecting vegetable crops

Common name of pest	Binomial name of pest	Vegetable host plants	Distribution
Carrot Cyst Nematode	*H. carotae*	Carrot	England, Italy
Brassica Cyst Nematode	*H. cruciferae*	Brussels Sprouts, Cabbage, Kale	Europe and U.S.A. (California)
	H. cajani	Pigeon-pea	India
Soybean Cyst Nematode	*H. glycines*	Snap Bean	Asia and U.S.A.
Pea Cyst Nematode	*H. goettingiana*	153 varieties of pea including Chick Pea (Garbanzo)	Europe
Golden Nematode	*H. rostochiensis*	Eggplant, Tomato	Cosmopolitan including North and South America, Europe, Mediterranean
Sugar-beet Nematode	*H. schachtii*	Table Beets, Broccoli, Brussels Sprouts, Cabbage, Cauliflower, Swede (Turnip)	Cosmopolitan throughout the temperate zones in the world

After hatching the nematodes are attracted to the rhizospheres of host plants where larvae usually penetrate host tissue in metabolically active zones near the root-cap. Once feeding commences and syncytia are initiated larvae undergo rapid changes to assume the characteristics of various succeeding stages. For those species of *Heterodera* having several generations per year, the eggs of early generations are deposited in a matrix attached to the posterior end of the body. Eggs deposited in this manner probably hatch in a relatively short time. Most commonly, the eggs remain within the female's body which, after fertilization and upon death, is transformed into a cyst which is an important protective device insuring survival of the species. Cysts of various ages often assume distinctive colors.

Two of the most common former methods of distributing cyst nematodes were with soil as in "tare dirt" and with the movement of seed.
Common dissemination agents now are machinery, livestock or any
other means of transporting contaminated soil. Certainly irrigation
water (tail water) or waste water from handling and processing contributes to dissemination of these pests.

4. Interrelationships

Interactions of various *Heterodera* spp. with other plant pathogens
apparently have not been a subject of intensive investigation. Some
workers consider cyst nematodes as agents predisposing plants to
fungus diseases, and other workers have noted cysts often contain
saprozoic nematodes and potential biological control agents (amoeba,
fungi and other nematodes).

5. Control

Protective cysts make these pests more difficult to control by nematicides and cultural practices. The narrow or specific host range is a condition which facilitates a cultural practice like crop rotation especially
if there is only one species of cyst nematode in the area. In such
instances alternate non-host crops and an effective weed control program is a very successful control method. Unfortunately, the period of
rotation becomes very long because of delayed larval emergence from
a high density of cysts in the soil. Some soil fumigant-nematicides,
however, are very effective against all stages of these pests. Some
attention is given to a search for resistant varieties of various susceptible crops. In some instances resistance has been found in wild type
relatives of host crops but difficulties have been encountered in transferring the resistance to commercially acceptable varieties.

D. Stem Nematodes (*Ditylenchus* spp.)

1. Occurrence

Bulb and stem nematodes, principally *Ditylenchus dipsaci*, were some
of the earliest known plant pests and commonly occur in temperate
zones of the world.

2. Important Species and Principal Hosts

One species is widely recognized in this group although *D. allii*,
synonymized by some with *D. dipsaci*, is a common parasite of garlic
and onions in several regions of Czechoslovakia and the U.S.S.R.
Differences in the host preferences of various collections of *D. dipsaci*
are responsible for various subdivisions of species into biotypes, patho-

types, populations, races or strains. Such differences have resulted in confusion about the behavior, distribution, and host range of species. The principal vegetable hosts of *D. dipsaci* are celeriac, garlic, leek, lettuce, onions, shallots and swedes.

3. Symptoms and Nematode Behavior

Although symptoms vary with the host plant, development of host and the climate, the most severe injury apparently occurs in the primary stem tissues. Thick, swollen and distorted shoots, deformed leaves, and spongy bulb tissue which separates from the basal plate, are the most common symptoms. In seedlings injury is sometimes so severe that infected plants fail to mature. In older plants there is frequently severe stunting and as the growing season progresses the foliage dies down prematurely. Brzeski and Rajewski (1966) report that nematodes migrate up the onion stalk to the inflorescence after the parent bulb has started to decompose. Rarely does a complete stand of host plants succumb at once as infected plants often are scattered, but the nematode population spreads rapidly through the soil and in a few years may infect several hectares of the crop. When plants are grown in close proximity, the nematode infection may spread more rapidly.

Nematodes usually enter host plants early in the growing season when stem tissues are developing. The pectic enzymes in the nematode's salivary secretion enable it to penetrate the plant and to feed on the tissues. Dissolution of the middle lamella is partly responsible for some pathogenic symptoms, but there is no agreement among investigators as to the source of this enzymatic activity. Although nematodes survive in fallow soil for a year or two, depending on soil type and climate, they can survive for several years in a quiescent form (fourth-stage larvae) in host tissue such as bulbs, leaves, seeds, crown areas of established plants and crop debris. Volunteer and weed hosts also are important reservoirs of infection. All stages of the nematode, including eggs, occur in developing host tissue.

Nematodes are spread from one area to another by a variety of agencies which move infested soil and infected plants or crop debris. Movement in irrigation water, wind, rain and floods, etc., also are important means of dissemination.

4. Interrelationships

In addition to role of a predisposer for various fungi and saprozoic nematodes, *D. dipsaci* is important in disease complexes involving other pathogens. According to Myuge (1960) the presence of this nematode in water produced 100% collar rot (*Botrytis allii*) of onion in comparison to a 30% infection without the nematode. Hawn (1963) has demonstrated that *D. dipsaci* has a significant role in dissemination of phytobacteria.

5. Control

Sanitary measures which destroy the nematode or restrict the distribution of infested soil and infected propagation stocks accomplish much in controlling this pest. Although regulatory procedures (quarantines and propagation stocks certified to be free of nematode pests) usually are effective preventive measures, the pest and subsequent disease problems often become well established before discovery. After discovery of the disease, the control program may involve cultural, physical or chemical approaches. Cultural approaches usually consist of roguing diseased plants, control of reservoir hosts (weeds and volunteer crop plants) and crop rotation with non-susceptible crops. Special consideration must be given in selecting rotation crops because of different host preferences of various subdivisions of this pest. Physical control usually consists of a hot-water treatment of bulbous crops (garlic, shallots, etc.). It is important to consult authorities or references for exact time and temperature durations; otherwise the nematodes may survive or the plant material may be severely injured. Two vegetable crops often treated in this manner are garlic (20 min at 49°C) and shallots (2 h at 43·5°C or 90 min at 45°C). To obtain maximum control for certain crops, e.g. garlic, it is necessary to employ a pre-soak (2 h at 25°C) and the incorporation of a wetting agent and formaldehyde in the hot-water bath (Lear and Johnson, 1962).

Chemical control with certain nematicides has been effective when applied to soil before planting. Dichloropropene nematicides are usually recommended; severe phytotoxicity occurs in most bulb crops when dibromochloropropane is used. Many of these crops are grown in organic or heavy soils and special attention should be given to soil preparation and other conditions that affect soil fumigation results in these conditions. Certain chemicals have been used to eliminate these pests from seed and roots of some crops but usage requires specific instructions.

E. Potato Rot Nematode (*Ditylenchus destructor*)

1. Occurrence

Although most information about *Ditylenchus destructor* is presented elsewhere, this nematode is better known as a parasite of ornamental bulbs and potatoes; it does attack other vegetables. This nematode is primarily a temperate zone pest and occurs sporadically throughout the world.

2. Important Species and Principal Hosts

Early literature citations referred to the potato rot nematode as a "strain" or "population" of *D. dipsaci* but it is now regarded as a distinct species. Most information on vegetable hosts has been obtained from greenhouse or controlled environmental studies. Safyanov (1965) studied the host range of a *D. destructor* population from Kazakh, U.S.S.R. and concluded that chick pea, cow pea, everlasting pea, and tomato were good hosts, cucumber and garlic poor hosts, and onion and radish non-hosts. Thorne (1961) made laboratory transfers from potato to carrot, beet, mint, onion, sweet potatoes and turnip. Faulkner and Darling (1961) noted that parsnip and radish also served as suitable laboratory hosts of *D. destructor*. Henderson (1951) reported that Baker had observed *D. destructor* on carrot in Canada.

3. Symptoms and Nematode Behavior

Symptoms vary with the host plant but usually consist of superficial, sunken discolored lesions. Apparently *D. destructor* will readily attack a number of hosts with little indication of host preference. Thorne (1961) reports that a significant reduction in overall size and some morphological changes in digestive and reproductive systems occur after transferring the nematode from one host to another. According to a number of workers the nematodes infect the plant via natural host openings. Once established, they develop colonies containing all stages, which enlarge with corresponding development of lesions.

4. Interrelationships

Associations of *D. destructor* and fungi are well known. For example, Faulkner and Darling (1961) demonstrated that this nematode fed and reproduced on 64 species of fungi representing 48 genera, 8 orders, and all major classes. Baker *et al.* (1954) studied relationships of mites, fungi and *D. destructor* in potatoes and concluded that mites and nematodes competed for fungus mycelia in potato tissue.

5. Control

The ability of this nematode to live on numerous hosts (crop and weed) and on fungal mycelia make cultural measures difficult or ineffective. Control therefore consists of local quarantine regulations to restrict movement of infected propagation stocks, hot-water treatment of certain propagation stocks, and of soil fumigation with standard nematicides.

F. Root lesion Nematodes (*Pratylenchus* spp.)

1. Occurrence

The endoparasitic root lesion nematodes are widely distributed in temperate zones. A few species are cosmopolitan and others are limited to specific areas.

2. Important Species and Principal Hosts

At least nine species of *Pratylenchus* have been reported to attack vegetables and among these *P. coffeae* and *P. penetrans* are most common. Most temperate zone vegetables are attacked by one or more species of root lesion nematode (Table III).

TABLE III. Distribution and hosts of root lesion nematodes (*Pratylenchus* spp.) affecting vegetable crops

Binomial name of pest	Vegetable host plant	Distribution
P. brachyrus	Mung Bean, Velvet Bean, Okra, Cowpea, Tomato and Watermelon	Australia, U.S.A. (Louisiana)
P. coffeae	Beans, Broccoli, Cabbage, Cauliflower, Onion, Pea, Sweet Potato, Tomato	Australia, Brazil, Japan, U.S.A., probably cosmopolitan
P. globulicola	Pea	U.S.S.R. (Ulyabrinsk)
P. penetrans	Most temperate zone vegetables	Cosmopolitan
P. pratensis	Cabbage	U.S.A. (Massachusetts, Washington)
P. scribneri	Lima Bean, Snap Bean, Tomato	U.S.A. (California)
P. thornei	Pea	Netherlands
P. vulnus	Many vegetables	U.S.A. (California)
P. zeae	French Bean	Australia

3. Symptoms and Nematode Behavior

Above-ground plant symptoms of distressed growth conditions such as chlorosis, stunting, and general lack of vigor are indicative of infection, but are non-specific. Usually symptoms involve more than one plant and occur in zones or spots in field planting. As the common name of this nematode group implies, these parasites produce lesions on roots and these are more reliable diagnostic symptoms. Although lesions occur on larger roots, preferred feeding sites seem to occur in the metabolically active tissues of feeder roots which are often destroyed. Frequently other subterranean host tissues are attacked, including crowns, rhizomes and tubers. All stages of the nematode usually occur in the marginal areas of lesions and healthy tissues. Acedo and Rohde (1968) reported that secondary roots of cabbage had many lesions that girdled lateral roots, exposed the stele and frequently caused these roots to break off. They also reported that portions of the stem base were attacked and developed lesions.

Most *Pratylenchus* spp. are polyphagous and feed on woody perennials as well as on succulent annuals including vegetables. The migratory nature of these pests should be noted as they continuously move to lesion perimeters, in and out of soil and from feeding site to feeding site. Usually all developmental stages can be found in infection sites; however, both larvae and adult can enter host tissue.

4. Interrelationships

Much attention has been given to interactions of root lesion nematodes and fungi, particularly *Verticillium* spp. McKeen and Mountain (1960) demonstrated a synergistic relationship between *P. penetrans* and *Verticillium albo-atrum*. At low and intermediate levels of *Verticillium* inoculum, the nematodes increased wilt and larger numbers of nematodes occurred in eggplant roots. Nematodes alone had no adverse effects upon eggplant but *Verticillium* alone was pathogenic. Mountain and McKeen (1962) obtained similar results with *Verticillium dahliae* and *P. penetrans* in tomato wilt. In greenhouse studies of *P. penetrans* and *V. dahliae* on pepper, the damage caused by these organisms was additive according to Olthof and Reyes (1969). Edmunds and Mai (1966) reported a significant increase in nematodes recovered from celery roots infected with *P. penetrans* and *Trichoderma viride* than from roots infected with only nematodes. The combined effect of these organisms significantly retarded growth of shoots and roots.

5. Control

Control of these pests is usually achieved with nematicides containing dichloropropene mixtures. In some instances, crop rotation may provide control but its efficiency is dependent upon the host range of a particular species and availability of economic crop alternatives. Nematologists in the Netherlands have shown that African marigolds (*Tagetes* spp.) produce a nematicidal material (terthienyl) which suppresses *P. penetrans* populations. In some instances, interplanting marigolds with susceptible crops has reduced injury by decreasing the population of nematodes. Several intercroppings of marigolds may be necessary to decrease very large populations of nematodes to a negligible level but the use of integrated control methods may be of more value.

G. Burrowing Nematodes (*Radopholus* spp. and *Hirschmanniella* spp.)

1. Occurrence

The genus *Radopholus* occurs in many tropical and semitropical areas of the world, but has not been reported from all warm climates.

2. Important Species and Principal Hosts

The only important species, *Radopholus similis*, is responsible for the "yellows disease" of black pepper (*Piper nigrum*) on the island of Bangka, Indonesia which is an outstanding example of devastation that can be caused by nematode pests. The following vegetables are reported to be hosts: asparagus, bean, beet, cabbage, cantaloup, carrot, cowpea, okra, pea, sweet potato, pumpkin, radish, rutabaga, squash, and tomato; non-hosts include broccoli, cauliflower, celery, lettuce and turnip, according to Esser (1963).

3. Symptoms and Nematode Behavior

"Yellow disease" of pepper is a spreading disease which initially involves a few vines but can spread to infest the entire planting. Infected vines cease to grow, foliage turns yellow, terminal growth dies back, leaves drop and then the vines die. Large roots are nearly destroyed. These pests are endoparasites which initially feed on cortical cells of roots usually causing necrotic spots and lesions on the feeder root system. The nematodes often migrate short distances to other feeding sites as decaying tissues and secondary organisms render feeding sites uninhabitable.

4. Interrelationships

Interactions with burrowing nematodes and *Fusarium* spp. are known for some crops but such relationships have not been reported for vegetables.

5. Control

Control of this pest varies from no known method for black pepper, unless standard pre-plant nematicides can be used, to application of such materials for other vegetables.

At least one pathogenic problem of vegetables is caused by a species of *Hirschmanniella*. Sher (1954) listed *H. gracilis* from lotus (*Nelumbrium nelumbo*) in Hawaii. The nematode produces lesions on the broad, hollow and sectioned rhizomes despite the fact that the rhizomes develop on the bottom of a bog often covered by 60–90 cm of mud and water. Lesions probably facilitate development of secondary pathogens which accentuate necrotic development; eventually the rhizome erodes through and mud and water penetrate into the interior of the rhizome.

The entire rhizome breaks down, decays, and the product is a total loss. No control procedures have been developed.

H. Reniform Nematode (*Rotylenchulus* spp.)

1. Occurrence

Reniform nematodes are pests of numerous plants including vegetables in tropical and semitropical areas and distribution varies with the species.

2. Important Species and Principal Hosts

Although 9 species of reniform nematode now occur, only *Rotylenchulus reniformis* is considered a vegetable pest. Linford and Yap (1940), who initially discovered and described this pest from Hawaii, listed the following hosts: artichoke, bean, beet, cabbage, carrot, cauliflower, chard, cucumber, eggplant, lettuce, okra, pea (including cowpea and pigeon pea), radish and squash. Birchfield and Brister (1962) added lima bean, muskmelon, pumpkin, sweet potato, and watermelon, and the following as non-hosts: mustard, onion, pepper, spinach and turnip. Decker *et al.* (1966) report that *R. reniformis* is a common parasite of cucumber, sweet potato and tomato in Cuba, and Ayala and Ramirez (1964) report that this nematode is a common pest in Puerto Rico. Caveness (1967) found *R. reniformis* to be a potential vegetable pest in the Western Region of Nigeria as shown in shade-house tests. The following were hosts: carrot, cowpea, sweet potato, tomato and water yam. It is also known to occur on tomato in Queensland, Australia (Colbran, 1964) and D'Souza and Screenwasan (1965) list this nematode as a pest of bush bean from South India.

3. Symptoms and Nematode Behavior

Damage varies from trace to severe depending on type of host, initial nematode populations and biological and physical factors affecting the seasonal increase of nematodes. Common symptoms of unthrifty growth only suggest a possible nematode association, but root injury indicated by stunting, discoloration and epidermal cell necrosis is more characteristic. According to Birchfield and Martin (1965) who investigated the disease in sweet potato, initial feeding by young females destroys root epidermal cells. As feeding progresses cortical parenchyma near the epidermis collapse and the nematode's migration towards the stele further disrupts cortical parenchyma with injury continuing into the phloem where maturing females feed. In some instances feeding may be limited to the cortex, as in cowpea, with the swollen posterior end of the nematode usually protruding. As implied by the common name, the mature female assumes a reniform or kidney shape, then produces eggs in a gelatinous matrix which usually occurs on the external surface of the host unless deeply embedded in tissue. Although vermiform males sometimes enter host tissue they are most frequently found in the rhizosphere or in the gelatinous matrix.

4. Interrelationships

Combinations with other pathogens are not well known. However, Jenkins and Taylor (1967) refer to an interaction in *Fusarium* wilt of cotton.

5. Control

Control can be accomplished usually with rotation as many economic non-susceptible crops are available but standard nematicides also are satisfactory.

I. Sting Nematodes (*Belonolaimus* spp.)

1. Occurrence

Sting nematodes are large, slender ectoparasites which probably are native in the Southeastern Coastal Plain areas of the U.S.A. None has been reported elsewhere.

2. Important Species and Principal Hosts

Only two species (*Belonolaimus gracilis* and *B. longicaudatus*) have been described and these occur in mixed populations which make it difficult to determine the extent of damage caused by each species. The collective host range includes a number of vegetables such as celery, bean and cowpea.

3. Symptoms and Nematode Behavior

Top growth of diseased plants is usually chlorotic, stunted, and yields are reduced in both quantity and quality. Celery seedlings may be severely stunted and in some instances the nematodes enter seed and thus reduce stand emergence. Characteristic symptoms are stubby roots which sometimes do not penetrate below the top 5 cm of soil, but necrosis and coarse roots also may occur.

According to Standifer (1959) lesions in injured bean roots consist of two distinct areas—a hollow cavity and an area of ruptured cells which encircle the cavity. Maturity of stelar tissues near the lesions was affected as were cells a considerable vertical distance from the nematode.

4. Interrelationships

Studies of sting nematodes and other pathogens in vegetable crops have not been reported but Holdeman and Graham (1954) have demonstrated a relationship between *Fusarium* wilt and sting nematode (*B. gracilis*) in cotton.

5. Control

Resistant plant varieties are not known for these nematodes but many non-hosts are available for rotation and many nematicides are successful control agents.

J. Awl Nematode (*Dolichodorus* sp.)

1. Occurrence

The ectoparasitic awl nematodes have been reported from scattered areas in many parts of the world. Most of the eight described species are limited to a continental area except for *Dolichodorus heterocephalus*, the only known vegetable pest, which probably occurs in three or four widely separated areas of the U.S.A.

2. Important Species and Principal Hosts

D. heterocephalus is associated with a "red root" disease of celery and is not known to be a pest outside of Florida. Other hosts include bean, Chinese water chestnut (*Eleocharis dulcis*), pepper and tomato.

3. Symptoms and Nematode Behavior

Growth of all known hosts is retarded and often there is decreased seedling emergence. Although this nematode is associated with restricted root growth of Chinese water chestnut, corm weight was not affected. Feeding usually occurs on root-tips, but may extend along other surfaces, causing necrotic lesions. Root-tip devitalization results in growth cessation and a stubby root condition. The nematode seems to prefer wet soil and is most destructive in such areas.

4. Control

Satisfactory control has been obtained with rotation of nonsusceptible crops or with application of standard nematicides.

K. Sheath Nematodes (*Hemicycliophora* spp.)

1. Occurrence

Many species of ectoparasitic sheath nematode have been described in this cosmopolitan genus but individual species usually have a limited or local distribution.

2. Important Species and Principal Host

Four species of *Hemicycliophora* have been reported as pests or potential pests of vegetables. Van Gundy (1957) reported that blackeye bean, celery, pepper, squash and tomato were hosts of *H. arenaria*. *H. parvana* has been associated with bean and celery. Klinkenberg (1963) observed *H. similis* feeding on roots of *Brassica oleracea* (variety not given) and lettuce. A field disease condition occurs in the marine, sandy soils of the Netherlands on carrots damaged by *H. typica* (Kuiper, 1959).

3. Symptoms and Nematode Behavior

Field symptoms are known only for *H. typica* as indicated by poor growth and "stubby root" of carrot. Most parasitic species feed on root hairs and often cause their disappearance or retard growth, but *H. arenaria* causes gall formation.

4. Control

Control measures have been used against *H. typica* which disappeared along with the symptoms after a nematicide treatment.

L. Pin Nematodes (*Paratylenchus* spp.)

1. Occurrence

The cosmopolitan genus *Paratylenchus* which occurs in most climatic zones contains some of the smallest plant parasites. Most species of these ectoparasites are limited in distribution to specific areas or localities.

2. Important Species and Principal Hosts

Although at least 50 species are known in this genus only three are regarded as important vegetable pests. *Paratylenchus hamatus* is widely distributed in the U.S.A. and has been reported as a celery and parsley pest in northeastern states and is a pest of mints in Washington State. *P. minutus* occurs in Hawaii and experimentally infects cabbage, cowpea, cucumber, okra and radish. *P. projectus* occurring in the eastern U.S.A. increased on a number of vegetables belonging to the Chenopodiaceae, Cruciferae and Leguminosae. Some kinds of vegetables, such as spinach, squash and tomato, have susceptible and non-susceptible varieties. Carrot, cucumber, eggplant, endive, parsley and sweet pepper were reported as non-hosts.

3. Symptoms and Nematode Behavior

Chlorosis and reduction of growth appear to be the most common above-ground symptoms. These ectoparasitic nematodes feed directly on roots which show a noticeable absence of feeder roots and poorly developed secondary roots, although some individuals have been found partly or completely enclosed in host tissue. Infection of mints can be detected by delayed flowering and in differences of fresh and dry weights between infected and non-infected plants (Faulkner, 1964). According to Jenkins and Taylor (1967) the fourth, or pre-adult stage, apparently does not feed and represents a survival stage.

4. Interrelationships

P. hamatus is associated with field symptoms of chlorotic and stunted celery, and produces similar symptoms in the greenhouse. Isolated from many root lesions was a species of Rhizoctonia which may have contributed to the disease according to Lownsbery et al. (1952).

5. Control

Both P. hamatus and P. minutus have been controlled successfully with halogenated hydrocarbon nematicides. Since host ranges are not extensive it should be possible to select economic non-host crops for rotation.

M. Spiral Nematodes (Helicotylenchus spp., Rotylenchus spp. and Scutellonema spp.)

1. Occurrence

At least three genera of spiral nematodes contain vegetable pests. Although species of Helicotylenchus and Rotylenchus are commonly found in temperate climatic zones, they seem to be abundant as are species of Scutellonema in the warmer climate of tropical and semi-tropical zones.

2. Important Species and Principal Host Crops

The eight species of Helicotylenchus in Table IV are or may be potential parasites of vegetables. Despite the fact that large populations have been found in association with various plant species, most parasitic evidence results from controlled greenhouse experiments. The following two species of Rotylenchus are vegetable parasites: R. uniformis damages carrot and pea in the Netherlands; R. robustus reduces lettuce yields, and carrot, cauliflower and Brussels sprouts support high populations in greenhouse tests in California. The genus Scutellonema contains the following three vegetable pests: S. bradys, sometimes called the "yam nematode", is an important yam parasite in India, Jamaica, and Nigeria; S. clathricaudatum has been shown in shade-house tests to be a potential pest of cassava, carrot, chilies, Bara melon, okra, cowpea and yams (water, white and yellow) in the Western Region of Nigeria (Caveness, 1967). Colbran (1964) records *S. brachyurum* as a pest of guar bean in Queensland, Australia.

TABLE IV. Distribution and hosts of spiral nematodes (*Helicotylenchus* spp.) affecting vegetable crops

Binomial name of pest	Vegetable host range	Distribution
H. cavenessi	Cassava, Chilies, Eggplant, Bara Melon, Onion, Indian Spinach, and Tomato	Nigeria (Western Region)
H. digonicus	Beans	U.S.A. (Wisconsin)
H. dihystera	Hyacinth Bean, Velvet Bean	Queensland, Australia
H. micro-cephalus	Cassava, Chilies, Garlic, and Onion	Nigeria (Western Region)
H. multicinctus	Bean, Beet, Cabbage, Cantaloup, Carrot, Garlic, Lettuce, Onion, and Pea	Cuba
H. microlobus	Bean, Carrot, Cucumber, Muskmelon, Onion, Radish, Rutabaga, Squash, Tomato, Turnip, and Watermelon	U.S.A. (Minnesota)
H. nannus	Lima Bean	U.S.A. (Maryland)
H. pseudo-robustus	Cassava, Carrot, Chilies, Efinrin, Eggplant, Cowpea, Indian Spinach, Tomato, numerous vegetables	Nigeria (Western Region) U.S.A. (Minnesota)

3. Symptoms and Nematode Behavior

Spiral nematodes, so named because of their configuration when killed by heat, are ectoparasites but frequently embed the anterior end of their bodies in host tissue. Above-ground symptoms such as chlorosis, stunted growth and reduction in yield are associated with this group of nematodes. Impaired functioning of the root system by destruction or partial destruction of feeder roots and lesions on larger roots and other underground plant tissue is more specific. In addition to reduced yield and size, *R. uniformis* also causes misshapen roots (Kuiper and Drijfhout, 1957). The growth of carrots improved when they followed potatoes and declined rapidly when the preceding crops were beets or carrots.

4. Interrelationships

All three genera are known to interact with other pathogens in disease complexes. *Helicotylenchus dihystera* increases the incidence of carnation wilt by providing entry for *Pseudomonas caryophylli* according to Stewart and Schindler (1956). The interaction of *Rotylenchus uniformis* and *Fusarium oxysporum* f. *pisi* race 3 and its importance in "early yellowing" disease of peas was investigated by Labruyere *et al.* (1959). Neither pest alone causes severe root rot in peas but together they cause an extensive decay of the root cortex which is the chief symptom in roots of peas with "early yellowing". Surface lesions caused by *Scutellonema bradys* are portals of entry for bacteria, fungi, mites, etc., which often destroy the entire root, a complex often referred to as "dry rot of yams".

5. Control

Control of these pests or diseases varies from improved storage conditions for "dry rot of yams", crop rotation for *Rotylenchus uniformis* affecting carrots, to soil fumigation for *Rotylenchus robustus* on lettuce. Yields of marketable heads of lettuce can be doubled when treated with pre-plant nematicides (Lear *et al.*, 1969). Pre-plant nematicides seem to be effective against all spiral nematode pests of vegetables.

N. Stunt Nematodes (*Tylenchorhynchus* spp.)

1. Occurrence

This genus is widely distributed with at least one species found in every agricultural area. Individual species, however, are usually limited to specific areas.

2. Important Species and Principal Hosts

Although many species have been described only a few are known to be pathogenic. In the southeastern U.S.A., *Tylenchorhynchus claytoni* causes a stunting of pea and *T. martini* damages sweet potato. In the Netherlands, the host range of *T. dubius* includes such potential hosts as cauliflower, pea, radish and turnip (Sharma, 1968).

3. Symptoms and Nematode Behavior

Stunt nematodes are ectoparasites and feed primarily on epidermal cells of roots in the region of elongation but occasionally they are observed partly or totally embedded in host tissue. This results in a stubby root condition and subsequently results in a loss of weight and decrease in size of host.

4. Control

Pre-plant nematicides are known to control stunt nematodes on other crops and should be satisfactory for control on vegetable hosts.

Host ranges of stunt nematodes are not extensive and several economic non-host crops should be available for a rotation program.

O. Stubby Root Nematodes (*Trichodorus* spp.)

1. *Occurrence*

Trichodorus spp. are common inhabitants of most soils (cultivated and non-cultivated) in tropical and temperate climatic zones. Most species tend to be limited to specific continents or specific areas within continents.

2. *Important Species and Principal Hosts*

Although several species of *Trichodorus* have been described, only a few are parasites of vegetables. *Trichodorus christiei* is widely distributed in the U.S.A. and is a polyphagous species associated with many

Fig. 1. Field of onions (var. Yellow danvers) showing area of poor stunted growth caused by *Trichodorus allius*.

vegetable crops but is most injurious on onion and tomato in the eastern U.S.A. *T. allius* injures onion in Oregon (U.S.A.) (Figs. 1 and 2) and has been found on pepper in California. *T. minor* has been reported to occur on garden pea and tomato in Millaro, and Nambour (Australia).

3. Symptoms and Nematode Behavior

Above-ground symptoms caused by *Trichodorus* spp., though non-specific, consist of patches or zones of stunted plants which often are chlorotic and may exhibit severe symptoms of water stress during the warmest time of the day. As the crop approaches maturity plants within infested areas often die prematurely. Roots of infected plants rarely show evidence of necrosis or galls but infected tomato roots lack a root-cap or a region of elongation and show a reduced meristematic region and protoxylem elements near the root apex (Jenkins and Taylor, 1967). Other workers have suggested that reduction or cessation of meristematic activity occurs or that cell maturation in diseased roots takes place nearer the meristem than in healthy ones. Although tap and primary roots may be infected in the seedling stage, or to some extent on older plants, secondary and feeder systems seem to be favorite

FIG. 2. *Trichodorus allius*-infected onions showing stunted plants with darkened, stubby roots in comparison with a healthy onion (right).

infection sites. Infected root systems are characterized by short "stubby" remnants of the secondary and feeder root system.

This group of ectoparasites do not possess a hollow stylet characteristic of other plant-parasitic nematodes and so feeding is accomplished by puncturing the cortex and cell walls with the "tooth" and withdrawing cell contents by powerful esophageal suction. The nematodes also possess an unusually thickened cuticle which probably protects them from adverse edaphic conditions.

4. Interrelationships

Several species of *Trichodorus* are vectors of certain soil-borne viruses (NETU). *T. pachydermus*, *T. teres* and *T. viruliferus* can transmit pea early browning virus; and *T. allius*, *T. christiei*, *T. primitivus*, *T. similis* and *T. teres* are vectors of tobacco rattle virus.

5. Control

The lack of ecological information on most species of this genus is a serious obstacle in the development of biological and cultural control programs. Asparagus is both resistant and nematicidal to *T. christiei* but it is not known if this crop has a similar effect upon other species in the genus. The host range of *T. allius* is small and thus crop rotation should be successful against this pest. Soil fumigation should be effective against all species and in particular against *T. christiei* and *T. minor* which are known to have a wide host range.

P. Dagger Nematodes (*Xiphinema* spp.)

1. Occurrence

The genus is distributed throughout the world in temperate and tropical zones alike. Individual species, however, tend to be limited to various regions.

2. Important Species and Principal Hosts

Although numerous species of *Xiphinema* have been described, only eight are known to cause plant injury. Most of these seem to prefer woody perennials although those species which transmit NEPO viruses feed on a variety of vegetable indicator plants (bean, cucumber, pea, spinach, watermelon, etc.). Evidence that *Xiphinema* spp. injure vegetable crops is meagre but *X. diversicaudatum* is pathogenic to tomato.

3. Symptoms and Nematode Behavior

Stunted growth in zone-like field patterns is a common indication of infection but, like most root ectoparasites, *Xiphinema* leaves hosts nearly devoid of secondary and feeder roots. *X. diversicaudatum*, however, on peanut, rose and other hosts causes galls containing giant cells and increases meristematic activity of the root tips (Davis and Jenkins, 1960).

4. Interrelationships

The *Xiphinema* were the first nematodes known to transmit certain soil-borne viruses. Although most of these viruses are injurious to crops other than vegetables, certain vegetables are used as indicator crops.

5. Control

Standard nematicides provide adequate control of these pests.

Q. Needle Nematodes (*Longidorus* spp.)

1. Occurrence

This genus (containing more than 40 species) is widely distributed in temperate and tropical zones of the world. Except for one species *Longidorus elongatus* which occurs in Canada, Europe and in the U.S.A., most species tend to be distributed in much smaller areas.

2. Important Species and Principal Hosts

Currently four species are considered as vegetable parasites. *L. africanus* causes serious injury to head lettuce in the Imperial Valley of California according to Radewald *et al.* (1969). Lamberti (1969) examined the host range and, in addition to lettuce, noted that eggplant, mint and spinach were injured. Yassin (1967) reports that *L. brevicaudatus* is associated with the stunting and yellowing of Jew's Mallow (*Corchorus olitorius*), a common vegetable crop of the Sudan. *L. elongatus* is a common pest of peppermint (*Mentha piperita*) in western Oregon (U.S.A.). In greenhouse trials this nematode severely damaged table beets and tomato. Cohn *et al.* (1968) reported that *L. vineacola* is an important pest of onions in Israel.

3. Symptoms and Nematode Behavior

Host plants usually are severely stunted with discolored foliage (e.g. peppermint). Most evidence of pathogenicity occurs in the root system where feeder roots are often lacking or consist of short, stubby remnants. Frequently tap-roots of young seedlings may be retarded, swollen or terminated and sometimes excessive secondary root formation or localized necrosis may develop near some feeding sites.

4. Interrelationships

Some species of *Longidorus* are known to vector certain soil-borne viruses (NEPO). *L. attenuatus* and *L. elongatus* are vectors of tomato black ring virus in England.

5. Control

Satisfactory control can be achieved with standard nematicides. *L. elongatus* has been controlled with dibromochloropropane in Oregon (U.S.A.) peppermint plantings. Cohn *et al.* (1968) reported that *L. vineacola* could be controlled with ethylene dibromide in Israel onion plantings with a 160% increase in yield without phytotoxicity.

VI. Vegetable Culture in Controlled Plantings

Many species of vegetables are grown in climatically controlled or partially controlled environments such as hot beds, cold or hot frames, glass-, green-, lath- and shade-houses and it is inevitable that various nematode pests will eventually become established in climatically controlled plantings. In most instances the pests will be introduced with propagation stocks (rhizomes, roots, transplants, tubers and occasionally seed), soil or water (unless it comes directly from a well). The most common nematode pests found in this type of culture are one or more species of root-knot nematode (*Meloidogyne*) which are injurious to most crops, particularly cucumber, melons and tomato. Occasionally other crops such as the leaf lettuces may be injured by these pests. Other nematodes including *Ditylenchus*, *Pratylenchus* and some of the ectoparasites may become established and cause problems. Sometimes the vigor of vegetables (chicory and rhubarb) forced in greenhouses may be impaired by the nematodes which accompany the "crown areas" to the greenhouse.

A wide assortment of vegetables are grown from seed to transplant size in controlled conditions and other plants may be cultured in beds for rhizomes, rooted cuttings, slips, etc. Rarely do such stocks show evidence of nematode infection at this early stage of growth. Since these types of propagation materials often are resold and distributed elsewhere, they are a major means of disseminating nematode pests and therefore maintenance of nematode-free sources is extremely important.

1. Control

The two major opportunities for nematode introduction are with propagation stocks and soil. Propagation stocks should be obtained from reliable sources and soil should be subjected to heat or steam pasteurization or to treatment by a standard nematicide recommended for this purpose.

VII. General Methods of Controlling Nematodes in Vegetables

There are many methods of controlling nematodes in vegetables but recommended procedures vary depending on climate, crop, cultural or production practices, economics, environmental restrictions, pests, soil, time of year, etc. Therefore, it is recommended that professional assistance be obtained in diagnosing nematode problems and developing a control program.

1. Preventive Methods

Obviously preventing establishment of nematode pests is the initial step of successful control. Preventive measures are of major importance when: new areas are brought under cultivation; crops are shifted from one area to another; pests are located in specific areas, preventing contamination or reintroduction following a control program. Unfortunately, nematodes usually are established by the time field or glasshouse problems are recognized. Nearly every agency which moves is capable of disseminating nematodes. Propagation stocks, including rhizomes, roots, seed, transplants and tubers, are of special importance to the vegetable industry because they may harbor nematode pests. All such stocks should be obtained from reliable sources which practise rigid nematode control.

2. Cultural Methods

Variations in cultural practices can be of considerable value in control but are limited by climate, crop, pest and production practices. The most successful cultural practice is crop rotation with non-susceptible or tolerant crops. Obviously a nematode pest with a wide host range can seldom be controlled in this manner.

3. Physical Methods

Application of heat (dry or steam) as a means of denematizing soil in limited areas of controlled environment is a very useful control method as is the hot-water treatment of infected propagation stocks.

4. Biological Methods

One of the most successful control methods is the use of resistant varieties which are available for some vegetables (bush type snap-, lima- and pole bean, cantaloup, carrot, garlic, gherkin, okra, bell- and hot pepper, sweet potato, tomato, watermelon, etc.). Control by this method is limited to those crops where resistance is available and also may be affected by climate and soil types, genera, species and biotypes of nematode pests.

Other biological control measures such as the use of antagonistic plants (asparagus and African Marigolds) are successful in specific instances but are not in general use. Attempts to combat nematodes with their enemies such as nematode-trapping fungi are being investigated.

5. Chemical Methods

Standard nematicides (halogenated hydrocarbons—bromine and chlorine) are usually applied to the soil before planting. Although a number of formulations and combinations with other materials are made, most treatments consist of injecting a liquid into the soil which volatilizes into a gaseous fumigant. Recently organophosphates and carbamates, usually formulated as granules, have been developed as nematicides and some are cleared for use on vegetable crops.

References

Acedo, J. R. and Rohde, R. A. (1968). *Nematologica* **14**, 1.

Anon. (1971). Special Publication No. 1, 7 pp. Suppl. *J. Nematol.*

Ayala, A. and Ramirez, C. T. (1964). *J. agric. Univ. P. Rico* **48**, 140–161.

Baker, A. D., Brown, G. L. and James, A. B. (1954). *Science* **119**(3081), 92–93.

Beijerinck, M. W. (1883). De oorzak der kroefziekte van de jonge ajuinplanten. Maanblad uitgegeven van wege de Hollandsche maatschappij van Landlouw, V.

Berkeley, M. J. (1855). *Gard. Chron.* **14**, 220.

Birchfield, W. and Brister, L. R. (1962). *Pl. Dis. Reptr* **46**, 683–685.

Birchfield, W. and Martin, J. M. (1965). *Phytopathology* **55**, 497.

Brzeski, M. W. and Rajewski, M. (1966). *Biul. warzyw* **8**, 195–206.

Caveness, F. E. (1967). *Pl. Dis. Reptr* **51**, 33–37.

Chatin, J. (1884). Recherches sur l'anguillule de l'oignon. Paris.

Chitwood, B. G. (1949). *Proc. helminth. Soc. Wash.* **16**, 90–104.

Christie, J. R. (1959). Plant nematodes; their bionomics and control. *Univ. Fla agric. Exp. Stn* (Gainesville).

Cohn, E., Krikun, J. and Yisraeli, U. (1968). *Pl. Dis. Reptr* **52**, 525–527.

Colbran, R. C. (1964). *Qd J. agric. Sci.* **21**, 77–123.

D'Souza, G. I. and Screenwasan, C. S. (1965). *Indian Coff.* **29**, 11–13.

Davis, R. A. and Jenkins, W. R. (1960). *Bull. Md agric. Exp. Stn* **A–106**.

Decker, H., Casamayor, G. R. and Gandoy, P. (1966). Mems. a Cent. Investnes Aprogec. Univ. Cent. Las Villas, Year 1966, 169–175.

Edmunds, J. E. and Mai, W. F. (1966). *Phytopathology* **56**, 1320–1321.

Esser, R. P. (1963). *State of Florida (U.S.A.) Depart. of Agric. Publication* **N-49**.

Faulkner, L. R. (1964). *Phytopathology* **54**, 344–348.
Faulkner, L. R. and Darling, H. M. (1961). *Phytopathology* **51**, 778–786.
Hawn, E. J. (1963). *Nematologica* **8**, 65–67.
Henderson, V. E. (1951). *Nature, Lond.* **167**, 952–953.
Holdeman, Q. L. and Graham, T. W. (1954). *Phytopathology* **44**, 683–685.
Jenkins, W. R. and Taylor, D. P. (1967). "Plant Nematology." Reinhold, New York, Amsterdam, London.
Klinkenberg, C. H. (1963). *Nematologica* **9**, 502–506.
Kuiper, K. (1959). *Meded. LandbHogesch. OpzoekStns Gent* **24**, 619–627.
Kuiper, K. and Drijfhout, E. (1957). *Meded. LandbHogesch. OpzoekStns Gent.* **22**, 419–426.
Labruyere, R. E., Ouden, H. and Seinhorst, J. W. (1959). *Nematologica* **4**, 336–343.
Lamberti, F. (1969). *Pl. Dis. Reptr* **53**, 421–424.
Lear, B. and Johnson, D. E. (1962). *Pl. Dis. Reptr* **46**, 635–639.
Lear, B., Johnson, D. E. and Miyagawa, S. T. (1969). *Pl. Dis. Reptr* **53**, 952–954.
Libman, G., Leach, J. G. and Adams, R. E. (1964). *Phytopathology* **54**, 151–153.
Liebscher, G. (1890). *Deuts. landw. Pr.* **56**, 436–437 and **84**, 627.
Linford, M. B. and Yap, F. (1940). *Proc. helminth. Soc. Wash.* **7**, 42–44.
Lownsbery, B. F., Stoddard, E. M. and Lownsbery, J. W. (1952). *Phytopathology* **42**, 651–653.
Lucas, G. B., Sasser, J. N. and Kelman, A. (1955). *Phytopathology* **45**, 537–540.
McKeen, C. D. and Mountain, W. B. (1960). *Can. J. Bot.* **38**, 789–794.
Mayol, P. S. and Bergeson, G. B. (1970). *J. Nematol.* **2**, 80–83.
Mountain, W. B. and McKeen, C. D. (1962). *Nematologica* **7**, 261–266.
Myuge, S. G. (1960). *Conference on Scientific Problems of Plant Protection, Budapest, Proceedings*, 333–338.
Olthof, Th. H. A. and Reyes, A. A. (1969). *J. Nematol.* **1**, 21–22.
Pitcher, R. S. (1965). *Helminth. Abst.* **34**, 1–17.
Powell, N. T. and Nusbaum, C. J. (1960). *Phytopathology* **50**, 899–906.
Radewald, J. D., Osgood, J. W., Mayberry, K. S., Paulus, A. O. and Shibuya, F. (1969). *Pl. Dis. Reptr* **53**, 381–384.
Safyanov, S. P. (1965). *Vest. sel'. -khoz. Nauki, Alma-Ata* **8**, 121–126.
Schuster, M. L. and Thorne, G. (1956). *J. Am. Soc. Sug. Beet Technol.* **9**, 193–197.
Sharma, R. D. (1968). *Neth. J. Pl. Path.* **74**, 97–100.
Sher, S. A. (1954). *Pl. Dis. Reptr* **38**, 687–689.
Standifer, M. S. (1959). *Pl. Dis. Reptr* **43**, 983–986.
Stewart, R. N. and Schindler, A. F. (1956). *Phytopathology* **46**, 219–222.
Thorne, G. (1961). "Principles of Nematology." McGraw-Hill, New York, Toronto and London.
Thorne, G. and Schuster, M. L. (1956). *Proc. helminth. Soc. Wash.* **23**, 128–134.
Van Gundy, S. D. (1957). *Pl. Dis. Reptr* **41**, 1016–1018.

Vecht, J. van der (1950). Op planten parasiterende aaltjes (Tylenchidae). *In* "Plagen van de cultuurgewassen in Indonesie", pp. 16–42.

Whitehead, A. G. (1968). *Trans. zool. Soc. Lond.* **31**, 263–401.

Yassin, A. M. (1967). *Pl. Dis. Reptr* **51**, 30.

17

Nematode Diseases of Flower Bulbs, Glasshouse Crops and Ornamentals

N. G. M. Hague

Department of Zoology
University of Reading
Reading, England

Introduction

These crops can be conveniently grouped together because they are grown in mostly temperate to subtropical regions in countries with highly developed urban economies where cut flowers, flowering shrubs, trees and glasshouse products are being used increasingly to decorate parks, gardens and homes. The acreages used for production of these crops are low in comparison with the total land area under cultivation, but the cash returns are high because of the intensive cropping. Monoculture is a common feature of these crops and consequently they suffer from a wide range of pests and diseases, but because of their high value the annual use of expensive phytosanitary measures are possible, even essential.

From the standpoint of the grower or nurseryman the most important loss is in stock which is unmarketable because of nematodes in the planting stock; infected bulbous plants are a good example of this type of loss. In the U.S.A., Anon. (1971) reports an average of 10% losses due to nematodes in 86,000 hectares of ornamental crops (including flowers, bulbs and flowering plants) which represents a financial loss of about $60 million annually. Other effects caused by nematodes are more difficult to assess, such as the influence of the burrowing nematode, *Radopholus similis*, on the exportation of nursery stock grown in infested soil. California and Canada have very strict regulations concerning the importation of planting material harbouring nematodes, and the North American quarantine restrictions for the potato cyst nematode, *Heterodera rostochiensis*, are well known, particularly with respect to the importation from western Europe of ornamentals and bulbs which may have been grown in infested soil. Every country and even different states within a country have different regulations, but wherever flowers and ornamentals are grown intensively the nurseryman is aware of nematodes both as actual and potential problems and he must take the necessary action to prevent spread of infestations.

I. Flower Bulbs

The latest available data on bulb acreages in different countries are shown in Table I. Gladiolus, narcissus and tulip are the most important bulb crops and of secondary importance are iris, hyacinth and lily. Except for gladiolus, grown in the southern part of the U.S.A., bulb crops are confined to temperate regions: the Netherlands, Britain, the U.S.A. and Japan are the principal producers.

A. Narcissus, Tulip and Hyacinth

The stem and bulb eelworm *Ditylenchus dipsaci* is the most important nematode pest and today is potentially as great a menace as it was in the 1920s when stem eelworm nearly wiped out the narcissus industry in Britain.

The main infective stage of the stem eelworm is the fourth stage larva which after penetrating the host develops rapidly inside the plant, the life-cycle being completed in about 20 days at 15°C (Yuksel, 1960). In narcissus a 15,000-fold multiplication during one season's growth has been reported and this would lead to complete and rapid destruction of the bulbs.

The symptoms on an actively growing narcissus plant are very specific (Figs 1a and 1b), the elongate, blister-like area (spikkel) on the leaf being readily detected when the leaf is passed between finger

TABLE I. Latest available data on number of hectares planted to ornamental bulbs in major producing countries (based on Gould, 1967); 1 acre = 0·405 hectares

Country	Year	Gladioli	Hyacinths	Bulbous Iris	Lilies	Narcissus	Tulips	Total
Canada	1962	36	1	4	0	168	25	234
Great Britain	1965					3029	1846	4875
France	1964					70		70
Guernsey	1965			13		154		167
Holland	1965	1912	851	501		1323	5725	10312
Ireland	1964					16		16
Italy	1965					66		66
Japan	1965	144	23	164	317	47	747	1442
Jersey	1965	34		21		158		213
Scilly Islands (G.B.)	1965					232	2	234
U.S.A.	1965	8449	6	412	304	902	119	10192

and thumb. Flowers from infected plants often have shortened deformed stalks that emerge from the soil at abnormal angles. Infestations are easily diagnosed in lifted narcissus by cutting the bulb transversely to reveal the characteristic brown ring symptoms. Eelworm multiplication in lifted narcissus may continue rapidly (Winfield and Hesling, 1966) and may lead to the formation of "eelworm wool" (Fig. 1b). Spikkel and brown ring symptoms are found in only very heavily infested plants, but for every obviously infected bulb in an infested stock there are many bulbs which, though apparently clean when examined, nevertheless contain large numbers of nematodes. It is for this reason that narcissus growers always treat bulbs before replanting.

The symptoms of *D. dipsaci* attack in tulips are more obvious than those on narcissus (Fig. 1c). Both leaves and flowers are damaged. The flower stem may be thickened and bent and may possess pale lesions just below the flower head which itself may be distorted. Flower petals often fail to colour completely and the leaves become fragile, showing transverse and longitudinal splits. Symptoms are not very clear in lifted bulbs but if bulbs appear dry or shrivelled ("corky") then the stock is likely to be generally infested.

In hyacinth *D. dipsaci* produces symptoms very similar to those found in narcissus, e.g. spikkels and brown rings. The root lesion nematode, *Pratylenchus penetrans*, is an important pest of narcissus particularly on light soils with a high organic content where it replaces *D. dipsaci* as the dominant pest species (Seinhorst, 1957). *P. penetrans* has been reported from the Scilly Isles (G.B.) and from the northwestern states of the U.S.A. where it attacks not only narcissus but also lilies.

1. Culture and Spread of Stem Eelworm

Narcissi and tulips are grown for two purposes: firstly to produce bulbs for retail or wholesale, and secondly for growing outdoors or in glasshouses for the flower trade.

In Britain narcissi are grown as a "two-year down" crop. The normal planting sized bulb, a "chip" or single-nozed bulb develops in the first year to a "round" which is the grade often used for forcing under glass. After a further year's growth the "round" develops into a double-nozed "mother" bulb to which are attached a number of "chips". These chips can be split off and replanted to start the cycle again. In Holland most narcissi are grown on a 1 year cycle, the replanted bulbs being hot-water treated to keep eelworm at a very low level: this is particularly important because narcissi and tulips are grown on the same land with no rotation, except that narcissus, tulip, hyacinth or some other bulb crop may be alternated.

FIG. 1. Stem eelworm, *Ditylenchus dipsaci*, infections in bulb crops.
(a) Symptoms in narcissus foliage; showing distorted foliage with thickened, yellow
spikkels. (By courtesy of the Glasshouse Crops Research Institute.)

Fig. 1. Stem eelworm, *Ditylenchus dipsaci*, infections in bulb crops.
(b) An infected narcissus bulb showing the "eelworm wool" in the basal disc region and
the soft appearance of the bulb. (Crown copyright, by permission of H.M.S.O.)

FIG. 1. Stem eelworm, *Ditylenchus dipsaci*, infections in bulb crops.
(c) Symptoms in tulip; showing the cracked flower stem, angled flower head with
   deformed, discoloured petals and the streaked foliage. (By courtesy of the Glasshouse
   Crops Research Institute.)

o*

Tulips are a 1 year down crop, often planted after narcissus, but never the other way round because the tulip race of *D. dipsaci* will breed in narcissus. The Dutch take particular care with their control of stem eelworm in narcissus to prevent it spreading to the economically more important tulip crop.

Both tulips and narcissi forced in glasshouses are cold-treated for 6 weeks prior to planting to encourage early growth, and under the ensuing moist warm conditions of the glasshouses they are extremely vulnerable to eelworm attack. The risk of economic loss when forcing narcissus bulbs is great because any eelworm infestation would be widely dispersed in the stock, since bulbs for forcing are not hot-water treated because of the risk of damaging the all-important flower crop. Hesling (1965) reviewed the "races" of *D. dipsaci* and came to the conclusion that there were perhaps 20 biological races. The host range of several of the races comprises many plants and recently it was shown that many of these races can interbreed (Webster, 1967), in which case it is time to reconsider the status of races as a pest problem. Hesling (1966) investigated the alternative hosts of the narcissus race of *D. dipsaci* and showed that the eelworm bred, sometimes very well, in 30% of the 200 kinds of plant species tested: 50% of the plants were not attacked and the remaining 20% were entered but breeding did not occur. The narcissus race of *D. dipsaci* will breed in broad, runner and french beans, shallot, chickweed, turnip, pea, onion, beet, *Brassica* seedlings and tulip. The last five are crops grown in rotation with bulbs in the fenlands of Britain where the majority of narcissus bulbs are grown.

Seinhorst (1956) has shown that in Holland *D. dipsaci* is a persistent menace on heavy soils even when the host crop is not grown frequently, whereas on the lighter soils of western Holland where bulbs are grown without rotation the eelworm only becomes important when the host crops are overcropped. In Britain reasonably wide rotations are practised but, as mentioned above, many of the alternate crops are good hosts to the eelworm.

In bulbous crops *D. dipsaci* is spread in the lifted bulb and in the soil; nematodes migrate rapidly from infested bulbs (Hague and Kondrollochis, 1969) and quickly attack surrounding uninfested bulbs (Webster, 1964).

2. Control of Stem and Bulb Eelworm

(a) *Hot-Water Treatment.* The hot-water treatment (HWT) of narcissus planting stocks over the last 50 years has undoubtedly saved the industry from time to time. The treatment damages flowers unless it is done at the right time in relation to the stage of development of the embryonic flowers; the time of year when the correct stage is reached depends on the cultivar and district. However, bulbs for forcing should not be hot-water treated except in an emergency.

In 1933 Staniland advocated immersion at 110°F (43·3°C) for 3 h but by the 1960s the standard treatment had risen to 112°F (44·4°C) for 3 h. Despite the recommendations of Weaving (1960) for better circulation of water in the tanks to improve the eelworm kill, it has been claimed in Britain that outbreaks of eelworm have been increasing. The indications were that strains of eelworm were becoming resistant to HWT and Woodville (1964) reported a rise in the "killing time" for active stages of *D. dipsaci* from 17 min in 1933 to 246 min in 1964.

Stem eelworms inside bulbs are more susceptible to HWT than when they are treated in water outside bulbs (Woodville and Morgan, 1961) where HWT at 112°F (44·4°C) for 3 h is lethal to only 38% of the eelworms (Green, 1963). In Canada and the U.S.A. the recommended treatment is 110°F (43·3°C) for 4 h while in Holland even higher temperatures (45°C+) are being considered for cleaning infested stocks.

Warm storing, i.e. conditioning the bulbs for one week at 85°–86°F (29·5°–30°C) prior to HWT reduces the damage to bulbs and flowers, but the eelworm control may be slightly inferior because eelworms also become conditioned (Winfield, 1968). "Eelworm wool", on the outside of the bulb, is particularly difficult to kill and thus the addition of a nematicide fungicide such as 0·5% formaldehyde to the hot water is advantageous. Organomercurial compounds, such as Aretan, are used but this has been discouraged in recent years because of their human toxicity. HWT will control also bulb scale mite and bulb fly.

In hyacinths, stem eelworm is controlled by HWT at 110°F (43·3°C) for 4 h. Except for a few varieties, tulips are not hot-water treated because of phytotoxicity.

(b) *Chemical Treatments.* In Britain the use of the organophosphorus compound thionazin (Nemafos = Zinophos) for tulips as a cold-water dip of dry bulbs, in a 0·23% solution (500 ml Nemafos 46% EC in 100 litres water) for 2 h, has been approved for the control of stem eelworm (Oliff *et al.*, 1969). However, the technique has not been used officially in Holland where they still prefer to destroy all infected stocks. A thionazin dip, causing only minor damage, i.e. a slight growth check to some varieties of tulips, has been a great boon to tulip growers in Britain because the only alternative control is rogueing out of obviously infested plants in the field, a very inefficient practice because symptomless bulbs contain nematodes.

A cold-water dip in 0·23% solution of thionazin for 2½ h is equally effective against stem eelworm in narcissus but only if the bulbs are to be down for 1 year (Oliff, 1966b). Thionazin is taken up by narcissus bulbs, and during the first year eelworms migrate from the bulb and reinvade adjacent bulbs in which the thionazin has been metabolized (Hague and Kondrollochis, 1969). Thionazin is a systemic compound but eelworms do not appear to be killed by feeding on cells containing the chemical. The most likely mode of action seems to be that eelworms are prevented from feeding as long as thionazin or its metabolites remain in the vicinity of the eelworms. Such a mode of action of systemic compounds is of great interest as many of the new organo-phosphate and carbamate (carbamyl oxime) chemicals seem to act as "contact" poisons by temporarily causing abnormal eelworm behaviour. If the eelworms are removed from the chemical they then behave and reproduce normally. This type of action has been called "nemato-static".

Thionazin is very effective for treating narcissus prior to forcing and in fact its major advantage over HWT is the minimal flower damage. Suitable fungicides such as formaldehyde and Aretan should be used together with thionazin which has no fungicidal action. Thionazin cold-water dips can be done in the same tanks as those used for HWT and circulation of the dip is essential. Bulbs should be planted as soon as possible after dipping and great care should be used in handling dipped bulbs as thionazin can be absorbed through the skin. Thionazin can be applied also as a foliar drench (Oliff, 1966a) to control stem eelworm in narcissus but not in tulips. A drench is not economic on a large scale but may be used to remove small loci of infection in a crop.

B. Iris

Bulbous iris, particularly the cultivar Wedgwood, are commonly attacked by *Ditylenchus destructor* and damage is often severe when iris is forced under glass. Characteristic symptoms are found in lifted bulbs (Fig. 2). The life-history of *D. destructor* (the potato-rot nematode or iris nematode) is similar to that of *D. dipsaci*, but unlike stem eelworm, there is no resistant stage (eelworm wool). *D. destructor* or at least some isolates of the species are known to feed on fungi and thus it may be able to overwinter in fallow ground. *D. destructor* closely resembles *D. dipsaci* and care is needed in identification, as both species vary greatly in size according to their hosts. In a heavy attack the nematode causes a grey-black rot on the roots and poorly developed, yellow-tipped leaves. *D. destructor* can be controlled by HWT at 110°F (43·3°C) for 3 h with added 0·5% formaldehyde, but some varieties may be damaged and so effective control can be obtained repeating the treatment every year with a shorter dipping time (Gould, private communication).

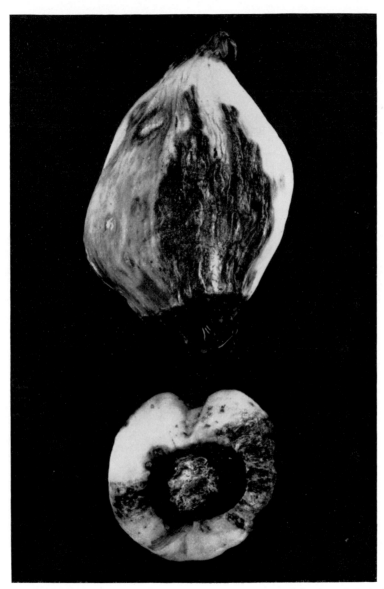

FIG. 2. Dark, rough lesions typical of a *Ditylenchus destructor* infection of the bulbous iris. (By courtesy of the Glasshouse Crops Research Institute.)

C. Lily

The most important eelworm pest of Bellingham hybrids (*Lilium aurelianense*) and Easter lilies (*L. longiflorum* cvs) is the leaf and bud eelworm *Aphelenchoides fragariae* which has an extremely wide host range including begonia, gloxinia, peony and violet. The life-cycle, similar to that of *A. ritzemabosi* on chrysanthemums, takes about 10–14 days from egg to adult. It becomes endoparasitic by passing through stomata and in the same way as the chrysanthemum eelworm can move over the surfaces of the plant in water films associated with the high moisture culture of these types of plant. Muller (1966) reported that HWT at 111°F (43·5°C) for 1 h, successful in Oregon, U.S.A., was phytotoxic in Holland. He suggested various temperature × time treatments ranging from 36°C for 6 h to 41°C for 1 h which were successful in controlling *A. fragariae*.

Various species of *Pratylenchus* are severe pests of lilies. *P. penetrans* severely damages Easter lilies (Jensen, 1961), *P. scribneri* is a serious pest of *Amaryllis* (Christie and Birchfield, 1958) and *P. convallariae* may eliminate the flower crop of the lily of the valley, *Convallaris majalis* (Cayrol and Ritter, 1962). *P. convallariae* is controlled by careful HWT at 110°F (43·3°C) for 1 h. *P. penetrans* is best controlled by planting healthy stock into fumigated soil, but, as with all bulb crops, it is very easy to advise this but more difficult to achieve in practice as it is almost impossible to be completely sure that one has healthy bulbs. D'Herde *et al.* (1960) reported good control of *P. penetrans* in roots of lily with HWT at 46°C for 20 min and in soil with DD at 4 litres/ha, which doubled yield. More recently the use of 10% granules of phorate applied at 112–170 kg/ha (11–18 kg ai/ha) gave excellent control of both *P. penetrans* and aphids (Jensen, 1966), while phorate dip treatments (0·5%) were also reported to be efficacious (Hart *et al.*, 1967).

D. Gladiolus

The U.S.A. grows about 80% of all gladioli produced in the world (Gould, 1967). Infection of gladiolus corms by root-knot nematodes (*Meloidogyne* spp.) is a major problem in the early autumn and spring plantings in Florida (Overman, 1969). The nematode invades roots, daughter corms and cormels which develop after flowering and nematodes will survive in corm tissue in the soil as a source of inoculum for next season.

For control of nematodes in soil, a pre-plant fumigation with Vorlex (80% chlorinated hydrocarbons and 20% methylisothiocyanate) at 327 litres/ha is recommended (Engelhard and Overman, 1968). To prepare nematode-free planting stocks HWT is recommended at 136°F (57·8°C) for 30 min (Magie *et al.*, 1966) and often it is necessary to warm-store corms for 2–3 months at 75°–85°F (23·8°–29·5°C) following harvest if grown at average temperatures below 70°F (21·1°C). Also dips in the organophosphates thionazin and fensulfothion at 0·5 g active ingredient per litre of water have reduced nematode infestations in corms (Overman, 1969).

E. Economics

In all bulb crops the spread of nematodes can occur in the planting stock and in the soil. In narcissus and tulips which are temperate crops it has proved difficult to control *D. dipsaci* in soil with conventional soil sterilants as they are not sufficiently effective to eliminate the nematode completely in order to produce "eelworm free" bulbs. As Seinhorst (1957) showed, even one eelworm per 500 g soil was sufficient to cause an infestation in onions on heavy soils, and thus it is a risk to plant even hot-water treated or chemical-dipped bulbs into soil despite treatment with soil fumigants. Control measures against nematodes in the planting stock are commercially satisfactory but nevertheless need repeating at frequent intervals or else populations will build up to damaging levels.

In Holland where tulips represent about 60% of the total bulb acreage, *D. dipsaci* does not exist officially but the authorities are sufficiently afraid of this eelworm to destroy all infested stocks, and it has thus proved difficult to introduce a dip treatment with thionazin, which has nevertheless been approved in Britain for *D. dipsaci* in tulips. The cost of a thionazin dip is approximately £7 per metric ton for both tulips and narcissi which is a relatively small price to pay for a crop worth at least £2500/ha.

In Britain less than 1% of the 3250 hectares of narcissi is rendered useless by eelworm; this is a loss of £60,000 on a crop worth about £6 million per annum (Wallis, private communication). However, even moderately infested crops are hot-water treated, and there is therefore a loss of income in the year of treatment because bulbs are unsaleable and the grower must wait 2 years for the crop. Wallis estimates that about half the area under narcissus is treated annually at a cost of approximately £7/metric ton which gives a total cost to the industry of about £150,000. Taking into account both the losses in crops rendered useless by the eelworm and the cost of phytosanitary measures, eelworm infestations cost about 3% of the total value of the crop. This figure is probably high because it does not take into account the value of the cut flower crop.

In bulb crops it is the potential build-up of the nematode to severe levels of infestation which forces growers to undertake routine phytosanitary measures against eelworms. Most bulb growing is done by specialists who have learned to live with pests and diseases, and because of the high value of the crops, expensive and effective control measures have been universally adopted. But in Britain where narcissus growing has expanded considerably in recent years, eelworm spread has increased significantly, largely because routine control measures have not been properly applied by many farmers.

II. Glasshouse Crops

In glasshouses, plants have a rapid and luxuriant growth as temperatures rarely fall below 10°C and they are fed and watered regularly. Usually only one crop is grown at a time thus providing a uniform environment for the rapid multiplication of the eelworm. Although as many as three crops per year may be grown, glasshouse cropping becomes virtually a monoculture which, under the ideal conditions of growth, can lead to the maximum multiplication of a pest problem. In glasshouses, eelworm diseases are almost always associated with fungal root rots. Root-knot eelworms may act as precursors of fungal infection and in some instances the fungus may compete with the eelworm for available food. This association between eelworms and fungi necessitates control measures for both types of pathogen.

A. Tomato and Cucumber

In the tropics and subtropics, including the Mediterranean areas, tomatoes are grown out of doors and are attacked by a wide range of nematodes (see Chapter 16). In temperate areas, tomatoes are grown under glass and are one of the most important cash crops.

The principal eelworm pest of tomatoes in western Europe is the potato cyst nematode, *Heterodera rostochiensis*, but the eelworm does not build up to the same extent on tomatoes as it does on potatoes. In Britain the potato cyst nematode is ubiquitous and many solanaceous weeds are good hosts to one or more of the biotypes. Thus it is not surprising that the nematode is a common glasshouse pest.

The potato cyst nematode has been spread with seed potatoes and its presence in glasshouse soils, while being a hazard to tomato production, is also a potentially serious contaminant of plants imported into other countries, especially North America, where stringent regulations exist to prevent the ingress of this nematode. The life-cycle of the nematode on potato and tomato is similar, taking about 8 weeks for development of a new generation of females after invasion by second-stage larvae. Under glasshouse conditions more than one generation may occur, and this would lead to a quick build-up of the population.

The tomato is hypersensitive to the potato cyst nematode; populations considered more or less harmless to potatoes cause severe stunting to tomato seedlings or young transplants. Young tomato plants attacked by the eelworm have a purplish hue similar to plants grown at low temperatures. Small swellings occur on infested tomato roots and they should not be confused with the galls caused by root-knot nematodes. The rate of population increase is dependent on the initial population level, the highest increases in population occurring at low initial levels when the largest cysts are produced, because competition within roots is minimal. At higher initial population levels, competition decreases the amount of food available to each invading larva, the "giant cell" becomes smaller and there is a tendency for more males to be produced. In potato growing, *H. rostochiensis* populations will decrease under rotation but in glasshouses where tomatoes are grown annually the interval is never long enough for a significant reduction in population to occur.

James (1966), reporting an association between the brown root rot fungus, *Pyrenochaeta lycopersici*, causing "corky root" of tomatoes and the potato cyst nematode, indicated that fungal exudates were interfering with the hatch of second-stage larvae. James (1968) also showed that the fungus competes for food in the root resulting in the depression of nematode populations to very low levels. *Rhizoctonia solani* and the potato cyst nematode have been reported to damage the host more severely when both pathogens are present together (Dunn and Hughes, 1964).

The root-knot eelworm, *Meloidogyne incognita*, is an important pest of glasshouse tomatoes although it is much commoner as a field pest in tropical and subtropical countries. The occurrence of root-knot eelworms is probably due to their importation in soil and planting material, particularly pot plants, which may originate in tropical or subtropical countries. Like *Heterodera* species, root-knot eelworms induce the formation of "giant cells" in the stele where the eelworm becomes sedentary and feeds. Attacked plants are generally unthrifty and tend to wilt. There are many records of disease complexes involving *Meloidogyne* spp., with fungal and bacterial wilts of tomato. Root-knot nematodes increase the incidence of *Fusarium* wilt and may break wilt resistance. The bacterial wilt caused by *Pseudomonas solanacearum* is increased in severity when tomatoes are attacked by root-knot nematodes.

H. rostochiensis populations in soil can be measured by floating cysts free from the soil, estimating their larval content and thereby obtaining an estimate of the eggs or larvae/g of soil. The Agricultural Development and Advisory Service in Britain have given a figure of 20–40 eggs/g as being dangerous depending on soil type. *M. incognita* populations are more difficult to estimate particularly for pre-planting purposes because the eggs are laid in an egg sac which becomes free in the soil, and the population can only be sampled when the larvae have hatched. Indicator plants can be used to estimate the population but it is a lengthy and time consuming process. The galling index on roots after harvesting the previous crop is the best guide but usually the population will have reached severely damaging levels before the grower becomes aware of the problem: this happens because during the build-up period of an infestation, external symptoms such as wilting are not evident in crops where adequate water is continuously applied.

Cucumbers are attacked also by *M. incognita*, on which crop it produces very large root galls, but root-knot nematodes are relatively minor pests compared with red spider, and fungal diseases, e.g. black root (*Phomopsis*) and mildew.

B. Control of Nematodes in Soil

Hussey *et al.* (1969) gave a detailed account of the methods used for controlling cyst and root-knot nematodes in glasshouses. In the soil environment fungi are more difficult to control than nematodes and it is common practice to apply sterilization techniques which control the fungi and the nematodes at the same time.

Steam sterilization is still regarded as one of the best methods of controlling both fungi and nematodes but it is becoming increasingly expensive because of the labour cost of moving pipes and sheets in the glasshouse, but in contrast with most chemical treatments it leaves no phytotoxic residues. Annual steaming does enable the growth of economic crops but the cost per hectare of contract steaming in Britain (£2000) is becoming prohibitive. Steam–air treatments may prove to be superior to the more conventional steaming techniques.

Liquid halogenated hydrocarbon fumigants such as DD (dichloropropane-dichloropropene) at 450 kg/ha and EDB (ethylene dibromide) at 225 kg/ha effectively control both root-knot and cyst eelworms, but only in the absence of root rotting fungi. Methyl isothiocyanate (Trapex) is a fungicide/nematicide and is preferable if both types of pathogen are present. These three chemicals are injected into soil which should be well cultivated before injection and then properly sealed after treatment. Under glasshouse conditions treatment may be done by hand-injector because pipes and purloins interfere with the used of motorized injectors. However, in modern glasshouses mechanical injection is the rule.

Two other chemicals used for treatment of glasshouse soils liberate methyl isothiocyanate after hydrolysis in soil. Vapam (metham-sodium) supplied as 32–35% solution is best applied as a soil drench at 120 ml/m² by diluting the concentrate to about 1 in 100 or 200. Dazomet (or mylone) is a powder formulation usually applied at 450 kg/ha and rotovated into the soil. Recently a dust-free formulation (prill) has been put on the market. Dazomet will only decompose to methyl isothiocyanate in the presence of moisture, and dry soils must therefore be watered after application. All these chemicals must be applied several weeks before planting and the timing depends on factors such as temperature, moisture content of soil, soil porosity, etc. There have been several instances of phytotoxicity with vapam applied at low temperatures, and tomatoes are sometimes tainted in DD treated soil.

In recent years in western Europe methyl bromide has been replacing other methods of sterilization, mainly because planting can follow very soon after treatment. The nematicidal properties of methyl bromide were investigated by Newhall (1947) and the chemical has been extensively used in the U.S.A. to control root-knot eelworm on tomatoes grown outdoors. Its fungicidal properties were thought to be poor but recent experiments (Galley and Hague, 1967) have shown that exposures of soil for 4 days or longer will give good control of the brown root rot and of *H. rostochiensis* on tomatoes. At normal temperatures methyl bromide (BP 4°C) is a volatile gas which is applied under elevated plastic sheets, the edges of which are buried in soil. It penetrates the soil by downward diffusion.

Methyl bromide is applied in three ways:

1. Using 1 lb canisters which are pierced under the sheet after being placed at the necessary intervals to give the required dose.

2. By vaporization from a cylinder using a dosimeter (Hague *et al.*, 1964), the gas being discharged through a "T" piece for distribution over the treated area.

3. By mechanical injection, the polythene sheet being laid down by the machine immediately after treatment. Although a satisfactory and efficient method for use outdoors in the U.S.A., this method has been found dangerous in glasshouses in western Europe and has not obtained official approval.

The second method is the safest, as the application can be done from outside the glasshouse and distribution is much quicker and more efficient than by the canister method. The generally accepted dosage is 50–100 g/m² of methyl bromide. Although 24–48 h exposure is sufficient to kill the nematodes, at least 4 days is preferred in order to rid the soil of both fungal and nematode problems. In glasshouses the operating personnel are recommended to wear respirators. The advantages of methyl bromide (see Chapter 18) over other fumigants are that planting can follow within a few days of treatment; it is a good weedicide and is an effective soil fungicide at high dosages (concentration × time products usually expressed as mg.h/litre).

All methods of soil sterilization cause increases in mineralization and the amount of nitrogen available to plants. Such soil amendment effects are most pronounced in glasshouse soils when no treatment has been given for several years and thus the nitrogen balance of fertilizers must be adjusted to prevent coarse growth.

C. Economics

Tomatoes and cucumbers are not exposed to nematode attack in their early stages of growth because the seedlings are produced in sterilized soil and it is only when they are planted out into the untreated, glasshouse soil that they are exposed to nematodes and soil fungi. Since they are protected during the early stages of growth, tomatoes and cucumbers are able to withstand all but the heaviest attacks by nematodes, even though nematode populations will build up to high levels by the end of the growing season.

Although it is difficult to obtain valuations of tomato and cucumber crops, Sheard (personal communication) has given a figure of £13·4 million as the nursery gate value for tomatoes in Britain, based on a range of £15,000–£37,000/ha off-take from the crop depending on various factors, such as whether the crop is grown in a cold or hot house or whether cropped continuously with tomatoes. Soil sterilization is normally done either annually or biannually at a cost per hectare of £1976 steam, £1088–£1235 methyl bromide or £370–£494 for chemicals such as dazomet and vapam, representing from $1\frac{1}{2}$ to 8% of the total value of the crop which is mainly at risk from soil fungi rather than nematodes. All treatments must control soil fungi, particularly brown root rot and *Rhizoctonia* and thus nematicides such as DD and ethylene dibromide would not be effective.

Cucumbers, another crop where fungal diseases are more important than nematodes, were valued at £3·6 million by Sheard and are probably worth about £50,000 per hectare. Stream sterilization is the normal means of control.

Resistant varieties are another way of controlling nematodes in tomatoes. The rootstocks KN and KNVR are resistant to *M. incognita* but not to the potato cyst nematode. However, Graham (1966) showed that these rootstocks are more tolerant to attack by *H. rostochiensis*, and good yields have been obtained in soil infested with up to 135 eggs/g, which would be considered an impossibly high population for normal commercial varieties. Recent work by Hesling and Ellis (private communication) has shown that all commonly grown tomato cultivars are susceptible to the various "biotypes" or "pathotypes" of *H. rosto-chiensis*, but there is some evidence of resistance in some of the crosses with *Lycopersicum peruvianum*.

III. Flowers and Ornamentals

A. Chrysanthemum

In recent years there has been a change from the traditional "early outdoor" and "late" varieties to the cultivation of the year-round chrysanthemum. Chrysanthemums are attacked by the bud and leaf nematode, *Aphelenchoides ritzemabosi*, in temperate regions and by various root-knot nematodes, *Meloidogyne* spp., the sting nematode, *Belonolaimus longicaudatus*, and stubby-root nematodes, *Trichodorus* sp., in parts of the southern U.S.A., e.g. Florida.

A. ritzemabosi is a severe pest for the traditional grower but year-round flowers are grown entirely under glass where conditions are not so favourable for the eelworm.

The chrysanthemum eelworm is always worse in wet, cool seasons because spread is encouraged by the surface water films in which the eelworms migrate (Wallace, 1959). The eelworms are introduced when cuttings are taken from infested "stools", and they spread rapidly throughout the crop. The life-cycle takes about 14 days, the female eelworms laying eggs where they feed either in bud axils or inside leaves. If the air humidity is high the nematodes may feed ectoparasitically in the terminal and auxiliary buds but the nematodes also enter the leaves via stomata and feed as they move through the parenchyma cells. In most chrysanthemum varieties the veins hinder nematode migration within leaves and thus the characteristic symptoms take the form of interveinal discoloration (Fig. 3). These are pale green at first, gradually becoming yellow and then dark brown or black as the leaf dies. As the season develops eelworms tend to move up the plants and the disease appears to spread up the plant from the base. The dead leaves do not drop off but hang in ragged clusters giving the plant a bedraggled look. In severe attacks eelworms may completely "blind" young plants, leading to the production of side growths before they are required, thus upsetting the timing of flowering.

Chrysanthemum cultivars vary in their susceptibility to eelworm, and the leaf-browning symptoms occur more rapidly on resistant varieties. Wallace (1961) suggested that "resistant" varieties are hypersensitive and the eelworm becomes isolated in a necrotic and unfavourable environment in which they cannot breed. The eelworm can be transported in the overwintering "stools" where it remains in the bud axils. Soil infestation from fallen leaves does not seem to be a problem in soil (French and Barraclough, 1961) but weeds which are good reservoir hosts, even though they may show no obvious symptoms, are a source of cross-infestation. Experimentally *A. ritzemabosi* has remained viable for two years in leaves which have been dried slowly but such conditions are unlikely to occur in nature.

Fig. 3. Chrysanthemum leaves showing the dark brown, interveinal lesions typical of an *Aphelenchoides ritzemabosi* infection. (By courtesy of the Glasshouse Crops Research Institute.)

A. ritzemabosi is one of the commonest and most widespread of plant nematodes, having been recorded from about 160 species of ornamentals, crop plants and weeds. The most important flower/ornamentals which could be cross-infested are aster, dahlia, delphinium, phlox, verbena, zinnia and african violet.

Meloidogyne incognita is occasionally found attacking chrysanthemums in glasshouses in temperate regions and root-knot nematodes are important pests mainly in Florida and the southern U.S.A.

1. Control of *A. ritzemabosi*

Bud and leaf eelworms occur in diseased stock from which they are spread during propagation, and they are also common in composts because of the large number of alternative hosts. Disinfestation of potting composts is best done by steam sterilization or methyl bromide fumigation.

Chrysanthemum "stools" are the main source of spread and Staniland (1950) showed that good control could be obtained by hot-water treatment at 115°F (46·1°C) for 5 min; but Hesling (1961) indicated that it was virtually impossible to treat at a constant temperature for so short a time. Further work has shown that chrysanthemum varieties vary widely in their susceptibility to hot water. A simple modification of the technique was recommended in which sufficient stools were added to the water at about 118°F (48°C) to bring the temperature down to about 116°F (46·6°C), no further heating being necessary, and stools were removed after 5½ min when the temperature was down to 115°F 46·1°C).

With the advent of all the year round chrysanthemums, three and a half crops per year being normal, *A. ritzemabosi* has tended to decline in severity because the cuttings, where the nematode is easily spread, can be treated with organophosphorus drenches such as thionazin at 0·03% applied as two applications three weeks apart (Hussey *et al.*, 1969). A single application of 0·02% is effective at a rate of 6·8 litres/m², which also controls leaf miner, red spider mite and the important aphid, *Myzus persicae*. The use of dazomet to control diseases in outdoor chrysanthemums has also led to a decline in eelworm in the more traditionally grown crop.

Economically, chrysanthemums are one of the most valuable crops grown. In all the year round crops grossing around £74,000/ha, control of eelworm is about £12·4 per hectare in the cutting stage. The incidence of eelworm in the traditional short season crop may be higher, the gross value less (£15,000–£22,000/ha) and the eelworm control more costly. Organophosphate and other systemic compounds have been used to control nematodes in relatively few plants but it is possible that they can be used on a wider range of ornamental species that are not amenable to hot-water treatment. Most of these crops are sufficiently high in value to support expensive control measures.

B. Roses

Roses are an increasingly important economic flower crop, the largest producers being in Holland, U.S.A., France and Britain, the value of this crop in the latter country exceeding £8 million (Sheard, private communication). The dagger nematodes (*Xiphinema* spp.) cause severe damage to roses under glass, the nematode being able to build up on the perennial crop. In North America the most important species is *X. americanum* which causes galling of the root tips somewhat similar to that caused by root-knot nematodes on roses. In western Europe *X. diversicaudatum* severely damages roses and also is the vector of the strawberry latent ringspot virus which is associated with malformation, chlorosis and dwarfing of roses in glasshouses (Harrison, 1967). The closely related needle nematode *Longidorus macrosoma* also is a pest of roses (Brown, 1965). The root lesion nematodes *Pratylenchus vulnus* and *P. penetrans* (Oostenbrink *et al.*, 1957) are important pathogens of roses, *P. vulnus* probably being introduced into rose culture via the *Rosa canina* rootstocks imported from various countries.

Soil fumigation is effective against these nematodes, *X. diversicaudatum* being particularly susceptible to treatment by DD at 900 kg/ha as a pre-plant treatment (Peachey and Brown, 1965). Post-planting drenches with nemagon (DBCP) controlled *P. vulnus* in container-grown roses (Sher and Bell, 1965). In Holland, Beuzenberg and den Ouden (1968) used nemagon with success against *P. penetrans*, a nematode which is controlled also by interplanting roses with African Marigold, *Tagetes* (Oostenbrink *et al.*, 1957).

C. Carnations

Glasshouse carnations are valued at nearly £4 million in Britain but fungi are the main pathogens. Esser and Overman (private communication) estimate carnations to be worth $1·5 million in Florida with a 5% loss due to nematodes.

Criconemoides xenoplax (Sher, 1959) and *C. curvatum* (Streu, 1960) cause growth decline in carnations and Streu also implicated *Tylenchorhynchus dubius* and *Paratylenchus dianthus*. Various species of *Meloidogyne* attack carnations and root-knot nematodes are implicated in a wilt disease caused by *Fusarium oxysporum* f. *dianthi* (Schlindler *et al.*, 1961).

In glasshouses, steam sterilization is still preferred to chemical methods because carnations are extremely sensitive to the bromine present in many of the effective soil sterilants. Methyl bromide can be used for carnations but thorough leaching must follow treatment to remove soluble bromides.

D. Ornamentals

In temperate and subtropical regions, the growth of shrubs in containers and pots has expanded enormously in recent years. Sheard puts the value in Britain at £12 million and in the U.S.A., Oregon alone grows about $15 million of trees and shrubs (Jensen, private communication), while Esser and Overman value Florida's woody ornamental industry at nearly $30 million. The above authors estimate that nematodes are responsible for about 5% of the loss in these crops and therefore growers are keen to take preventive measures.

Azaleas and rhododendrons are two of the most important crops and Esser (1967) reported 20 species of nematode attacking azalea, *Tylenchorhynchus* spp., *Trichodorus* spp., *Meloidogyne* spp., and *Helicotylenchus* spp. being the most important. Barker *et al.* (1965) reported *Tylenchorhynchus claytoni* as an important pathogen of azalea in the U.S.A., and in Belgium azaleas are attacked by both *T. claytoni* and *Trichodorus christiei*. Control is by pre-plant and post-plant treatments of nemagon.

Decorative foliar plants such as ferns and caladiums are menaced by nematodes, ferns being particularly susceptible to *Aphelenchoides fragariae* and *Pratylenchus penetrans*. The latter nematode is controlled by 4·5–9·0 kg/ha of a mixture of thionazin and phorate granules (Rhoades, 1968). *Meloidogyne incognita* attacking caladiums is controlled by treating the tubers in hot water at 50°C for 30 min or by dipping the tubers in 1000 ppm emulsions of organophosphates such as thionazin and Mocap (O-ethyl S,S-dipropyl phosphodithioate) (Rhoades, 1970).

The commonest nematode pest of begonias and gloxinias is *Aphelenchoides fragariae* which can be controlled by a dip or drench in organophosphates such as meta-systox or thionazin.

Infestations of *Meloidogyne* spp., and *Pratylenchus penetrans* on begonia can be very damaging (D'Herde *et al.*, 1962), steam sterilization or DD fumigation both being efficient methods of controlling *P. penetrans*.

Meloidogyne incognita attacks container grown plants such as creeping bugleweed (*Ajuga reptans*), flowering dogwood (*Cornus florida*) and Japanese boxwood (*Buxus microphylla japonica*) which is also attacked by *Rotylenchus buxophilus* (Golden, 1956). *M. incognita* on dogwood is controlled by bare root dips with various organophosphates at 1000 ppm for 15–30 min (Johnson *et al.*, 1970).

Infestations of the same nematode on boxwood (Johnson, 1969a) and creeping bugleweed (Johnson, 1969b) were decreased with drench and granular applications of various phosphate and carbamate compounds.

The cactus cyst nematode, *Heterodera cacti*, and *Meloidogyne* spp. are found often on imported cactus (Southey, 1957). Langdon and Esser (1969) gave a list of cacti infested with *H. cacti* in Florida. Some plants, particularly *Zygocactus truncatus*, are very heavily infested and show wilt symptoms with the tops turning a reddish colour. Control is either by cutting off existing roots (Southey, 1965) or by the use of aqueous organophosphate drenches (O'Bannon and Esser, 1970).

IV. Conclusions

On a "per hectare" basis, the crops dealt with in this chapter are some of the most expensive, ranging from tulip and narcissus grown out of doors worth at least £2500/ha, tomatoes £25,000/ha and all the year round chrysanthemums which have been valued at up to £74,000/ha. Estimates of the loss attributable to the direct effect of nematodes in producing unmarketable crops vary from about 0·1% in narcissus in Britain to 5% for bulb crops and lilies in Oregon, U.S.A. (Jensen, private communication) and a 10% overall loss in the U.S.A. for ornamentals (Anon., 1971).

It is difficult to get accurate estimates of the total cost to the nurseryman but a fair average figure would probably be in the order of 5–10%, depending on crop and standard of cultivation, etc. Johnson (private communication) has suggested that losses due to nematodes in container-grown ornamentals in Georgia, U.S.A. may be as high as 20–25% of production potential, and it is in the field of container or pot-grown shrubs and ornamentals that the greatest expansion is occurring with all its attendant chances of spreading nematodes during potting procedures.

The absolutely devastating effects of *D. dipsaci* to tulip and narcissus and *D. destructor* to iris are well known to bulb growers but the nematode perils of growing ornamentals and nursery shrubs, etc., are only just beginning to be realized.

All these crops are grown in almost continuous monoculture and it is significant that the most effective control of nematodes has been obtained in these crops because the value of the crops warrants expensive phytosanitary measures. Crops grown in protected cultivation are subject to many fungal root rots as well as nematodes and thus control measures should always take into account the control of both types of pathogen.

References

Anon. (1971). Special Publication No. 1, 7 pp. Suppl. *J. Nematol.*

Barker, K. R., Worf, G. L. and Epstein, A. H. (1965). *Pl. Dis. Reptr* **49**, 47–49.

Beuzenberg, M. P. and den Ouden, H. (1968). *Meded. Dir. Tuinb.* **31**, 299–304.

Brown, E. B. (1965). *Pl. Path.* **14**, 45–46.

Cayrol, J. C. and Ritter, M. (1962). *Annls Epiphyt.* **13**, 301–328.

Christie, J. R. and Birchfield, W. (1958). *Pl. Dis. Reptr* **42**, 873–875.

D'Herde, J., Brande, J. van den and Gillard, A. (1960). *Nematologica* (Supplement) **2**, 64–67.

D'Herde, J., De Maeseneer, J. and Van den Brande, J. (1962). *Revue Agric. Brux.* **15**, 1471–1481.

Dunn, E. and Hughes, W. A. (1964). *Nature, Lond.* **201**, 413.

Engelhard, A. W. and Overman, A. J. (1968). *Phytopathology* **58**, 727.

Esser, R. P. (1967). *Pl. Dis. Reptr* **51**, 46–49.

French, N. and Barraclough, R. M. (1961). *Nematologica* **6**, 89–94.

Galley, D. J. and Hague, N. G. M. (1967). *Proc. 4th Br. Insectic. Fungic. Conf.* 1967. 56–62.

Golden, A. M. (1956). *Bull. Md agric. Exp. Stn* **A-85**, 28 pp.

Gould, C. J. (1967). *Florist's Review* **140** (3633), 14–16, 70–71.

Graham, C. W. (1966). *Pl. Path.* **15**, 76–85.

Green, C. D. (1963). *Nature, Lond.* **198**, 303.

Hague, N. G. M. and Kondrollochis, M. (1969). *Proc. 5th Br. Insectic. Fungic. Conf.* 1969, 284–289.

Hague, N. G. M., Lubatti, O. F. and Page, A. B. P. (1964). *Hort. Res. Edinb.* **3**, 84–101.

Harrison, B. D. (1967). *Ann. appl. Biol.* **60**, 405–409.

Hart, W. H., Maggenti, A. R. and Lenz, J. U. (1967). *Pl. Dis. Reptr* **51**, 978–980.

Hesling, J. J. (1961). *Pl. Path.* **10**, 139–141.

Hesling, J. J. (1965). Rep. Glasshouse Crops Res. Inst. 1965 (1966), 132–141.

Hesling, J. J. (1966). *In* "The Daffodil & Tulip Year Book, 1966".

Hussey, N. W., Read, W. H. and Hesling, J. J. (1969). "The Pests of Protected Cultivation." Edward Arnold, London.

James, G. L. (1966). *Nature, Lond.* **212**, 1466.

James, G. L. (1968). *Ann. appl. Biol.* **61**, 503–510.

Jensen, H. J. (1961). *Stn Bull. Ore. agric. Exp. Stn* **579**, 34 pp.

Jensen, H. J. (1966). *Pl. Dis. Reptr* **50**, 923–927.

Johnson, A. W. (1969a). *Pl. Dis. Reptr* **53**, 128–130.

Johnson, A. W. (1969b). *Pl. Dis. Reptr* **53**, 295–298.

Johnson, A. W., Ratcliffe, T. J. and Freeman, G. C. (1970). *Pl. Dis. Reptr* **54**, 952–955.

Langdon, K. R. and Esser, R. P. (1969). *Pl. Dis. Reptr* **53**, 123–125.

Magie, R. O., Overman, A. J. and Waters, W. E. (1966). *Fla Agric. Exp. Stn Bull.* No. 664A.

Muller, P. J. (1966). *Meded. LandbHogesch. OpzoekStns Gent* **31**, 666–671.

Newhall, A. G. (1947). *Agric. Chem. Balt.* **2**, 30–31, 63.

O'Bannon, J. H. and Esser, R. P. (1970). *Pl. Dis. Reptr* **54**, 692–694.

Oliff, K. E. (1966a). *Hort. Res. Edinb.* **6**, 79–84.

Oliff, K. E. (1966b). *Hort. Res. Edinb.* **6**, 85–90.

Oliff, K. E., Hague, N. G. M. and Kondrollochis, M. (1969). *Pl. Path.* **18**, 6–9.

Oostenbrink, M., S'Jacob, J. J. and Kuiper, K. (1957). *Tijdschr. PlZiekt.* **63**, 345–360.

Overman, A. J. (1969). *Proc. Fla St. hort. Soc.* **82**, 362–366.

Peachey, J. E. and Brown, E. B. (1965). *Expl Hort.* **13**, 45–48.

Rhoades, H. L. (1968). *Pl. Dis. Reptr* **52**, 383–385.

Rhoades, H. L. (1970). *Pl. Dis. Reptr* **54**, 411–413.

Schindler, A. F., Stewart, R. N. and Semeniuk, P. (1961). *Phytopathology* **51**, 143–146.

Seinhorst, J. W. (1956). *Nematologica* **1**, 159–164.

Seinhorst, J. W. (1957). *Tech. Committee, Southern Regional Nematology Project (S-19), Alabama* **3**, 39 pp.

Sher, S. A. (1959). *Phytopathology* **49**, 761–763.

Sher, S. A. and Bell, A. H. (1965). *Pl. Dis. Reptr* **49**, 982–985.

Southey, J. F. (1957). *Nematologica* **2**, 1–6.

Southey, J. F. (1965). "Plant Nematology." Tech. Bull. No. 7, H.M.S.O., London.

Staniland, L. N. (1933). *J. Min. Agric.* **40**, 343–355.

Staniland, L. N. (1950). *Ann. appl. Biol.* **37**, 11–18.

Streu, H. T. (1960). *Phytopathology* **50**, 656.

Wallace, H. R. (1959). *Ann. appl. Biol.* **47**, 350–360.

Wallace, H. R. (1961). *Nematologica* **6**, 49–58.

Weaving, G. S. (1960). *Commercial Grower* No. **3349**, 619–625.

Webster, J. M. (1964). *Ann. appl. Biol.* **53**, 485–492.

Webster, J. M. (1967). *Ann. appl. Biol.* **59**, 77–83.

Winfield, A. L. (1968). *C.r. VIII-ème Symp. int. Nématologie*, Antibes 1965: 128.

Winfield, A. L. and Hesling, J. J. (1966). *Pl. Path.* **15**, 153–156.

Woodville, H. C. and Morgan, H. G. (1961). *Expl Hort.* No. 5, 19–22.

Woodville, H. C. (1963). *Expl Hort.* **10**, 90–95.

Yuksel, H. S. (1960). *Nematologica* **5**, 289–296.

18
Nematode Pests of Mushrooms

J. J. Hesling

Glasshouse Crops Research Institute
Littlehampton
Sussex, England

Introduction

Mushroom growing is a young industry, for although mushrooms have been grown by enthusiasts for centuries, it is only in the last 20 years that scientific methods of growing have permitted crops to be harvested with certainty on a very large scale. Mushroom growing has an aura of mystery, which results from the uncertainties of cropping. These uncertainties arose because of earlier ignorance of the mushroom's needs in terms of temperature, moisture, ventilation, nutrition and ecology, and because of diseases, pests, and unreliable spawn. Nematode pests have been associated with crop losses for over half a century, and although nematode problems now are fewer they are particularly prevalent where growing methods are unhygienic. When harmful nematodes occur on large farms their effects are proportionately more serious, while their control may be disproportionately difficult.

Near-perfect, hygienic mushroom growing is possible only in purpose-built mushroom houses, but these are not universal. Hence, disease and pest problems are common, and their incidence can be related to the standards of growing, and especially to the sophistication of the growing houses and to the facilities available to control the pathogens. Modern growing methods facilitate controlled mushroom production that has changed the industry's aura from one of "muck and magic" to one of factory farming, where the aim is to understand and control all aspects of production (and marketing) to get the biggest returns.

There have been and still are many different systems of mushroom growing, and these affect the incidence of diseases and pests, and especially nematodes. The following chapter outlines the basic processes of mushroom growing and their influence on mushroom nematodes.

I. Mushroom Growing and Nematology of Production Processes

Mushroom mycelium is grown on a suitable substrate (compost) after inoculating it with a culture of mycelium (spawn). When the mycelium has grown through the compost the latter is covered ("cased") in a layer of soil or of peat and chalk, and the mushrooms later appear through the "casing". This sequence of cultural processes, each of which may influence the nematodes, is elaborated in Table I.

A. Compost Collection

Nematodes are ubiquitous, and are invariably found in moist, well-aerated environments. Hence, manures, straw, corn cobs, hay and even sawdust naturally harbour nematodes. Cottonseed meal and proprietary compost-activators are usually dry and contain few or no nematodes, but if wetted and contaminated they become a favourable substrate and support many. Nematodes in a dry, inactive "dormant" state (anabiosis) on compost materials revive and multiply rapidly when these materials are wetted. Impure water is often used to wet compost stacks and is a source of nematodes (Jankowska, 1964). The abundance of eelworms in raw compost material depends on its nature, age and condition, and on how it became contaminated.

Every compost stack is a unique environment, but three main types of nematode invariably occur: saprophagous, fungal-feeding, and predaceous. Additionally, in composts containing animal manures, the infective larvae of many animal-parasitic nematodes occur, and in some areas of the world these may present a health hazard to those working with the infested manures.

B. Composting (Phase I)

Composting is a microbial fermentation which produces considerable heat (140°–180°F or 60°–82°C), and heat coupled with aerobic conditions

TABLE I. The processes involved in mushroom production

Process	Notes
A. Compost collection	e.g. Horse manure, cereal, straw, corn cobs, other manures (pig, poultry), additives (gypsum, cottonseed meal, molasses, leafy hay, proprietary activators).
B. Composting (Phase I)	Controlled fermentation of compost components heaped in a stack usually 1·0–1·5 m high and 1·0–1·8 m wide. May take from one to several weeks.
C. Composting (Phase II) peak-heat, compost pasteurization	Final fermentation in a closed environment (e.g. a shed) with temperature control (not to exceed 140°F (60°C)). Has two functions: (i) removal of gases (e.g. ammonia) in the compost, and "finishing off" composting; (ii) eradication of diseases and pests.
D. (i) Spawning	When compost has cooled to about 80°F (27°C) it is inoculated with "spawn". There are three common methods (Flegg, 1962): a. Spot-spawning. Spawn in lumps buried 5 cm deep at intervals throughout the compost. b. Surface-spawning. Spawn scattered on surface of compost and "ruffled in". c. Through-spawning. Spawn distributed through entire depth of compost by hand or machine.
D. (ii) Spawn-run	Growth of mycelium through the compost, recommended temperature 70°–75°F (21·1°–23·8°C). Usually requires about 14 days.
E. Casing	Covering of soil, or peat and chalk, about 2–4 cm deep, on the surface of the compost.
F. Cropping	Mushrooms tend to appear in "flushes", the first about 3 weeks after casing; often the beds are watered between flushes; recommended temperature about 60°–65°F (15·6°–18·3°C).
G. Cook-out	The disinfection of the contents of the cropping house.

favour the thermophilic micro-organisms whose activities produce the best substrate for later mushroom growth. Water, nitrogenous additives, and (very recently) soluble carbohydrates (e.g. molasses) are used to accelerate microbial action, and soon thermophiles predominate. A compost stack is self-insulating and the temperature in the centre becomes lethal to nematodes. However, in the cooler, moist, peripheral regions of the stack, enormous multiplication of nematodes, especially of saprophagous nematodes, occurs.

The composting process takes either about 7 days ("short composting") (Sinden and Hauser, 1952; Hayes and Randle, 1968a and b) or, sometimes, several weeks ("long composting") and during these times the stack is turned to achieve uniformity and to assist aerobic fermentation. In "short composting" the stack is turned about every 2 days and may then be "finished", i.e. little further fermentation occurs and this ensures that any particular environment in the stack is ephemeral. The only organisms that can increase without interruption are those which thrive at high temperatures ($140°–180°F$ or $60°–82°C$). Such temperatures are lethal to nematodes. No nematode is known with a life-cycle of less than 2 days, so in "short composting" there is little opportunity for the development of enormous populations of nematodes. With "long-composting", however, where the compost turning may be at weekly, or longer, intervals, the saprophagous nematodes have ample opportunity for increase as many of them have life-cycles of 3–7 days. There is a temperature gradient from the centre (hot) to the outside (cool) of a fermenting compost stack, thus providing zones where the temperature is optimal for nematode breeding. Long intervals between turns permit weed fungi to become established, and if stylet-bearing mycophagous nematodes are present they may thrive on these fungi. The life-cycles of mycophagous nematodes are usually longer than a week, and efficient turning of the stack at weekly intervals kills many of them. However, because stack turning is never a perfect "sides to middle" some nematodes survive and continue to multiply on the surviving fungus. Thus with "long composting", both saprophagous and mycophagous nematodes have better opportunities for survival and increase than they have with "short composting".

As fermentation proceeds many physical, chemical and biological changes occur in composts and affect the eelworm populations. The relative dominance of nematode species in a compost stack may change as conditions in the stack change. Where conditions permit rapid breeding of saprophagous forms, many of the larvae swarm on top of the compost stack and congregate in glistening flame-like clusters. These swarming nematodes attach themselves to flies, mites, hands, implements, etc., and in this way they are transferred to other mushroom beds. Most of the flies associated with compost carry nematodes on their legs and bodies. The mycophagous nematodes and the saprophagous forms (suspect pathogens (Hesling, 1966a) and carriers of pathogenic micro-organisms) (Steiner, 1933), are eliminated at the end of composting by peak-heat or pasteurization (Composting Phase II, below).

C. Composting (Phase II)

When composting (Phase I) is complete, the compost is placed in boxes or trays, and placed usually in an insulated room. In some parts of the world compost is made up into beds on the ground, or on floors of caves, cellars, etc. There it continues to ferment, and the heat generated (peak-heat or "sweat-out"), frequently with much ammonia, "finishes" the compost. The temperature of the compost in peak-heat ideally rises to $135°–140°F$ ($57°–60°C$); higher temperatures are deleterious because they encourage weed fungi. The air temperature in the room housing the compost during Phase II is often only $100°F$ ($37·8°C$), and although the centre of the compost is hot, the temperature of the boxes, trays, shelves and house structure scarcely exceeds the air temperature. Thus, nematodes which often are driven by heat from the centre of the compost and come to lie on the compost surface, or on boxes, are not killed, as neither the temperatures nor the common pesticides applied (mainly to control arthropods) at these sites are lethal. During Phase II the amount of fermentation and hence of heat generation, of carbon dioxide and ammonia evolution, and the moisture content of the compost, are controlled as much as possible. Some composts fail to peak-heat naturally, because fermentation finishes in Phase I, and under these circumstances few nematodes will be killed unless artificial heat is used. To ensure that all the benefits of Phase II are achieved, and especially the control of pests, many growers prefer to heat the compost artificially until compost and air temperatures are high enough (up to $140°F$ or $60°C$) to kill all pests including nematodes ("Pasteurization"). Exposure to such temperatures for only a few minutes is lethal to nematodes, but the insulation properties of the compost, etc., dictate that hours rather than minutes are needed in practice. Generally, the deeper and wetter the compost and the thicker the wood of trays and boxes, the longer will be the time needed to achieve uniform heating, and hence, uniform nematode kill. Boxes of compost should be raised off the floor and a fan used to distribute the heat uniformly around them so as to preclude draughts.

Different methods of pasteurization/peak-heating are used and many

P

result in uneven heat distribution (Moreton and John, 1962); therefore eelworms, doubtless, survive. Nematode control during Phase
II can only be assured by the controlled heating of the compost and
its containers until all the contents of the peak-heat room reach temperatures of 135°–140°F (57°–60°C). Sometimes it is culturally disadvan-
tageous to pasteurize compost for a long time, and growers may be
tempted to end the treatment before all nematodes have been eradicated,
with resultant crop losses.

Pasteurization facilities on mushroom farms throughout the world
vary greatly and the incidence and severity of nematode problems can
be related to the quality of the pasteurization. Standards of mushroom
growing throughout the world are improving rapidly, and old methods
of composting, with reliance on residual fermentation in Phase II to
effect a "peak-heat", are rapidly being replaced by new methods of
"short-composting" followed by controlled pasteurization by artificial
heat. Pasteurization is then needed more for pest control than for
"finishing" the compost. The capital and running costs of heat pasteurization are great, and the heat is damaging to the house structures.
Methyl bromide fumigation, after a compost is "finished" in Phase I,
recently has been shown to replace the pesticidal action of heat pasteurization, and it causes little or no structural damage.

D. Spawn, Spawning and Spawn-run

Before the advent of "pure culture spawn" made under sterile conditions, the spawns available harboured many pests including nematodes. Spawns supplied today are nematode-free.

The three spawning methods described in Table I evolved largely
because of different local needs on mushroom farms, and because it is
beneficial to get the quickest possible colonization (spawn-run) of all
the compost with a vigorous mycelium. Generally, a speedy spawn-run
gives less chance to organisms which compete with mushroom in the
compost. Spot- and surface-spawning perhaps give slower mycelial
colonization than through-spawning. Both spot- and surface-spawning
are done by hand, and there is a risk of the worker spreading nematodes
with his hands or with "bats" sometimes used to scatter spawn
between the closely-stacked compost boxes.

Through-spawning ensures that spawn is distributed throughout the
mass of compost and is often done by machine. Mixing of spawn with
compost aids nematode dispersal because the machine may become
contaminated with nematodes and spread them to different batches of
compost. Saprophagous nematodes and their associated bacteria compete with mushrooms, and when they are present it is advisable to use

more spawn so that the compost is colonized by the mycelium as rapidly as possible.

In "healthy" well-prepared compost, initial eelworm numbers as high as 1500 per ml compost failed to affect mycelial growth (Hesling, 1966b). Saprophagous nematodes will, however, multiply in such compost during the few days before it is colonized, and also in imperfect composts where the physical, chemical or biological environment is unsuitable for mycelial growth. The reason for the dominance of mushroom over such nematodes in good compost is not clear, but it may be due to the uptake of moisture by the mushroom, thus denying the eelworms the moisture film essential for their movement and breeding; also, healthy mycelium appears to have a hydrophobic surface, for it is not easily wetted. Furthermore, the mycelium is believed to secrete bactericidal substances (Jackson, 1965), and the scarcity of certain bacteria in the mycosphere may limit the number of bacteria-feeding eelworms (Hirschmann, 1952).

E. Casing

1. Casing as an Environment

Few or no mushrooms appear on beds not cased (Lambert, 1938; Pizer and Leaver, 1947) but the functions of the casing layer are not completely understood. The micro-organisms present in the casing differ from those occurring in the compost and recently Hayes *et al.* (1969), and Park and Agnihotri (1969a and b) showed that metabolites of certain bacteria in casing seem to trigger sporophore formation. These beneficial and vitally important bacteria increase in the casing layer, making nutritional use of volatile metabolites of the mushroom mycelium (Hayes *et al.*, 1969), and their role and numbers may be affected by nematodes in the casing.

In general, substances used for casing have been selected because they give a reliable casing layer which is "self-cleansing" from the mushroom, has a near neutral pH, and a physical structure that permits good water retention and good aeration and at the same time withstands watering without loss of structure. Both loam and peat/chalk casing media also provide a favourable environment for nematodes (Wallace, 1963). In such an acceptable environment the availability of suitable food is an important factor determining the species of nematode present. Most casing materials contain much organic matter, and when on mushroom beds, bacteria become abundant in them. Organic matter and bacteria are important foods of saprophagous (microbivorous) nematodes, so many thrive in the casing. However, not all flourish, as the ecological optima vary for different species. Thus, those nematodes normally common in the saline moisture of raw manure would be osmotically uncomfortable in hypotonic casing moisture. As cropping proceeds and conditions change, e.g. decaying bases of cut mushrooms become abundant, and diseases and pests locally alter the environment so the relative sizes of the populations of saprophagous nematodes change.

Ideally, casing soil contains very little mycelium, and so the adverse effects of mushroom on nematode movement and feeding are not important eelworm-limiters in the casing layer. Sometimes, however, the casing is overgrown with a dense mycelial mat ("stroma" or "overlay"), and in this condition casing contains few saprophagous nematodes. Some fungal-feeding nematodes are found in the casing wherever there is mycelium.

2. Nematodes Associated with Casing Materials

(a) *Casing from Loam.* Loam contains many potential mushroom diseases and pests, and the nematodes in it may be fungal-feeding, saprophagous and predaceous forms. Unless the loam is sterilized, these nematodes will be transferred to mushroom beds, and, alone or with other deleterious organisms, may cause crop losses. There is no evidence that during the normal preparation of the loam, numbers of potentially pathogenic nematodes are decreased. Clearly, loam should be sterilized before use (see later).

(b) *Casing from Peat and Chalk.* These are mixed in approximately equal quantities and used extensively. The chalk is commonly direct from quarries and may be considered to be eelworm free. Deeply dug, uncontaminated peat that has been processed and kept dry harbours few nematodes, but shallow-dug peat, because of its proximity to plant life, may harbour many nematodes of different species. If the peat becomes contaminated and wet, the nematodes in it, including the fungal-feeding forms, breed rapidly. Therefore, peat should be stored dry and uncontaminated, and preferably should be sterilized before use. The need for sterilization is greater on those mushroom farms where nematodes cannot be eradicated at the end of cropping, since the chance introduction of a few fungal-feeding nematodes—with parthenogenetic species only one—may lead to chronic infestation that lowers yields on every succeeding crop. Rothwell (1969) has warned that a brown mould *Plicaria fulva*, which interferes with mushroom production, is encouraged if peat is heat sterilized above 137°F (58·3°C).

F. Cropping

1. General

Records of nematode species in mushroom crops are scattered in the literature (Paesler, 1957; Choleva, 1966; Brzeski and Jankowska, 1966) and a survey of nematodes that breed in mushroom beds probably would show that the more common species were from the Rhabditida, Aphelenchoidea, and Tylenchidae. Other forms, such as larvae of entomophagous nematodes, are resident in compost but do not breed in it. Both predaceous (Diplogasterids) and microbivorous forms of the Rhabditida breed in the mushroom beds as well as mycophagous and predaceous (e.g. *Seinura*) forms of the Aphelenchoidea. The genus *Ditylenchus* in the Tylenchidae includes at least one common and very destructive species.

It is convenient now to discuss nematode species belonging to the three groups—fungal-feeders, predators, and saprophagous (microbivorous) forms—because the feeding habits of nematodes greatly influence their importance as pests.

2. Fungal-feeding Nematodes

These feed by piercing the mycelial wall with their hollow stylet and sucking out the contents of the pierced cells. The damaged mycelium "bleeds" and the liquid lost in this way produces a moist film (often absent on undamaged mycelium) which may facilitate nematode movement. Hundreds of cells are pierced, and as the nematodes breed the damage rate increases. Depending on the time of their introduction to the beds, fungal-feeding nematodes may attack the mycelium at any time from spawning onwards. Where cropping lasts only 6–8 weeks and initial numbers of nematodes are low, or if they get into the beds at a late stage in cropping, the crop loss will be negligible. However, greater initial numbers, or increased numbers of nematodes owing to a longer cropping time, may cause a steady decline in mushroom yield and production may finally stop. A decline in yield is often the first outward symptom of attack by fungal-feeding nematodes. Prior to this time, however, mycelium in infested compost will have become sparse and stringy, but this will not normally be seen because growers are reluctant to disturb the beds to check the health of the mycelium (Figs. 1a and b). In the final stages of attack, no mushrooms appear, and the compost surface sinks, often in patches, and has a peculiar medicinal odour. Although this odour is associated with nematode damage, it is not specific to it, and is probably a product of the microbial breakdown which follows the nematode attack on the mycelium, and which causes the sogginess and collapse of the compost. A greyish mould, the "Cephalothecium disease" (Lambert *et al.*, 1949), often appears on the surface of the sunken patches and is composed of the fruiting bodies of an eelworm-trapping fungus, *Arthrobotrys* sp.

Peak populations of fungal-feeding nematodes occur shortly before the last of the mycelium is killed, and are followed by a rapid population

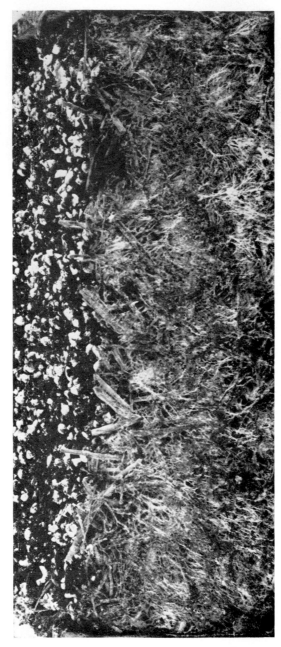

Fig. 1. (a) Profile of eelworm-free compost and casing showing abundant mushroom mycelium. (By courtesy of Glasshouse Crops Research Institute.)

FIG. 1. (b) Profile of *Ditylenchus myceliophagus*-infested compost and casing showing almost complete destruction of mycelium. (By courtesy of Glasshouse Crops Research Institute)

decline as the nematodes die or leave their fouled environment. When their migration is arrested by dry surfaces the nematodes often congregate in masses on the boxes, on the casing, or they may even hang like miniature stalactites beneath the beds (Ritter, 1957; Hesling, 1966b) (Fig. 2). There is no evidence that they wilfully move to fresh feeding sites (Cayrol and Ritter, 1967). This escape reaction with subsequent aggregation in drying conditions has been recorded in many species (Cairns, 1954; Hirschmann, 1962; Evans, 1970). Aggregation in such cases is the inevitable result of restriction of migration. Although aggregation may help survival (Ellenby, 1968), it seems unlikely that it is a form of behaviour evolved for this purpose.

Damage by fungal-feeding forms is usually patchy because of uneven eelworm distribution. By the time damage is noticed and the cause identified the peak nematode population may have passed and extensive spread may have occurred. Migrating nematodes may be transported to other mushroom beds on hands, implements, or by insects; they may be washed to lower beds during watering. There is no method of controlling fungal-feeding nematodes in mushroom beds without destroying the mycelium. Even if there were, it is unlikely that the mushroom could recolonize zones in compost where the mycelium had been destroyed because in these places the compost is toxic. When eelworm damage is detected during cropping all that can be done is to remove or physically isolate the offending box or patch of compost and disinfest it. The safest and surest way to disinfest is by heating throughout to a minimum temperature of 160°F (71°C).

3. Predaceous Nematodes

When compost is healthy and mycelial growth good these nematodes are more abundant in the casing than in the compost.

The numbers of predaceous nematodes probably depend on the abundance of their prey. Their population dynamics, and the predator/prey relationships have received little study, but there is no evidence to suggest that predaceous nematodes could usefully control harmful nematodes, or that they are harmful to cropping.

4. Saprophagous Nematodes and Bacterial Problems of Mushrooms

The experimental evidence showing the pathogenicity of saprophages is scanty and of doubtful validity. They probably are not primary pathogens, but thrive, sometimes in astronomical numbers, where other pests and diseases have inhibited mushroom growth. Thus they are often associated with poor crops, and a possibly wrong correlation is made between them and crop failure. Several workers have claimed that saprophagous nematodes are harmful (Kux and Rempe, 1954; Van

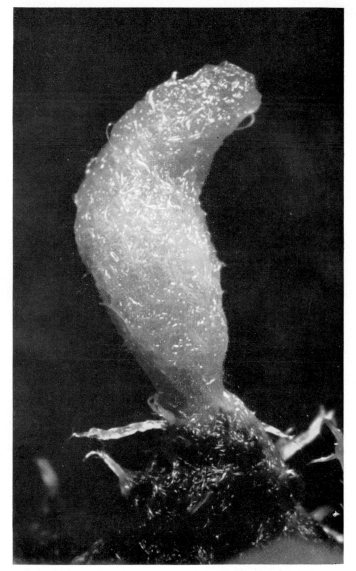

Fig. 2. An aggregation of *Ditylenchus myceliophagus* on casing after destruction of the mushroom mycelium in compost below.

Haut, 1956; Blake and Conroy, 1959; Conroy and Blake, 1960). However, experiments proving this have often involved the inoculation of spawned compost or spawn cultures with mixtures of nematodes and other pathogenic organisms and their noxious by-products, in a way that rarely, if ever, occurs on a good, hygienic mushroom farm.

P*

It appears, however, that an abundance of saprophagous nematodes may encourage spread of bacterial diseases such as "blotch" or "pit" because bacteria are spread as surface contaminants of nematodes or in their faeces. Possibly bacteria and eelworms may act synergistically, and the resulting damage is often in the form of a readily visible blemish on the mushroom cap which decreases its market value.

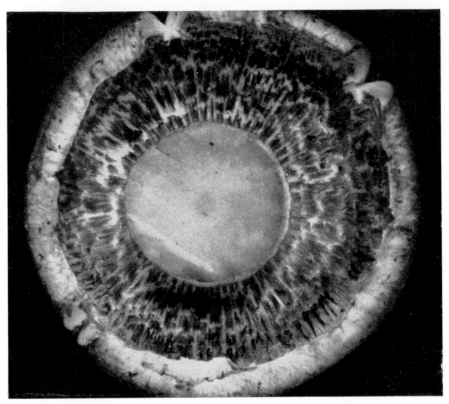

FIG. 3. The underside of a mushroom showing almost complete lysis of gill tissue caused by bacteria spread (*inter alia*) by nematodes. (By courtesy of Glasshouse Crops Research Institute.)

In one case of "pit", a bacterium (believed to be systemic in mushroom) initially caused lysis of the gill tissue of closed button mushrooms, with the bacterial colonies first forming minute glistening droplets on the gills. Later, the gill tissue was completely lysed, and pits developed on the mushroom cap (Fig. 3). The pits were moist and acted as collecting sites favourable for a rapidly-breeding parthenogenetic *Rhabditis* sp. Eventually the whole mushroom was covered in bacterial slime containing abundant nematodes which helped bacterial spread.

Saprophagous nematode/bacterial problems are perhaps less common in countries with high standards of growing, but as little is known of the conditions that invoke these problems, their relative rarity where growing standards are good cannot be equivocally explained by concomitant good hygiene. It has not been possible to reproduce "pit" experimentally by using large inocula of bacteria and of nematodes, together or separately. Methods of composting and growing change, and so the future may bring conditions that favour nematode/bacterial diseases, and "pit" may return as a serious problem, as it was in the 1950s (Atkins, 1954).

G. "Cook-out"

When cropping ends mushroom beds contain many pathogens which are readily spread, especially during the removal of the spent compost. Any debris left in the growing house provides disease inoculum for the next crop. To prevent disease spread and carry-over, the house and its undisturbed contents are often sterilized *in situ*. It is impractical to use chemical sprays or drenches as sterilants, for none is available that fulfils all the needs. Thus, a suitable sterilant needs to be: cheap, a broad-spectrum killer, sufficiently penetrating to reach pests hidden in crevices, non-persistent, and non-toxic to humans. A 2% solution of formalin is sometimes used to wash down shelves and floors after the growing house has been emptied, but it rarely eliminates all nematodes.

Another treatment, unlikely to eliminate nematodes, is used in cavern cellars of Budapest where floors are washed with a drenching spray of 10% commercial sodium hypochlorite followed by a sprinkling of slaked lime on floors and walls. After treatment the cellars are sealed for 2 days, then ventilated and left undisturbed for 2–3 months (Heltay, 1957).

Experience has shown that nematodes are among the most difficult pathogens to eradicate, and the most reliable method of controlling them is to heat the house and contents until every part of the compost, shelves, boxes, etc., has reached a temperature of 160°F (71°C), which must then be held for at least 12 h (Hayes, 1969a).

Efficient "cook-out", like efficient compost pasteurization by heat, is possible only in purpose-built houses with facilities for controlling and evenly distributing the heat. Unlike pasteurization, however, there is no cultural reason for keeping the temperature at 135°–140°F (57°–60°C), and theoretically, the hotter the cook-out, the better the kill of pathogens. High temperatures for long durations will ensure complete eradication of nematodes and other pathogens. The treatment is more successful if heat transfer is improved e.g. by fans, and cold spots near badly fitting doors, at floor level, etc., avoided.

Some nematodes are more difficult to kill by heat when in a dry anabiotic state (Cairns, 1954; Klingler and Lengweiler-Rey, 1969), also, dry compost material is a good insulant and will protect nematodes enclosed in it. Therefore, before cook-out by heat it is always advisable to damp down any dry compost, woodwork and floors.

An effective cook-out is impossible on many mushroom farms in the world because the cropping houses have no facilities for it, or will not tolerate the structural stresses that the fierce heat creates. Where mushrooms are grown in caves or cellars, cook-out is often impractical or dangerous. In Taiwan (Formosa) many mushroom houses are built of rice straw on a bamboo framework. These houses are "disinfected" annually, and are demolished after 2–3 years to break the cycles of pests and diseases (Luh, 1967).

In recent years, methyl bromide has been used increasingly as an alternative to cook-out. Properly used, it is extremely effective and causes no structural damage, though it is sometimes more expensive (see later).

II. Nematodes of Real or Probable Importance

A wide range of nematode groups occurs in "raw" compost, but fortunately only a few of the species encountered cause crop loss. It seems that modern methods of composting and growing permit few species to survive, and at mushroom growing temperatures (65°F or 18·3°C) some of these reproduce too slowly to be a problem.

Ditylenchus myceliophagus and *Aphelenchoides composticola* have been recorded as mushroom pathogens in almost every country where mushrooms are grown. Other recorded pathogens are seemingly universal, but become a problem only infrequently, e.g. *Aphelenchus avenae*. Many other fungal-feeding species are rare in mushroom culture and so are infrequently recorded as the cause of crop loss. Fungal-feeding nematodes survive weeks of starvation, and if dried slowly in peat or compost they will survive for many years in a cryptobiotic state until revived by moisture. On revival they are able to feed and breed upon a suitable host. Cryptobiotic eelworms survive best at low temperatures, e.g. 35°F (1·8°C).

Aphelenchus and *Aphelenchoides* feed briefly at each attacked hyphal cell (Fisher and Evans, 1967; Siddiqui and Taylor, 1969), but *Ditylenchus* may remain at a feeding site for over an hour. In this time it passes digestive juices into the cell through the stylet, and later withdraws and ingests the products of this extra-corporeal digestion. Thus

D. myceliophagus affects cells adjacent to the one it has punctured (Doncaster, 1966) and is considered to be more destructive than *A. composticola* (Cayrol, 1967a), but both are sufficiently destructive to cause crop loss.

A. *Aphelenchoides*

Many species of *Aphelenchoides* are found in mushroom beds and their identification is outside the scope of this book. The following selected references are helpful: Baranovskaya (1963); Eroshenko (1967, 1968a, b); Franklin (1955, 1965); Husain and Khan (1967); Riffle (1970); Sanwal (1961, 1965); Shavrov (1967); Siddiqi and Franklin (1967); Siddiqui and Taylor (1967); and Thorne and Malek (1968). Species of *Aphelenchoides* found associated with mushroom are listed in Table II.

TABLE II. *Aphelenchoides* species recorded from mushroom beds or experimentally shown to be harmful to mushroom

Species	Remarks	References
A. composticola	Widely distributed and has a 10-day life-cycle. A major pest of mushrooms.	Franklin (1957), Arrold and Blake (1966), Cayrol (1967b)
A. bicaudatus	Widely distributed, feeds on many common fungi. Lower reproductive rate than *A. composticola* and unlikely to be such a serious pest.	McLeod (1967), Siddiqui and Taylor (1967)
A. coffeae	Slow reproduction on mushroom, no evidence that it causes crop loss.	McLeod (1968), McLeod and Conroy (1968)
A. limberi (syn. *Paraphelenchoides limberi* (Khak, 1967)	Rarely a problem, adversely affected by spawn-run temperature which is high enough to limit its breeding. Thrives on *Botrytis cinerea*; parthenogenetic.	Cayrol (1967a)
A. saprophilus	Similar to *A. bicaudatus* in its effects on mushroom. Common in rotting plants, probably lives on many soil fungi.	Juhl (1966), McLeod (1968)
A. cyrtus *A. dactylocercus* *A. helophilus* *A. parietinus* *A. sacchari* *A. spinosus* *A. winchesi* (syn. *Seinura winchesi*)	Have been found in mushroom beds, or experimentally shown to be harmful to mushroom. Little known of extent of damage, or of biology.	Paesler (1957)

1. Aphelenchoides composticola

Described by Franklin in 1957, this species was first reported in 1956 (Moreton *et al.*, 1956) and has been recorded in most countries that have a mushroom industry. Cayrol (1967a) found that 13% of mushroom compost samples from French farms were infested with this species, while investigations in Britain showed it was usually present in samples of mushroom compost (Hooper, 1962a). Arrold and Blake (1968) showed that infestations of 1, 10 and 50 *A. composticola* per 100 g compost at spawning reduced mushroom yields by 26, 30 and 42% respectively over a 12–14 week cropping period. With 10 or 50 *A. composticola*/100 g, cropping ceased after 12 weeks.

A. composticola is small (0·5 mm long), varying in size and shape according to its food supply. It thrives on many common fungi, the life-cycle from egg to adult occupies 10 days at 64°F (18°C), and 8 days at 73°F (28°C) (Cayrol, 1967b). Arrold and Blake (1966) showed that reproduction was more rapid at 25°C than at 15, 20 or 30°C, the sex-ratio being unbalanced in favour of females (Brun and Younes, 1969). In water, *A. composticola* tend to aggregate which probably helps to ensure that several eelworms are available to initiate a fresh infestation.

B. *Ditylenchus*

1. Ditylenchus myceliophagus

Until *D. myceliophagus* was described by Goodey in 1958, *Ditylenchus* nematodes in mushroom beds were usually identified as *D. destructor*; since then most reports of *Ditylenchus* in mushroom beds cite *D. myceliophagus*. This species has a world-wide distribution and is common. Cayrol (1967a) found that 85% of the samples of mushroom compost contained this nematode, which he considered to be the most damaging of all mycophagous species. Goodey (1960) demonstrated that populations as low as 3/100 g compost at spawning caused a yield loss of about 30%; Arrold and Blake (1968) found that 20 or more *D. myceliophagus*/100 g compost at spawning prevented cropping, and that 20, 100 and 300 eelworms/100 g at casing caused yield reductions of 50, 68, and 75% respectively.

D. myceliophagus is a bisexual species that thrives on many common fungi. It increases optimally at mushroom-growing temperatures 64°F (18°C), but at 78·8°F (26°C) reproduction almost ceases. The life-cycle is completed in 40 days at 55·4°F (13°C), 26 days at 64°F (18°C) and 11 days at 73°F (23°C) (Cayrol, 1962; Evans and Fisher, 1969). Cairns (1954) showed that the increase of *D. myceliophagus*, and hence damage by it, was much less below 55°F (13°C), and he suggested that mushrooms should be raised under cooler conditions (45°–50°F or 7·2°–10°C) when *D. myceliophagus* is present. However, this would not be economically acceptable for intensive mushroom production.

When *D. myceliophagus* is feeding well on mushroom it is about 1 mm long, but conditions of starvation, other hosts, and different temperatures affect its size, so that care is needed in identification (Cayrol, 1965; Brzeski, 1967; Cayrol and Legay, 1967). Like *A. composticola*, *D. myceliophagus* aggregates in water.

2. Ditylenchus destructor

Before *D. myceliophagus* was described in 1958 *Ditylenchus* from mushroom beds were usually identified as *D. destructor* which is very similar in appearance to *D. myceliophagus*, and some believe that the latter is merely a strictly *mycophagus* race of *D. destructor* (Smart and Darling, 1963). However, *D. destructor* and *D. myceliophagus* will not interbreed, and consistently differ serologically and morphologically (Webster and Hooper, 1968; Evans and Fisher, 1970a). There is no doubt that *D. destructor*, or some races of it, feed and multiply on mushroom and destroy it (Seinhorst and Bels, 1951; Faulkner and Darling, 1961). *D. destructor* has a wide host range amongst fungi and higher plants, and it is common and widespread. Thus there is some risk of its being introduced into mushroom beds. Reports of crop losses due to *D. destructor* seem rare, suggesting that the race(s) of *D. destructor* that feed on mushroom are less common than their counterparts which destroy higher plants, or, as with *Aphelenchus avenae*, that some factor(s) limit(s) their pathogenic potential. Little is known of the optimal temperature for reproduction of *D. destructor*, or of the nematode's persistence. *D. destructor* is said not to withstand drying, and there are no reports of its aggregation.

C. Other Mycophagous Nematodes

1. Aphelenchus avenae

Although this is a common cosmopolitan species with a large number of fungal hosts, and often has been shown to rapidly reproduce parthenogenetically on mushroom, and destroy it, it seems to occur in mushroom beds much less frequently and in smaller numbers than *A. composticola* or *D. myceliophagus* (Hooper, 1962a; Townshend, 1964; Goodey and Hooper, 1965). The reason that it is relatively scarce and harmless in mushroom beds is not clear but it may be because spawn-running and cropping temperatures are too low, since the optimum temperature for reproduction in *A. avenae* is about 30°C (Fisher, 1968, 1969; Evans and Fisher, 1970b).

A. avenae from different isolates vary in size, internal morphology, fecundity and sex-ratio; males are usually extremely rare. Different hosts, temperature and starvation stress affect all of these characteristics (Goodey and Hooper, 1965; Juhl, 1966; Dao, 1970; Evans and Fisher, 1970b).

2. Paraphelenchus myceliophthorus

This species was discovered in mushroom compost and destroys mushroom on which it increases 50,000-fold in 8 weeks at 65°–70°F (18·3°–21·1°C) (Goodey, 1960).

Fortunately *P. myceliophthorus* is uncommon in mushroom beds. Little is known of its biology and of the optimal conditions for reproduction. It is known to feed on soil fungi, and it can be assumed that it normally exists in association with fungi colonizing rotting organic matter. Species related to *P. myceliophthorus* are primarily fungal feeders; some are parthenogenetic and must be considered a potential menace to mushroom growing. Taylor and Pillai (1967), Khak (1967) and Husain and Khan (1967) recently described species and produced identification keys.

III. Control of Mushroom Nematodes in Houses and in Compost

A. Chemical Control: Special Problems

The chemical control of mushroom nematodes has some difficulties which are common to all pest control, and others that are unique to this crop and its mode of production. In common with pest control generally, there must be no toxicity to the crop or the consumer or grower; the best time and site must be found for the chemical's application, and the treatment must be economic.

In mushroom growing, cropping is often continuous for 6–8 weeks, and the mushrooms are harvested and consumed soon after they appear. Therefore any chemical applied during cropping must be non-toxic and/ or non-persistent. Growers and others are frequently in contact with compost and its containers, and if toxic chemicals are added to the compost there is a risk to the operators. On the other hand, if protective clothing is used, work output (and hence wages) is often lowered—an unpopular event. Compost-filled boxes closely packed into pasteurization rooms and growing rooms are difficult to treat with nematicides owing to access problems. Also, once it has been spawned, compost should not be disturbed.

Mushroom is sensitive to many chemicals, and the sporophore on the surface is sometimes more sensitive than the mycelium. Chemicals that are safe in compost may, when applied to the surface of mushroom beds, or incorporated in the casing, induce abnormal or unsaleable mushrooms. Thus the application of some chemicals must be limited to compost only (Hussey, 1967).

A well-prepared compost contains about 70% by weight of moisture, and only a slight deviation from this is permissible. Nematicidal drenches must not add more than 2% moisture, for an excess will make the compost too "soggy". The alternative use of a low volume of concentrated nematicide, with the concentrate diluting itself in the compost moisture, may create difficulties of application and handling. To

FIG. 4. Custom-composting. This yard has a weekly output of about 1000 tons.   The incorporation of nematicides into vast stacks of fermenting compost is a major undertaking. (By courtesy of Bushell's Ready Mixed Ltd.)

ensure complete control of nematodes in compost those nematicides which are contact-killers must be finely and evenly dispersed in compost, and must be soluble (or solubilized) in compost moisture. Only then will they reach the relatively immobile nematodes in the water films in which they live. The thorough admixture of substances into compost stacks as a separate process from the routine "turning" may add greatly to the expense of compost preparation (Fig. 4). The obvious time and place to incorporate nematicides into compost is during the last turn in Phase I, or during machine spawning. However, any nematicide used in Phase I must withstand the heat, etc., of Phase II, or else the amount of its degradation must be allowed for by adding more chemical—an additional expense. Injection of nematicides with fumigant action seems to offer good prospects for nematode kill throughout a properly prepared stack, but fumigation would be ineffective on the outside of the stack unless it were sealed with a gas-proof sheet. Sheeting, however, might prevent beneficial aerobic fermentation, trap and concentrate noxious gases, and/or change the composting process from aerobic to anaerobic. Heat and gases in a sheeted stack might neutralize the nematicide. Fumigation usually requires several hours, and this delay might upset carefully timed composting procedures and cropping schedules.

Clearly, there are many real and potential difficulties in treating large volumes of compost with nematicides. Even if these were overcome, the grower would have to face the fact that, in practice, nematode kill is never complete, and that recontamination, especially by saprophages, seems to occur very frequently before the crop is finished. As most mushroom nematodes increase very rapidly, a nematode problem might still occur towards the end of cropping. Thus, the ideal nematicide needs to be not only an *eradicant*, but also a *protectant*. The latter implies persistence, which, in turn, implies non-toxicity to mushroom and man and, with this, every nematicide so far marketed seems ruled out. One substance, however, Thionazin = Nemafos = Cynem = Zinophos = EN18133 (OO-diethyl O–2 pyrazinyl phosphorothioate) has shown considerable promise because it is effective against the destructive fungal-feeding nematodes at very low concentration. It has no detectable residues in mushrooms grown in correctly treated compost, and is non-fungitoxic at concentrations of 80 ppm in mushroom compost (Oliff, 1965; Hesling, 1966b). It is, however, very toxic (acute oral L.D. 50 in rat 11–16 mg/kg) and considerable risk is involved in its use because it needs to be mixed with compost at high concentrations to avoid overwetting compost. Fortunately, the spawning process in mushroom cultivation provides a convenient stage at which this chemical can be mixed with compost with minimum risk and little delay. The emulsifiable concentrate (e.c.) formulation may be applied (diluted in water) as a coarse drench to the compost as it passes through the spawning machine. During compost treatment, and when boxes are handled afterwards, suitable protective clothing must be worn. The concentrate is diluted in the compost-moisture, and an acceptably even distribution is achieved.

One advantage of treating compost with Thionazin during spawning is that it is unlikely to be disturbed (handled) during the immediately-following 2 weeks (spawn-run), or even during the further 10–12 weeks until the end of cropping. The half life of Thionazin in compost is about 14 days (Hesling and Kempton, 1969), so that when picking commences

the toxic hazard is small. Limited experiments using the granular formulation of Thionazin (which is probably safer to use) have shown this
to be nearly as effective as the emulsifiable concentrate (Hesling, 1969).
Despite claims of systemic activity of Thionazin, adequate control of
fungal-feeding nematodes has been achieved only by mixing the e.c.
or granules as uniformly as possible in the compost; "spot" placement
of either is ineffective. Little is known of the effects of Thionazin on
saprophagous nematodes, but their control by Thionazin may be hin-
dered because the high pH that is often associated with vigorously
breeding populations of them in compost may "neutralize" the chemical.

B. Chemical Control and Prevention of Mushroom Nematodes

In the last 20 years several chemically different nematicides have
been marketed, but for many different reasons (see below), none is
completely suitable for use in mushroom growing. To control nematodes before and during cropping a non-fungitoxic nematicide is needed.
An all-purpose sterilant could theoretically be used to control nematodes
in compost at the end of cropping, but there would be great practical
difficulties in its application. Infested empty boxes, shelf-boards, etc.
may be washed or dipped between crops in 2% commercial formalin
(a 40% solution of formaldehyde in water) or in proprietary sterilants
such as cresylic acid, "santobrite" (sodium pentachlorophenate), and
"Sudol" (a mixture of selectively distilled xylenols and ethyl phenols),
but none is completely effective against nematodes, and residual traces
of some of these chemicals may cause deformation of mushrooms.
Moreton and John (1957) showed that dips of cold 0·4% xylenol
solubilized with Teepol, and of cold 2% cresylic acid did not control
Aphelenchoides sp. in wooden units assembled to simulate the joints and
cracks in the mushroom tray. However, if the wooden units were immersed in boiling water or boiling cresylic acid for only one minute, kill
was complete. These results emphasize the ineffectiveness of cold dips
for controlling nematodes.

C. Heat

Efficient pasteurization, and post-cropping "cook-out" should effectively control nematodes before spawning and after cropping. However,
for heat to be effective it must reach the nematode. Edwards (1969)
reported a test which showed a temperature difference of 10°F (6°C)
through 1 in. (2·5 cm) of wood, and a time lag of 12 h in the rise of
temperature between the inside, as compared with the temperature
outside, a mushroom tray. This example indicates how "flash-steaming"

to disinfest trays and structures, which involves merely "hosing" these with a jet of steam, is inadequate to eradicate nematodes. Complete immersion of trays and boxes for 1–5 min (depending on size) in boiling water seems the speediest and most effective way of eradicating contaminants. Where cook-out or pasteurization are not done, or not done well enough, alternative methods of controlling nematodes are needed. Control by methyl bromide is effective, but this is a dangerous gas and in some countries its use is restricted, so the need for other control methods remains.

D. Methyl Bromide Fumigation

The properties and use of methyl bromide were usefully reviewed by Thompson (1966). Many mushroom houses lend themselves to gas-proofing, and hence to methyl bromide fumigation, while the advent of expendable, cheap, plastic sheeting to cover compost stacks makes their fumigation an inexpensive, simple procedure. Methyl bromide fumigations can be completed in about 3 days, in which time the stack is sealed for only 24 h. Thus there is little risk of adversely affecting compost that has been treated before cropping. The gas disperses readily at temperatures of 50°F (10°C) and above, and if houses or stacks are properly aired after treatment there is little risk from residual gas. To date there is little information about bromine residues in the compost or in mushrooms, but analogy with the fumigation of other commodities suggests that these residues are insignificant. Lawson (1955), Blake and Conroy (1959), and Kohn (1962) treated compost with methyl bromide in sealed houses and under thin polythene sheets, and obtained acceptable high kills of pest and disease organisms without the damage to the house structure and contents that steam sterilization incurs. The dosages of methyl bromide Kohn used were large (C.T.P.s* of 2880 in the mushroom house and 5760 under sheeted compost stacks). Research in Britain shows that a C.T.P. of 600 gives practical control of mushroom nematodes and other pests (Hussey et al., 1962; Hesling, 1963). At temperatures higher than 67°F (19·4°C), this C.T.P. may be just adequate to control diseases caused by viruses and *Verticillium malthousei* (Hayes, 1969b).

The dosage of fumigant applied is usually given as a weight of fumigant per volume of the entire enclosed space. During fumigation some

* C.T.P. = Concentration Time Product. Within limits, a brief exposure to a large dose of fumigant has the same effect as a long exposure to a small dose. Thus, a vapour concentration of 1 g/litre for 1 h has the same effect as 0·1 g/litre for 10 h. C.T.P. is usually expressed as milligrams × h/litre, or oz × h/1000 ft³, or g × h/m³, these three products being equivalent for all practical purposes.

methyl bromide is always lost by sorbtion and solution in the materials being fumigated, and, unless sealing is perfect, leakage also occurs, and this may be dangerous. The methyl bromide contains 2% chloropicrin as a warning lachrymatory agent. The concentration of the fumigant must be maintained if control is to be successful, and it should be checked from time to time. Only in this way can the extent of any loss by sorbtion or leakage be assessed, and the correct "topping up" dose be added. Growers are advised to demand a record of the concentration of the fumigant during treatment, and a statement of the C.T.P. achieved (Flegg, 1968).

In 1967 it was announced that commercially-conducted field trials of methyl bromide fumigation on mushroom farms in Britain had been satisfactory (Anon., 1967a). Today, methyl bromide fumigation of mushroom houses and compost is a common and increasing practice, and several firms offer fumigation services. Because methyl bromide fumigation is hazardous, official policy in many countries is that it should be done only by the trained personnel of contract servicing companies. The sales policy of manufacturing firms in Britain is to supply methyl bromide only to approved servicing companies, and so the safe use of the substance is assured. Such a policy is wise, but it inevitably incurs high fumigation costs.

Fumigation costs at each site are based on the work and distance involved. In Britain the cost ranges from £2 to £3 per 28·3 m³ of fumigated space. Methyl bromide costs about £0·94 per kg. A dose of 900 g per 28·3 m³ for 24 h theoretically gives a C.T.P. of 768 which controls all nematodes and allows a margin for losses due to adsorption, etc.

It is important that methyl bromide be properly used if it is to be effective and safe. Several detailed publications are available to guide growers and operators, and all those involved in methyl bromide fumigation in mushroom growing are urged to read them (Burns Brown, 1959; Anon., 1960; Heseltine, 1961; Monro, 1961; Anon., 1968a; Anon., 1968b).

1. Methyl Bromide Fumigation in Mushroom Houses

The house must be effectively sealed. Gaps around doors, ventilators, cracks, and other sources of leaks must be covered in gas-proof materials and sticky plastic tape. Sometimes houses cannot be sealed, so the contents are fumigated under a gas-proof sheet within the house itself. This method cannot control pests and disease organisms outside the sheet, and a residuum of infection will be left on the inside walls and ceiling, etc., of the house.

For gassing mushroom houses methyl bromide is most conveniently and economically used from bulk containers of the gas liquefied under pressure. Halide warning lamps should be used to indicate any leakage, and respirators with suitable canisters should be worn or be ready for use by the operators. Warning notices should be posted, and all unauthorized access by humans and animals to the fumigation area should be prohibited and prevented. The methyl bromide is vaporized in a coil immersed in a water bath at about 170°F (76·6°C), and then released through perforated or multi-outlet pipes into the mushroom house, the temperature of which should be at least 67°F (19·4°C). The dispersion of the gas may be aided at first by fans, provided that these do not encourage leaks, so that an early uniform dispersion of the gas can be obtained. The required concentration should be maintained, or the fumigation extended until the (minimal) desired C.T.P. of 600 g/h/1000 m³ is reached; 24 h is a convenient duration for fumigation on most mushroom farms.

2. Methyl Bromide Fumigation of Sheeted Stacks of Compost

"Finished" compost may be fumigated with methyl bromide as an alternative to peak-heat. The compost stack should be sealed with poly-thene sheet (minimum thickness 0·1 mm) beneath which gas may be released either from small canisters placed at intervals under the sheet or from perforated pipes from a bulk container. Ideally the area of a section through the stack should not exceed 2·8 m² and the stack should neither be compressed nor excessively wet. If these conditions are not fulfilled the gas may fail to penetrate the entire stack. It is preferable for the compost stack to have a level, firm base, and the lower edges of the sheeting must be well sealed to the substrate.

Where stacks are fumigated out of doors at a C.T.P. exceeding 600 g/1000 m³, the pest kill is good, but a complete kill of disease organisms cannot be expected unless the temperature under the sheet exceeds 67°F (19·4°C). If the gas concentration under the sheeted stack cannot be measured, a dose of 170 mg of methyl bromide per cubic metre of closely sheeted compost for 24 h will provide adequate control of nema-todes and pests.

When fumigation of mushroom houses or sheeted compost stacks is complete, the houses must be opened and sheets removed to permit thorough airing and gas dispersion. During the airing time the compost is necessarily completely open to recontamination, and flies and other arthropod vectors may reintroduce nematodes.

E. Resistant Strains

This is an ideal method of nematode control, if a resister can be found. However, unless the resister is also economically acceptable, the method is more theoretical than practical, for knowledge of mushroom genetics is scanty, and the stability of a selection cannot be guaranteed. Investigations of several commercial strains of mushroom showed that none was resistant to *Aphelenchoides composticola*, and that slight resistance to *Ditylenchus myceliophagus* (manifest by lower reproductive rate on resisters) would be of little practical value (Hesling, 1964). Spawn types seem to change autonomously, spawn manufacturers often continuously select improved versions of their products. One problem of research on resistant spawns is the uncertainty of supplies of unchanged strains.

F. Chemical Control in Casing Materials

Unlike compost, which quickly deteriorates if not used soon after Phase II, casing can be prepared in bulk and stored. Thus, where adequate hygienic storage is available, large quantities can be chemically sterilized, and used when all traces of harmful residual sterilant have disappeared. Every precaution is needed to avoid recontamination of treated casing by nematodes carried by flies and other vectors. Prolonged storage of casing in large closed bins or sheeted heaps may encourage changes, e.g. due to anaerobic conditions in the centre of the material, that could be detrimental to cropping.

Several chemicals have been used to sterilize casing; experimentally, Vapam (sodium methyl dithiocarbamate) has been used at a rate of 500 ml/m³, and treated casing was sealed under plastic for 4 weeks before use. Yields of beds covered with Vapam-treated casing were inferior to the untreated controls, but where the treated casing was subsequently peak-heated before use, yields were not depressed. These effects were attributed to the fungitoxicity of residual Vapam (Bukowski, 1967). Beds covered with peak-heated casing (no Vapam) gave yields of about 10% more. Bech and Rasmussen (1967) used Vapam at a rate of 0·5 litres in 3 litres water/m³ casing. Treated casing was enclosed in polythene sheet for 4 days at 30°C, and then aired for a further 4 days at 40°C. No fungitoxicity was observed.

Shell "DD" has been used at a rate of 100 ml DD per m³ casing. It was mixed into the casing as this was turned. After treatment the casing was covered with a plastic sheet for 15 days; it was then vigorously turned to release residual DD vapours. A temperature of 64°–72°F (18°–22°C) is needed to ensure that DD is effective and disperses at the end of treatment (Cayrol, 1967a).

Chloropicrin at 0·5 litres/m³ and Basamid = Dazomet at 500 g/m³ have been used in the same way as Vapam (see above, Bech and Rasmussen, 1967) without fungitoxicity problems.

For best results with most fumigants, high temperatures 60°–80°F (16°–27°C) are advisable. Inferior results from fumigation at suboptimal temperatures may limit the use of these substances to mushroom farms in warmer climates. It is advisable to use these chemicals out of doors, for all emit harmful, unpleasant, lachrymatory vapours on contact with soil, and a respirator may be needed. Dazomet is least harmful in this respect for it is in granular form (the others being liquids) and releases unpleasant fumes less rapidly. For best results with fumigants, casing should be merely moist and uncompacted, and not wet or compressed.

Formalin has been used, with mixed success, to sterilize casing. Cairns and Thomas (1950) claimed that it was ineffective, but unfortunately did not record treatment details. Bech and Rasmussen (1967) used 6 litres of 50% formalin/m^3 and found no fungitoxic effects.

It is unlikely that such chemical treatments of casing will have much use on those modern mushroom farms with facilities for heat-pasteurization or a methyl bromide fumigation service. Also, the time involved in treating and airing casing may be prohibitive, as may be the high temperatures needed for success. Nevertheless the treatments are inexpensive.

IV. Economics

In all countries with a mushroom-growing industry there has been a startling increase in production during the last 20 years (Table III).

TABLE III. World production of mushrooms (metric tons) (Data mainly from Longhill, 1969)

Country	1950	1960	1963	1966	Recent
U.S.A.	30,000	50,000	75,000	77,400	82,000 (1968)
France	10,000	30,000	43,000	49,200	74,000 (1970)
Formosa	—	6,000	22,000	36,800	
U.K.	5,600	17,500	20,500	33,000	45,000 (1969)
Holland	300	3,000	6,500	16,000	29,000 (1970)
Canada	1,500	3,000	6,000	7,500	
Denmark	700	2,500	4,500	6,000	
W. Germany	600	2,500	4,000	13,000	
Belgium	800	2,000	3,000	4,000	
Switzerland	500	1,500	2,500	2,750	
Sweden	350	1,500	2,000	2,750	
Eire	220	750	1,500	1,625	
Hungary	—	1,100	—	1,350	

This increase is due to the greater number of growers and to improved yields (Table IV). At the same time competition for sales has increased, prices have fallen (Table V) and the cost of production (Table VI) has risen by about 5% per annum (Ganney, 1970). Declining profitability has accelerated the trend towards more scientific and controlled methods of growing, with the minimization of factors that adversely affect production. There is no doubt that where these trends have been followed, nematode problems have been rarer. But problems (such as nematodes) that are of little consequence when profits are high, assume increasing importance when profits decline. The industry must continually improve its efficiency in order to survive and with this improvement nematode problems may become still rarer.

TABLE IV. Holland: Number of growers and crop yields 1940–1965 (Anon., 1967b)

Year	No. of Growers	Yield kg/m²
1940	20	3·0
1950	50	4·0
1959	520	4·0
1960	530	4·0
1963	675	8·0
1964	800	9·0
1965	825	11·0
1970	1100	?

TABLE V. Net return to grower per kilogramme of mushroom 1945–1968 (based on Longhill, 1969; Ganney, 1970)

Year	£
1945	0·72
1949	0·45
1955	0·39
1965	0·34
1968	0·30

The efficiency of mushroom growing is differently expressed because of different outlooks and methods of growing: (1) yield per unit area of bed surface, (2) yield per unit weight of fresh compost, (3) yield per unit weight of compost dry matter, (4) yield per unit volume of growing space. On the basis of (2) above, efficiency is currently (though very rarely!) maximal at 1:3, i.e. about 360 kg of mushrooms from one ton of fresh compost (Rasmussen, 1970). Commonly, a ratio of 1:6 is

achieved and (averaging out price fluctuations) is economically acceptable, but this is reliably and regularly obtained only on the best
equipped mushroom farms where there has been a large capital investment in purpose-built growing rooms with ventilation, air filtration,
and peak-heat/cook-out facilities. A ratio of 1:8 may make mushroom
growing profitless.

TABLE VI. Production costs (£/m² mushroom bed) on two farms (based on
Ganney, 1970)

	Farm 1	Farm 2
Average Output (kg/m²)	14·65	11·72
Item costed		
Materials		
Compost	0·55	0·40
Spawn	0·16	0·15
Casing	0·09	0·16
Packing	0·33	0·25
Fuel	0·14	0·11
Sundries	0·08	0·03
Labour		
Picking	0·42	0·31
Packing	0·07	0·05
Operational	0·39	0·42
Other costs	1·35	0·98
Total costs per m²	3·56	2·89
Total costs per kg	0·24	0·25

An obvious way to improve efficiency is to maintain yields for a
longer cropping period, for this saves the cost of frequent composting.
But extended cropping is unwise when pathogenic nematodes are
present, for it permits their increase so that they, and not inadequacies
of the compost or "old age" of mycelium, are the factor limiting
cropping. Where nematodes are a chronic problem, their effects will be
lessened and efficiency will be improved by encouraging early yields
of mushroom and shortening cropping time.

Fungal-feeding nematodes undoubtedly lower production efficiency.
However, there are few data from mushroom farms on the extent of
their effects, but experimental results show that losses can be great
enough to make growing uneconomic (Hesling, 1966b; Arrold and
Blake, 1968). Cayrol (1963) demonstrated the problems caused by nematodes in crops grown in ridge beds on the infested floors of caves in

France, and Bukowski (1967) claimed that, in Poland, about 20% of crop failures were due to nematodes, causing some growers to abandon mushroom production.

References

Anon. (1960). Fumigation with methyl bromide. "Precautionary Measures." H.M.S.O. London.

Anon. (1967a). *MGA Bull.* **206**, 92.

Anon. (1967b). *Mushr. Sci.* **6**, 553.

Anon. (1968a). "Chemical Compounds used in Agriculture and Food Storage. Recommendations for Safe Use in Great Britain. *Methyl bromide* (containing 2% chloropicrin as warning agent), a Soil Fumigant (agricultural and horticultural use)." Recs/453. Issued 3.9.68. Minist. Agric. Fish. Fd, London.

Anon. (1968b). "Safe and Efficient Fumigation Practice." Minist. Agric. Fish. Fd, London.

Arrold, N. P. and Blake, C. D. (1966). *Nematologica* **12**, 501–510.

Arrold, N. P. and Blake, C. D. (1968). *Ann. appl. Biol.* **61**, 161–166.

Atkins, F. C. (1954). *Mushr. Sci.* **2**, 92–95.

Baranovskaya, J. A. (1963). *In* "Helminths of Man, Animals and Plants, and their Control: Papers on Helminthology presented to Academician K. I. Skrjabin on his 85th Birthday", *Moscow: Izdatelstvo Akad. Nauk SSSR*, 480–483.

Bech, K. and Rasmussen, C. R. (1967). *Mushr. Sci.* **6**, 515–521.

Blake, C. D. and Conroy, R. J. (1959). *J. Aust. Inst. agric. Sci.* **25**, 213–218.

Brun, J. L. and Younes, T. (1969). *Nematologica* **15**, 591–601.

Brzeski, M. W. (1967). *Bulletin de L'académie polonaise des sciences* **15**, No. 3, 147–149.

Brzeski, M. W. and Jankowska, J. (1966). *Roczniki nauk rolniczych—Series A—roslina* **91**, 245–254.

Bukowski, T. (1967). *Mushr. Sci.* **6**, 485–493.

Burns Brown, W. (1959). *Pest Infestation Research, Bull.* No. 1. H.M.S.O., London.

Cairns, E. J. (1954). *Mushr. Sci.* **2**, 161–164.

Cairns, J. and Thomas, C. A. (1950). *Mushr. Sci.* **1**, 89–91.

Cayrol, J. C. (1962). *Mushr. Sci.* **5**, 480–496.

Cayrol, J. C. (1965). *90th Congrès des Sociétés Saventes.* Nice, 1965. *Tome* **2**, 533–541.

Cayrol, J. C. (1967a). *Mushr. Sci.* **6**, 475–482.

Cayrol, J. C. (1967b). *Nematologica* **13**, 23–32.

Cayrol, J. C. and Legay, J. M. (1967). *Ann. Épiphyties* **18**, 193–211.

Cayrol, J. M. and Ritter, M. (1967). *Rev. Zool. agric. appl.* **66**, nos. 7–9, pp. 92–102.

Choleva, B. (1966). *Pl. Sci.* **3**, 97–102.

Conroy, R. J. and Blake, C. D. (1960). *Agric. Gaz. N.S.W.* **70**, 644–648.

Dao, F. D. (1970). *Meded. LandbHogesch. Wageningen.* **70-72**, 181 pp.

Doncaster, C. C. (1966). *Nematologica* **12**, 417–427.

Edwards, R. L. (1969). *MGA Bull.* **240**, 529.

Ellenby, C. (1968). *Proc. R. Soc. Lond.* **169**, 203–213.

Eroshenko, A. S. (1967). *Zool. Zh.* **46**, 617–620.

Eroshenko, A. S. (1968a). *Soobshch. dal' nevost. Kl. V.L. Komarova sib. Otdel. Akad. Nauk SSSR*, No. 26, 58–66.

Eroshenko, A. S. (1968b). *In* "Parasites of Animals and Plants", *Moscow. Nauka*, No. 4, 224–248.

Evans, A. A. F. (1970). *J. Nematol.* **2**, No. 1, 99–100.

Evans, A. A. F. and Fisher, J. M. (1969). *Nematologica* **15**, 395–403.

Evans, A. A. F. and Fisher, J. M. (1970a). *Nematologica* **16**, 113–123.

Evans, A. A. F. and Fisher, J. M. (1970b). *Nematologica* **16**, 295–305.

Faulkner, L. R. and Darling, H. M. (1961). *Phytopathology* **51**, 778–786.

Fisher, J. M. (1968). *Aust. J. biol. Sci.* **21**, 161–171.

Fisher, J. M. (1969). *Nematologica* **15**, 22–28.

Fisher, J. M. and Evans, A. A. F. (1967). *Nematologica* **13**, 425–428.

Flegg, P. B. (1962). *MGA Bull.* **146**, 68–70.

Flegg, P. B. (1968). *MGA Bull.* **228**, 582–584.

Franklin, M. T. (1955). *J. Helminth.* **29**, 65–76.

Franklin, M. T. (1957). *Nematologica* **11**, 306–313.

Franklin, M. T. (1965). *In* "Plant Nematology" (J. F. Southey, ed.) *Tech. Bull.* No. **7**, *Minist. Agric. Fish. Fd* 2nd ed., 131–141.

Ganney, G. W. (1970). *MGA Bull.* **241**, 4–17.

Goodey, J. B. (1958). *Nematologica* **3**, 91–96.

Goodey, J. B. (1960). *Ann. appl. Biol.* **48**, 655–664.

Goodey, J. B. and Hooper, D. J. (1965). *Nematologica* **11**, 55–65.

Hayes, W. A. (1969a). *MGA Bull.* **240**, 520.

Hayes, W. A. (1969b). *MGA Bull.* **234**, 240–243.

Hayes, W. A. and Randle, P. E. (1968a). *MGA Bull.* **218**, 81–97.

Hayes, W. A. and Randle, P. E. (1968b). *Rep. Glasshouse Crops Res. Inst.* 1968, 142–147.

Hayes, W. A., Randle, P. E. and Last, F. T. (1969). *Ann. appl. Biol.* **64**, 177–187.

Heltay, I. (1957). *Mushr. Sci.* **3**, 41–49.

Heseltine, H. K. (1961). "The Use of Thermal Conductivity Meters in Fumigation Research and Control." Pest Infestation Research Bulletin No. 2.

Hesling, J. J. (1963). Nematology. *Rep. Glasshouse Crops Res. Inst.* 1962, 79–82.

Hesling, J. J. (1964). Nematology. *Rep. Glasshouse Crops. Res. Inst.* 1963, 83.

Hesling, J. J. (1966a). *Ann. appl. Biol.* **58**, 477–486.

Hesling, J. J. (1966b). *Pl. Path.* **15**, 163–167.

Hesling, J. J. (1969). Nematology. *Rep. Glasshouse Crops Res. Inst.* 1968, 93–94.

Hesling, J. J. and Kempton, R. J. (1969). *Proc. 5th Brit. Insectic. Fungic. Conf.* 1969. 185–188.

Hirschmann, H. (1952). *Gewäss. Zool. Jb. (Syst.)* **81**, 313.

Hirschmann, H. (1962). *Proc. helminth. Soc. Wash.* **29**, 30–43.

Hooper, D. J. (1962). *Nature, Lond.* **193**, 496–497.

Husain, S. I. and Khan, A. M. (1967). *Proc. helminth. Soc. Wash.* **34**, 167–174.

Hussey, N. W. (1967). *Proc. 4th Brit. Insectic. Fungic. Conf.* 1967. **1**, 76–81.

Hussey, N. W., Wyatt, I. J. and Hesling, J. J. (1962). Entomology. *Rep. Glasshouse Crops Res. Inst.* 1961, 63.

Jackson, R. M. (1965). *In* "Ecology of Soil Borne Plant Pathogens, Prelude to Biological Control" (K. F. Baker and W. C. Snyder, eds), pp. 363–373. University of California Press.

Jankowska, J. (1964). Obserwacje nad wystepowaniem i szkodliwoscia micieni w pieczarbamiach *Praca magisterska SGGW. Skierniwice.*

Juhl, M. (1966). *Saertryk af Tidsskrift for Planteavl* **69**, 511–531.

Khak, M. M. (1967). *Zool. Zh.* **46**, 1251–1253.

Klingler, J. and Lengweiler-Rey, V. (1969). *Z. PflKrankh. PflPath. PflSchutz* **76**, 193–208.

Kohn, S. (1962). *MGA Bull.* **152**, 329–332.

Kux, M. and Rempe, H. (1954). *Mushr. Sci.* **2**, 175–177.

Lambert, E. B. (1938). *Bot. Rev.* **4**, 397.

Lambert, E. B., Steiner, G. and Drechsler, C. (1949). *Pl. Dis. Reptr* **33**, 252–253.

Lawson, M. (1955). *MGA Bull.* **69**, 746.

Longhill, D. J. (1969). *Mushr. Sci.* **7**, 497–484.

Luh, Chi-Lin (1967). *Mushr. Sci.* **6**, 441–448.

McLeod, R. W. (1967). *Nature, Lond.* **214**, 1163–1164.

McLeod, R. W. (1968). *Nematologica* **14**, 573–576.

McLeod, R. W. and Conroy, R. J. (1968). *Agric. Gaz. N.S.W.* **79**, 237–238.

Monro, H. A. U. (1961). Manual of Fumigation for Insect Control. FAO Agricultural Studies No. 56. Rome.

Moreton, B. D. and John, M. E. (1957). *Mushr. Sci.* **3**, 99–101.

Moreton, B. D. and John, M. E. (1962). *MGA Bull.* **147**, 91–96.

Moreton, B. D., John, M. E. and Goodey, J. B. (1956). *Nature Lond.* **177**, 795.

Oliff, K. E. (1965). *Hort. Res. Edinb.* **5**, 36–37.

Paesler, F. (1957). *NachrBl. dt. PflSchutzdienst, Berl.* **7**, 129.

Park, J. Y. and Agnihotri, V. P. (1969a). *Antonie van Leeuwenhoek* **35**, 523–528.

Park, J. Y. and Agnihotri, V. P. (1969b). *Nature, Lond.* **222**, 984.

Pizer, N. H. and Leaver, W. E. (1947). *Ann. appl. Biol.* **34**, 34–44.

Rasmussen, C. R. (1970). *MGA Bull.* **243**, 138–151.

Riffle, J. W. (1970). *Proc. helminth. Soc. Wash.* **37**, 78–80.

Ritter, M. (1957). *Mushr. Sci.* **3**, 90–99.

Rothwell, J. B. (1969). *MGA Bull.* **232**, 162.

Sanwal, K. C. (1961). *Can. J. Zool.* **39**, 143–148.

Sanwal, K. C. (1965). *Can. J. Zool.* **43**, 987–995.

Seinhorst, J. W. and Bels, J. (1951). *Tijdschr. PlZiekt.* **57**, 167–169.

Shavrov, G. N. (1967). *Zool. Zh.* **46**, 762–764.

Siddiqi, M. R. and Franklin, M. T. (1967). *Nematologica* **13**, 125–130.

Siddiqui, I. A. and Taylor, D. P. (1967). *Nematologica* **13**, 581–585.

Siddiqui, I. A. and Taylor, D. P. (1969). *Nematologica* **15**, 503–509.

Sinden, J. W. and Hauser, E. (1952). *Mushr. Sci.* **1**, 52–59.

Smart, G. C. Jr. and Darling, H. M. (1963). *Phytopathology* **53**, 374–381.

Steiner, G. (1933). *J. agric. Res.* 1933, **46**, 427–435.

Taylor, D. P. and Pillai, J. K. (1967). *Proc. helminth. Soc. Wash.* **34**, 51–54.

Thompson, R. H. (1966). *J. Stored Prod. Res.* **1**, 353–376.

Thorne, G. and Malek, R. B. (1968). *Tech. Bull. S. Dak. agric. Exp. Stn.* No. 31. 111 pp.

Townshend, J. L. (1964). *Can. J. Microbiol.* **10**, 728–737.

Van Haut, H. (1956). *Nematologica* **1**, 165–173.

Wallace, H. R. (1963). "The Biology of Plant Parasitic Nematodes." Edward Arnold, London.

Webster, J. M. and Hooper, D. J. (1968). *Parasitology* **58**, 879–891.

19

Nematodes and Biological Control

J. M. Webster

Pestology Centre
Department of Biological Sciences
Simon Fraser University
Burnaby, Vancouver, B.C., Canada

I. Introduction to Beneficial Associations

Nematode and insect pests are commonly controlled by the application of highly efficient chemical pesticides, many of which have high mammalian toxicity and some of which are general biocides. It is this latter feature which has encouraged scientists to attempt to find alternative means of controlling these pests. The object of this chapter is to review recent attempts in the biological control of plant-parasitic nematodes and the use of nematodes to control insect pests of agricultural and forestry crops.

The aim of studies in biological control is to be able to modify the ecological environment of the pest so as to restrict its activities to below an economically important level. A biological change in the pest's environment can be achieved by one of four ways, namely (i) the regular release of large numbers of parasites or predators of the pest into the troubled area, (ii) introduction and establishment of parasites or predators of the pest from an epidemic area, (iii) environmental manipulation so as to increase the numbers of parasites and predators of the pest that are already present and, (iv) integration of these biological means with chemical means of control. In order for these techniques to be successful a very close study has to be made of the ecological and physiological relationships between the parasite/predator and its host/prey.

II. Parasites and Predators of Plant-parasitic Nematodes

There are many reports of parasites and predators attacking soil inhabiting nematodes and there has been speculation, supported by some experimental evidence, that these parasites/predators regulate the nematode populations. This section reviews the progress made in understanding the physiology and ecology of these relationships, and indicates the potential of the parasites/predators as agents in the biological control of plant-parasitic nematodes.

A. Microbial Diseases of Nematodes

Dollfus (1946) reported many protozoan infections of nematodes but it is probably only the sporozoa that cause major, lethal diseases of nematodes. Since Micoletzky (1925) first described *Duboscqia*, several species have been reported parasitizing plant parasitic nematodes in various parts of the world: *D. penetrans* from *Pratylenchus brachyurus* in the U.S.A. (Thorne, 1940), from *Meloidogyne javanica* in Japan (Steinhaus and Marsh, 1962) and from several species of plant-parasitic nematodes in the Netherlands (Kuiper, 1958). Williams (1960) found 34% of 174 *Meloidogyne* females examined from sugar-cane in Mauritius to be filled with spores of a sporozoan that resembled *D. penetrans*. Esser and Sobers (1964) reported a population of *Helicotylenchus microlobus* to be heavily infected with a sporozoan, probably *Duboscqia* sp. About 25% of the infected nematodes were dead and filled with spores or had them adhering to the cuticle. There is uncertainty as to the form of the complete life-cycle of this sporozoan. Possibly the spores are carried passively in the soil water, adhere to passing nematodes and then produce an infection plug and penetrate the cuticle (Boosalis and Mankau, 1965). The effectiveness of *Duboscqia* in regulating nematode populations is unknown, but the fact that it is relatively common and that it sterilizes and kills its host suggests that its effect may be considerable. The sporozoan *Legerella helminthorium* was recorded as parasitizing the free-living nematode *Mononchus composticola* (Canning, 1962) but there is no information as to whether this parasite attacks also plant-parasitic nematodes. In a glasshouse experiment an unidentified sporozoan decreased a population of *Pratylenchus scribneri* by 50% in 55 days, and also severely attacked the larval root-knot nematodes *Meloidogyne javanica* and *M. incognita* (Prasad and Mankau, 1969).

Bacteria can be lethal agents to nematodes (Dollfus, 1946). Most populations of *Xiphinema americanum* in West Virginia (U.S.A.) were reported (Adams and Eichenmuller, 1963) to contain nematodes infected with *Pseudomonas denitrificans*. The bacteria were particularly common in the intestinal cells and the ovary of the nematodes, which suggests possible transovarial transmission. However, there was no record of it affecting nematode movement. Iizuka *et al.* (1962) recorded that cell-free extracts of a *Pseudomonas* sp. caused a high mortality of *Meloidogyne* sp. In some interesting experiments the free-living nematodes *Caenorhabditis briggsae*, *Rhabditis oxycerca* and *Panagrellus* sp. were lysed by two isolates of myxobacters (Myxobacterales) in liquid and solid medium, but the plant-parasitic nematodes *Aphelenchus avenae* and *Heterodera trifolii* were not (Katznelson *et al.*, 1964). Within 24 h most of the free-living nematode larvae and adults were dead and by 72 hours many of the nematodes had decomposed. However, the authors emphasized that such potent activity on agar plates probably does not occur in the complex soil environment which may be so well buffered that toxins do not accumulate.

Since the demonstration by Loewenberg *et al.* (1959) of the existence of a virus in larval *Meloidogyne incognita* that caused larval sluggishness and an inability to cause galls, there has been little further evidence for such a potentially important phenomenon. Recently, however, Das and Raski (1969) showed that specimens of the virus vector, *Xiphinema index*, that were carrying the grapevine fanleaf virus survived significantly longer than did the virus-free individuals but there was no effect on rate of reproduction. As a result of the recent advances in such areas as electron microscopy and serology we can expect to hear more of virus/nematode associations and their effect on the nematode.

B. Fungal Parasites and Predators

Many workers believe that fungal parasites and fungal predators of nematodes are potentially the most useful biological control agents against plant-parasitic nematodes. These fungi are ubiquitous and have been found in environments ranging from Finnish saunas (Salonen and Ruokola, 1968) to Devonshire moss (Duddington, 1951). Duddington (1957) reviewed the role of these nematophagous fungi in his book, and Dollfus (1946), Dreschsler (1950) and Soprunov (1958) contributed major works on the structure, taxonomy and biology of the predacious fungi. However, despite some interesting early observations it is only

Q

recently that progress has been made in studying their ecological and physiological relationships. There are useful descriptive keys to the most common genera of the predacious fungi, namely *Arthrobotrys*, *Dactylella* and *Dactylaria* (Cooke and Godfrey, 1964; Sachchidananda and Swarup, 1966). *Arthrobotrys oligospora* is one of the most common of the nematophagous fungi in Canada and it was isolated 108 times from 175 collection sites in Quebec (Estey and Olthof, 1965). The predacious fungi capture nematodes by one of two methods: adhesive processes and mechanical ring traps. The former method is either an anastomosing network of hyphae some of which are sticky (e.g. *Arthrobotrys* sp.) or a hyphal network from which grow adhesive loops or knobs (e.g. *Dactylella ellipsospora*). The nematodes stick to these adhesive hyphal processes during their random migrations. Within 2 hours of the nematode adhering to an adhesive hypha it is immobilized and outgrowths from the adhesive hypha penetrate the nematode cuticle. Once inside, the hypha swells into an infection bulb from which are produced numerous trophic hyphae. These trophic hyphae absorb the body contents and so kill the nematode.

The fungi with mechanical ring traps are divided into those with constricting rings and with non-constricting rings. In *Dactylaria candida* a three-celled non-constricting ring is attached to the parent hypha by a stalk. Nematodes during their migrations pass into these rings and become firmly wedged. Hyphae quickly grow from the ring cells, penetrate the nematode cuticle and absorb the body contents. The captured nematode is held securely but occasionally some break free. In doing so they usually take the hyphal ring collar with them and in a short time hyphae penetrate the nematode cuticle from the collar and kill the nematode. The constricting ring, as in *Dactylella doedycoides* and *Dactylaria brochopaga*, is similar in general form to that of the non-constricting type but when the nematode enters the ring the cells suddenly enlarge to about three times their original volume and so hold the nematode until the trophic hyphae from the ring cells have penetrated the nematode and absorbed its contents (Fig. 1). The process of ring constriction has stimulated several investigations (Muller, 1958; Lawton, 1967). The internal surface of the ring is thigmotactic and when sensitized by touch the cell wall structure suddenly changes and increases the permeability of the cell wall to water: the ring cells expand and constrict the ring.

These sophisticated capturing devices are surprisingly efficient in situations of high nematode density. When a large nematode is caught in one of the traps, it struggles vigorously to free itself but the energy is dissipated effectively along the many very flexible hyphae, and in the course of the struggle the nematode frequently becomes even more firmly trapped. Olthof and Estey (1963) showed that in addition to traps and adhesive hyphae the predacious fungi may produce a toxin that kills the nematode before the fungal hyphae penetrate the cuticle.

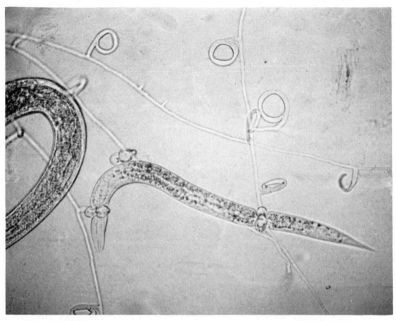

Fig. 1. *Dactylaria brochopaga*, a constricting-ring predacious fungus, trapping a stylet-bearing nematode. Note the expansion of the constricting ring cells. (By courtesy of G. L. Barron.)

The non-trap-forming endoparasitic fungi usually enter the nematode by a germ tube that penetrates the cuticle from a sticky spore. Once inside the nematode, the hyphae ramify throughout the body, absorb the contents and spore-bearing hyphae, then emerge from the nematode cadaver. Thus, with *Caternaria vermicola*, which frequently attacks nematode pests of sugar-cane, mobile zoospores adhere to the nematode cuticle near the mouth, anus or vulva and, if not dislodged within a few hours, produce germ tubes that penetrate the nematode's body (Birchfield, 1960). The fungus *Phialophora heteroderae* enters the cyst of *Heterodera rostochiensis* via the oral or vulval openings and then penetrates the egg shell, so decreasing the number of viable larvae (van der Laan, 1956). Conidia of endoparasitic hyphomycetes that are ingested by saprophytic nematodes usually lodge in the buccal cavity, oesophagus or, as in the fungus *Harposporium helicoides*, in the intestine (Fig. 2) before penetrating into the body of the nematode (Barron, 1970).

Sayre and Keeley (1969) showed in culture a linear relationship between increasing number of zoospores of *Catenaria anguillulae* and incidence of the fungus disease in *Panagrellus redivivus* and in the stem eelworm, *Ditylenchus dipsaci*. However, massive doses of zoospores would be necessary in the field to obtain a mortality equivalent to that which occurred in the laboratory.

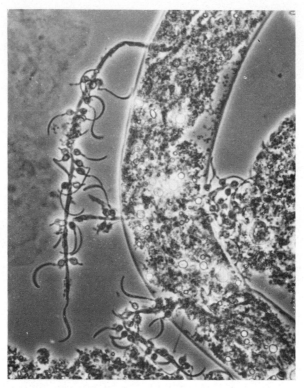

Fig. 2. The germinating conidia of *Harposporium helicoides* bursting out of a parasitized nematode. Reproduced with permission from Barron (1970).

Other species of fungi are potential biological control agents of plant-parasitic nematodes in that they retard development by chemical secretions. A fungus which produces a grey, sterile mycelium in culture and is the cause of brown-root rot of tomatoes, adversely influences the development of the potato root eelworm, *Heterodera rostochiensis*. In the laboratory James (1966, 1968) showed that exudates produced by the grey sterile fungus significantly inhibited hatch of *H. rostochiensis*, the invasion of the host and the number of cysts produced. Recently, Jorgenson (1970) observed that the damage to sugar-beet was less when the fungus, *Fusarium oxysporum*, and beet cyst nematode, *Heterodera schachtii*, were present together than when the nematode occurred alone. The fungus inhibited nematode invasion and development in the sugar-beet seedlings and thereby decreased the number of maturing nematodes about threefold. The converse nematode–fungal relationship was reported by Nonaka (1959) who stated that the severity of rice stem rot, caused by the fungus *Leptosphaeria salvinii*, was decreased when the rice plant was attacked by *Aphelenchoides besseyi*, causing "white tip". Our knowledge of disease complexes is increasing greatly but our understanding of the physiological relationships is not sufficient as yet to develop these "mutual inhibition" phenomena in a biological controlled programme.

1. Trap Production and Fungal Nutrition

Predacious fungi do not produce traps all the time but require a stimulus to do so. The nutritional requirements of the fungi are the primary factors affecting fungal growth and the initiation of trap production, and much work has been done in this area.

Cooke (1962) suggested that during the decomposition of organic matter in the soil the nematode-trapping fungi are capable of trapping nematodes only while suitable carbon sources are available to provide energy for growth. Hence, Satchuthananthavale and Cooke (1967a) tested seven common nematode-trapping fungi in culture to ascertain their carbohydrate requirements. All fungi grew well on the hexose and pentose. All network-forming fungi grew better on polysaccharides than did the ring formers, which showed a reduced ability to utilize these carbohydrates. *Monacrosporium bembicodes* cannot efficiently utilize maltose, sucrose, cellulose, starch or glycogen and this is correlated with its low trap producing ability. Little or no predacious activity occurs in untreated soils, but the addition of cabbage tissue or sucrose to the soil increases the nematode populations and fungal trapping activity. However, the fungi trap nematodes for only a short period after the onset of decomposition of the cabbage tissue. The competition of nematode-trapping fungi and other fungi for these substrates is intense, but as soon as the nematode-trapping fungus begins to grow actively the young growing hyphae are stimulated into predacious activity and so can utilize the extra nematode substrate. Thus, in general, the predacious activity of nematode-trapping fungi is often independent of nematode populations, but dependent on the release of certain soluble carbohydrates.

Many of the common nematophagous fungi have thiamine requirement for growth and also, except for *Dactylaria thoumasia*, for biotin

(Satchuthananthavale and Cooke, 1967b). The fungi having an adhesive network (*Arthrobotrys oligospora, Dactylaria clavispora* and *D. thoumasia*) utilize nitrite, nitrate, ammonium and organic nitrogen for growth. However, *A. anchonia, A. dactyloides, Monacrosporium bembicodes* and *M. doedycoides*, which form constricting rings, cannot utilize nitrite and have a reduced ability to grow when nitrate is the sole nitrogen source. Under such conditions they obtain essential nitrogen from nematode prey (Satchuthananthavale and Cooke, 1967c). The effects of substrate on the predacious activity of two isolates of *Arthrobotrys* were measured by expressing the number of non-survivors of a free-living nematode population exposed to the fungi for 4 days as a percentage of a comparable population of nematodes not exposed to predation (Hayes and Blackburn, 1966). They called this percentage the "predacity number". Thus, when the energy source in the substrate was varied the nutrient requirements for mycelial growth were supplemented by nutrients from trapped nematodes and this was visible as a change in predacity. The predacity of *A. oligospora* was less when nitrogen was supplied as ammonium than when it was supplied as nitrate. *A. robusta* predacity was greatest on media deficient in biotin and with nitrate as the nitrogen source.

Broths in which *Neoaplectana glaseri* had developed axenically caused the mycelium of the predacious *A. conoides* to produce traps. The active principle in the broth was named "nemin" (Pramer and Stoll, 1959). Later, Lawton (1967) showed that *A. dactyloides* and *Dactylella deodycoides* produced rings after the addition of a liquid containing living sterile nematodes and also if placed on a medium containing horse serum. Both fungi produced rings on the β- and γ-globulin fractions near the negative pole after separation of the horse serum by gel electrophoresis.

In order to ascertain the success of the predacious fungi in trapping nematodes it was important to know of their competitiveness and how long they would persist in the soil. Hence, Cooke and Pramer (1968) used a model system in which nematode-trapping fungi and the mycophagous nematode, *Aphelenchus avenae*, were studied in dual culture. The presence of *A. avenae* in the culture did not significantly affect the rate of radial growth of the fungal colonies on 2% cornmeal agar. After the agar surface was completely covered by the fungus there was a brief period of trapping activity but the nematode populations continued to increase and reached levels which eventually killed the fungus. This model showed that the population growth of the nematode was not significantly influenced by the fungus under these experimental conditions, in which the substrate was finite.

Cooke (1962) suggested that for nematode-trapping fungi to be successful in the biological control of nematode pests the following two requirements would have to be satisfied. Firstly, an increase in the period during which the fungi produce traps and secondly, an intensification of nematode-trapping activity during this period. Hence, perhaps the addition of an appropriate rich energy source and an active fungus to the soil could successfully control plant-parasitic nematodes.

2. Use of Soil Amendments to enhance Fungal Predation

Many attempts have been made to use predacious fungi in a biological control programme. The approach usually has been to inoculate the soil with a predacious fungus and then the soil has been supplemented with organic matter in order to stimulate fungal growth and trapping. Linford (1937) was one of the first to use nematophagous fungi in a biological control programme. Incorporation of fresh plant material into the soil was followed by a rapid rise in the total population of free-living nematodes. However, within 21 days the nematode population was greatly diminished. Linford *et al.* (1938) reported that the decomposition of large amounts of organic matter (chopped pineapple plants) in the soil was associated with a reduction in numbers of *H. marioni* (= *Meloidogyne* sp.) galls on the roots of cowpea indicator plants. Addition of organic matter at rates of about 50, 100, and 150 metric tons per hectare of soil gave progressively fewer nematode galls on the cowpea but this reduction of infestation was short-lived. The authors hypothesize that the sudden increase in organic matter greatly increases the total population of nematodes (plant-parasitic and free-living) in the soil and also greatly increases, in turn, the numbers of nematode predators such as nematode-trapping fungi, predacious nematodes and predacious mites. These collectively then decrease the number of nematodes during the early weeks of decomposition. Thus nematodes are necessary to initiate the formation of trapping organs, but the fungi are incapable of remaining in an actively predacious state in the absence of an organic energy source other than nematodes. The organic factors appear to have a direct effect on the fungi apart from the indirect one of influencing the nematode populations (Fig. 5, p. 491).

Since that time numerous other authors have described various combinations of fungal and organic additives to the soil. Duddington *et al.* (1956) treated microplots of beet heavily infested with beet cyst eelworm, *Heterodera schachtii*, with *Dactylaria thoumasia* and with organic matter (bran). Both the fungus and the organic matter increased the yield of beet. The fungus had no significant effect on the final cyst population whereas organic matter caused a significant depression. However, Duddington *et al.* (1961) decreased the number of cereal cyst eelworm (*Heterodera avenae*) larvae invading oat seedlings by green manure treatments of chopped cabbage leaves, and showed that this was due to the stimulation of the predacious fungi. The fact that the green manure without the addition of fungi to the plots failed to decrease eelworm larval invasion appreciably, confirms the fact that it is the fungus rather than the manurial value of the treatment which was effective.

Tarjan (1961) failed to decrease populations of the burrowing nematode, *Radopholus similis*, in the roots of citrus seedlings or trees by applications of nematophagous fungi, organic material and lime. Various organic materials supported fungal (*Arthrobotrys musiformis*) growth but did not induce nematode predation. Reductions in nematode disease incidence have been obtained by adding to the soil castor pomace (Mankau and Minteer, 1962), finely divided oil cakes (Singh and Sitaramaiah, 1966), sawdust and urea (Singh and Sitaramaiah, 1967), cellulose (Walker *et al.*, 1967) and chitin (Mankau and Das, 1969). The mode of action of these amendments is probably to increase the microbial activity which in turn increases the numbers of nematodes and of their predators and parasites. Unfortunately, the large quantities of these amendments that are required often make them commercially impractical.

The experiments reported by Patrick *et al.* (1965) support another explanation as to how soil amendments decrease plant-parasitic nematodes. They purified aqueous extracts of rye, adjusted the pH to 5·0 and then exposed the root-knot nematode, *Meloidogyne incognita*, and lesion nematode, *Pratylenchus penetrans*, to the extract and observed their activity (over 3 h) and survival (after 24 h). Concentrations of 380–440 ppm immobilized 50% of the plant-parasitic nematodes but concentrations of 410–540 ppm were necessary to kill them. In soil, concentrations of 1350–2100 ppm were needed to kill 50% of the plant-parasitic nematodes.

In the development of an integrated control programme it would be advantageous for the nematode-trapping fungi to remain alive in spite of soil treatments with chemical nematicides. Mankau (1968) showed that the common nematicide 1,2-dibromo-3-chloropropane (DBCP) was not fungitoxic to seven species of predacious fungi in outdoor microplot tests or in the laboratory. Ethylene dibromide (EDB) stopped fungal growth but was not lethal. However, 1,3-dichloropropene-1,2-dichloropropane (DD), chloropicrin, sodium methyldithiocarbamate, 1,3-dichloropropene and a mixture of DD and methyl isothiocyanate were all lethal. Some fungi utilize low concentrations of DBCP, and of EDB, as carbon sources and show increased growth at concentrations of 21 ppm of DBCP in laboratory culture conditions. In laboratory soil tests fungi were recovered from soil treated with 250 ppm DBCP and up to 700 ppm of EDB; the conidia tolerated higher concentrations. In California citrus culture where predacious fungi are important in the population dynamics of *Tylenchulus semipenetrans* the fungus persists in the citrus rhizosphere after DBCP treatment and may account in part for the gradual decline in population of citrus nematode larvae. This is a good example of integrated control in nematology.

C. Predacious Nematodes

Predacious nematodes are one of the most important natural agencies attacking plant-parasitic nematodes. Many of the large dorylaims and mononchids and also some *Seinura* spp. are efficient nematode predators. The dorylaims *Discolaimus*, *Dorylaimus* and *Actinolaimus* pierce nematodes and nematode eggs with their stylet and suck out the body contents by rhythmic oesophageal contractions (Linford and Oliveira, 1937). The small *Seinura christiei* kills nematodes much larger than itself because of its ability to paralyse the prey. The prey becomes inactive almost immediately the stylet is inserted, and repeated secretions (presumably of toxic saliva) of the dorsal oesophageal gland pass into the oseophagus during feeding (Linford and Oliveira, 1937). In the mononchids the flexible buccal cavity walls and the strong oesophageal sucking action enable the rasping and cutting action of the teeth to penetrate nematode cuticle and permit removal of the body contents. Nematodes such as *Tripyla monhystera* that have a very flexible buccal cavity can swallow whole nematode prey half as wide as they themselves.

Cobb (1924) reported a mermithid as parasitizing a mononchid and, more recently, an unidentified mermithid was found parasitizing 36% of 36 *Aporocelaimellus obscurus* females and larvae (Tjepkema, 1969). Both these records are of mermithids killing predatory nematodes. Nevertheless, in some instances species of these invariably lethal parasites may attack plant-parasitic nematodes and could significantly decrease the populations. More information is needed on this type of association.

Populations of predacious nematodes vary greatly with soil type, and Thorne (1927) records populations of mononchids as large as 121,416,194 per hectare in light soils. Of four mononchids tested *Mononchus papillatus* fed most readily on the adult and larval *Heterodera schachtii* (Thorne, 1927). The number of parasitic nematodes killed per day increased with the age of *M. papillatus* up to a maximum of 65 *Heterodera radicicola* larvae (Steiner and Heinly, 1922). *M. papillatus* survives well in dry soil, and is a hermaphrodite which may facilitate the rapid build-up of soil populations of this species. However, it probably would be of economic importance only if large populations could be maintained in the soil at the time of occurrence of the plant-parasitic nematodes.

Q*

Thorne (1927) sampled sugar-beet fields for *H. schachtii* and mononchids for over 2 years and concluded that the mononchids were of doubtful economic value as biological control agents of the sugar-beet pest because of their unstable populations.

Esser (1963) observed the prey preference of predacious dorylaims in water agar cultures. Many plant-parasitic nematodes such as *Hoplolaimus tylenchiformis*, *Belonolaimus longicaudatus* and *Criconemoides* sp. were resistant to predation, but *Meloidodera floridensis*, *Pratylenchus penetrans*, *P. vulnus*, *Paratylenchus curvitatus* and a *Meloidogyne* sp. were very susceptible; 1000 *P. penetrans* in culture were devoured in 2 days by 40–50 dorylaims. The rate of predation on nematodes of *Mononchoides potohikus* in the laboratory was independent of the prey species but dependent on the prey density (Yeates, 1969).

In some citrus orchards in California large numbers of the dorylaimid, *Thornia* sp., feed on the citrus pest, *Tylenchulus semipenetrans*. Large numbers of *Thornia*, which is cultured easily on fungus, were added to soil to ascertain its effect as a biological control agent of the citrus nematode (Boosalis and Mankau, 1965). Groups of potted seedlings were infested with *Tylenchulus* and *Thornia*, with *Thornia* alone, with *Tylenchulus* alone, and, in the control group, without any nematodes. After 29 months there was a seven-fold increase of *Thornia* in the presence of *Tylenchulus*, although the number of *Tylenchulus* did not significantly decline. The authors suggest that under certain conditions the predation of *Thornia* and of other organisms on *Tylenchulus* may significantly decrease the population of this citrus pest.

Despite the apparent success of predacious nematodes they probably do not occur in the soil in sufficiently large numbers in the crucial early phase of build-up of a plant-parasitic nematode population for them to be efficient natural regulators of these pests. More research is needed on the factors affecting the population dynamics of predatory nematodes.

D. Other Predacious Soil Fauna

Collembola, especially *Onychiurus armatus*, are active predators of nematodes but it is not known whether they significantly decrease soil populations of plant-parasitic nematodes. In a pot experiment with a 1:5 ratio of *O. armatus* and *Heterodera cruciferae* cysts, Murphy and Doncaster (1957) estimated that about 6·9% of the cysts were damaged by *O. armatus*. This collembolan is a voracious feeder and devours large nematodes, including mature female *Heterodera* within a few hours. In experiments where organic matter was added to the soil the numbers of Collembola increased greatly in association with an increase in total nematode numbers.

Rodriguez *et al.* (1962) observed that when given an equal choice, the adult mite *Macrocheles muscaedomesticae* preferred house fly eggs over nematodes while proto- and deutonymphs under the same conditions preferred nematodes. An oribatid mite, *Pergalumna* sp., is reported to eat *Pelodera lambensis* and *Tylenchorhynchus martini* in large numbers (Rockett and Woodring, 1966). A culture of 30 of the free-living nematode, *Pelodera*, was eliminated within 24 h by three adult and four nymphal *Pergalumna*. As *Pergalumna* kills large numbers of nematodes and is able to survive on fungi in the absence of nematodes it seems that this and other species of mite may be important under some conditions in regulating populations of nematode pests in the soil.

Tardigrades have been observed feeding on several nematode species (Hutchinson and Streu, 1960; Doncaster, 1962). The tardigrade *Hypsibius myrops* was cultured on *Panagrellus redivivus* for several months (Sayre, 1969), and it significantly decreased populations of the root-knot nematode, *Meloidogyne incognita*, and the stem nematode, *Ditylenchus dipsaci*, in moist Sphagnum over 3 days.

Schaeffenberg and Tendl (1951) reported that the enchytraeid worms *Fridericia* and *Enchytraeus* controlled populations of the beet cyst eelworm, *Heterodera schachtii*, in sugar-beet roots. However, enchytraeids are usually saprophagous and probably would devour endoparasitic *Heterodera* larvae only when they are embedded in extensively damaged host plant tissue: this suggests that enchytraeids are of little value in regulating natural populations of nematodes.

Turbellarians, however, show rather more promise as they consume significant numbers of nematodes in certain soil types; although under normal conditions they are unlikely to be present in sufficient numbers to regulate the populations of plant-parasitic nematodes. In a pot experiment (Sayre and Powers, 1966) in which 350 *Meloidogyne incognita* larvae and 30 turbellarians, *Adenoplea* sp., were inoculated into soil 48 h before tomato seedlings were planted, the turbellarians caused a significant decrease in the root-knot galling. Sayre (1970) records that 20 *Adenoplea* in culture that had been starved for 24 h greatly decreased a population of 300 *Panagrellus redivivus*.

Amoeboid organisms feed on various nematode species. Despite its small size and relatively lethargic action *Theratromyxa weberi* is known to engulf a nematode in less than 2 h (Zwillenberg, 1953) (Fig. 3).

E. Antagonistic Higher Plants

Over the years there have been numerous reports of crops which produce a root secretion that is toxic to nematodes. Ellenby (1945, 1951) confirmed early reports by showing that exudate from white or black

mustard seedlings and dilute solutions of allyl isothiocyanate, the
mustard oil of black mustard, decreased the hatch of *Heterodera rostochiensis* larvae. A dose of 3 ml of allyl isothiocyanate per 10 plants
increased the yield of potatoes by 100% (Ellenby, 1951). In a small
scale field trial in which the mustard oil was applied at a rate equivalent
of 0·12 metric tons per hectare, with peat as the carrier, the potato
yield increased by 100%.

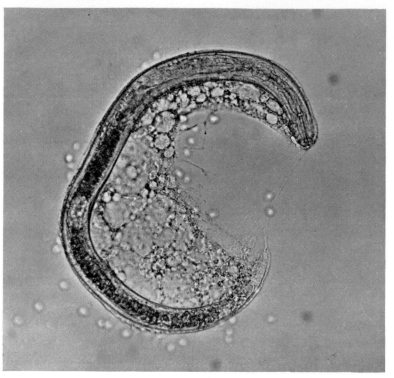

Fig. 3. The amoeba *Theratromyxa weberi* engulfing a second-stage larva of *Heterodera
schachtii*. The contractile strands of cytoplasm bridging the gap between the nematode's
head and tail eventually cause the nematode to become coiled in a digestive cyst. (By
courtesy of C. C. Doncaster.)

Of all the plants most studied for their nematicidal effects, *Tagetes*
spp. are probably the best known and also appear to have the greatest
success in nematode control. Winoto Suatmadji (1969) reviewed much
of the work done on *Tagetes* spp. Oostenbrink (1957) showed that
T. patula decreased *Pratylenchus* populations 90% more than did
fallow, and populations of *P. pratensis*, *P. penetrans* and *Tylenchorhynchus dubius* in soil were suppressed. It was shown that 16 varieties

of *Tagetes patula* and *T. erecta* were effective when the nematodes were exposed to these plants in the soil for 3–4 months. The cultivation of *Tagetes* spp. between the rows or round the stem base of the host plant crop suppresses the *Pratylenchus* population both in the plant roots and in the soil. However, some ectoparasites, such as *Criconemoides mutabilis*, were reported to be unaffected by *Tagetes* spp. Oostenbrink (1957) and Hesling *et al.* (1961) stated that the exudates of *Tagetes* sp. did not decrease the hatch of *Heterodera rostochiensis*. Nevertheless, this plant caused a slight reduction in the nematode population when grown for 4 months in soil infested with *H. rostochiensis* (Omidvar, 1962). In field experiments *Tagetes* sp. increased the yield of most crops on sandy and peaty soils by 10–40%, indicating that *Tagetes* sp. in a 2-year rotation are more effective in decreasing *Pratylenchus* populations than is fallow (Oostenbrink, 1960). These results are supported by those of Kerr (1963) in decreasing *P. loosi* and *Meloidogyne* populations on tea in Ceylon.

Rohde and Jenkins (1958) showed that asparagus would not support *Trichodorus christiei* for more than 40–50 days and that tomatoes, which were normally a good host, supported only very low populations when asparagus also was grown in the plot. The juices from asparagus roots were toxic to many nematode species at 1:10 dilution, and the number of *T. christiei* on tomato roots were decreased by a leaf spray or root drench at 1000 ppm. Atwal and Mangar (1969) described laboratory experiments in India which showed that root exudates from sesame (*Sesamum orientale*) have nematicidal properties against *Meloidogyne incognita*. When *Abelmoschus esculentus* was grown in *M. incognita*-infested soil it was only slightly attacked and there were fewer nematodes compared with when *A. esculentus* was grown in the absence of sesame.

Ochse and Brewton (1954) and Colbran (1957) report that *Crotalaria* spp. are good cover crops for decreasing populations of *Meloidogyne* spp., but a report (Anon., 1965) from the southern U.S.A. where both *Crotalaria* and *Tagetes* spp. have been used in trials, suggest that *Tagetes* is more effective. In the U.S.A. populations of *Pratylenchus brachyurus* built up and those of *Xiphinema americanum* decreased under *Crotalaria spectabilis* (Good, 1968).

III. Nematode Parasites of Insect Pests

In addition to the large numbers of parasitic nematodes that are economically important pests there are many nematode species which attack insects that are themselves pests. In so doing, these nematodes are frequently beneficial to man as they decrease the harmfulness of insects in one of several ways, namely by (i) killing the immature or adult insect, (ii) sterilizing the insect or decreasing its fecundity, (iii) changing the insect's behaviour such that it fails to perform the essential functions of life and/or (iv) slowing development of the insect.

Although the occurrence of natural regulation of insect populations by nematodes has been known for many years there has been a lack of understanding of the physiological and ecological relationships between the nematode parasite and the insect host, and so the practical use of nematodes as biological control agents has been minimal. The remainder of this chapter considers those nematode parasites of insect pests of crops that are potential biological control agents.

A. Mermithids

This group of nematodes are obligate parasites of insects, to which they are invariably lethal. The insect is infected either by active penetration of the nematode larva through the cuticle into the haemocoel or by ingestion of the embryonated egg. In the latter instance the larva hatches in the gut and migrates rapidly to the haemocoel. Once in the haemocoel the mermithid larva grows and moults until the pre-adult stage is reached, when it leaves the host either intersegmentally or through one of the natural openings (Fig. 4). The final moult, mating and oviposition occur outside the host. This aggressive method of exit from its host by a nematode that is frequently many times the length of the insect host, is always lethal to the host. However, prior to emergence the mermithid usually induces sterility and frequently causes morphological abnormalities of the host (Fig. 4). Usually it is the larval or nymphal host that is infected but Welch (1963) records infective larvae invading the adult Bogong moth, Agrotis infusa. It seems likely that the nature of the host's physical environment and the behaviour of the host determine the developmental stage of the insect that is infected.

Couturier (1963) described two species of mermithid parasitic in the common chafer, Melolontha melolontha. The density of this temperate climate insect pest is cyclic and Couturier demonstrated that the population density of the mermithids followed very closely that of the host. It is repeatedly recorded that the incidence of mermithids is very variable and, frequently, areas that have high mermithid incidence in one year have a lower incidence the following year. Also, sites of high mermithid incidence are often localized. These facts support the theory that these parasites are probably very effective locally in decreasing host insect populations from one year to the next and also that they are not readily dispersed. In a survey of the grasshoppers of western Canada, Smith (1958) recorded a considerable range of incidence of parasitism and the incidence declined westwards. However, in early observations in the northeastern U.S.A. Glaser and Wilcox (1918) found that the incidence of mermithids in grasshoppers was as high as 76%. Various species of grasshopper are hosts of mermithids and much has been written about the ecology and bionomics of this relationship.

Fig. 4. *Mermis nigrescens* escaping from the cadaver of an adult desert locust, *Schistocerca gregaria*. Note the malformed wings resulting from picking up the infection at the nymphal stage. (By courtesy of R. Gordon.)

More recently the nutritional relationships of the desert locust, *Schisto-cerca gregaria*, and *Mermis nigrescens* have been investigated (Gordon and Webster, 1972) and these show the type of food *M. nigrescens* requires and its source within the insect host.

Many other insect pests suffer the depredations of mermithid infections, such as the rice borer *Chilo suppressalis* in Japan (Kaburaki and Imamura, 1932), forest lepidoptera in Voronezhsk oblast (Artyukhovskii, 1955), cutworm larvae in Russia (Artyukhovskii and Negrobov, 1959) and India (Khan and Hussain, 1964), alfalfa weevil in the U.S.A. (Poinar, 1962), cucumber beetle in the southern U.S.A. (Cuthbert, 1968) and the earwig in Canada (Wilson, 1971).

B. *Neoaplectana* spp. and other Rhabditids

Many rhabditids have harmless associations with insects (they are generally bacterial and detritus feeders), but some *Neoaplectana* species are lethal to insects. The DD-136 strain of *N. carpocapsae* has a bacterium (*Achromobacter nematophilus*) in mutualistic association with it. These bacterial cells are carried in the ventricular portion of the intestinal lumen of the infective larval DD-136 which is taken up by the feeding insect. The larval nematode penetrates the gut wall and enters the insect haemocoel, at which stage the bacteria are released through the nematode's anus (Poinar, 1966). The bacteria feed and multiply in the haemocoel, kill the insect and create within the cadaver a "bacterial soup" upon which the nematodes feed and multiply. The infective nematode larvae either remain in the cadaver or migrate into the surrounding environment to be picked up by another host. This lethal nematode/bacterium combination was first reported by Dutky (1959) who recognized its potential as a biological control agent.

Other species of *Neoaplectana* have been reported, but their effect on insect populations is unknown; 33% infection of the pronymphs of the sawfly *Acantholyda nemoralis* was reported in Poland (Weiser and Koehler, 1955) and a 25% infection of larval tussock moths, *Crombus simplex*, in New Zealand (Hoy, 1954). Recently, Turco *et al.* (1970) found that the sugar-cane borer, rice water weevil, cucumber beetle and white grub were killed by *N. glaseri*.

Most other rhabditids, including other *Neoaplectana* species associated with insects, are found frequently in decomposing material or in soil where they feed on bacteria. Consequently, in such habitats these nematodes and bacteria are found also in insect guts. *N. affinis* and *N. bibionis* live in the gut of the "host" insect without causing apparent injury until the insect dies, when they invade the haemocoel and reproduce in the decaying cadaver (Welch, 1965). Recently two rhabditids from leatherjackets, *Tipula paludosa* larvae, were shown not to be pathogens, but rather opportunists that remain in the gut until death or debilitation of the insect; they then feed and reproduce in the cadaver (Lam and Webster, 1971). However, Sandner and Stanuszek (1967) describe a diplogasterid nematode that killed hibernating adult Colorado potato beetles, *Leptinotarsa decemlineata*, and Niklas (1960) considered that *Diplogasteroides berwigi* played an important role in the biological control of the chafers, *Melolontha* spp.

C. Aphelenchids and Tylenchids

An early report of a tylenchid affecting insect populations was that of Goodey (1930) who described the nematode, *Howardula oscinellae*, causing sterility of frit flies, *Oscinella frit*. Another species of *Howardula* and one of *Parasitylenchus* sterilize fruit flies, *Drosophila* sp. (Welch, 1959). Phorid flies, *Megasalia halterata*, on mushrooms are commonly infected with a *Bradynema* sp. which sterilizes the female fly (Hussey, 1965). Several forest insect pests are attacked by this group of nematodes. Massey (1958) showed that *Parsitylenchus elongatus* sterilized and was sometimes lethal to the fir engraver beetle, *Scolytus ventralis*, while *Contortylenchus elongatus* in the haemocoel of *Ips confusus* decreased the egg-laying capacity of the beetle by 50% (Massey, 1960). Another nematode of potential value to the forestry industry is *Deladenus* sp. The female of this nematode lives in the haemocoel of the wood-boring wasp, *Sirex* sp., and causes sterilization (Bedding, 1967). The juveniles produced by the parasitic female are released into the galleries where they feed for several generations on a fungus, *Amylostereum*, before again entering a *Sirex* larva. This nematode may thus survive long periods in the absence of the insect host.

D. The Use of Nematodes in Insect Control

Several nematode species have been tested for their efficiency as agents for insect control, but the difficulties of producing large numbers and of their successful application has restricted the species of nematode tested in the field. Nearly all practical applications of nematodes to control insects have utilized species of *Neoaplectana* (Table I). Larvae of *Neoaplectana glaseri* applied to turf at 292,000 nematodes/m² caused high mortality of the Japanese beetle, *Popillia japonica* in the field (Glaser, 1932). The DD-136 strain of *N. carpocapsae* has been tried in the field most often and with varying degrees of success (Table I). Niklas (1967) reviews the use of DD-136 in controlling insects and Poinar (1971) tabulates the details of field trials with *N. glaseri* and DD-136, and the results obtained. Dutky (1959) reported on applying DD-136 in aqueous sprays to apple orchards for several years to control populations of the codling moth larva, *Cydia pomonella*. Insect mortalities in excess of 60% occurred and the nematode was reported to be superior to many chemical insecticides, in that it persisted from one generation of the *C. pomonella* to the next by penetrating the cracks in the apple tree bark to infect the cocoons. Jacques (1967) did not obtain a significant decrease in the population of Winter moth (*Operophtera brumata*) on apple trees after spraying the trees with DD-136 suspension. However, DD-136 sprayed on to pine bark, together with a wetting agent, entered the galleries of the pine bark beetle *Dendroctonus frontalis*, and killed the brood and adults (Moore, 1970). The high, though rather impractical, dose equivalent of $55 \cdot 5 \times 10^9$ infective

TABLE I. Successful tests with the DD-136 strain of *Neoaplectana carpocapsae* against insect pests of different crops

Pest insect	Notes	Reference
VEGETABLE AND FIELD CROPS		
Leptinotarsa decemlineata (Colorado potato beetle)	Significant reduction in plot tests but no control in field tests	Welch and Briand (1961)
Heliothis virescens (Tobacco budworm) *Protoparce sexta* (Tobacco hornworm)	80–85% reduction in field population at dose of 50,000 per plant	Chamberlin and Dutky (1958)
Heliothis zea (Corn earworm)	High mortality, but death of host occurred after crop damage had been done	Tanada and Reiner (1962)
Heliothis virescens (Cotton boll worm) *Mescinia peruella* *Pococera atramentalis* *Anthonomus vestitus* (Cotton weevil) *Dysdercus peruvianus* (Peruvian cotton stainer)	Infected in laboratory tests	Tang (1958)
Hylemya brassicae (Cabbage maggot) *Ostrinia nubilalis* *Pieris rapae* (Imported cabbageworm)	Field populations significantly decreased and, for *O. nubilalis*, better than by chemical insecticides	Welch and Briand (1961)
Trichoplusia ni (Cabbage looper) *Heliothis zea* (Corn earworm)	27% and 46% mortality respectively in field tests	Moore (1965)
Leptinotarsa decemlineata (Colorado potato beetle)	Successful population decrease	Scognamiglio *et al.* (1968)
Diabrotica balteata (Banded cucumber beetle)	Reduction of damage	Creighton *et al.* (1968)
ORCHARD CROPS		
Cydia pomonella (Codling moth)	In the field a consistent 60% mortality	Dutky (1959)
FOREST CROPS		
Rhyacionia buoliana (European pine shoot moth) *Choristoneura fumiferana* (Spruce budworm) *C. pinus* (Jack-Pine budworm) *Symmerista albifrons*	Insects were infected by DD-136	Schmiege (1963)

Table I (*continued*)

Pest insect	Notes	Reference

Hylobius radicis
  (Pine root-collar weevil)
Monochanus scutellatus
Diprion similis
  (Introduced pine sawfly)
Neodiprion lecontei
  (Red-headed pine sawfly)
Pikonema alaskensis
Pristiphora erichsonii
  (Larch sawfly)		

Pristiphora erichsonii (Larch sawfly)	Significant kill in laboratory tests to which "thick water" and an evaporation retardant added	Webster and Bronskill (1968)
Rhyaciona frustrana (Nantucket pine tip moth)	Field population decreased but not economic	Nash and Fox (1969)
TURF AND FORAGE CROP *Tipula paludosa* (European marsh cranefly)	The mortality of DD-136 increased if mixed with *Bacillus thuringiensis* var. *thuringiensis*	Lam and Webster (1972)

DD-136 larvae per hectare caused a 100% mortality of leatherjackets, *Tipula paludosa* larvae (Lam and Webster, 1972). However, by combining a β-exotoxin preparation of *Bacillus thuringiensis* var. *thuringiensis* with a low inoculum of DD-136 these authors obtained a 90% mortality of *T. paludosa* larvae.

Many insect pests of vegetable crops and orchard trees have been the target of DD-136 spray programmes but in general the results have been disappointing (Table I). This failure of a potentially lethal nematode to control an insect population is due to a number of ecological and physiological factors which influence the insect–nematode relationship. Recently work has been done on the factors which mitigate against successful control of harmful insect populations by nematodes. There are three main factors: (i) resistance of insects to nematode parasitism, (ii) intolerance of the free-living stage(s) of the nematode to low humidities and extremes of temperature and, (iii) inability to produce large numbers of infective nematodes for use in a control programme.

(i) Poinar (1969) reviews the insect's defence mechanisms to nematodes. Various forms of encapsulation may isolate a pathogenic nematode that has penetrated the insect's haemocoel. However, multiple infection may overcome this resistance and the insect then will succumb to the nematode. A humoral form of resistance may occur in some insects, as nematodes present in the haemocoel have been known to cease development in the absence of cellular encapsulation. Some insects are more susceptible to nematode attack at certain stages of their development, as in the instance of DD-136 infection of *Leptinotarsa decemlineata* (Webster, unpublished) where the mortality of DD-136 infected nymphs is greater than that of adults. If susceptibility of an insect to a nematode changes with the insect's physiological state the possibility should be considered of inducing physiological change at the time of infection by the nematode control agent.

(ii) Schmiege (1962) reported that above 32°C the activity of the infective stage of neoaplectanids decreased and the mortality increased to 100%. Similarly, Weiser (1966) showed that the survival time of infective *N. carpocapsae* larvae was less than 1 h at 38°C. Welch and Briand (1961) emphasized the importance of water in the longevity of infective nematode larvae and stated that infective DD-136 larvae probably survived only about 1 h when sprayed on foliage. Webster and Bronskill (1968) used a Gelgard M/evaporation retardant/surfactant mixture to extend the survival period of the nematode and so increase the mortality of the larch sawfly, *Pristiphora erichsonii*.

(iii) To control insect populations with nematodes large numbers of infective nematodes must be available at short notice. It is therefore essential to be able to produce them at low cost in either *in vitro* culture or *in vivo* in insectary populations of host insects. DD-136 is readily cultured on the wax moth larva, *Galleria mellonella* (Dutky *et al.*, 1964), and can be stored at about 5°C in 0·1% formalin for several years providing sufficient oxygen is available. Mass breeding of this nematode is now done commercially in the U.S.A. on a medium based on dog food (House *et al.*, 1965). Cuthbert (1968) reported rearing several generations of *Filipjevimermis leipsandra* in two species of cucumber beetle, *Diabrotica*. To date there have been no successful methods described for culturing *in vitro* the highly pathogenic mermithids or sphaerularids. These endoparasites are dependent on the host's physiology for their own development and so a greater knowledge of the nutritional requirements of the nematode is a necessary prerequisite to their *in vitro* culture. The recent research of Gordon and Webster (1971, 1972) should increase the chances of successful mermithid culture in the near future.

Providing that these limiting factors can be removed the mermithids and some tylenchids and aphelenchids show considerable potential as biological control agents of agricultural and forestry insect pests (Fig. 5).

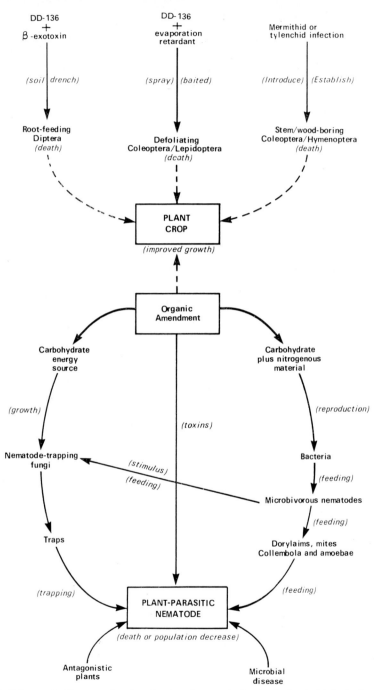

Fig. 5. Schematic diagram showing the involvement of nematodes in biological control of crop pests.

IV. Conclusions

This review emphasizes that the common use of parasites and predators to decrease populations of nematode pests, and of nematodes to decrease populations of insect pests, is currently not an economical or practical proposition. However, some of these biological control agents show promise as alternatives to chemical biocides (Fig. 5). A more intensive study of predatory mites and of microbial diseases of nematodes would be useful, since many of the known examples appear to be highly successful in decreasing nematode populations. The complex interaction between soil flora and fauna following the addition of organic amendments to the soil deserves much more investigation, particularly as this method of controlling plant-parasitic nematodes is the only economical method in some of the developing countries. Mermithids are potentially good control agents both of pest insects and of plant-parasitic nematodes, but more research on them is needed before one can hope to establish them in the field as regulators of pest populations. The initial success of DD-136 as a biological insecticide is already being improved upon by combining it with other pathogens, and its efficacy could be further enhanced by using it in conjunction with a bait to improve uptake. Some experiments have shown that the susceptibility of a nematode or of an insect pest to a parasite may be increased during periods of physiological change, e.g., at moulting. Hence, the addition of a developmental hormone at the time of parasite application may enhance control of the pest. In general, more research is urgently needed on (i) the physiology of the host–parasite relationships so as to facilitate the culture of the parasites, and (ii) on the ecology of these relationships to improve parasite application and so increase their effectiveness.

References

Adams, R. E. and Eichenmuller, J. J. (1963). *Phytopathology* **53**, 745.
Anon. (1965). *Agric. Res. Wash.* **14**, 12.
Artyukhovskii, A. K. (1955). *Voronezhsk. Lesokhozyaystevennyy Inst.* 37–40.
Artyukhovskii, A. K. and Negrobov, V. P. (1959). *Tr. Khopersk. Gos. Zapovednika* **3**, 268–270.
Atwal, A. S. and Mangar, A. (1969). *Indian J. Ent.* **31**, 286.
Barron, G. L. (1970). *Can. J. Bot.* **48**, 329–331.
Bedding, R. A. (1967). *Nature, Lond.* **214**, 174–175.
Birchfield, W. (1960). *Mycopath. Mycol. appl.* **13**, 331–338.
Boosalis, M. G. and Mankau, R. (1965). *In* "Ecology of Soil-borne Plant Pathogens" (K. F. Baker and W. C. Snyder, eds), pp. 374–389. University of California Press, Los Angeles.
Canning, E. U. (1962). *Arch. Protistenk.* **105**, 455–462.
Chamberlin, F. S. and Dutky, S. R. (1958). *J. econ. Ent.* **51**, 560.

Cobb, N. A. (1924). *J. Parasit.* **11**, 120–121.

Colbran, R. C. (1957). *Qd agric. J.* **83**, 499–501.

Cooke, R. C. (1962). *Trans. Br. mycol. Soc.* **45**, 314–320.

Cooke, R. C. and Godfrey, B. E. S. (1964). *Trans. Br. mycol. Soc.* **47**, 61–74.

Cooke, R. C. and Pramer, D. (1968). *Phytopathology* **58**, 659–661.

Couturier, A. (1963). *Annls Épiphyt.* **14**, 203–267.

Creighton, C. S., Cuthbert, F. P. and Reid, W. J. (1968). *J. Invert. Path.* **10**, 368–373.

Cuthbert, F. P. (1968). *J. Invert. Path.* **12**, 283–287.

Das, S. and Raski, D. J. (1969). *J. Nematol.* **1**, 107–110.

Dollfus, R. P. (1946). *Encycl. Biol.* **27**, 1–481.

Doncaster, C. C. (1962). *Parasitology* **52**, 19P.

Dreschler, C. (1950). *Mycologia* **42**, 1–79.

Duddington, C. L. (1951). *Trans. Br. mycol. Soc.* **34**, 322–331.

Duddington, C. L. (1957). "The Friendly Fungi." Faber and Faber, London.

Duddington, C. L., Everard, C. O. R. and Duthoit, C. M. G. (1961). *Pl. Path.* **10**, 108–109.

Duddington, C. L., Jones, F. G. W. and Moriarty, F. (1956). *Nematologica* **1**, 344–348.

Dutky, S. R. (1959). *Adv. appl. Microbiol.* **1**, 175–200.

Dutky, S. R., Thompson, J. V. and Cantwell, G. E. (1964). *J. Invert. Path.* **6**, 417–422.

Ellenby, C. (1945). *Ann. appl. Biol.* **32**, 237–239.

Ellenby, C. (1951). *Ann. appl. Biol.* **38**, 859–875.

Esser, R. P. (1963). *Proc. Soil Crop Sci. Soc. Fla* **23**, 121–138.

Esser, R. P. and Sobers, E. K. (1964). *Proc. Soil Crop Sci. Soc. Fla* **24**, 326–353.

Estey, R. H. and Olthof, Th. H. A. (1965). *Phytoprotection* **46**, 14–17.

Glaser, R. W. (1932). *New Jers. Dept Agric. Bureau Plant Industry, Circular* **211**, 3–34.

Glaser, R. W. and Wilcox, A. M. (1918). *Psyche* **25**, 12–15.

Good, J. M. (1968). *In* "Tropical Nematology" (G. C. Smart and V. G. Perry, eds), pp. 113–138. University of Florida Press, Gainesville.

Goodey, T. (1930). *Phil. Trans. R. Soc. Lond.* **218**, 315–343.

Gordon, R. and Webster, J. M. (1971). *Exp. Parasit.* **29**, 66–79.

Gordon, R. and Webster, J. M. (1971). *Parasitology,* **64**, 161–172.

Hayes, W. A. and Blackburn, F. (1966). *Ann. appl. Biol.* **58**, 51–60.

Hesling, J. J., Pawelska, K. and Shepherd, A. M. (1961). *Nematologica* **6**, 207–213.

House, H. L., Welch, H. E. and Cleugh, T. R. (1965). *Nature, Lond.* **206**, 847.

Hoy, J. M. (1954). *Parasitology* **44**, 392–399.

Hussey, N. W. (1965). *Proc. XII Int. Congr. Ent. Lond.,* 1964. p. 752.

Hutchinson, M. T. and Streu, H. T. (1960). *Nematologica* **5**, 149.

Iizuka, H., Komagata, K., Kawamura, T., Kunii, Y. and Shibuya, M. (1962). *Agric. biol. Chem., Tokyo* **26**, 199–200.

Jacques, R. P. (1967). *J. econ. Ent.* **60**, 741–743.

James, G. L. (1966). *Nature, Lond.* **212**, 1466.

James, G. L. (1968). *Ann. appl. Biol.* **61**, 503–510.

Jorgenson, E. C. (1970). *J. Nematol.* **2**, 393–398.

Kaburaki, T. and Imamura, S. (1932). *Proc. Imp. Acad. Japan* **8**, 109–112.

Katznelson, H., Gillespie, D. C. and Cook, F. D. (1964). *Can. J. Microbiol.* **10**, 699–704.

Kerr, A. (1933). *Rep. Tea Res. Inst. Ceylon*, 95–102.

Khan, M. Q. and Hussain, M. (1964). *Indian J. Ent.* **26**, 124–125.

Kuiper, K. (1958). *Tijdschr. PlZiekt.* **64**, 122.

Laan, P. A. van der (1956). *Tijdschr. PlZiekt.* **62**, 305–321.

Lam, A. B. Q. and Webster, J. M. (1971). *Nematologica* **17**, 201–212.

Lam, A. B. Q. and Webster, J. M. (1972). *J. Invert. Path.* In press.

Lawton, J. R. (1967). *Trans. Br. mycol. Soc.* **50**, 195–205.

Linford, M. B. (1937). *Science, N.Y.* **85**, 123–124.

Linford, M. B. and Oliveira, J. M. (1937). *Science, N.Y.* **85**, 295–297.

Linford, M. B., Yap, F. and Oliveira, J. M. (1938). *Soil Sci.* **45**, 127–141.

Loewenberg, J. R., Sullivan, T. and Schuster, M. L. (1959). *Nature, Lond.* **184**, 1896.

Mankau, R. (1968). *Pl. Dis. Reptr* **52**, 851–855.

Mankau, R. and Das, S. (1969). *J. Nematol.* **1**, 15–16.

Mankau, R. and Minteer, R. J. (1962). *Pl. Dis. Reptr* **46**, 375–378.

Massey, C. L. (1958). *Proc. helminth. Soc. Wash.* **25**, 26–30.

Massey, C. L. (1960). *Proc. helminth. Soc. Wash.* **27**, 14–22.

Micoletzky, H. (1925). *Mem. Acad. R. Sci. Lettres Danemark (Copenhagen).* *Sec. Sci.* **8** Sec. t. X, no. 2, 55–310.

Moore, G. E. (1965). *J. Kans. ent. Soc.* **38**, 101–105.

Moore, G. E. (1970). *J. Nematol.* **2**, 341–344.

Muller, H. G. (1958). *Trans. Br. mycol. Soc.* **41**, 341–364.

Murphy, P. W. and Doncaster, C. C. (1957). *Nematologica* **2**, 202–214.

Nash, R. F. and Fox, R. C. (1969). *J. econ. Ent.* **62**, 660–663.

Niklas, O. F. (1960). *Mitt. biol. BundAnst. Ld-u. Forstw.* **101**.

Niklas, O. F. (1967). *Mitt. biol. BundAnst. Ld-u. Forstw.* **124**.

Nonaka, F. (1959). *Sci. Bull. Fac. Agric. Kyushu Univ.* **17**, 1–8.

Ochse, J. J. and Brewton, W. S. (1954). *Proc. Fla St. hort. Soc.* **67**, 218–219.

Olthof, T. H. and Estey, R. H. (1963). *Nature, Lond.* **197**, 514–515.

Omidvar, A. M. (1962). *Nematologica* **7**, 62–64.

Oostenbrink, M. (1960). *Meded. LandbHogesch. OpzoekStns Gent* **25**, 1065–1075.

Oostenbrink, M., Kuiper, K. and s'Jacob, J. J. (1957). *Nematologica* **2**, Suppl. 424–433.

Patrick, Z. A., Sayre, R. M. and Thorpe, H. J. (1965). *Phytopathology* **55**, 702–704.

Poinar, G. O., Jr. (1962). Ph.D. Thesis, Cornell University.

Poinar, G. O., Jr. (1966). *Nematologica* **12**, 105–108.

Poinar, G. O., Jr. (1969). *In* "Immunity to Parasitic Animals" (G. J. Jackson and R. Herman, eds), Vol. 1, 173–210. Appleton-Century-Crofts, New York.

Poinar, G. O., Jr. (1971). *In* "Microbial Control of Insects and Mites" (H. D. Burges and N. W. Hussey, eds), 181–203. Academic Press, London.

Pramer, D. and Stoll, N. R. (1959). *Science, N.Y.* **129**, 966–967.

Prasad, N. and Mankau, R. (1969). *J. Nematol.* **1**, 301–302.

Rockett, C. L. and Woodring, J. P. (1966). *Ann. ent. Soc. Am.* **59**, 669–671.

Rodriguez, J. G., Wade, C. F. and Wells, C. N. (1962). *Ann. ent. Soc. Am.* **55**, 507–511.

Rohde, R. A. and Jenkins, W. R. (1958). *Phytopathology* **48**, 463.

Sachchidananda, J. and Swarup, G. (1966). *Indian Phytopath.* **19**, 279–285.

Salonen, A. and Ruokola, A. (1968). *Maatalous Aikakausk* **40**, 142–145.

Sandner, H. and Stanuszek, S. (1967). *In* "Insect Pathology and Microbial Control" (P. A. van der Laan, ed.), pp. 210–212. North-Holland Publ. Co., Amsterdam.

Satchuthananthavale, V. and Cooke, R. C. (1967a). *Nature, Lond.* **214**, 321–322.

Satchuthananthavale, V. and Cooke, R. C. (1967b). *Trans. Br. mycol. Soc.* **50**, 221–228.

Satchuthananthavale, V. and Cooke, R. C. (1967c). *Trans. Br. mycol. Soc.* **50**, 423–428.

Sayre, R. M. (1969). *Trans. Am. microsc. Soc.* **88**, 266–274.

Sayre, R. M. (1970). *The American Biology Teacher* **32**, 487–490.

Sayre, R. M. and Keeley, L. S. (1969). *Nematologica* **15**, 492–502.

Sayre, R. M. and Powers, E. M. (1966). *Nematologica* **12**, 619–629.

Schaeffenberg, B. and Tendl, H. (1951). *Z. angew. Ent.* **32**, 476–488.

Schmiege, D. C. (1963). *J. econ. Ent.* **56**, 427–431.

Scognamiglio, A., Giandomenico, N. and Talame, M. (1968). *Boll. Lab. Ent. agr. "Filippo Silvestri" di Portici* **26**, 191–204.

Singh, R. S. and Sitaramaiah, K. (1966). *Pl. Dis. Reptr* **50**, 668–672.

Singh, R. S. and Sitaramaiah, K. (1967). *Indian Phytopath.* **20**, 349–355.

Smith, R. W. (1958). *Can. J. Zoo.* **36**, 217–262.

Soprunov, F. F. (1958). *Akad. Nauk Turkmenskoi, USSR Ashkhabad.*

Steiner, G. and Heinly, H. (1922). *J. Wash. Acad. Sci.* **12**, 367–386.

Steinhaus, E. A. and Marsh, G. A. (1962). *Hilgardia* **33**, 349–490.

Tanada, Y. and Reiner, C. (1962). *J. Insect Path.* **4**, 139–154.

Tang, J. L. (1958). *Revista Peru. Entomol. Agricola* **1**, 19–22.

Tarjan, A. C. (1961). *Proc. Soil Crop Sci. Soc. Fla* **21**, 17–36.

Thorne, G. (1927). *J. agric. Res.* **34**, 265–286.

Thorne, G. (1940). *Proc. helminth. Soc. Wash.* **7**, 51–53.

Tjepkema, J. P. (1969). *J. Nematol.* **1**, 29.

Turco, C. P., Hopkins, S. H. and Thames, W. H. (1970). *J. Parasit.* **56**, 277–280.

Walker, J. T., Specht, C. H. and Mavrodineau, S. (1967). *Pl. Dis. Reptr* **51**, 1021–1024.

Webster, J. M. and Bronskill, J. F. (1968). *J. econ. Ent.* **61**, 1370–1373.

Weiser, J. (1966). *Nemoci Hmyzu.* Academia, Prague, 554 pp.

Weiser, J. and Koehler, W. (1955). *Roczniki Nauk Lesnych* **11**, 93–110.

Welch, H. E. (1959). *Parasitology* **49**, 83–103.

Welch, H. E. (1963). *Parasitology* **53**, 55–62.

Welch, H. E. (1965). *A. Rev. Ent.* **10**, 275–302.

Welch, H. E. and Briand, L. J. (1961). *Can. Ent.* **93**, 759–763.

Williams, J. R. (1960). *Nematologica* **5**, 37–42.

Wilson, W. A. (1971). *Can. Ent.* **103**, 1045–1048.

Winoto Suatmadji, R. (1969). "Studies on the Effect of *Tagetes* Species on Plant-parasitic Nematodes." Veenman and Zonen N.V., Wageningen, The Netherlands.

Yeates, G. W. (1969). *Nematologica* **15**, 1–9.

Zwillenberg, L. O. (1953). *Antonie van Leeuwenhoek* **19**, 101–116.

20
Evaluation and Integration of Nematode Control Methods

M. Oostenbrink

Laboratorium voor Nematologie
Landbouwhogeschool
Wageningen
The Netherlands

I. Introduction

A. General

Control of a plant nematode is essentially an economic method of population regulation. It is well known that nematode communities composed of populations of several species occur in nearly every soil, and that populations in cultivated or in untilled soil nearly always contain 30–50% plant parasites in addition to nematodes of predatory and saprozoic species. An established nematode community in cultivated or uncultivated soil usually harbours populations of more than ten plant-parasitic species from more than five different genera, and some of them become numerous and possibly noxious when suitable host plants are grown intensively. In a field with constant soil type and climatic conditions the crops grown would determine to a large extent the population density of the nematode species present. Nematode damage is often associated with short rotations and therefore with the culture of main crops.

Great persistence, low reproductive rate and little active locomotion account for the modest but regular and predictable fluctuations in population densities of plant-parasitic nematodes. These characteristics explain why the appearance of nematode damage often lags several years behind the initial infection, its original occurrence as patches which gradually spread over the fields, and also the non-specific disease symptoms which, at first, are often difficult to distinguish from other types of poor growth. Scrutiny of the symptoms may reveal that many infestations are, in fact, specific for a particular plant–nematode relationship. In most instances of nematode damage, and of nematode control, the quantitative relation between the nematode population density and the symptoms has to be considered.

B. Nematode Density/Plant Yield Relationship

Most plant nematodes, including very noxious species, are polyphagous. This explains why one species can be involved in many host–nematode relations and corresponding nematoses.* Host plant species differ in their ability to support nematode reproduction and in their tissue response to the nematode. These two characters may vary independently. They can be measured and may then give a quantitative picture of the plant–nematode relationship. Each nematose is a specific relationship between a nematode species, or trophotype†, and a particular plant species, or variety.

The parasitic behaviour and the damage caused by plant-parasitic nematodes can be demonstrated and illustrated in various ways, such as the marked population fluctuations caused by plants, the dome-shaped nematode reproduction curves resulting from damage at increasing pre-plant densities, the significant regressions of crop yields on pre-plant nematode densities in many plant–nematode relationships, and the results of inoculation experiments (cf. Oostenbrink, 1966). The negligible mobility and slow reproduction are the main reasons why the pre-plant population density of nematodes largely determines the degree of damage.

---

* Nematose, meloidogynose, pratylenchose, etc., are considered useful words to indicate infestation by nematodes, by *Meloidogyne* species, by *Pratylenchus* species, etc.

† Trophotypes are strains or biological races with differences in host preference or other feeding behaviour.

Empiricism has shown that the rectilinear regression of the final plant yield Y on the logarithmically transformed pre-plant nematode densities P_i, as illustrated in Fig. 1A, is a useful model and is comparable to the well-known dose–response curve used in animal toxicology. Natural populations in infested fields, however strongly affected by the different preceding crops, appear to fit such straight regression lines. The line is characterized by a linear equation, i.e. by the two constants of such an equation. Differences in the constants may measure

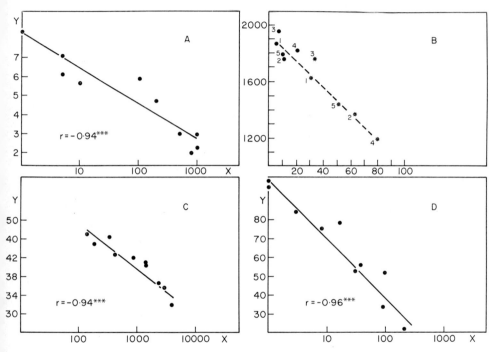

FIG. 1. Four regressions of test plant yield Y (ordinate) on pre-plant nematode densities $P_i = X$ (abscissa), r = regression coefficient, *** = significant at 0·1% level.

A. Pea seed yield in kg per 16 m² on *Meloidogyne hapla* larvae per 100 ml of soil, logarithmic scale (after Oostenbrink, 1966).

B. Tobacco yield in lb per acre on root-knot (*Meloidogyne javanica*) index. The nematode population is differentiated by the cultivation of different crops as well as by fumigation of the soil (based on figures published by Daulton, 1964).
★ = soil not fumigated; ● = soil fumigated. Preceding crops of tobacco are: 1 = *Eragrostis curvula* var. Ermelo, 2 = *Chloris gayana* var. Giant, 3 = *Chloris gayana* var. Katambora, 4 = *Setaria sphacelata* var. Kazungala, 5 = *Panicum maximum* var. Sabi.

C. Percentage of potato plants escaping *Phoma solanicola*-infestation on *Ditylenchus dipsaci* per 500 ml of soil, logarithmic scale (after Hijink, 1963).
The fungus infestation occurs in proportion to the nematode density.

D. Oat seed and straw as kg per 12 m² on mixed population of *Pratylenchus crenatus* + *Tylenchorhynchus dubius* + *Rotylenchus robustus* per 300 ml of soil, logarithmic scale (after Oostenbrink, 1966).

differences in nematode activity, in the condition of the plant, or in environmental influences. Deviations of one or more points from the line may help to determine and measure the effect of other factors on plant growth: side effects of soil treatments, nitrogen effects in rotation experiments and additional effects of other pathogens. The formula is useful also in establishing the basis for use in advisory soil sample examination (cf. Figs 2 and 3) and in completing the reproduction curve of nematodes on damaged plants, because the yield may now also be expressed in terms of the pre-plant nematode density P_i (Oostenbrink, 1966).

The differences of P_i may be due to inoculation, or to irregular distribution under natural circumstances in the field, or to differences caused by rotations or soil treatments. Figure 1B is a rectilinear regression of a yield on a population which is differentiated by the cultivation of different crops as well as by fumigation of the soil. Yield may also be related to two or more parasitic nematodes which damage the plant simultaneously (Fig. 1D) and it may be necessary to give one species more weight than the other (Hijink, 1964). Furthermore, it is remarkable, and important for controlling the damage, that the regression even holds where the nematodes do not act directly as pathogens but only as vectors or initiators of other diseases (Fig. 1C; Hijink, 1963).

The regression line reflects the growth reduction of the sick plant. In the soil a certain percentage of roots may escape infection because of irregular distribution of the nematodes and the fact that a thin surface layer and the subsoil are normally not or only slightly infested. This may cause a rest yield based on escape and therefore a horizontal line at very high densities. In instances where a nematode causes no damage at all, the whole curve is horizontal. Also with noxious nematodes a horizontal curve must theoretically precede the sloping regression line at increasing densities, because the plant can sustain or compensate for a very light infection. In pot experiments such low densities may be established but in naturally infested fields populations are normally too high. There are records of pot experiments in which light inoculations initially have promoted plant growth, a phenomenon which has not been confirmed in natural conditions. Deviations from the straight line regression are, therefore, seldom found under practical conditions.

Some workers combine the horizontal graphs of the root regeneration and escape due to irregular distribution with that of the damage graph, into one sigmoid curve for which an exponential formula is derived (Seinhorst, 1965, 1968). Such a curve combines fundamentally different phenomena and extrapolation from it is difficult as compared with the straight regression line.

D. = Ditylenchus
A. = Aphelenchoides
M. = Meloidogyne
H. = Heterodera
P. = Pratylenchus
R. = Rotylenchus
Tr. = Trichodorus

Nematode reproduction

0   nil
1   little
2   moderate
3   strong

Susceptibility to damage

not known
nil
little
moderate
strong

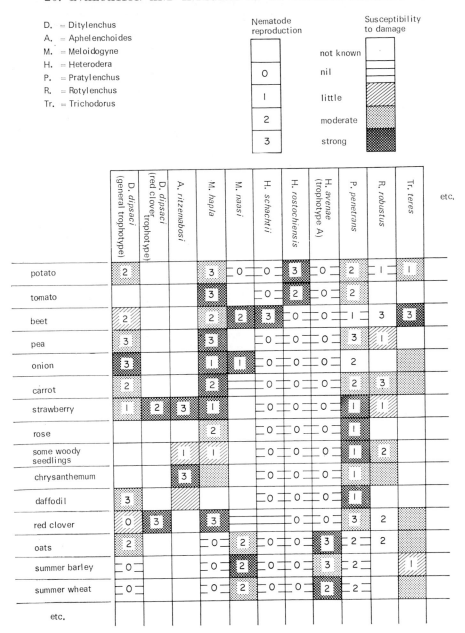

Fig. 2. Scheme of main crops versus main nematode infestations. For each crop–nematode combination known data about damage are indicated by shading and nematode reproduction is characterized by means of a number.

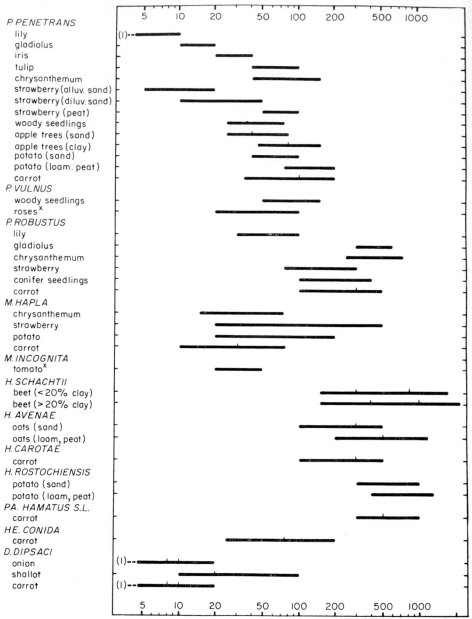

FIG. 3. Indication of field infestations expected to cause moderate crop damage; in some instances further specification into categories is indicated by dots on the infestation lines in the figure. Lower densities than indicated are considered acceptable, higher densities are too risky for normal crop production. Ordinate: nematode–plant relationship; * = glasshouse crops, P = *Pratylenchus*, R = *Rotylenchus*, M = *Meloidogyne*, H = *Heterodera*, Pa = *Paratylenchus*, He = *Hemicycliophora*, D = *Ditylenchus*. Abscissa: nematode density per 100 ml of soil; except for *D. dipsaci*, per 1000 ml; logarithmic scale.

C. Application of Population Density Figures for Nematode Control

The relative ease with which population density figures can be obtained launched a quantitative approach in plant nematology, which proved to be valuable for the development of experimental, inventory and advisory work, and therefore of control.

1. Experiments

It is, or should be, common practice that nematode control experiments are preceded and accompanied by quantitative evaluations of the nematode populations. Identification of the species is normally required and must often be complemented by a test about the potential pathogenicity of the trophotypes. Sampling, extraction and methods of estimating nematode populations are developed sufficiently well to furnish useful, reproducible figures despite the fact that they are usually not absolutely quantitative.

2. Inventories

It is difficult to obtain meaningful quantitative figures in inventories unless the variability due to sampling time, plant, soil and other environmental influences can be reduced or taken into account. A qualitative, faunistic list of the species present is valuable for some aspects of nematode control but is usually insufficient. A detailed inventory of populations and corresponding yield losses, and therefore of nematode density/plant yield relationships, is needed as a basis for a nematology control programme in every agricultural area. There are several types of inventories (cf. Dao *et al.*, 1970). One efficient method is the quantitative analysis of plant nematode communities in each field of a few typical farms per area, and evaluation of the nematode figures with the help of rotation and fumigation trials on these farms. The quantitative results of inventories, corresponding field trials, and observations based on practical experience may for each area lead to a general scheme of main crops versus main nematode infestations, as indicated under *3* and to a more precise base for advice for a particular field as indicated under *4*.

3. Plant–Nematode Scheme

Figure 2 is an example, comprising part of a scheme published by Hijink and Oostenbrink (1968) and extended by Stemerding and Kuiper (1968) showing the main crops versus the main nematode infestations for a particular area. Known data about susceptibility to damage and host suitability of the plant for nematode reproduction are indicated. A scheme of this type summarizes the knowledge about important

R

nematode infestations in an area and grows in size, detail and accuracy with time.

4. Advisory Base

Pre-plant nematode population densities are used in practice to help to select safe fields, to avoid heavily infested fields for particularly susceptible crops, to decide upon the necessity to disinfect the soil or to choose a resistant crop. In this type of advisory work the fields are often put in a few (2–5) categories according to the infestation rates found. Figure 3 records the density of several nematode species on different hosts which would cause moderate damage in the Netherlands. Such data are used on a large scale for practical advisory work by the Bedrijfslaboratorium voor Grond-en Gewasonderzoek at Oosterbeek. Such advisory work evaluates the degree of infestation of individual fields and it allows the optimum exploitation of knowledge of populations in relation to damage. The experience gained up to now indicates that an overall nematode inventory as indicated under Section 2, every 3–5 years, is rewarding for nearly every farm or nursery.

II. Evaluation of the Main Methods of Control

The control of nematode damage may be aimed at preventing nematode entry, suppressing its population, mitigating its effect, or a combination of these principles. There is no easy and cheap treatment for general use, and plant nematode control is generally difficult or expensive. This results in varying attempts at obtaining satisfactory control, and if this fails, basic changes of the cropping system or considerable nematode damage often has to be accepted. In practice several control methods may play a role, either alone or combined and these are briefly discussed and evaluated below (Fig. 6).

A. Quarantine to prevent Entry

Quarantine, in the general sense, comprises all officially organized activities to keep the population density in the country, area, or field at the zero level. The complete arrest or deterrence of a plant nematode is seldom if ever the result of human efforts: the chances of preventing an infection are low, owing to the efficient natural spread of nematodes, their good chances for establishment, their latent occurrence and the difficulty in eradicating them.

Quarantine measures should be practical and in accord with the specific factors appertaining to the particular situation, with the biological information about the significance of the parasite, and also with the expectancy of being effective. Half-grown knowledge often leads to unbalanced measures of quarantine requirements. It makes little sense to take strict measures against harmful nematodes, unless a country is willing to sacrifice much of its trade and traffic. It has been shown (Dao, 1970) that tropical nematodes cannot thrive in cool climates (except in glasshouses) and that the nematodes from temperate zones do not thrive or even establish themselves in tropical regions (except at high altitudes), and it would therefore be unnecessary for a western European country to make a regulation against the sting nematode, *Belonolaimus longicaudatus* and the lesion nematode, *Pratylenchus zeae*, or for a Caribbean country to make a regulation against *Heterodera avenae* or *Pratylenchus penetrans*.

It is, on the other hand, practical for the Netherlands to maintain regulations against the potato cyst nematode, *Heterodera rostochiensis*, and not against the beet cyst nematode, *H. schachtii*, whereas Britain does the reverse. In both instances the regulations are against the nematode which is still local in distribution but could spread further, and not against the nematode which has already reached its potential distribution area. If expert estimates promise no more than a delay of the introduction or slowing of the spread, it may nevertheless be justifiable to take quarantine measures against a nematode especially if it concerns a major pathogen of an important crop and is apparently spreading from a local source. This may hold for a country, an area, a farm or even a field.

B. Eradication of Established Foci

Complete eradication of an established plant nematode population is difficult to obtain unless the cropping system is also drastically altered, so as to facilitate a natural decline. Usually a residual population builds up again from a small residue, or infections spread from untreated adjacent areas. Campaigns to eradicate *Heterodera rostochiensis* on potato in the U.S.A., the Netherlands and other countries, or the burrowing nematode, *Radopholus similis*, in citrus in the U.S.A., or *Rhadinaphelenchus cocophilus* on oil palm in Venezuela have in fact failed, and there are many other examples. It has been claimed that *Ditylenchus destructor* in some foci in the U.S.A. has been eradicated by soil fumigation. This may be true, but in the same years the nematode disappeared as a pest of potato in some areas of the Netherlands without any direct control activity. If treatments are repeated or if strict hygiene is maintained, eradication of a nematode pest to unnoticeable densities is not an exceptional occurrence in infested glasshouses or mushroom nurseries. Chances of eradicating an established population in the field, however, are low. There are nevertheless cases where attempts to eradicate foci are necessary or justified, even when the result is merely to decrease the level of the population in the known focus to that of the surrounding area.

C. Farm Hygiene and Cultural Measures in General

In infected crops a particular nematode population may be kept at a tolerably low level by a variety of practical measures. Such integrated or harmonious control may comprise elements of more specific methods. They are: the use of uninfested seed, the disinfection of seed, the removal or destruction or disinfection of infested plants and materials, the application of organic manures, special soil tillage techniques or curative treatments with nematicides. Also the methods discussed hereafter under Sections D–J may be included.

Selection of uninfested plants or rejection of infested plants or crops is a common method for obtaining nematode-free seed or planting material. Some methods which are recorded as successful to denematise infested seed are: cleaning wheat seed from *Anguina tritici* by flotation in salt solutions, disinfection of flower bulbs, strawberry plants, chrysanthemum sets, banana rhizomes, woody and other plants of several nematode species by hot-water treatment and disinfection of onion and clover seeds from *Ditylenchus dipsaci* by treatment with methyl bromide. Disinfestation of seedbeds and curative treatment of growing plants with non-phytotoxic nematicides are related techniques with the same purpose. Early destruction of palm trees infested with *Rhadinaphelenchus cocophilus* and sanitation of glasshouse cultures by removing those plants with visible symptoms of the foliar nematode, *Aphelenchoides*, infestation belong to this chapter. Application of organic manure, compost and mulch are known to mitigate the increase as well as the effect of certain plant nematodes and are sometimes applied against nematode infestations. The same holds for certain soil tillage techniques; cf. also Section D.

D. Biological and Ecological Control

Control of a species by the introduction or regulation of a specific repressive or stabilizing organism or by modifying the nematode's environment is possible in principle, although few practical successes are known. Control of nematodes by manuring, tillage, rotation, resistant varieties and by modifying planting dates, could probably be included under this heading as soon as an analysis of the mechanism reveals the exact biological, chemical and physical influences involved.

The main objective in all instances is to keep the nematode population at a tolerable level. Noticeable effects are obtained against populations of some plant nematodes in soil under experimental conditions by the inoculation of nematophagous fungi (Duddington, 1957) and of predatory mites and springtails (Sharma, 1971). Several other organisms have been found to be potentially effective against nematodes. In general, however, plant nematode populations are very persistent, and there are no instances where biological control of plant nematode damage has been of practical significance.

Some examples of success by means of ecological control are the control of *H. rostochiensis* on potato by very early planting and lifting (Grainger, 1959), reduction of the stubby-root nematode, *Trichodorus teres*, by rototilling the soil (Oostenbrink, 1964), and suppression of the root-knot nematodes, *Meloidogyne* spp., infestation of tomato by lowering the temperature and compensating for the growth stagnation by excess of light (Brande and Gillard, 1957).

The possibilities for establishing specific methods of biological and ecological control are potentially present, but results to date are meagre, owing to the poor knowledge of the plant–nematode relationships.

E. Escape due to Selection of Particular Fields

Escape from damage by the selection, based on soil sample results, of lightly infested or non-infested fields for high value crops is an established method of minimizing nematode infestations in some countries. It is, in fact, a special type of rotation procedure, but it may lead to a large increase in the population of parasitic nematodes in the fields concerned unless the initial density is zero.

Examples of this approach are the selection of fields free from the stem eelworm, *Ditylenchus dipsaci*, for the cultivation of silver onions or from *Pratylenchus penetrans* for the cultivation of woody stock or from *Heterodera rostochiensis* for the cultivation of potatoes.

F. Direct Chemical or Physical Control

Soil disinfestation with nematicides, heat and other radical treatments normally results in immediate suppression of the pre-plant population to 20–40%; 10% or in exceptional instances 1% may be reached. The treatment dosage or temperature is, for economic reasons, usually adjusted to the minimal kill which is safe for the next crop, and the crop may consequently restore the population to its initial noxious density. The course of the population density is different if slowly working nematicides or nematicidal manures are used, or if plants comprising nematicidal principles are grown. This is also the case if non-phytotoxic nematicides are applied to the soil or to the plant during the growing season.

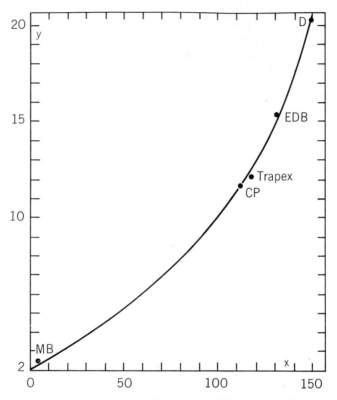

FIG. 4. The relation between boiling point of the soil fumigant and the percentage of experiments with phytotoxic effect on the crop, when more than 2400 experiments were evaluated; cf. text. MB = methyl bromide, CP = chloropicrin, EDB = ethylene dibromide, D = dichloropropane. Ordinate: Y = percentage of toxic cases. Abscissa: X = boiling point of the soil fumigant (after Eissa, 1971).

The effect on plant growth of partial soil sterilization by means of suitable chemicals or heat is drastic. A recent review of more than 2400 plant-treatment combinations was recently made in the Netherlands (Eissa, 1971). The yield of treated soils was 141% of untreated soils, or 154% if instances of phytotoxicity were excluded; these figures were 172% and 188% respectively if the best average partial soil sterilization treatment for each crop was evaluated. Taylor (1961) recorded an average crop yield of 187% for 853 experiments in the U.S.A. The figures for both countries, however, were obtained in experiments on traditionally problem fields. An overall average crop yield for all soils treated with nematicides is estimated at 120–130% of untreated soils. It is difficult to know with certainty how much of the effect of partial soil sterilization is due to elimination of nematodes, but the author estimates that this is about half.

The review by Eissa (1971) also revealed that 20% of all treatments were toxic to the crop and caused yields less than these from untreated soil. The incidence of phytotoxicity resulting from different nematicidal fumigants appeared to be strongly correlated with the boiling point of the chemical (Fig. 4). This stresses the difficulty of the partial soil sterilization technique. It is probable that phytotoxicity often decreases the beneficial effect of soil fumigation.

The most widely used nematicides are based on ethylene dibromide, dichloropropene or dibromochloropropane. Heat, chloropicrin or methyl bromide are used if problems other than nematodes also have to be suppressed. Other good nematicides are based on methylisothiocyanate but their effects against fungi and other organisms is problematic. Among the new nematicides with low phytotoxicity and apparently with some systemic effect are formulations of thionazin, aldicarb, and fensulfothion. They are technically promising, but their application to food crops is often discouraged due to the risk of residues which are toxic to mammals. The fact, however, that powerful new nematicides are coming to the fore indicates the possibility that a simple, safe, curative treatment will become available eventually. It should, of course, be economic, for the high cost of nematicide application is the main barrier for general application.

G. Fallow

Fallow is a widely used technique in certain countries for suppressing nematode populations by starvation. With dry fallow the percentage kill may be very high due to the lethal effect of drought (cf. Fig. 5), and may then be considered natural soil disinfestation. The method is slow and may require a full crop-growing season. Dry fallow is practised in many parts of the world and must be considered an important method of nematode control in some areas.

H. Crop Rotation

The aim of rotation is to decrease nematode population density, by using a non-host crop, to an acceptable level for growing a susceptible main crop. Rotation of crops in its various forms is still the most

important method for the prevention and suppression of plant nematode problems.

Historically, rotation was used to solve problems of soil fertility and
structure, water supply, weed and disease control and economic problems of crop husbandry and labour management. The main function for
crop rotation in modern agriculture is to prevent and suppress noxious
populations of soil-borne pathogens. However, as soon as soil disinfestation becomes economic, rotation is usually abandoned.

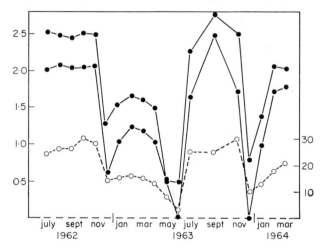

FIG. 5. Relation between the soil moisture content (broken line) and the population
density of *Hoplolaimus* sp. (lower solid line) or of *Hoplolaimus* sp. + *Helicotylenchus* sp.
+ *Hemicriconemoides* sp. (upper solid line) in a non-irrigated mango orchard at Abdullah
Hall (U.P., India). The drought extremes (December 1962, June 1963, December 1963)
coincide with minimum population densities (after Khan *et al.*, in press).

Ordinate at left: nematode numbers per 100 ml of soil, logarithmic scale.
Ordinate at right: percentage moisture in the soil.
Abscissa: time scale.

Escape from nematode attack by selecting uninfested fields (E),
fallow (G), the ley system, cultivation of cover or green manure crops,
and cultivation of nematode-trapping or resistant plant crops (I), deep-
digging of soil, refreshening of pot soils, amelioration of plant sites
by the addition of fresh soil and abandoning of soils for new areas are
essentially rotations. Rotation and soil disinfestation are complemen-
tary techniques, because each of them may be effective and so make the
other superfluous, but it is often economic to integrate both methods.

The study of the effect of rotation as a nematode control method is
far from complete, but a number of introductory studies are available
(cf. Anon., 1968; Good, 1968; Hijink and Oostenbrink, 1968).

I. Resistant Varieties

The cultivation of a resistant variety may suppress a nematode's population to 10–50% of its harmful density. It is in fact a strong rotation effect. A resistant variety of the susceptible plant species may be more effective than other unsuitable host plants in decreasing a nematode population. Current research in breeding and selection of new resistant varieties against several major plant nematodes appears to be promising. Resistant gene material can usually be found also against polyphagous nematodes, and crossing with existing crop varieties is often successful. The basis of the resistance can often be indicated in terms of genes, and in a few instances extreme resistance is known to be related to the presence of nematicidal substances in the plant. Most nematode species appear to comprise or to be able to develop aggressive trophotypes. New resistant plant varieties are usually damaged within one or two decades to the same degree as the old varieties owing to accumulation of new trophotype populations.

Resistance has been derived, for example, from *Solanum andigenum* and other wild *Solanum* species against *Heterodera rostochiensis* in potato, from *Avena sterilis* against *H. avenae* in oats, from *Lycopersicum peruvianum, Gossypium barbadense* and several *Nicotiana* species against *Meloidogyne* spp. in tomato, cotton and tobacco respectively, from *Citrus aurantium* against *Tylenchulus semipenetrans*. In other crops special varieties within the plant species harboured resistance, e.g. in barley against *H. avenae*; in soy-bean against *Heterodera glycines*; in white clover against *H. trifolii*; in peach, alfalfa, red pepper, beans and other crops against *Meloidogyne* species; in red clover, oats, rye and alfalfa against *Ditylenchus dipsaci*; in *Chrysanthemum* against *Aphelenchoides ritzemabosi*.

J. Tolerant Varieties

Non-susceptibility to damage (tolerance) is often not correlated with host unsuitability (resistance) of plants to nematodes. Tolerant varieties of suitable host species may grow well in infested soil despite the infestation, but they may increase the nematode population up to a harmful level.

III. Summary and Integration

The array of cumbersome control techniques listed above would be removed and the research priority in nematology would change, if an effective, cheap, nematicide were found. It could probably be used with profit on most crops if the treatment cost was less than 10% of the crop

R*

value. The chances that a paramount nematicide without practical
limitations, for example a safe systemic nematicide for treating seed
or growing plants, will be discovered in the near future are small. Until

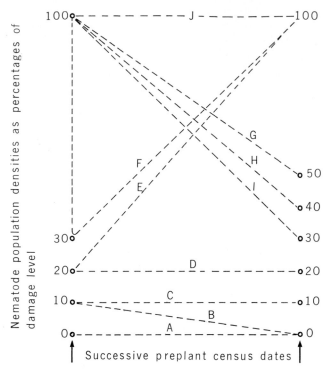

FIG. 6. Principal characteristics of ten methods of nematode control, A–J, in relation
to the nematode's population density. Each method is represented by a graph.

A = quarantine to prevent entry;
B = extermination of established foci;
C = farm hygiene and cultural measures in general;
D = biological and ecological control;
E = escape due to selection of fields;
F = direct chemical or physical control;
G = fallow;
H = crop rotation;
I = resistant varieties;
J = tolerant varieties.

that moment, the above methods or integrations of them will have to
be used.

  The principal characteristics of the ten methods A–J in relation to a
nematode's population density is illustrated in Fig. 6. The effects illustrated are approximations of the result which generally could be
expected.

In practice integration of different methods is the rule rather than the exception. Two cases of true integrated control will be mentioned as examples.

When a soil is heavily infested with *Pratylenchus penetrans*, one fumigation treatment may reduce the population to a very low level. An advisable cropping sequence (cf. Fig. 2) could then be:

1. strawberry or woody seedlings (which are susceptible but do not cause a large increase of the residual nematode infection),
2. potato (moderately susceptible and causing moderate nematode reproduction),
3. oats or barley (not susceptible but causing nematode reproduction),
4. beet (not susceptible and suppressing the population), after which moderately susceptible crops can be grown again successfully. It would be impractical to grow cereals, potato or peas immediately after the soil treatment. In this case intelligent combination of soil fumigation and rotation is desirable.

A well-known integration of methods can control *Heterodera rostochiensis* in potato in a two-year rotation combining soil fumigation, cultivation of resistant potato varieties and cultivation of susceptible varieties in a systematic programme. The effects of the different treatments and crops on the nematode population density are variable, but they can be presented approximately in a 4-year period as follows: 100% (high initial density) $\times \frac{3}{10}$ (reduction caused by a moderately successful soil fumigation) $\times \frac{1}{2}$ (reduction caused by a non-host) $\times \frac{3}{10}$ (reduction caused by a resistant potato variety) $\times \frac{1}{2}$ (reduction caused by a non-host) $\times 44$ (reproduction factor $= R$, on a susceptible crop) $= 100\%$. In instances where the non-hosts would only cause population decline to $\frac{2}{3}$, the result could be $100\% \times \frac{3}{10} \times \frac{2}{3} \times \frac{3}{10} \times \frac{2}{3} \times 25 = 100\%$. R is usually less than 25 and so the scheme in both cases would finally lead to extermination of the nematode population. Early potatoes, for which R is often 5–10, may perhaps be grown safely in a 2-year rotation when either the soil treatment or the resistant variety is included.

It is seen that the integration of the different methods offers effective control possibilities which are of great practical value in the suppression of harmful nematode population densities as well as to prevent the build-up of measurable populations of nematodes. The tendency in modern agriculture to grow crops as a monoculture and to grow more crops per year, makes the prevention and suppression of nematode damage increasingly imperative for obtaining good yields.

References

Anon. (1968). *Natu. Acad. Sci. Wash.* **1696**, 172 pp.
Brande, J. van den, and Gillard, A. (1957). *Nematologica* **2**, 398–404.

Dao, F. (1970). *Meded. LandbHogesch. Wageningen* **70-2**, 181 pp.

Dao, F., Oostenbrink, M. and Viets, H. A. (1970). *Versl. Meded. plziektenk. Dienst Wageningen Sep. Ser.* **415**, 84 pp.

Daulton, R. A. C. (1964). *Biokemia* **5**, 10–15.

Duddington, C. L. (1957). "The Friendly Fungi: A New Approach to the Eelworm Problem." Faber and Faber, London.

Eissa, M. F. M. (1971). *Meded. LandbHogesch. Wageningen* **71-14**, 129 pp.

Good, J. M. (1968). *In* "Tropical Nematology" (G. C. Smart and V. G. Perry, eds), pp. 113–138. University of Florida Press, Gainesville.

Grainger, J. (1959). *Eur. Potato J.* **2**, 184–198.

Hijink, M. J. (1963). *Neth. J. Pl. Path.* **69**, 318–321.

Hijink, M. J. (1964). *Meded. LandbHogesch. OpzoekStns Gent* **29**, 818–822.

Hijink, M. J. and Oostenbrink, M. (1968). *Versl. Meded. plziektenk. Dienst Wageningen sep. Ser.* **368**, 7 pp.

Oostenbrink, M. (1964). *Nematologica* **10**, 49–56.

Oostenbrink, M. (1966). *Meded. LandbHogesch. Wageningen* **66-4**, 46 pp.

Seinhorst, J. W. (1965). *Nematologica* **11**, 137–154.

Seinhorst, J. W. (1968). *C.r. 8 Symp. int. Nématologie, Antibes* 1965, 83.

Sharma, R. D. (1971). *Meded. LandbHogesch. Wageningen* **70-1**, 154 pp.

Stemerding, S. and Kuiper, K. (1968). Tjeenk Willink, Zwolle, 178 pp.

Taylor, A. L. (1961). *Abstr. 6. int. Nematology Symp. Gent 1961*, 80.

Appendix

Weights and Measures Equivalents

1 millimetre (mm)	0·039 inch
1 centimetre (cm)	0·394 inch
1 metre (m)	1·094 yards
1 square centimetre (cm^2)	0·155 square inch
1 square metre (m^2)	1·195 square yards
1 hectare (ha)	2·471 acres
1 cubic centimetre (cm^3)	0·061 cubic inch
1 litre	1·760 pints
1 gramme (g)	0·035 ounce
1 kilogram (kg)	2·205 pounds
1 metric ton	0·984 ton

Approximate Monetary Equivalents of One U.S. Dollar*

Dutch guilder	3·15
English pound	0·38
German mark	3·15
Indian rupee	7·91
Japanese yen	303·03
Swedish krone	4·71

* (approximate values as of January, 1972)

Author Index

Numbers in italics indicate those pages where references are given in full.

Subject Index

Note that the names of the crop hosts to plant-parasitic nematodes are not indexed as they are already suitably indexed at the beginning of each chapter.

S

S*